T0271989

XAFS for Everyone

XAFS for Everyone provides a practical, thorough guide to x-ray absorption fine structure (XAFS) spectroscopy for both novices and seasoned practitioners from a range of disciplines. It's enhanced with more than 200 figures as well as cartoon characters who offer informative commentary on the different approaches used in XAFS spectroscopy.

This second edition now includes chapters on spatial and temporal resolution, alternative measurement modes including resonant inelastic x-ray scattering (RIXS) and high-energy resolution fluorescence detection (HERFD), and an expanded chapter on experimental design. In addition, this edition adds new sections on wavelet transforms, blind source separation, free electron lasers, and theoretical XANES standards, as well as three new case studies.

XAFS for Everyone covers sample preparation, data reduction, tips and tricks for data collection, fingerprinting, linear combination analysis, principal component analysis, and modeling using theoretical standards. It describes both near-edge (XANES) and extended (EXAFS) applications in detail. Examples throughout the text are drawn from diverse areas, including materials science, environmental science, structural biology, catalysis, nanoscience, chemistry, art, and archaeology. In addition, eight case studies from the literature demonstrate the use of XAFS principles and analysis in practice.

The text includes derivations and sample calculations to foster a deeper comprehension of the results. Whether you are encountering this technique for the first time or looking to hone your craft, this innovative and engaging book gives you insight on implementing XAFS spectroscopy and interpreting XAFS experiments and results. It helps you understand real-world trade-offs and the reasons behind common rules of thumb.

Scott Calvin has been using x-ray absorption fine structure (XAFS) since 1998 to study systems as diverse as solar cells, magnetic nanoparticles, soil samples, battery cathodes, analogues to atmospheric dust particles, and pigments used in 18th-century paintings. Dr. Calvin is currently the director of the prehealth program at Lehman College of the City University of New York, helping students their achieve their dreams of becoming physicians, dentists, pharmacists, veterinarians, and more.

$$\chi(k) = S_o^2 \sum_i N_i \frac{f_i(k)}{kD_i^2} e^{-\frac{2D_1}{\lambda(k)}} e^{-2k^2\sigma_i^2} \sin\left(2kD_i + \delta_i(k)\right)$$

The EXAFS Equation

XAFS for Everyone

Second Edition

Scott Calvin
Illustrated by Kirin Emlet Furst

CRC Press
Taylor & Francis Group
Boca Raton London New York

CRC Press is an imprint of the
Taylor & Francis Group, an **informa** business

Front cover image: © Kirin Emlet Furst

MATLAB® and Simulink® are trademarks of The MathWorks, Inc. and are used with permission. The MathWorks does not warrant the accuracy of the text or exercises in this book. This book's use or discussion of MATLAB® or Simulink® software or related products does not constitute endorsement or sponsorship by The MathWorks of a particular pedagogical approach or particular use of the MATLAB® and Simulink® software.

Second edition published 2025

by CRC Press
2385 NW Executive Center Drive, Suite 320, Boca Raton FL 33431

and by CRC Press
4 Park Square, Milton Park, Abingdon, Oxon, OX14 4RN

CRC Press is an imprint of Taylor & Francis Group, LLC

First edition published 2013

ISBN: 978-1-032-74548-0 (hbk)
ISBN: 978-0-367-34567-9 (pbk)
ISBN: 978-0-429-32955-5 (ebk)

DOI: 10.1201/9780429329555

Typeset in Adobe Garamond
by Deanta Global Publishing Services, Chennai, India

For Bruce, Matt, John, and Ed
without whom this book
would be very different.

Contents

Foreword to the Second Edition... xvii
Preface to the Second Edition xviii
Preface to the First Edition xx
Author xxiii
About the Panel xxiv

Introduction 1
XAFS for My Birthday 1
Reference 3

SECTION I   THE XAFS EXPERIMENT

1   XAFS in a Nutshell..7
   1.1   X-Ray Absorption Spectra 8
         1.1.1   X-Ray Absorption Spectroscopy..8
         1.1.2   Background 8
         1.1.3   Edge 9
         1.1.4   X-Ray Absorption Near-Edge Structure......................................9
         1.1.5   Extended X-Ray Absorption Fine Structure 9
   1.2   Basics of EXAFS Theory...9
         1.2.1   Fermi's Golden Rule 9
         1.2.2   EXAFS Is Due to Interference of an Electron with Itself 9
         1.2.3   Relationship of k to Photon Energy..11
         1.2.4   EXAFS and Structure 11
         1.2.5   Scattering Probability..12
         1.2.6   Multiple Neighbors...12
         1.2.7   Multiple Scattering 13
         1.2.8   Phase Shifts..13
         1.2.9   Spherical Waves 13
         1.2.10 Incomplete Overlap...14
         1.2.11 Mean Free Path 15
         1.2.12 EXAFS Is an Average 15
         1.2.13 Enough for Now 17
   1.3   Some Terminology 17
         1.3.1   Edge..17
         1.3.2   XANES 17
         1.3.3   NEXAFS 18
         1.3.4   EXAFS 18
         1.3.5   White Line..18
         1.3.6   Preedge 18
         1.3.7 E_o..19

1.4   Data Reduction 19

      1.4.1   From Raw Data to χ(*k*) 19

      1.4.2   Fourier Transform 20

1.5   XAFS Is Not a Black Box 24

1.6   Overview of Approaches to XAFS Analysis 25

      1.6.1   Fingerprinting 25

      1.6.2   Linear Combination Analysis 25

      1.6.3   Principal Component Analysis 26

      1.6.4   Curve Fitting to a Theoretical Standard 27

References 28

2   **The Hardware: Light Sources, Beamlines, and Detectors** **29**

2.1   X-Ray Sources 29

      2.1.1   Synchrotron Light Sources 30

      2.1.2   Energy-Recovery Linacs 30

      2.1.3   Tabletop Sources 30

2.2   Bending Magnets and Insertion Devices 32

      2.2.1   Brilliance 32

      2.2.2   Wigglers 32

      2.2.3   Undulators 33

      2.2.4   Free Electron Lasers 33

2.3   Monochromators 35

2.4   Measurement Modes and Detectors 35

      2.4.1   Measurement Modes 35

      2.4.2   Ion Chambers 36

      2.4.3   Energy-Discriminating Fluorescence Detectors 36

      2.4.4   Current-Mode Semiconductor Detectors 36

      2.4.5   Wavelength-Dispersive Fluorescence Detectors 37

      2.4.6   Electron Yield 37

      2.4.7   Other Detectors 37

      2.4.8   Beamline Equipment 37

      2.4.9   Simultaneous Probes 38

      2.4.10   Microprobe 38

2.5   Getting Beamtime 38

References 39

3   **Spatial and Temporal Resolution** **41**

3.1   Time-Resolved Experiments 42

      3.1.1   Evolution on a Scale of Months or Years: Just Wait! 42

      3.1.2   Freezing Time: Quenched Samples 42

      3.1.3   Evolution on a Scale of Tens of Minutes 42

      3.1.4   Quick XAFS 43

      3.1.5   Energy-Dispersive XAFS 43

      3.1.6   Pump-Probe 44

3.2   Microprobes 45

References 47

4   **High-Energy-Resolution Techniques** **49**

4.1   Energy Resolution 50

4.2   High-Energy-Resolution Fluorescence-Detected X-Ray Absorption Spectroscopy 52

4.3   Resonant Inelastic X-Ray Scattering 53

References 54

5 Experimental Design **55**
  5.1 Identifying Your Questions 55
  5.2 Automation 56
  5.3 Experimental Plan 58
  5.4 *Ex Situ* Experiments 59
  5.5 *In Situ* Experiments 59
  5.6 *Operando* Experiments... 59
  5.7 Multimodal Measurements... 60
  References 60

6 Sample Preparation **61**
  6.1 XAFS Samples.. 62
  6.2 Absorption.. 62
      6.2.1 Undesirable Photons 62
      6.2.2 The Absorption Coefficient 63
  6.3 Sample Characteristics for Transmission 64
      6.3.1 Thickness, Mass, and Absorption Lengths 64
      6.3.2 Uniformity.. 66
      6.3.3 A Short Digression on Noise 67
      6.3.4 Optimum Thickness for Transmission 68
      6.3.5 Edge Jump 69
  6.4 Sample Characteristics for Fluorescence 73
      6.4.1 From Incidence to Detection 73
      6.4.2 Understanding Your Sample 76
      6.4.3 The Mathematics of Self-Absorption........................... 77
      6.4.4 Geometrical Factors 78
      6.4.5 How Thick Is 'Thick'?.. 80
      6.4.6 How Concentrated Is 'Concentrated'? 81
  6.5 Which Technique Should You Choose? 81
  6.6 Preparing Samples 82
      6.6.1 Powder 82
      6.6.2 Thin Films 88
      6.6.3 Solid Metals 88
      6.6.4 Solutions and Liquids... 88
      6.6.5 Gases 89
      6.6.6 Environmental 89
      6.6.7 *In Situ* and *Operando* 89
  References 91

7 Data Reduction **92**
  7.1 Preprocessing.. 92
      7.1.1 Rebinning.. 92
      7.1.2 Selecting Channels and Scans...................................... 93
      7.1.3 Calculating Unnormalized Absorption 93
      7.1.4 Truncation 94
  7.2 Calibration and Alignment... 94
      7.2.1 Aligning Reference Scans.. 94
      7.2.2 Merging 94
      7.2.3 Calibrating.. 96
  7.3 Finding Normalized Absorption 97
      7.3.1 Deadtime Correction 97
      7.3.2 Deglitching 97
      7.3.3 Choosing E_o... 97

    7.3.4 Normalization 98
    7.3.5 Self-Absorption Correction 103
  7.4 Finding $\chi(K)$ 104
    7.4.1 Background Subtraction 104
    7.4.2 $\chi(E)$ 107
    7.4.3 Converting from E to k 108
    7.4.4 A Second Chance at Self-Absorption Correction 108
    7.4.5 Weighting $\chi(k)$ 108
  7.5 Finding the Fourier Transform 108
    7.5.1 About Fourier Transforms 108
    7.5.2 Data Ranges Are Finite 109
    7.5.3 Windows 114
    7.5.4 Zero Padding 115
    7.5.5 Choice of Windows 116
    7.5.6 Fourier Transforms Are Complex 117
    7.5.7 'Corrected' Fourier Transforms 120
    7.5.8 Back-Transforms 121
  7.6 Wavelet Transforms 124
    7.6.1 Comparison of Fourier and Wavelet Transforms ... 125
    7.6.2 The Uncertainty Principle 127
    7.6.3 Choosing Parameters for the Parent Function 128
  References 131

8 **Data Collection** **133**
  8.1 Noise, Distortion, and Time 134
  8.2 Detector Choice 134
    8.2.1 Predicting Signal-to-Noise Ratio in Transmission ... 135
    8.2.2 Signal-to-Noise Ratio in Fluorescence 138
    8.2.3 Energy-Discriminating Fluorescence Detectors ... 139
    8.2.4 Wavelength-Dispersive Fluorescence Detectors ... 140
  8.3 Mode Choice 140
  8.4 Before You Begin 140
    8.4.1 Plan Your Beamtime 140
    8.4.2 Get to Know Your Beamline 141
  8.5 Optimizing the Beam 142
    8.5.1 Aligning the Beam 142
    8.5.2 Choosing Pre-I_0 Vertical Slit Width 143
    8.5.3 Reducing Harmonics 145
  8.6 Ion Chambers 149
    8.6.1 Physics of Ion Chambers 149
    8.6.2 Limitations 150
    8.6.3 Choosing Fill Gases 150
    8.6.4 Amplifiers 151
  8.7 Suppressing Fluorescent Background 152
    8.7.1 Suppressing Scatter Peaks 153
    8.7.2 Suppressing Low-Energy Peaks 156
    8.7.3 Making the Choice 157
  8.8 Aligning the Sample 158
  8.9 Scan Parameters 161
    8.9.1 Scan Regions 161
    8.9.2 Number of Scans 163
    8.9.3 Time-Resolved Studies 164
    8.9.4 Making the Most of Your Beamtime 165

8.10 'What's That?'.. 165
    8.10.1 Noise... 165
    8.10.2 Monochromator Glitches... 166
    8.10.3 Other Edges.. 167
    8.10.4 Other EXAFS... 168
    8.10.5 Multielectron Excitations.. 170
    8.10.6 Bragg Peaks.. 171
    8.10.7 Monotonic Time-Dependent Effects..................................... 172
    8.10.8 Oscillatory Time-Dependent Effects..................................... 173
    8.10.9 Electronics Out of Range... 174
    8.10.10 Sample Motion 175
    8.10.11 Loss of Beam.. 177
References... 177

SECTION II   XAFS ANALYSIS

9   Fingerprinting

9   **Fingerprinting**.. **183**
9.1 Matching Empirical Standards... 183
9.2 Fingerprinting Spectral Features.. 185
9.3 Semiquantitative Fingerprinting... 186
    9.3.1   Example: Vanadium XANES... 187
    9.3.2   Fitting Features... 189
9.4 Theoretical XANES Standards... 192
References... 193

10   **Linear Combination Analysis**... **194**
10.1 When LCA Works.. 194
    10.1.1 A Simple Example... 194
    10.1.2 Intimate Mixtures in Transmission....................................... 196
    10.1.3 Intimate Mixtures in Fluorescence.. 196
10.2 When LCA Doesn't Work.. 197
    10.2.1 Nonuniform Samples in Transmission.................................. 197
    10.2.2 Surface Gradients in Thick Fluorescence Samples................. 197
10.3 An Example of LCA... 197
10.4 Statistics of Linear Combination Fitting.. 200
    10.4.1 Normalization: A Source of Systematic Error....................... 200
    10.4.2 Degrees of Freedom and Statistically Distinguishable Fits 202
    10.4.3 Quantifying Fit Mismatch... 203
    10.4.4 Degrees of Freedom 204
    10.4.5 The Hamilton Test... 204
    10.4.6 Uncertainties... 205
10.5 Combinatoric Fitting... 206
10.6 Sources of Systematic Error 207
    10.6.1 Energy Alignment.. 207
    10.6.2 Background 208
    10.6.3 Attenuation: Self-Absorption, Inhomogeneous
            Transmission Samples, Harmonics, Dead Time, and So On. 208
    10.6.4 Energy Resolution.. 209
    10.6.5 Glitches.. 209
    10.6.6 Noise.. 209
10.7 Choosing Data Range and Space For LCA..................................... 210
    10.7.1 XANES in Energy Space 210
    10.7.2 XANES in Derivative Space 210

10.7.3 EXAFS in Energy Space 211
10.7.4 EXAFS in $\chi(k)$ 211
10.7.5 The Back-Transform of EXAFS 216
References 216

11 Principal Component Analysis 217
11.1 Introduction 218
11.1.1 An Example from the Literature 218
11.1.2 Isosbestic Points 218
11.2 The Idea of PCA 219
11.3 How Many Components? 222
11.3.1 Appearance of Components 223
11.3.2 Fourier Transform of Components 224
11.3.3 Compare to Measurement Error 224
11.3.4 Scree 225
11.3.5 Objective Criteria 226
11.4 How Many Constituents? 228
11.4.1 Relationship to Number of Components 228
11.4.2 Energy Misalignment 228
11.4.3 Other Structural Free Parameters 229
11.4.4 Coupled Constituents 230
11.5 PCA Formalism 230
11.6 Cluster Analysis 231
11.7 Target Transforms 233
11.8 Blind Source Separation 235
11.8.1 Transformation Matrix 236
11.8.2 Iterative Target Transform Factor Analysis (ITTFA) 236
11.8.3 Evolving Factor Analysis (EFA) 237
11.8.4 Simple-to-Use Interactive Self-Modeling Mixture Analysis
(SIMPLISMA) 238
11.9 PCA of EXAFS 239
11.10 How PCA Is Used 239
References 241

12 Curve Fitting to Theoretical Standards 243
12.1 Fitting 244
12.2 Theoretical EXAFS Standards 245
12.2.1 Muffin-Tin Potentials 246
12.2.2 Final State Rule 247
12.2.3 Losses 247
12.3 The Path Expansion 248
12.3.1 Convergence 248
12.3.2 Full Multiple Scattering 250
12.4 EXAFS Fitting Strategies 252
12.4.1 Bottom-Up Strategy 252
12.4.2 Top-Down Strategy 253
12.5 Theoretical XANES Standards 254
12.5.1 Real-Space Multiple-Scattering 254
12.5.2 Finite Difference Method 254
References 256

SECTION III MODELING

13   **A Dictionary of Parameters** **261**
    13.1   Common Fitting Parameters 262
        13.1.1   Half Path Length 262
        13.1.2   Degeneracy 266
        13.1.3   Mean Square Relative Displacement 269
        13.1.4   Amplitude Reduction Factor 271
        13.1.5 E_0 275
    13.2   Less Common Fitting Parameters 277
        13.2.1   Cumulants 277
        13.2.2   Third Cumulant 278
        13.2.3   Fourth Cumulant 280
        13.2.4   Fifth and Higher Cumulants 282
        13.2.5   Mean Free Path 283
    13.3   Scattering Parameters 285
    References 290

14   **Identifying a Good Fit** **292**
    14.1   Criterion 1: Statistical Quality 293
        14.1.1   Number of Independent Points in EXAFS 293
        14.1.2   Measurement Uncertainty 296
        14.1.3   Reduced χ^2 298
    14.2   Criterion 2: Closeness of Fit 299
        14.2.1 R Factor 299
        14.2.2   Hamilton Test 300
    14.3   Criterion 3: Precision 302
        14.3.1   Finding Uncertainties in Fitted Parameters 302
        14.3.2   Calculations of Uncertainties by Analysis Software 302
        14.3.3   How Precise? 303
        14.3.4   Correlations 303
    14.4   Criterion 4: Size of Data Ranges 303
    14.5   Criterion 5: Agreement Outside the Fitted Range 304
    14.6   Criterion 6: Stability 306
    14.7   Criterion 7: Are the Results Physically Possible? 307
    14.8   Criterion 8: How Defensible Is the Model? 307
    14.9   Evaluating a Fit 308
    References 310

15   **The Process of Fitting** **311**
    15.1   Identify Your Questions 311
        15.1.1   Example: Which Ligand? 311
        15.1.2   Example: Where's the Dopant? 312
    15.2   Prepare Your Data 312
        15.2.1   Transform to $\chi(k)$ 312
        15.2.2   Choose k Weighting 313
        15.2.3   Choose k Range 314
        15.2.4   Choose k Window 316
    15.3   Plan Your Strategy 316
        15.3.1   Example: Which Ligand? 316
        15.3.2   Example: Where's the Dopant? 316
    15.4   Fit! 318
        15.4.1   Choice of R Range 318
        15.4.2   Art of Fitting 322

15.4.3 Perfecting Your Fit 323

15.4.4 Stressing Your Fit 324

References 325

16 Starting Structures **326**

16.1 Crystal Structures 326

16.1.1 Cluster Size and EXAFS 326

16.1.2 Cluster Size and XANES 328

16.1.3 Sources for Crystal Structures 329

16.2 Calculated Structures 329

16.3 Mixtures 329

16.4 Inequivalent Absorbing Sites 332

16.5 Histogram Methods 332

16.6 Multiple-Edge Fits 333

16.7 Site Occupancy 334

16.7.1 Vacancies 334

16.7.2 Treating as a Mixture 334

16.7.3 Creating a Mixed Model 334

16.7.4 Creating Multiple Mixed Models 335

References 335

17 Constraints **337**

17.1 Rigorous Constraints 338

17.2 Constraints Based on a Priori Knowledge 338

17.3 Constraints for Simplification 339

17.3.1 Constraints Based on Grouping 339

17.3.2 Constraints Based on Estimates 341

17.3.3 Constraints Based on Standards 341

17.4 Some Special Cases 344

17.4.1 Lattice Scaling 344

17.4.2 Correlated Debye Model 344

17.5 Multiple-Scattering Paths 346

17.5.1 Focused Paths 346

17.5.2 Double Paths 348

17.5.3 Conjoined Paths 348

17.5.4 Triangles, Quadrilaterals, and Other Minor Multiple-Scattering Paths 349

17.6 Alternatives for Incorporating A Priori Knowledge 355

17.6.1 Restraints 355

17.6.2 Bayes–Turchin Analysis 356

References 357

SECTION IV   XAFS IN THE LITERATURE

18 Communicating XAFS **361**

18.1 Know Your Audience 361

18.2 Experimental Details 362

18.3 Data 362

18.3.1 What to Include 362

18.3.2 Labeling Graphs 363

18.3.3 Estimate of Noise 363

18.4 Data Reduction 363

18.5 Models and Standards 364

18.5.1 Curve Fitting to Theoretical Standards 364
18.5.2 Linear Combination Analysis and Principal Component
Analysis 364
18.6 Results..364
18.6.1 Graphs of Fits 365
18.6.2 Closeness of Fit 365
18.6.3 Uncertainties...365
18.7 Conclusions 366
Reference 366

19 Case Studies..367
19.1 Introduction to the Case Studies 367
19.2 Lead Titanate, a Ferroelectric 368
19.2.1 The Paper 368
19.2.2 The Scientific Question..368
19.2.3 Why XAFS?...369
19.2.4 The Structure 369
19.2.5 Experimental Considerations 371
19.2.6 The Model 373
19.2.7 Drawing Conclusions..374
19.2.8 Presentation 375
19.3 An Iron–Molybdenum Cofactor Precursor..................................376
19.3.1 The Paper 376
19.3.2 The Scientific Question..377
19.3.3 Why XAFS?...377
19.3.4 A Challenge and a Solution..377
19.3.5 Possible Structures 377
19.3.6 Experimental Considerations 378
19.3.7 Fingerprinting..379
19.3.8 The Models for EXAFS..379
19.3.9 Drawing Conclusions..381
19.3.10 Presentation 382
19.4 Manganese Zinc Ferrite, an Example of Fitting Site Occupancy 383
19.4.1 The Paper 383
19.4.2 The Scientific Question..383
19.4.3 Why XAFS?...383
19.4.4 The Structure 384
19.4.5 Experimental Considerations 384
19.4.6 The Model 384
19.4.7 Drawing Conclusions..387
19.4.8 Presentation 387
19.5 Sulfur XANES from the Wreck of the Mary Rose 388
19.5.1 The Paper 388
19.5.2 The Scientific Question..388
19.5.3 Why XAFS?...389
19.5.4 Experimental Considerations 389
19.5.5 Principal Component Analysis...390
19.5.6 Linear Combination Analysis...390
19.5.7 Drawing Conclusions..391
19.5.8 Presentation 391
19.6 Identification of Manganese-Based Particulates in Auto Mobile
Exhaust 392
19.6.1 The Paper..392

19.6.2 The Scientific Question 392
19.6.3 Why XAFS? 392
19.6.4 Experimental Considerations 392
19.6.5 Principal Component Analysis and Target Transforms 393
19.6.6 Linear Combination Analysis.. 393
19.6.7 Fingerprinting... 393
19.6.8 Drawing Conclusions.. 394
19.6.9 Presentation 394
19.7 *In Situ* Investigation of Cobalt Molybdenum Sulfide Catalysts 395
19.7.1 The Paper 395
19.7.2 The Scientific Question 395
19.7.3 Why XAFS? 395
19.7.4 Experimental Considerations 396
19.7.5 Principal Component Analysis and Blind Source Separation 396
19.7.6 Column-Wise Augmentation 397
19.7.7 EXAFS Modeling 398
19.7.8 Drawing Conclusions.. 398
19.7.9 Presentation 399
19.8 Speciation of Gold in Hydrothermal Fluids.................................... 401
19.8.1 The Paper 401
19.8.2 The Scientific Question 401
19.8.3 Why XAFS? 401
19.8.4 Experimental Considerations 401
19.8.5 HERFD-XANES Results 402
19.8.6 Theoretical XANES Standards 402
19.8.7 EXAFS Data 402
19.8.8 EXAFS Analysis 402
19.8.9 Drawing Conclusions.. 404
19.8.10 Presentation 404
19.9 Local Structure of an Entropy-Stabilized Oxide............................... 405
19.9.1 The Paper 405
19.9.2 The Scientific Question 405
19.9.3 Why XAFS? 406
19.9.4 The Structure 406
19.9.5 Experimental Considerations 406
19.9.6 The Model 406
19.9.7 Drawing Conclusions.. 407
19.9.8 Presentation 408
19.10 The Next Case Study: Yours 409
References 409

Index **411**

Foreword to the Second Edition

I really like this book. Not just because Scott is my long-time friend—we were post-docs together oh-so-long ago—but because he is a gifted teacher. Pedagogy is really important to Scott. And his commitment to teaching well is really important to me. I started working at beamlines, writing data analysis software, and teaching in XAFS workshops over two decades ago. I always needed a book like this and, more to the point, young scientists new to XAFS always needed a book like this. Now we have a second edition. Delightful!

If you think a science text needs to be a stodgy affair filled with long equations and long words, *XAFS For Everyone* is probably a bit off-putting. With colorful characters engaged in intertextual dialog and cute cartoon depictions scattered throughout, this book might not look serious enough for you. Phooey, I say! Give this book a go. You'll learn some XAFS and you'll learn it well.

The running conversations of Dysnomia, Carvaka, Kitsune, and the rest of the gang are the breezy charm that makes this volume very readable. They are the heavy pedagogy going on in the margins, expanding upon and explaining the difficult concepts running through the main text. The supporting cast also provides the spark of life that extends well beyond the page. There is ongoing debate among the users of my beamline about which flesh and blood scientist each character is meant to represent. Now, Scott assures anyone who asks that each character is an archetype and not a caricature of any real person. My users want none of that and most think that I am Mr. Handy. That can't be right, of course. Simplicio—perpetually slack-jawed in confusion—is who I most feel like. But Scott says 'Archetype!' and I believe him. Now, however, I need to get back to the beamline. There are at least eight things that need to be done right away …

Bruce Ravel

Physicist, Materials Measurement Laboratory, National Institute of Standards and Technology

Lead Beamline Scientist at the Beamline for Materials Measurement at the National Synchrotron Light Source

Preface to the Second Edition

It has now been ten years since the first edition of *XAFS for Everyone* was published. That almost sounds like an intentional interval, but truth be told I had intended to update *XAFS for Everyone* well before that. As is often the case, any number of events, from the personal to the global, intervened to delay those plans.

In retrospect, I am glad for the delay. The past few years have seen a burst of activity in XAFS and related fields; no fewer than 12 of the references in this edition are from the 2020s, including one of the new case studies in Chapter 19. Those new developments and examples would not have been included if the new edition had come out in 2020, as originally planned.

This new edition is expanded, revised, and reorganized. The first part of the book now includes separate chapters on hardware, spatial and temporal resolution, high energy-resolution experiments, and experimental design, all areas which have seen rapid development in the last decade. In addition, a section on waevelet transforms has been added to the chapter on data reduction. In Parts II and III, while the chapter titles remain the same, recent developments in principle component analysis (particularly blind source separation) and theoretical standards for XANES have been incorporated. Finally, three new case studies based on research conducted since the publication of the first edition have been added to Part IV.

If you're familiar with the first edition of this book, you might be surprised to find that my panel of cartoon characters have also been busy. Simplico and Kitsune are featured in *Cartoon Physics: A Graphic Novel Guide to Solving Physics Problems*. Rather than just commenting from the margins, they star in a full-length story, joined by several new characters including the jaguar Alfa, a malevolent department chair, who is the villain of the piece. Alert readers will also spot cameos by the other panelists! That book is aimed at undergraduates taking general physics, bringing them to a new audience.

Speaking of *Cartoon Physics*, it is worth noting that my coauthor for that book is Kirin Emlet Furst, the illustrator of *XAFS for Everyone*. I first met Kirin when as an undergraduate she joined my XAFS research group at Sarah Lawrence College. Kirin, after receiving her doctorate from Stanford University, is now a professor of environmental engineering at George Mason University.

For the second edition, I must again acknowledge Bruce Ravel of the National Institute of Standards and Technology. He has been a staunch supporter of my efforts in publishing, even when they seemed a bit unconventional. I am honored that he has provided the forward to this edition, as well as feedback on some of the new material and a tour of the XAFS beamlines at the National Synchrotron Light Source II.

As always, the editorial staff at Taylor and Francis, including my editor Carolina Antunes, have been first rate, and shown great patience as I blew past deadline after deadline. In my opinion, I feel the final product was worth it, and I hope you feel the same.

Finally, I would like to thank my wonderful wife, Erin, for her support and patience as I worked on this new edition, as well as our cats Watson and Gilgamesh who, like my panelists, provided their own commentary as I wrote, although most of it consisted of 'stop writing and give us a snack.'

Preface to the First Edition

Long after its discovery in 1920, x-ray absorption fine structure (XAFS) was the domain of physicists who explored and theorized about the mysterious energy-dependent structures that appeared at energies above the absorption edge in x-ray spectra. In 1971, Dale Sayers, Ed Stern, and Farrel Lytle finally provided a satisfactory description of the physical process that created these features. This description, which incorporated the application of Fourier transforms, provided a readily understood connection to the geometry of the material being measured, immediately suggesting that XAFS could be a powerful tool for characterizing a wide range of materials.

A short time thereafter, x-rays from synchrotron light sources began to be used to generate x-ray absorption spectra, dramatically improving the speed and accuracy of data collection. The next two decades featured a rapid development of the theory connecting structure to spectrum. This, along with the rapidly increasing computing power available to the typical scientist, spurred the development and dissemination of software that could aid in the analysis of XAFS.

But even with the most powerful software, XAFS is not a black box. The Fourier transform provides an evocative connection to structure, but it does not provide a method by which the structure of a substance can be read directly. The analysis of XAFS is a skill that has to be learned.

XAFS is no longer the sole domain of physicists. It is used as a tool in fields as diverse as materials science, synthetic chemistry, environmental science, structural biology, and cultural fields such as archaeology and art conservation.

The 'Everyone' in *XAFS for Everyone*, therefore, includes physicists and archaeologists, undergraduates and mid-career scientists, front-line researchers and referees and heads of research programs. I wrote this book expecting that many of my readers would have no more knowledge of physics than what is covered in a first-year undergraduate class, and I used calculus only sparingly. However, I did not hesitate to borrow examples and terminologies from multiple disciplines, so that you will find 'dopants' and 'cyclic voltammetry' jostling side by side with 'ligands,' soil samples, and *operando* experiments in catalysis.

Most likely, however, the first thing that struck you when you leafed through this book was the cartoon characters. Have we really come, you may have wondered, to the point that sophisticated scientific ideas have to be livened up by talking kangaroos and monocle-wearing ducks?

Actually, using colorful characters to disseminate scientific ideas is not new. Galileo, for instance, in his preface to the *Dialogue Concerning the Two Chief World Systems*, writes:

> I have thought it most appropriate to explain these concepts in the form of dialogues, which, not being restricted to the rigorous observance of mathematical laws, make room also for digressions which are sometimes no less interesting than the principle argument.

The question of whether it is better to prepare a sample for XAFS measurements by spreading powder on tape or by diluting it and pressing it into a pellet may not be as weighty as whether the Sun circles the Earth or vice versa, but it lends itself to a similar treatment. While XAFS experts usually agree on the result of an analysis, there is great variety in the strategies they use to get to that result. Rather than clutter this book up with a formal discussion of each approach, I put them in the mouths of seven characters, each of whom is then free to speak without fear of derailing the primary narrative. The characters can also use asides to speak to different slices of the broad audience for which the book is aimed; a physicist might want to see the details of a mathematical derivation, a biologist particular tricks for dealing with fragile samples, an undergraduate something about the culture of scientific research, and a graduate student writing his or her first article a little about conventions in publication.

Galileo's approach, however, is not without its perils. One of his characters was the hapless Simplicio, an earnest follower of the old, Aristotelian ideas. His role in the dialogues was to advance common but incorrect arguments that could then be knocked down by the others. Unfortunately, Galileo's enemies suggested to Pope Urban VIII that Simplicio was in fact modeled after His Holiness the Pope. This did not prove helpful to Galileo's subsequent career.

While I don't expect this work to lead to a trial for heresy, I am alert to the possibility that some readers may attempt to identify the characters with particular experts in the field. It seems wise, therefore, to briefly discuss how they were developed and how they are used.

The characters were created by three of my undergraduate research assistants: Blaine Alleluia, Sydney Alvis-Jennings, and Lauren Glowzenski. These three had a practical knowledge of XAFS and used it to develop the list of characters, including the basic approach each would express and the animal that would represent them. Another undergraduate student, Kirin Furst (now the book's illustrator), then wrote an early draft of what is now Chapter 6, including comments from the characters. With their personalities thus established, I went on to develop them in the remaining chapters. Since these students had little knowledge or contact with others in the larger XAFS community, the characters are based on ideas, not individuals.

Of course, they sometimes express sentiments that have been emphasized by one member or another of the XAFS community. But real experts incorporate traits from all of the characters. Whenever one of us emphasizes physical processes, we are like Carvaka the owl; when we choose to begin with a fixed protocol, we are like Robert the duck; when we think about XAFS in the context of a particular research problem, we are like Kitsune the lemur, and so on. I have personally said things that come out of the mouths of each of the characters and that includes my own incarnation of Simplicio, now a cartoon dinosaur, who often expresses ideas that I held at one point or another in my career and have since abandoned.

I have occasionally been asked about the names of the characters. Besides Simplicio, a tribute to Galileo's character of the same name, they are named after the late mathematician Benoît Mandelbrot, whom I once had the pleasure of seeing speak; Brigadier General Henry Martyn Robert, the American military engineer who authored *Robert's Rules of Order*; the Cārvāka (pronounced 'Charvaka') school of ancient materialistic philosophy in India; the ancient Greek goddess Dysnomia; and Kitsune (pronounced Kit-soo-nay), the fox of Japanese folklore. Mr. Handy's name needs no explanation.

A work of this scope is impossible to write without help, and I have had plenty of it. Pride of place goes to Bruce Ravel, a world-renowned XAFS expert, who provided detailed reviews of many of my chapters as I wrote them. Additional reviewers for one or more chapters included Dalton Abdala, Blaine Alleluia, Leslie Baker, Agnieszka and Krzysztof Banas, María Elena Montero Cabrera, Soma Chattopahyay, Jason Gaudet, Sanjeev Gautam, Mihail Ionescu, Julian Kaiser, Regina Kirsch, Lia Klofas,

Andy Korinda, Rajesh Kumar, Wei Li, Jing Liu, Shanshan Liu, Lisa Van Loon, Todd Luxton, Lachlan MacLean, Vladimir Martis, Richard Mayes, Jasquelin Pena, Michael Peretich, Brandon Reese, John Rehr, Alexander Riskin, Will Vining, Jasen Vita, Van Vu, Michael Weir, Darius Zając, and Peter Zalden. Authors, when acknowledging reviewers, often thank the reviewers for how much they have improved the book while taking personal responsibilities for any errors that remain. I had thought that to be a platitude, but I now know it to be true: *XAFS for Everyone* would not be what it is without the thoughtful effort put in by all my reviewers. Many of their suggestions cut Gordian knots that I had no way to unravel on my own. But the responsibility for any errors that remain is mine and mine alone.

I also must acknowledge Lu Han, my editor at Taylor & Francis. Before meeting Lu, I had concerns that this book, with its diverse audience and its unusual format, would make for a difficult editorial process. But Lu 'got it' immediately and has time and again shown that he can express the purpose and audience of *XAFS for Everyone* better than I can! From small details to big ideas to finding the balance between giving me space and keeping me moving, Lu has far surpassed what I had hoped for in an editor.

Of course, I would never have considered writing this book if it wasn't for the support I have received over the years from the XAFS community. An incomplete list: Johnny Kirkland and Kumi Pandya, beamline scientists at the National Synchrotron Light Source; John Bargar, Britt Hedman, and Joe Rogers at the Stanford Synchrotron Radiation Lightsource; Matt Newville, Bruce Ravel, and Sam Webb, who have volunteered countless hours writing the analysis software I've used to learn my craft; Marten denBoer and Vince Harris, advisors and mentors for my PhD and postdoc.

Software used in the writing of this book includes the following:

- ATHENA, ARTEMIS, FDMNES, FEFF, IFEFFIT, and SixPACK for XAFS analysis
- HEPHAESTUS for x-ray data
- EXCEL and NUMBERS for general computation
- PAGES for preliminary layout
- MATHTYPE for equations
- PRO FIT for graphing
- AVOGADRO for visualizing chemical structures

Data sources for examples were drawn from my own measurements over the years supplemented by the XAFS Model Compound Library. The latter collection of XAFS spectra, described by Newville et al. (1999, *J. Synch. Rad.* 6:276–277), can be accessed at xafs.org by choosing the 'databases' option. Other sources of data are identified in the text at the point that they are used.

Author

Scott Calvin has been using x-ray absorption fine structure (XAFS) since 1998 to study systems as diverse as solar cells, magnetic nanoparticles, soil samples, battery cathodes, analogues to atmospheric dust particles, and pigments used in eighteenth century paintings. He was a member of the principal research team for beamline X-11B at the National Synchrotron Light Source, was a National Academies Research associate specializing in synchrotron science at the Naval Research Laboratory, and provided beamline support at the Stanford Synchrotron Radiation Lightsource. He has shared his expertise in XAFS spectroscopy in short courses across the world, including California, Illinois, New York, Belgium, and Thailand. Less conventionally, he coauthored the open-source pop-up book *National Synchrotron Light Source II: Long Island's State of the Art X-Ray Microscope.*

Photo by Josh Feldman
@cooperedtot

With a major in astronomy and physics and a minor in classics from the University of California, Berkeley, followed by a PhD in physics from Hunter College of the City University of New York, Dr. Calvin has taught in a wide variety of settings, including the Hayden Planetarium, Lowell High School, Southern Connecticut State University, the University of San Francisco, Sarah Lawrence College, Lehman College of the City University of New York, Examkrackers, and through Mirare Services, an educational company he ran in the 1990s. The courses he has taught have run the gamut from standard subjects such as General Physics to innovative general education courses including Crazy Ideas in Physics, Rocket Science, and Steampunk Physics. His published books include *Beyond Curie: Four Women in Physics and Their Remarkable Discoveries, 1903-1963,* and *Cartoon Physics: A Graphic Novel Guide to Solving Physics Problems.*

Dr. Calvin is currently the director of the prehealth program at Lehman College of the City University of New York, helping students their achieve their dreams of becoming physicians, dentists, pharmacists, veterinarians, and more.

About the Panel

 Carvaka is a professor of physics at a major research university. She has studied XAFS for more than two decades. Her favorite quote is from Richard Feynman: 'I have only to explain the regularities of nature—I don't have to explain the methods of my friends.'

 **Dysnomia** is a materials chemist working for a major industrial company. In addition to her own research in catalyst materials, she has become her company's go-to gal for using XAFS to characterize a wide variety of materials. She's been known to leap before she looks, but sometimes that gets her places no one else has been.

 Mr. Handy is a member of the beamline staff at a user facility synchrotron. A bachelor's degree in physics and a master's in material science have prepared him well for his current position, which he has held for 15 years. Through supporting users, he has seen just about every imaginable problem and found ways of fixing most of them. His motto: 'If it ain't broke, don't fix it, but just because you don't know it needs fixing, doesn't mean it ain't broke.'

 Kitsune is an environmental biologist and a professor at a liberal arts college. Generally good natured, she bristles at the notion that she is an XAFS expert: 'I use XAFS as a tool to study systems I am interested in, and I try to use it well. But my expertise is in those systems. I am no more an XAFS expert than a plumber is a wrench expert!'

 Mandelbrant is a geochemist working for a government laboratory. They believe the best way to attack a problem is to start from simple building blocks, understand their behavior thoroughly, and only then move on to more complicated systems. Their favorite proverb: 'A journey of a thousand miles begins with a single step.'

 Robert is a curator at a major museum, who uses XAFS to help understand, restore, and occasionally authenticate historical treasures. He believes in following a protocol whenever possible, even if at times it later becomes necessary to abandon it. There is a rumor among his staff that he wrote out a step-by-step plan before proposing to his wife, a rumor Robert has never denied.

 Simplicio is a first-year graduate student with an under-graduate degree in paleontology. He is new to XAFS and is excited about learning how to use the technique effectively. 'I try to learn from my mistakes,' says Simplicio, 'which is good, because I make a lot of them.'

Introduction

XAFS for My Birthday

Aside from the Prefaces and a brief word at the end, this is the only time in this book that I will speak to you directly as the author. For the rest of the book, you will see conventional text accompanied by a panel of cartoon characters providing a running commentary.

Each of those characters represents one facet of my own approach to x-ray absorption fine structure (XAFS)—actually one facet of the approach of any successful practitioner in the field. When I sit at the beamline trying to make as much of my limited time as I can and I find myself tempted to take a shortcut, it is Dysnomia, the irreverent kangaroo, hopping on my shoulder to whisper in my ear. When I find myself wondering, as I analyze the spectra from a sample, what is physically happening in the sample, that is Carvaka, my grumpy owl. Mr. Handy, the excitable octopus, warns me when I am thoughtlessly about to make a mistake that has been made a hundred times before and reminds me how to correct it if I do. And so on down the line—you will become familiar with these voices as you work your way through this book, and before you know it, they will be voices in your head, too.

But while they can spin tales that bring out particular features of XAFS collection and analysis, I thought it would be best to start with a story from my own career, something that gives you a taste of what this technique can—and can't—do.

I was a postdoc; it was my birthday, and somehow, I found myself in the laboratory I had worked in as a graduate student, working feverishly on a problem whose solution had eluded me for years.

Years before, when I was a graduate student, one of the faculty members in my department (let's call him S.) had been researching various lithium-ion batteries, including ones based on pyrite (FeS_2). While S. was not my advisor, the two had good relations, and our groups often helped each other out on projects. This particular project initially sounded rather manageable. An electrochemist overseas would measure the electrochemical properties of a lithium–pyrite electrochemical cell. We would then measure the iron XAFS at various voltages and after various numbers of cycles, hopefully confirming the speciation in each case.

As is often the case in electrochemical work, the list of likely phases for the iron atoms in the cells was fairly small: FeS_2, FeS, Li_2FeS_2, Li_xFeS_2, and metallic iron.

At first, things went smoothly. Metallic iron exhibits XAFS spectra that are very different from the oxidized phases. For one thing, the fact that the iron atoms are reduced in comparison to the other phases results in an earlier onset of the *x-ray absorption edge*. In addition, scattering of ejected photoelectrons off of neighboring iron atoms produces a characteristic set of wiggles in the x-ray absorption spectra well above the edge (the *EXAFS region*). I quickly identified metallic iron as present in a few

DOI: 10.1201/9780429329555-1

of the most highly cycled samples. In Chapter 9 of this book, we'll learn more about *fingerprinting* approaches such as this.

The other phases, however, were not so easy to untangle. If I had possessed good specimens of each of the phases in their pure form (*standards*), I could have used linear combination analysis (Chapter 10) to determine how much of each was present. But although I had good standards for FeS_2, FeS, and metallic iron, those for the lithiated phases proved too hard to come by. My standard for Li_2FeS_2, for instance, had not had its composition verified by other techniques—the researcher who provided it was essentially guessing that it might be the desired material. In addition, the literature suggested that the structure of Li_2FeS_2 found in electrochemical cells was quite different from the bulk crystalline form. In fact, when I compared an extended x-ray absorption fine structure (EXAFS) spectrum calculated using the published structure of bulk Li_2FeS_2 to the spectrum of my standard, I found them to be considerably different. But without more information, it wasn't clear whether this was because the structure of electrochemically generated Li_2FeS_2 was that different from the bulk or whether my standard was not Li_2FeS_2.

I now know that *principal component analysis* (Chapter 11) would have been a good tool for this problem, but at the time I was unfamiliar with that technique.

And so I resorted to *modeling* (Chapter 12), a powerful but time-consuming technique that was my specialty. To model the spectra, I had to make guesses as to what each sample contained. For instance, I might presume a sample contained some metallic iron, some FeS, and some Li_2FeS_2. I allowed for modifications to the structure relative to the bulk crystalline forms, such as different coordination numbers and bond lengths. And I then used computer software to simulate the EXAFS spectra for my model and compare it to the experimental data.

This approach soon allowed me to rule out FeS in all of the samples—it just would not fit the data, no matter how I modified the structure. This was satisfying, as earlier work in the literature had argued against the presence of FeS as an intermediate. FeS_2 was also absent in all of the measured cells. (This may seem surprising at first, as these were nominally FeS_2 cells! But they had all been cycled at least 40 times, and it was well known that significant irreversible changes occurred during the first few cycles, suggesting that the starting material of FeS_2 was never regenerated.)

The approach also revealed that some of the spectra were consistent with either Li_2FeS_2 (an electrochemical form somewhat different from the bulk crystalline structure) or mixtures of metallic iron and Li_2FeS_2. But about half of the electrodes showed evidence of an additional phase, both in the charge–discharge curves and in the XAFS. This additional phase was not any of the initial suspects.

And thus I was stumped. I would return to the problem again and again, only to be continually defeated. S. took to crying out one word whenever he saw me in the halls: 'PYRITES!' This was his subtle way of reminding me I should continue to work on the problem.

Eventually, I wrote my dissertation (which thankfully had nothing to do with pyrites), graduated, and secured a postdoc hundreds of miles away. But still, the calls and e-mails came: 'PYRITES!' This project was not going to die a quiet death.

At last, I happened to be passing through the city where I had done my graduate work at the same time as the electrochemist from overseas. S. seized the opportunity to get us to agree to spend an afternoon sitting in my old lab discussing the problem. (He is a very persuasive fellow!) It just so happened that this one available day was my birthday, and that's how it turned out that I spent my birthday back in my old laboratory, working on a problem that had hounded me for years.

I had never met the electrochemist in person before. And we were both tired of this project. I had reached a dead end in my analysis, as had she. We weren't sure what we were supposed to be talking about.

So I began to describe my dead end. I explained how the samples with the mystery phase appeared to include fully reduced iron, but that neither conventional iron metal nor any of its allotropes (face-centered cubic iron, for instance) were consistent with the EXAFS. It was as if there were neutral iron atoms without iron neighbors, but of course that didn't make any sense.

Except that the electrochemist thought it might make sense. She remembered seeing work in the electrochemical literature suggesting that isolated neutral atoms of metal could form and coordinate to materials like oxides or sulfides.

So I tried modeling the unknown phase as neutral iron atoms coordinated to the oxygen of the polyethylene oxide (PEO) we used as a matrix in our electrodes. And it worked! The model could reproduce the features of the measured spectra quite well.

I also tried a model where the neutral iron atoms coordinated to the sulfur in the lithium sulfide formed by the cell. This model also worked—with all the freedom that I was allowing my models (bond lengths, phase fractions, coordination numbers, etc.), the two models could not be distinguished solely based on the ability to reproduce the measured spectra.

But my colleague and I had additional knowledge about the system. When I used the model with iron bonded to oxygen, a good fit required a coordination number of 8 ± 1 oxygen atoms. When I used sulfur instead, the fit required 7 ± 1 sulfur atoms coordinated to the iron, a similar result. The PEO, however, was a preexisting matrix in our electrode, and the amount of iron was significant. Requiring each iron to have even six nearest-neighbor oxygen atoms would completely disrupt the PEO matrix, a result that seemed to us to be implausible. The Li_2S, on the other hand, was forming and being consumed every time the cell was cycled anyway. It was not unreasonable that it could end up in an atomic scale mixture with neutral iron atoms, so that an iron–sulfur–lithium nanocomposite was formed.

The result made physical and chemical sense, was in accord with the electrochemical measurements, and was consistent with the XAFS. We soon published the results (Strauss et al., 2003), and I no longer had to worry about hearing the cry of 'PYRITES!' in the halls. All in all, it was not a bad way to spend my birthday.

Most XAFS analyses don't stretch on for years, but my experience with the pyrites project includes many aspects that are endemic to the technique. In *XAFS for Everyone*, we will see that the technique is not a 'black box,' providing definite answers to narrow questions. Rather it is a powerful scientific tool, requiring judgment and knowledge to wield effectively. Although nothing substitutes for the experience of applying XAFS to your own projects, this book is designed to shorten the learning curve, incorporating real-world examples, helpful tips, and general principles into a practical introduction to the technique.

Reference

Strauss, E., S. Calvin, H. Mehta, D. Golodnitsky, S. G. Greenbaum, M. L. denBoer, V. Dusheiko, and E. Peled. 2003. X-ray absorption spectroscopy of highly cycled Li/composite polymer electrolyte/FeS$_2$ cells. *Solid State Ion.* 164:51–63.

THE XAFS EXPERIMENT

1

Chapter 1

XAFS in a Nutshell

Something for Simplicio

That was a nice story from our author. I appreciate how the electrochemistry and XAFS worked together to answer a real-world question.	
Note how it was necessary to think about what was actually happening within the system physically to distinguish between the two models.	
I just like that it worked!	
Pardon me, but I'm new to all of this. I know some chemistry and a little physics, but I don't know anything about XAFS. I got the gist of the story, but only in a fuzzy way. Could you help me learn more?	
Friends! We must not leave Simplicio behind. Let us help him out by providing a simple overview of XAFS now, leaving the more complicated details for later.	
Thanks, Mandelbrant!	

DOI: 10.1201/9780429329555-3

1.1 X-Ray Absorption Spectra

1.1.1 X-Ray Absorption Spectroscopy

In an x-ray absorption experiment, a sample of interest is bombarded with x-rays of definite energy. Some of these x-rays are absorbed by the atoms in the sample, causing the excitation or ejection of a core electron. This absorption can be quantified by comparing the intensity of the incident beam to that of the transmitted beam, by measuring the fluorescence given off by the excited atoms as an electron fills the now-vacant core orbital (the *core hole*) or by measuring the ejected electrons as the core hole is filled (*Auger-Meitner electrons*). Once the absorption is determined for one energy of incident x-rays, the energy is changed slightly and the process repeated. By stepping through a range of energies in this way, a spectrum is created. An example of an x-ray absorption spectrum is shown in Figure 1.1.

Several types of features can be seen in this spectrum:

■ An initial gradual, nearly linear trend toward lower absorption with energy
■ A sharp rise in absorption above incident energies of about 7100 eV
■ Several small peaks and shoulders near or on that sharp rise
■ A gradual oscillation up and down relative to the general trend downward, seen from about 7150 eV to at least 7400 eV, and perhaps beyond. The amplitude of this effect decreases with increasing energy, and the oscillations also increase in breadth.

These features, or something like them, are common to most x-ray absorption spectra.

1.1.2 Background

The gradual trend toward lower absorption seen in Figure 1.1 is a basic property of the interaction of photons and matter: as a general rule, the probability of absorption of photons with atoms decreases as the energy of the photons increases.

Of course, there are many exceptions to this basic rule. And those exceptions tell us something about the material being examined.

Since the exceptions are what's interesting, the slow trends seen in x-ray absorption spectra are usually referred to as the *background*.

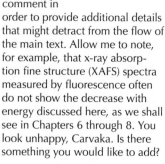

Welcome, gentle reader! Allow me to introduce myself. My name is Robert, and I shall often comment in order to provide additional details that might detract from the flow of the main text. Allow me to note, for example, that x-ray absorption fine structure (XAFS) spectra measured by fluorescence often do not show the decrease with energy discussed here, as we shall see in Chapters 6 through 8. You look unhappy, Carvaka. Is there something you would like to add?

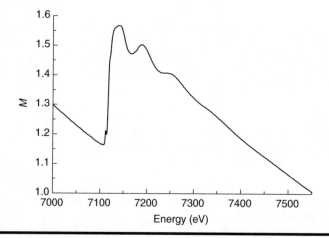

Figure 1.1 X-ray absorption spectrum of an electrode incorporating the iron–sulfur–lithium nanocomposite discussed in the Introduction. The *y*-axis is closely related to the absorption of x-rays by the sample (see Chapter 7), and the *x*-axis is the energy of the incident x-ray photons.

1.1.3 Edge

The sharp rise seen in Figure 1.1 is called an *edge*. It occurs because at energies below the edge x-ray photons do not have enough energy to excite electrons from some particular orbital, while above it they do. This leads to a sharp increase in absorption. For example, it takes approximately 7112 eV to excite an electron in the 1s orbital of iron, leading to a sharp increase in the absorption of iron-containing materials around 7112 eV, known as the *iron K edge*. (Excitations of orbitals from $n = 1$ states are referred to as K edges, $n = 2$ states as L edges, $n = 3$ states as M edges, and so on.)

1.1.4 X-Ray Absorption Near-Edge Structure

The peaks, shoulders, and other features near or on the edge are known as *x-ray absorption near-edge structure* (XANES). The interpretation of XANES is addressed primarily in Chapter 9.

I would like to make it clear that when you say that fluorescence spectra often don't 'show' a decrease with energy, it's an artifact of the way the measurement is performed, and not a sign of the fundamental physics of absorption. An atom doesn't 'know' whether you're measuring in fluorescence or transmission mode, and in fact you may be measuring in both.

1.1.5 Extended X-Ray Absorption Fine Structure

The gradual oscillations above the edge are known as *extended x-ray absorption fine structure* (EXAFS). This phenomenon is discussed in more detail in Section 1.2.

Come now, friends, we are getting ahead of ourselves! All those details will come in good time. Be like Mandelbrant (your friend the ant), start with the basics, and work from there.

1.2 Basics of EXAFS Theory

1.2.1 Fermi's Golden Rule

According to quantum mechanics, systems, when measured, exist in definite states. An atom may be in its ground state, for instance, or one of its electrons may be excited to a higher orbital, or an electron may be missing from the bound states of the atoms altogether resulting in an ion. A measurement designed to find the energy states of electrons in an atom will always yield definite answers: the atom either is in its ground state or is not; it has either 0, 1, or 2 electrons occupying its 1s orbital; and so on.

But when a system is disturbed, it can exist for a time as a superposition of these possibilities. Only when it is measured in some way is one of the possibilities selected.

For instance, when an x-ray enters the region occupied by an atom, several things could happen: the x-ray could pass through unchanged, it could scatter off of the atom, it could be absorbed by an electron in the 1s orbital of the atom, and so on. Until a measurement is made, the system exists as a superposition of these possibilities, each with its own probability. According to *Fermi's Golden Rule*, the probability of each possibility coming to pass depends on the similarity of the proposed final state to the indeterminate state prior to measurement—the more similar, the more likely (Dirac 1927).

In this context, 'measurement' doesn't necessarily imply human intervention. If, for instance, the x-ray goes on to interact with another atom farther down the beam path that provides indirect evidence that it was not absorbed by the first atom; a 'measurement' has occurred.

1.2.2 EXAFS Is Due to Interference of an Electron with Itself

If an x-ray ejects an electron from an atom, you can think of that electron (called a *photoelectron*) as a wave radiating out in all directions, with a wavelength λ given by

$$\lambda = \frac{h}{p} \tag{1.1}$$

where h is Planck's constant and p is the momentum of the electron.

The electron wave can scatter off of nearby atoms, returning to the original, absorbing atom. Since it's a wave, it could interfere constructively or destructively at that

Welcome, all! My name is Kitsune. We're not going to go into the mysteries of quantum mechanics deeply in this book. If you'd like to know more, you might want to consult an introductory text on the subject.

point. If it interferes constructively, then there is more electron density at the absorbing atom, and the system looks more like the (neutral atom + photon) state that occurs before measurement, than if it were to interfere destructively. That means that constructive interference increases the likelihood of finding this state after measurement rather than, say, the state where the photon continues on without being absorbed.

BOX 1.1 Mandelbrant's Plan

	So the bottom line is that the probability of the absorption of an x-ray is enhanced if it leads to constructive interference of the ejected electron at the location of the absorbing atom, and the probability is reduced if it would lead to destructive interference there. Oh, sorry, I don't think I introduced myself. My name's Dysnomia, and I like knowing the bottom line. I never liked derivations much; I leave that stuff to folks like Mandelbrant.
	But friend Dysnomia, we can learn so much from thinking things through! Derivations are nothing more than ways of starting from simple assumptions to learn more complicated things. We can begin to learn about EXAFS, for instance, by thinking of the electron scattering off of a nearby atom as if it were a plane wave bouncing off of a soft boundary.
	Mandelbrant, the photoelectron produced in EXAFS is not a plane wave, it's spherical! And 'soft boundary' is a little vague.
	True, friend Carvaka. But even though air resistance prevents a feather from falling as fast as a cannon ball, we are first taught in introductory physics that they fall at the same rate. Only later do we add in adjustments for phenomena such as air resistance, important though they may be. In just such a way, let us begin by treating the EXAFS photoelectron as a plane wave bouncing off of a soft boundary that does not change the phase of the wave and then add back in complications one by one.

If we treat the scattering of the ejected photoelectron off of a nearby atom as if it were a plane wave bouncing off of a soft boundary, we find that constructive interference at the location of the absorbing atom is achieved if the round-trip distance is a whole number of wavelengths, that is,

$$2D = n\lambda \tag{1.2}$$

where D is the distance from the absorbing atom to the scattering atom and n is an integer. (The number 2 is there because the electron needs to travel from the absorbing atom to the scattering atom and back.)

Since we are modeling the photoelectron as a plane wave (Box 1.1), and a plane wave is sinusoidal, the interference pattern should be as well. This leads to the probability of absorption being modulated by a factor χ:

$$\chi \propto \cos\left(2\pi \frac{2D}{\lambda}\right) \tag{1.3}$$

Note that this function gives a maximum when Equation 1.2 is satisfied, as we expect.

Wavenumber k is a parameter defined by

$$k \equiv \frac{2\pi}{\lambda} \tag{1.4}$$

which means Equation 1.3 can be rewritten in the compact form as

$$\chi \propto \cos(2kD) \tag{1.5}$$

1.2.3 Relationship of k to Photon Energy

Equations 1.1 and 1.4, along with the definition $\hbar \equiv \dfrac{h}{2\pi}$, can be combined to yield the useful relationship:

$$p = \hbar k \tag{1.6}$$

Using basic physics, the momentum p of the photoelectron can, in turn, be related to its kinetic energy T:

$$T = \frac{p^2}{2m_e} \tag{1.7}$$

where m_e is the mass of an electron.

Finally, the kinetic energy of the photoelectron is equal (in our simple model) to the energy E of the incident photon, less whatever energy was necessary to remove the photoelectron from the absorbing atom, which we'll call E_o:

$$T = E - E_o \tag{1.8}$$

Combining Equations 1.6 through 1.8 yields the result:

$$k = \frac{1}{\hbar}\sqrt{2m_e(E - E_o)} \tag{1.9}$$

1.2.4 EXAFS and Structure

We can already see why EXAFS can give us information about the structure of the material. As we scan through photon energy in the EXAFS region of an x-ray absorption spectrum, we are also scanning through photoelectron wavenumber. According to the proportionality in Equation 1.5, the spectrum should be modulated by an oscillatory factor. In our simple model, the oscillations will be spaced regularly in k, which means, according to Equation 1.9, that they'll be more and more spread out as we go to higher energies. An examination of Figure 1.1 suggests that this is indeed the case.

But what makes these oscillations useful for characterization is that the spacing of the oscillations also depends on D, the distance between the absorbing and scattering atoms. The smaller the value of D is, the slower the oscillations occur as a function of k (and E).

You can perhaps see how this might be useful. For instance, suppose you are trying to determine whether a metallic element in your sample is oxidized. If it were oxidized, the distance between the absorbing atom and nearby scatterers would be smaller than in the metallic form, leading to EXAFS oscillations that are more spaced out.

1.2.5 Scattering Probability

The heuristic derivation in Section 1.2.2 is based on the model that the photoelectron scatters elastically off of a nearby atom at a distance D from the absorber. Of course, many other things could happen to the photoelectron. It could, for instance, scatter inelastically off of a nearby atom. If that happens, some of the energy of the photoelectron is lost, its wavelength changes, and its interference pattern is different. Or it could fail to scatter off of the atom at all. In the usual manner of quantum mechanics, each of these possibilities has a probability. We can, therefore, write the proportionality in Equation 1.5 as

$$\chi(k) = f(k)\cos(2kD) \tag{1.10}$$

Here, $f(k)$ is a proportionality constant that is itself proportional to the possibility of scattering elastically off of the atom, but may also include other factors, such as geometrical ones. We've noted explicitly that the probability of scattering elastically depends on the wavenumber k (and thus the momentum) of the photoelectron.

A big atom with lots of electrons, such as lead, will usually cause a photoelectron to scatter with much higher probability than a small atom with few electrons, such as oxygen. In addition, the dependence of $f(k)$ on k is different for different elements, as is explained in Chapter 13. Thus, EXAFS can yield information about the type of atoms nearby the absorbing atom, as well as the distances to them.

Equation 1.10 is our first stab at 'the EXAFS equation,' but it is a very, very preliminary one. We'll spend the following several sections refining it.

 Before we go on, think of as many ways as you can in which Equation 1.10 is a simplified version of reality. Even as we continue to fill in details, see if you can think of more. By doing that, you will come to better understand the EXAFS equation and its limitations.

1.2.6 Multiple Neighbors

One correction to Equation 1.10 that may have occurred to you is that there isn't only one neighboring atom the photoelectron can scatter off of. Unless you are working with a diatomic gas, there are likely to be several. Some may be closer to the absorbing atom, and some farther. They may be of the same species, or not.

Each of these scattering events contributes separately to modulating the absorption probability. In other words, we can describe the modulation as a sum of the modulations from scattering off of each neighboring atom (we'll label the different scattering possibilities by the subscript i):

$$\chi(k) = \sum_i f_i(k)\cos(2kD_i) \tag{1.11}$$

Of course, sometimes several of those atoms will be of the same species at the same average distance. For example, consider a block of copper metal. Almost every copper atom is surrounded by 12 identical copper atoms, each the identical distance away (on average, anyway). We can group those 12 atoms together, by noting that they have 12 times the effect a single atom would have. This *degeneracy* is noted by the symbol N (see Chapter 13 for a more thorough discussion of degeneracy). Our equation now reads:

$$\chi(k) = \sum_i N_i f_i(k)\cos(2kD_i) \tag{1.12}$$

Thus, EXAFS also provides information on the number of scattering atoms, as well as their identities and distances.

1.2.7 Multiple Scattering

So far, we've discussed events where the photoelectron scatters elastically off of a nearby atom and then returns to the absorbing atom. This is known as *single scattering* or *direct scattering*. We could also, however, imagine the photoelectron scattering elastically off of one nearby atom, then off of another, and only then returning to the absorbing atom. This is known as *multiple scattering*. Multiple scattering does not require us to rewrite Equation 1.12, as long as we generalize the meaning of the factors. For a single scattering path, D_i was the distance from the absorbing atom to the scattering atom. Since the photoelectron has to get from the absorber to the scatterer and back again, it is half of the total distance traveled by the photoelectron; that is, it is half of the *path length*.

For multiple scattering, we'll simply define D_i in the same way as half the total distance traveled by the photoelectron when following that path. For instance, suppose the photoelectron begins at absorber A and then scatters off of atoms X and Y before returning to the absorber A. To get the value of D_i for that path, we would add up the distance from A to X, from X to Y, and from Y back to A, and then divide the total by 2.

For a discussion of how the degeneracy N_i is defined for multiple scattering paths, see Chapter 13.

Multiple scattering can include more complicated sequences. Even diatomic gases exhibit some multiple scattering, such as the path that travels from the absorber to the scatterer, then back to the absorber, back to the scatterer again, and finally once more to the absorber.

1.2.8 Phase Shifts

Equation 1.12 treats scattering atoms as if they're some sort of ideal 'soft boundary,' which do not change the phase of the photoelectron wave and yet reverse its direction instantaneously. The reality is, of course, more complicated; we'll examine it in detail in Chapter 13. For now, we'll simply observe that there should be a phase shift introduced into our equation. As long as we're putting in a phase shift anyway, we'll also switch from cosine to sine. (A sine function is, after all, just a phase-shifted cosine function.) While we could equally well continue to use cosine, for historical reasons it is traditional to use sine in the EXAFS equation.

Yeah, but scattering back and forth and back and forth like that doesn't usually contribute much to your spectrum. We'll talk more about this in Section 12.3.1 of Chapter 12 and Section 17.5 of Chapter 17.

$$\chi(k) = \sum_i N_i f_i(k) \sin\left(2kD_i + \delta_i(k)\right) \tag{1.13}$$

1.2.9 Spherical Waves

The photoelectron wave is not a plane wave, of course. It spreads out isotropically; that is, it is a spherical wave. One important ramification of this is that as the wave spreads out the scattering probability drops as the square of the distance. We acknowledge that explicitly by writing:

Like $f(k)$, the phase shift $\delta(k)$ depends on the species of the scattering atom.

$$\chi(k) = \sum_i N_i \frac{f_i(k)}{kD_i^2} \sin\left(2kD_i + \delta_i(k)\right) \tag{1.14}$$

This means we are defining $f_i(k)$ differently than we did before. In this form, and properly accounting for spherical wave effects on scattering, it is sometimes given the additional subscript 'eff,' short for 'effective.' Rather than clutter up the equation more than necessary, we'll follow common practice and leave it as $f_i(k)$, with context making it clear that it is the form appropriate for spherical waves.

BOX 1.2   There Is No Such Thing as 'The EXAFS Equation'

	I get the D_i^2 part of Equation 1.14, but where did that k in the denominator come from?
	If you don't like it, it doesn't have to be there. I mean, $f_i(k)$ is already a function of k, so we could get rid of it by just redefining $f_i(k)$. Or we could redefine $f_i(k)$ so that there is a k^2 in the denominator. Or a k in the numerator. All those things have been done, and they've all appeared in the literature at one time or another.
	What Dysnomia is trying to say is that, unlike the other factors we've discussed, k in the denominator doesn't correspond to a physical effect. It was chosen to make the definition and units of $f_i(k)$ consistent with those used in some other contexts. But depending on what you try to make the definition of $f_i(k)$ consistent with, you'll end up with different powers of k out front.
	Perhaps, Carvaka. But it is good to have symbols mean the same thing to everyone. Thus, the International Union of Crystallography has stepped in (IUCr 2011) and made the version with k in the denominator the standard one, meaning that $f_i(k)$ has units of length.
	But then, the IUCr follows that up by talking about 'other definitions' that have a k^2 in the denominator instead.
	As we'll see, friends, the idea that there is such a thing as 'the EXAFS equation' is a bit of a myth. There are many versions of the equation, written using somewhat different conventions, and emphasizing different aspects of EXAFS phenomena. While we will continue to talk about 'the EXAFS equation,' we should keep in mind that different versions are in use.

1.2.10 Incomplete Overlap

Regardless of what the photoelectron does, the final state of the absorbing atom is not the same as the initial state. The photoelectron has been ejected from the atom, leaving behind a core hole. All of the other electrons thus feel more positive charge from the nucleus (i.e., it is not as well shielded), and the orbitals adjust to this change. The incomplete overlap can be phenomenologically modeled by an element-dependent constant, the *amplitude reduction factor* S_0^2:

$$\chi(k) = S_o^2 \sum_i N_i \frac{f_i(k)}{kD_i^2} \sin\left(2kD_i + \delta_i(k)\right) \tag{1.15}$$

This effect is modest but not negligible; S_0^2 typically has a value between 0.7 and 1.0.

See Chapter 13 for a discussion of the limitations of the S_0^2 approximation.

1.2.11 Mean Free Path

Aside from scattering elastically off of nearby atoms, other fates can befall the photoelectron. It could scatter inelastically, perhaps exciting a valence electron from one of the scattering atoms or a *phonon* (vibration) in the crystal. Any event of this sort will remove energy from the photoelectron, thus changing its wavelength and the resulting interference condition. Because the number of different specific outcomes for this kind of event is very great, and because some will result in constructive interference at the absorbing atom and some destructive interference, the net result is to suppress part of the *main channel* of the EXAFS signal; that is, suppress $\chi(k)$ in Equation 1.15. The reason we can't just fold this suppression into our S_0^2 factor is that the effect has a strong D dependence; the farther the electron has to go, the more likely something other than elastic scattering will happen to it.

In addition, the core hole isn't going to wait around forever. Sooner or later, an electron in a higher orbital will fall into the core hole, likely resulting either in fluorescence or in the ejection of another electron (i.e., an Auger-Meitner electron). Either way, the absorbing atom will be in a different state, and thus, the overlap between initial and final states used for Fermi's Golden Rule will be different. Once again, signal will be removed from the main channel.

Section 1.2.11 provides a reasonable mental model of why we use a mean free path. But it's taking some liberties with the actual quantum mechanics. The photoelectron and the core hole are entangled; that is why Fermi's Golden Rule applies to this process in the first place. Some processes, such as core-hole decay, can destroy that entanglement.

Like the suppression due to inelastic scattering of the photoelectron, suppression due to decay of the core hole will also depend on the length of the photoelectron's path: the farther it has to travel, the longer it takes, and the longer it takes, the more likely the core hole won't be around anymore when it gets back. So both of the effects can be lumped together into a single factor:

$$\chi(k) = S_o^2 \sum_i N_i \frac{f_i(k)}{kD_i^2} e^{-\frac{2D_i}{\lambda(k)}} \sin\left(2kD_i + \delta_i(k)\right) \tag{1.16}$$

$\lambda(k)$ is called the *mean free path* of the photoelectron. It is primarily because of this term that EXAFS is a local phenomenon; it ensures that scattering off of atoms more distant than around 10 Å is usually negligible.

Why do we call it 'mean free path' if it also depends on the core-hole lifetime?

1.2.12 EXAFS Is an Average

So far, we've been discussing an EXAFS experiment as if a single photon is absorbed by a single atom, resulting in a single photoelectron. But in an actual experiment, there are millions of x-rays being absorbed by millions of atoms, and we're collecting the average. (In fact, that's how we measure a probability!)

If every absorbing atom were in an identical environment, then Equation 1.16 would be fine. But in a real material, the environments may differ. We'll divide those differences into four categories:

Think of it as short for 'mean free path of the photoelectron before something happens that messes up the main EXAFS channel.'

1. The absorbing element may be in more than one crystallographic environment. This may be because more than one phase is present (e.g., zinc metal and zinc oxide) or because a single phase has different kinds of sites (e.g., spinel crystal structures have both tetrahedrally and octahedrally coordinated cation sites).

2. There may be local differences in the environment of absorbing atoms. For example, a crystal might have defects that are scattered essentially at random throughout the material. Amorphous materials and liquids, where we can no longer discern a long-range crystal at all, are extreme cases of this. Local differences of these kinds are called *static disorder*.

3. There may be gradients within a material; for example, the composition of an amorphous film might change gradually with depth. These may be modeled as different crystallographic environments (e.g., for a nanoparticle, surface atoms might be distinguished from core atoms), or as if it were static disorder, or in some other way.

4. Even the environment of a single atom is not constant over time. At room temperature, most chemical bonds have vibration frequencies on the order of 10^{13} Hz. Typical core-hole lifetimes for x-ray spectroscopy are about 10^{-15} seconds, or about 1% of the time it takes a chemical bond to vibrate. By sampling millions of atoms, we will thus be sampling the variations in absorber–scatterer distance caused by those thermal vibrations as well. This effect is called *thermal disorder*.

When the absorbing atom is present in different crystallographic environments (case 1 from the list), each environment can be modeled as including its own set of scattering paths, and all sets of paths are then included in the sum in Equation 1.16. Rather than introduce a new factor to indicate the relative prevalence of each site, that weighting is just absorbed into N_i.

If thermal (case 4) and static (case 2) disorders are small, that is, not nearly large enough to change the interference from completely constructive to completely destructive (or vice versa), then they can be modeled by something called the *cumulant expansion* (see Chapter 13). To a first approximation, that leads the EXAFS equation to be modified with an additional factor dependent on the *mean square radial displacement*, symbolized by σ^2:

'Crystallographic environment,' in this context, doesn't necessarily mean the material has to be crystalline. An amorphous spinel-like structure, for instance, still has tetrahedral and octahedral sites. And while the protein you're working with might never have been successfully crystallized, if it has two zinc sites, it is clear that if it ever were crystallized those sites would be crystallographically distinct.

$$\chi(k) = S_o^2 \sum_i N_i \frac{f_i(k)}{kD_i^2} e^{\frac{2D_i}{\lambda(k)}} e^{-2k^2\sigma_i^2} \sin\left(2kD_i + \delta_i(k)\right) \tag{1.17}$$

We will discuss this factor, and other aspects of the cumulant expansion, in more detail in Chapter 13 (also see Box 1.3).

BOX 1.3 A Little More about σ^2

$e^{-2k^2\sigma_i^2}$ is a pretty complicated looking factor, and Chapter 13 is a long way off. Can't we talk a little more about it now?

Sure thing, Simplicio. σ_i^2 is the variance in D_i due to disorder. That means you could think of there being a parameter called σ_i that would be the standard deviation of that distance, although almost nobody ever actually does that in the literature. Since k is proportional to the number of photoelectron wavelengths per angstrom, $k\sigma_i$ is proportional to the spread of the distribution of D_i measured in wavelengths. For the approximation to work at all, $k\sigma_i$ needs to be much less than one; in other words, the disorder should shift the path lengths by much less than

> one wavelength of the photoelectron. From there, it's not surprising (to a mathematical physicist, anyway) that a first-order approximation involves a suppression with a Gaussian form.

If either thermal or static disorder is not small, then we can use a strategy somewhat like that used for different crystallographic environments, summing over paths that represent scatterers at various differences from the absorber (see, for example, Section 16.5 of Chapter 16).

Finally, the case of a gradient (case 3) can be handled in either manner or by a combination of the two. If the effect of the gradient is modest, then treating it as static disorder is reasonable. If it is more significant, then modeling it as a sum over different paths (e.g., 'core' paths and 'surface' paths) sometimes works well.

1.2.13 Enough for Now

Except for a few differences in notation, Equation 1.17 is identical to the equation recommended by the IUCr (2011). In this book, we will treat Equation 1.17 as 'the' EXAFS equation and have therefore enshrined it on the inside front cover.

That does not mean, however, that there are not further refinements that are sometimes made. For instance, the cumulant expansion can be taken further, allowing for somewhat more disorder to be accounted for. Another common modification is to include the effects of the polarization of the x-ray beam. We will discuss these considerations in future chapters.

1.3 Some Terminology

Certain aspects of XAFS terminology are not standardized well, and can be confusing. While most terms will be introduced as we need them, there are a few worth clarifying up front.

Note

Section 1.3 is based, in part, on material in the National Synchrotron Light Source Online Orientation, which is licensed under a Creative Commons Attribution-Share Alike 3.0 United States License.

1.3.1 Edge

The IUCr Dictionary defines an absorption edge as 'the energy at which there is a sharp rise (discontinuity) in the (linear) absorption coefficient of X-rays by an element' (IUCr 2011). This is, intentionally, not a very precise definition, as the sharp rise occurs over a range of energy—perhaps 10 or 20 eV.

'Edge' is often used to describe the sharp rise itself, rather than just the energy at which it occurs. For example, it could be said that the edge in Figure 1.1 exhibits a small shoulder about two-thirds of the way up.

'Edge' can also be used to identify the core hole corresponding to a given spectrum. Thus, Figure 1.1 can be described as 'the iron *K*-edge spectrum of the sample.'

Finally, 'edge' can be used as a synonym for E_o (see Section 1.3.7).

1.3.2 XANES

XANES stands for *x-ray absorption near-edge structure*. In Section 1.1.4, we mentioned that features 'near or on the edge' are known as XANES. But what exactly qualifies as 'near or on the edge?'

Section 1.2 focused on rough physical interpretations of the parameters in the EXAFS equation. Those who would like to see a somewhat more formal derivation could consult Bunker (2010).

You may have noticed that we already used 'edge' in all three senses in Section 1.1.3!

In actuality, there is not a sharp boundary between XANES and EXAFS, so choosing an energy of demarcation is somewhat arbitrary. The IUCr Dictionary suggests around 50 eV above E_o (IUCr 2011), but 30 eV above E_o is also commonly used.

1.3.3 NEXAFS

NEXAFS stands for *near-edge x-ray absorption fine structure*. Technically, NEXAFS is a synonym for XANES. In practice, the term NEXAFS is generally only used for low-energy edges, typically those below 1000 eV.

So you can either call the near-edge region XANES for both low- and high-energy edges, or you can call it NEXAFS for low-energy edges and XANES for high-energy edges.

1.3.4 EXAFS

EXAFS stands for *extended x-ray absorption fine structure*; that is, the oscillations present at energies above the XANES region. Notice that only the oscillations themselves are properly referred to as EXAFS, and not the gradual background. If you want to talk about the entire behavior of the spectrum at energies higher than those for XANES, including the background, glitches in the data, and so on, refer to 'the EXAFS region' rather than just 'the EXAFS.'

1.3.5 White Line

Often, a spectrum will exhibit a sharp feature at the top of the edge, showing absorption much higher than what is seen in the EXAFS region (see Figure 1.2). This sharp feature is called the *white line*.

The term 'white line' comes from the days when x-ray spectra were collected on film. Strong absorption at a given energy left the photographic negative unexposed, creating a literal white line.

1.3.6 Preedge

Preedge is used in at least two distinct ways:

1. To identify the fairly featureless part of the spectrum before the sharp rise associated with the edge.
2. To identify small features below the midpoint of the rising portion of the spectrum, such as the small sharp peak near 7115 eV shown in Figure 1.3. For this purpose, it's often used as an adjective: 'preedge feature.'

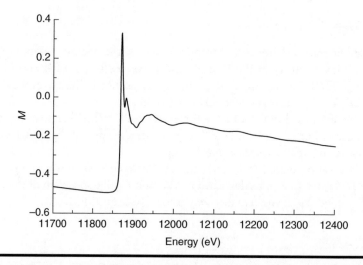

Figure 1.2 The arsenic *K*-edge spectrum of a partially oxidized gallium arsenide sample. The sharp feature around 11,870 eV is called the white line.

It is perhaps lamentable that the multiple ways in which the terms 'edge' and 'pre-edge' are used can lead to a 'preedge feature' occurring at a higher energy than a point described as the 'edge.' In practice, however, this strange terminology rarely causes confusion.

1.3.7 E_o

We introduced the parameter E_o in Section 1.2.3, calling it the energy 'necessary to remove the photoelectron from the atom.' This definition is neither precise (remove to where?) nor easy to directly measure. In fact, E_o is used to symbolize a variety of related quantities:

- It may be defined operationally, that is, in terms of the shape of the spectrum. For example, the first inflection point in the XANES region might be chosen, or the steepest inflection point, or a point halfway up the edge (see Section 7.3.3 of Chapter 7).
- It may be taken from a table (E_o for the iron K edge is 7112 eV).
- It may be chosen so as to make the EXAFS equation for a model structure match the data as well as possible (see Section 13.1.5 of Chapter 13).

These ways of defining E_o typically result in values that differ from each other by several electron volts. Failure to pay attention to which definition of E_o is being used can cause considerable confusion.

1.4 Data Reduction

Section 1.2 outlined the theory of EXAFS: given a material with a known arrangement of atoms, how would we predict what the spectrum looks like? But of course, we're usually more interested in the inverse problem: given a spectrum, how can we extract information about the atomic arrangement? For the rest of this chapter, we'll provide an outline of how that is done, once again leaving the details to chapters that follow.

The initial steps toward extracting information from XAFS spectra are to manipulate the data so as to present them in a standardized form—hopefully, one which emphasizes the features we are most interested in analyzing.

1.4.1 From Raw Data to χ(k)

When presented with a spectrum such as that shown in Figure 1.1, the first step is to normalize it, as has been done in Figure 1.4; that is, rescale it so that the rise associated with the edge (the *edge jump*) is 1.0. This facilitates direct comparison of the spectra of different samples, as well as the comparison of data with theory.

For XANES, data reduction is now done. But for EXAFS, additional steps are generally necessary.

Since EXAFS refers only to the oscillatory part of the spectrum, and not the gradual trends, a smooth background function is drawn through the EXAFS portion of the data, as has been done in Figure 1.5.

Next, the background is subtracted, yielding just EXAFS, symbolized by $\chi(E)$. This is shown in Figure 1.6.

Notice how the gradual curve from Figure 1.5 has been removed.

The next step is to convert $\chi(E)$ to $\chi(k)$, using Equation 1.9. The result is shown in Figure 1.7.

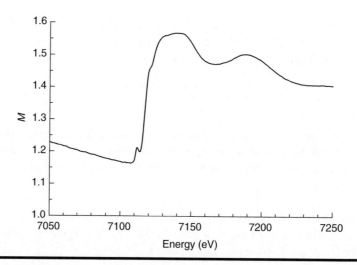

Figure 1.3 The same spectrum as in Figure 1.1, with the scale on the *x*-axis changed so as to see detail in the XANES region.

Some folks, particularly those who started working in the field in the 1970s and 1980s, informally refer to $\chi(k)$, or maybe the *k*-weighted $\chi(k)$, as 'the EXAFS.' If you are showing someone your data and they ask to 'see the EXAFS,' that's what they mean. A more recent colloquial term is '*k*-space graph,' and that's the informal term we'll use in this book.

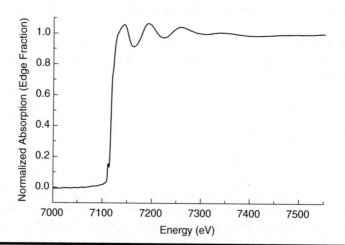

Figure 1.4 The data from Figure 1.1 after normalization.

Since the $e^{-2k^2\sigma_i^2}$ factor in the EXAFS equation causes $\chi(k)$ to drop off in amplitude with k, $\chi(k)$ is often multiplied by k, k^2, or k^3. This is called *k weighting* and results in a plot with more uniform amplitude, as can be seen in Figure 1.8.

It is also worth noting that the *k*-weighted data goes to zero at $k = 0$, which means that the XANES region is deemphasized. Since consistent background subtraction within the XANES region is difficult to achieve, this is desirable.

For some purposes, the data are now ready for analysis. But in many cases, an additional transformation is employed.

1.4.2 Fourier Transform

An inspection of Figure 1.8 shows a signal that is roughly sinusoidal, albeit with an amplitude that is a function of k. Given the system described in the Introduction, that makes sense; isolated neutral iron atoms correlated to a disordered lithium sulfide matrix will exhibit a fairly consistent distance to the neighboring sulfur atoms but are unlikely to have a consistent structure (i.e., a narrow distribution of absorber–scatterer distances) beyond those nearest neighbors.

k weighting makes the small signal at high *k* more visible, but it also enhances the noise at high *k*, resulting in the jagged appearance above *k* = 7 Å⁻¹ in Figure 1.8. Figures 1.7 and 1.8 reflect the reality that the signal-to-noise ratio is worse at high *k* than at low: in Figure 1.7 that manifests as a smaller signal; in Figure 1.8 as greater noise.

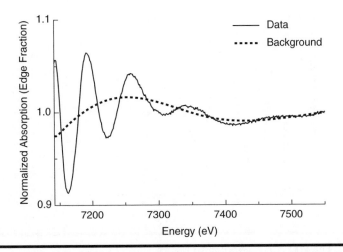

Figure 1.5 **The EXAFS region of the data from Figure 1.4. A smooth background function has been drawn through the data. When comparing to Figure 1.4, please observe the scale on each axis.**

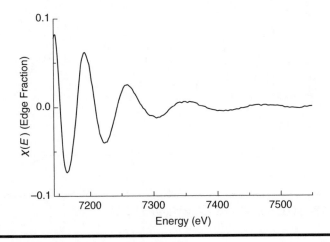

Figure 1.6 **The $\chi(E)$ extracted from Figure 1.5.**

So let's look at Figure 1.9, which shows the k^2-weighted $\chi(k)$ of a more ordered material, an iron metal foil.

This looks considerably less like a modulated sinusoid. A dominant spacing is still visible, with very large peaks spaced a bit more than one inverse angstrom apart. (Note the different scales on the y-axes of Figures 1.8 and 1.9!) But there are smaller peaks and shoulders visible, and the large peaks don't look much like sine waves. What's going on?

Iron metal, unlike the highly disordered substance which yielded the data in Figure 1.8, is an ordered crystal. Every iron atom is in a nearly identical environment, modified only by occasional defects and thermal vibrations. Every atom, therefore, has not only a similar near-neighbor environment but also similar next-nearest neighbors, next-next-nearest neighbors, and so on. Multiple terms in the EXAFS equation will, therefore, be significant. According to the EXAFS equation, each individual term is a modulated sinusoid. But when adding up multiple sine waves of different amplitudes, periods, and phases, the result can look something like Figure 1.9.

Fortunately, the problem of decomposing a function into constituent sine waves is an old one, with solutions going back to Joseph Fourier's work in the early nineteenth century (Fourier 1822). In honor of Fourier, the modern method is known as the

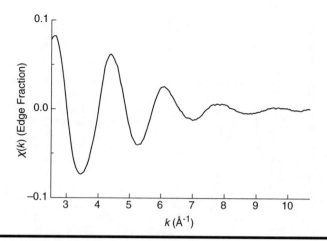

Figure 1.7   The χ(*k*) corresponding to Figure 1.6.

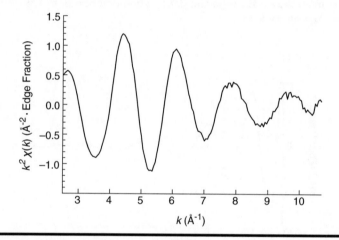

Figure 1.8   The χ(*k*) from Figure 1.7, weighted by *k²*.

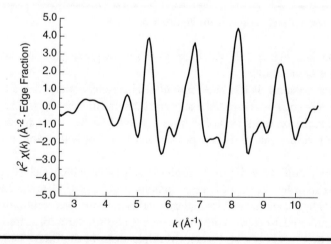

Figure 1.9   *k²*χ(*k*) for an iron metal foil.

Fourier transform. The magnitudes of the Fourier transforms of the data in Figures 1.8 and 1.9 are shown in Figure 1.10.

The differences we saw in the *k*-space graphs are reflected in the magnitudes of the Fourier transforms. The spectrum that is approximately sinusoidal in Figure 1.8 yields a Fourier transform magnitude with one main peak. The spectrum from Figure 1.9, on the other hand, yields multiple peaks.

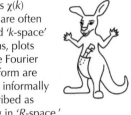

Just as $\chi(k)$ plots are often called 'k-space' graphs, plots of the Fourier transform are often informally described as being in 'R-space.'

BOX 1.4   Fourier Transforms Are Not Radial Distribution Functions

	The x-axis in Figure 1.10 is measured in angstroms, like bond distances are. Does that mean it's a graph of how many atoms are at each distance? A *radial distribution function*, I think it's called. That would be great! But if I'm wrong, I'm sure Mr. Handy will set me straight. He always seems to show up when I'm about to make a big mistake.
	I'm here, Simplicio! And you are about to make a really big, and common, mistake! *The magnitude of the Fourier transform is not a radial distribution function*! The peaks aren't at the right distances, they're not the right sizes, and they're not the right shape.
	Right. *The magnitude of the Fourier transform is not a radial distribution function.* But it's related to one. For example, the first big peak in the magnitude of the Fourier transform of the iron foil is at higher R than the first big peak in the sample with the iron–sulfur composite because the iron–iron nearest neighbor distance is larger than the iron–sulfur distance.
	For the magnitude of the Fourier transform to be a radial distribution function, the EXAFS equation would have to be simply $$\chi(k) = \sum_i N_i \sin(2kD_i),$$ with no multiple-scattering paths, no k weighting, no disorder, no dependence of scattering amplitude on species or k, an infinite mean free path, and an infinite amount of data with which to work. None of those things are true. If one compares the equation I just gave with the EXAFS equation developed in Section 1.2, one can see why there is some similarity between the magnitude of the Fourier transform and a radial distribution function. But *the magnitude of the Fourier transform is not a radial distribution function*.
	OK. But I have another question. We wrote the EXAFS equation using the symbol D for the distance between an absorbing atom and a scattering atom, but I notice the Fourier transform figures use the symbol R instead. Why?
	We have done that, friend Simplicio, to remind us that *the magnitude of the Fourier transform is not a radial distribution function*. Our notation is not a standard one in the literature, but we find it useful for avoiding confusion.

With the discussion of our panel (Box 1.4) in mind, we can now return to Figure 1.10. The transform of the metal foil shows multiple peaks because there are several shells of scattering atoms at fairly well-defined distances, unlike the transform of the

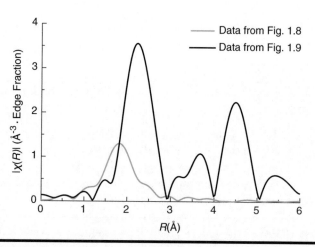

Figure 1.10 **Magnitude of the Fourier transforms of the data from Figures 1.8 and 1.9.**

sample with the composite. But the positions of those peaks in the R-space of the Fourier transform are not the same as the absorber–scatterer distances D_i from the EXAFS equation.

1.5 XAFS Is Not a Black Box

Section 1.2 outlines the theory of EXAFS spectra. The state of the art is a bit more advanced than what we have shown—see, for instance, Rehr and Albers (2000). Given an atomic-level structure of a material, we can predict its spectrum in k-space reasonably well. (See Chapter 13 for a discussion of some of the limitations of current theory.)

The theory of XANES is not quite as well developed, but great strides have been made in recent decades. We can now predict the general characteristics of XANES resulting from most structures, and in some cases, get good quantitative agreement between theory and experiment (see Section 12.5 of Chapter 12).

The problem is that, except as a matter of scientific curiosity, we are not particularly interested in predicting the spectrum corresponding to a given atomic arrangement. Instead, we are interested in the inverse problem: given a spectrum which we have measured, what is the structure of the material? It turns out that this question is difficult.

An examination of the EXAFS equation reveals why. For a completely unknown structure, there would be many unknown parameters. Each term in the EXAFS equation contains a D_i, σ_i^2, and N_i. What's more, there could be many important terms, corresponding to many important scattering paths. If the scattering elements are unknown, then $f_i(k)$ and $\delta_i(k)$ are also unknown, and those are functions, not just values! Finally, S_o^2 and E_o represent two more parameters that are unknown for a completely unknown structure.

The simple model that we have developed in this chapter suggests that S_o^2 and E_o are the same for every path. This is a reasonable assumption, but we will examine its limitations in Chapter 13.

That's a lot of undetermined parameters, but is it too many? Look at the EXAFS equation again. If S_o^2 is unknown and all the N_is are unknown, the task is hopeless: different combinations of S_o^2 and the N_i's could result in exactly the same spectrum, and thus, those parameters cannot be individually determined.

In Section 14.1.1 of Chapter 14, we can see that only limited information can be extracted from an EXAFS spectrum—perhaps 10 or 20 parameters. That's not enough to characterize a completely unknown structure.

And there's no way around that. *If you thought that you could put XAFS data through some procedure—a black box—which would yield the structure of your material, you were mistaken.*

So, if XAFS can't give you the structure of a completely unknown material, what good is it? How are people who use the technique getting anything out of it at all?

The answer is that they take advantage of our ability to calculate fairly well the spectra corresponding to given structures. That means that XAFS is very good at ruling out structures. Suppose we think a material is probably either pure goethite or pure lepidocrocite (two allotropes of FeOOH), but we don't know which. We can calculate what the spectrum of each of the proposed structures looks like, and see if either looks like our data. If only one does, that's a pretty good clue that it's the right structure, since we already had reasons (whether experimental or theoretical) for thinking our material was one of the two. If, on the other hand, neither calculated structure looks like the data, then our initial assumption was probably wrong, and we should think about other possibilities.

The EXAFS equation allows us to go a little beyond just choosing between possible structures. If, for instance, we have a biological compound the structure of which is generally known, but where a key question is the length of a copper–sulfur bond at an active site, it might be possible to use the known information about the material to reduce the number of unknown parameters in the EXAFS equation to a handful, including the desired bond distance. Those parameters could then be varied in a systematic way and the resulting spectra examined for agreement with the data. If only certain copper–sulfur bond lengths produce spectra similar to the data, then we have our answer.

Thus, while XAFS is not a black box, it can be a very useful tool for materials characterization. Section 1.6 provides an overview of the most common XAFS analysis techniques, each of which is then detailed in later chapters.

In the case studies in Chapter 19, we'll include some examples of published work where the authors attempted to choose between proposed structures, and *none* of them matched the data well! That's a valuable result that advances scientific knowledge, even if EXAFS doesn't end up telling us the actual structure.

1.6 Overview of Approaches to XAFS Analysis

1.6.1 Fingerprinting

The most straightforward approach to analysis is *fingerprinting*; that is, directly comparing measured data to a calculation or to other known data. In either case, the spectrum to which the data is compared is called a *standard*. The example of distinguishing between goethite and lepidocrocite given in Section 1.5 involves comparing the data to two different calculations; that's fingerprinting using *theoretical standards*. Instead of comparing to calculations, the data could have been compared to the spectra of known samples of goethite and lepidocrocite: fingerprinting using *empirical standards*.

This concept can be extended to circumstances where no single standard matches the data by looking at a series of standards and seeing where the data seem to fall in the series. For example, XANES spectra of sulfur in many different local environments and with various valences might be collected, creating a *library* of standards. It's often the case that examination of the spectra within the library, perhaps informed by theoretical considerations, can yield patterns that can be used to identify valence, symmetry, or other characteristics for an unknown spectrum, even when the unknown spectrum doesn't match any individual standard from the library.

Chapter 9 provides more information about this technique.

1.6.2 Linear Combination Analysis

If 20% of the chlorine atoms in a sample are in one local environment, and 80% are in a second local environment, then the resulting normalized XAFS spectrum will be a sum of 20% of the normalized spectrum for a sample where all the chlorine atoms are in the first environment and 80% of the normalized spectrum for a sample where all the chlorine atoms are in the second environment. (This is also true in *k*-space but is not true for the magnitude of the Fourier transform.)

The concept of linear combination analysis is therefore simple: assemble a library of standards and allow a computer to sum them in various linear combinations, reporting those that match the data well. Linear combination analysis is thus very effective at finding the relative amount of known constituents that are present.

For example, imagine a reaction in which the initial material and the end product are known, and it is suspected that there are no intermediates. It might be important to understand the kinetics: how fast does the reaction occur? At what point is it 50% complete? 90% complete? This could be discovered simply by measuring spectra at different periods and using linear combination analysis to determine the fraction of initial material and final product that are present at each time.

Or, suppose you are investigating a soil containing iron minerals, but it is unknown which minerals and in what proportions. If you have a library of common iron minerals, you could use linear combination analysis to discover which mixtures are consistent with your data.

If a linear combination analysis fails to find a good match to the data, it suggests that there is a constituent in the sample that is not in the library of standards. That can be useful information, but it also reveals the weakness of linear combination analysis: you have to have standards for everything that is in your mixture!

Linear combination analysis is covered in more depth in Chapter 10.

1.6.3 *Principal Component Analysis*

Since XAFS spectra of mixtures can be expressed as a linear combination of the pure spectra of the constituents, the machinery of linear algebra can be brought into play.

Suppose you have collected a series of four soil samples, all from the same site. The first is from a depth of 5 cm, the second from 10 cm, the third from 15 cm, and the last from 20 cm. From each, you measure the XAFS for the iron edge.

Those four spectra can be mixed together in a number of ways. For example, you're probably familiar with the idea of a *moving average*.

Spectra 1 and 2 can be averaged to form spectrum A, spectra 2 and 3 to form spectrum B, and spectra 3 and 4 to form spectrum C. By doing this, signal to noise can be improved, but you pay a price: where you started with four spectra, you now have only three. That means that you've lost information. If you somehow misplaced the original spectra 1–4, you couldn't reproduce them from your new spectra A–C.

But suppose you had made one more combination—it doesn't matter what, as long as it was independent of the other combinations. For argument's sake, let's say you averaged spectra 1, 2, and 3 and called the result D. You started with four spectra, and you still have four. Does that mean you could recover the original information?

Yes. Spectrum 1 can be recovered from (D – B). Spectrum 2 is (A + B – D). Spectrum 3 is (D – A). And spectrum 4 is (C + A – D).

A, B, C, and D can now be thought of as *components* of the original data set, analogous to the components of a vector. And just as many different coordinate systems could be used to represent the same vectors, many different choices of components could be used to represent the same set of spectra.

A particularly useful set of components to choose are what are called *principal components*. Under this system, the first component is chosen to be the average of all of the spectra, perhaps scaled by some constant. The second component is chosen so as to account for as much of the variability of the individual spectra from that average as possible; that is, linear combinations of the first and second component do as good a job of recovering each of the original spectra as any pair of components possibly could. The

Principal component analysis (PCA) is pretty abstract and hard to describe in a short section like this. Turn to Chapter 11 and the case studies in Chapter 19 for more!

third component is chosen so as to account for as much of the remaining variability as possible and so on until there are as many components as there were spectra originally.

The advantage of this kind of analysis is best seen when considering a fairly large set of spectra that are related in some way. Instead of just four soil samples, consider 20—perhaps they were not only collected at different depths but were also powdered and exposed to air for different periods.

It is very likely that the variations between all of the samples can be described by a small number of free parameters. Perhaps some have more of the iron mineral hematite, and some more goethite. Perhaps in some cases the goethite is highly disordered, and in others more crystalline. But since the samples are all related, all 20 can probably be described pretty well in terms of only a few variables—maybe 3, 4, or 5. And because of the way principal components are chosen, it would only take the same number of principal components, plus one for the initial average, to reproduce all of the original spectra quite well. If the samples can be described in terms of three variables, then the most important four components will do the job. If it takes five variables, then six components would be needed. Components beyond that will just account for tiny variations, such as those caused by noise in the measurement.

One of the great strengths of *principal component analysis* (PCA) is that, unlike other XAFS analysis techniques, it does not require you to have any kind of standards, whether theoretical or empirical. It can tell you how many parameters are needed to account for the variation of a set of spectra without knowing anything about them at all. It can also tell you which samples are relatively similar to each other and which are quite different. Finally, if you do have standards, you can use them to identify constituents that account for part of the variation without having to figure out what every constituent in the system is.

In short, the power of PCA is its ability to work with a system for which much is unknown. For this reason, it is often used as a preliminary technique, and then followed up with one or more of the other three (fingerprinting, linear combination analysis, or curve fitting to a theoretical standard) described in this book.

PCA is covered in more depth in Chapter 11.

It is not *quite* true that each component accounts for as much of the remaining variability as possible; it must also be orthogonal to the previous components. See Chapter 11, and the references it cites, for more information.

1.6.4 Curve Fitting to a Theoretical Standard

The final technique goes by the rather unwieldy name *curve fitting to a theoretical standard*. Many people just refer to it as *modeling* or *fitting*, although technically linear combination analysis is also a kind of fitting process.

In this technique, a candidate structure is first chosen by means of an educated guess—that is, the investigator's knowledge about the system identifies it as a likely possibility. (The candidate 'structure' may in some cases be a mixture of materials.) The theoretical spectrum is then computed for the candidate structure, creating a theoretical standard.

Next, the investigator decides how to describe modifications to the candidate structure in terms of a small number of parameters—perhaps the fraction of different phases that are present, or a thermal expansion, or the extent of some particular kind of disorder. The investigator then translates these parameters into the parameters of the EXAFS equation. For example, an investigator might be trying to find an unknown thermal expansion coefficient, and therefore wants to use it as a free parameter. According to the definition of 'thermal expansion coefficient,' that parameter multiplied by the temperature difference from the baseline temperature is the percentage change in all of the D_is. Finally, a computer program is used to vary the free parameters in such a way as to create the best fit between the modified theoretical standard and the data.

If the fit is poor, then the model is likely wrong—that is, either the candidate structure is not correct or the free parameters are not appropriate for the sample. If the fit is good, then that lends credence to the choice of candidate structure and yields best-fit values for the free parameters.

Of all the techniques, curve fitting to a theoretical standard is the most complicated to learn. Doing it well takes careful thought, good judgment, experience, and a combination of scientific and common sense. For that reason, not only does Chapter 12 discuss curve fitting to a theoretical standard in more detail but all of Part III is dedicated to it as well.

While it is the most difficult of the techniques to gain proficiency with, it also has the potential to be the most powerful and is what most scientists mean when they talk about EXAFS analysis. A combination of this book and a willingness to experiment and practice will be enough to get you started.

WHAT I'VE LEARNED IN CHAPTER 1, BY SIMPLICIO

- *EXAFS* is due to interference of an electron with itself: the probability of the absorption of an x-ray is enhanced if it leads to constructive interference of the ejected electron at the location of the absorbing atom, and the probability is reduced if it would lead to destructive interference there.
- The *EXAFS equation* attempts to incorporate modifications to the basic picture such as disorder, multiple atoms nearby the absorber, multiple scattering paths, and phenomena that remove signal from the main channel.
- Some common terms related to XAFS have more than one definition. Context can usually make it clear which is meant.
- For both XANES and EXAFS analysis, spectra must be *normalized*. For EXAFS, they are usually converted to *k*-space, and perhaps Fourier transformed as well.
- XAFS is not a black box.
- *Fingerprinting* means comparing data to a standard. Standards may be calculated (*theoretical*) or measured (*empirical*).
- *Linear combination analysis* is a simple and effective analysis technique when you have standards for all the possible constituents in your sample.
- Frankly, I'm not sure I understand *PCA* yet, but that's OK. It is explained in more detail in Chapter 11.
- *Curve fitting to a theoretical standard* means making an educated guess as to the structure of the sample, choosing a few parameters that are free to vary, and then using a computer to match that model to the measured data as well as possible.

References

Bunker, G. 2010. *Introduction to XAFS*. New York: Cambridge University Press.
Dirac, P. A. M. 1927. The quantum theory of the emission and absorption of radiation. *Proc. R. Soc. Lond. A.* 114:243–265. DOI:10.1098/rspa.1927.0039.
Fourier, J. 1822. *The Analytic Theory of Heat*. Paris: F. Didot.
IUCr. IUCr Dictionary CXAFS Contribution. Accessed December 12, 2011. http://www.iucr.org/__data/assets/pdf_file/0003/58845/IUCr-Dictionary-CXAFS-Contribution.pdf.
Rehr, J. J. and R. C. Albers. 2000. Theoretical approaches to x-ray absorption fine structure. *Rev. Mod. Phys.* 72:621–654. DOI:10.1103/RevModPhys.72.621.

Chapter 2

The Hardware: Light Sources, Beamlines, and Detectors

Consider Your Toolkit

I like tools! At home, I have a very nice hammer!	
Wonderful! I bet that's great if you'd like to pound a nail into a piece of wood. But what if you want to work with something that is not a nail? There's an old saying, that 'if all you have is a hammer, everything looks like a nail.'	
That's why it's important to have access to a wide variety of tools: hammers, wrenches, screwdrivers, etc. That way you'll choose the right tool for the job!	
The same thing is true in XAFS. You should become familiar with different light sources, beamlines, and detectors, so you can choose ones that will work well for your experiment.	

2.1 X-Ray Sources

To measure x-ray absorption fine structure (XAFS), you need x-rays, and you need to be able to scan them through different energies.

DOI: 10.1201/9780429329555-4

2.1.1 Synchrotron Light Sources

Currently, the most common way to produce x-rays for XAFS experiments is to use a *synchrotron light source* (often called either a *synchrotron* or a *light source* for short). These light sources inject electrons traveling near the speed of light into a *storage ring* to produce broad-spectrum x-rays (*synchrotron radiation*). The ring features straight sections alternating with curved sections. In the curved sections, magnetic fields created by *bending magnets* accelerate the electrons in a circular path, creating the desired broad-spectrum x-rays. Arrays of magnets may also be arranged in the straight sections to form *insertion devices*; for third-generation sources (see below), the insertion devices in the straight sections are central to the design.

The x-rays generated by bending magnets or insertion devices then travel down *beamlines*. The term 'beamline' encompasses not only the tube through which the x-rays travel but also all the equipment used for measurement. Even the chairs experimenters sit in are considered part of the beamline!

Storage-ring light sources (SLS) are classed by generation (Robinson 2009):

It used to be that light sources would have to interrupt the beam a couple of times per day to inject a new set of electrons, and some still do. But many now operate in *top-up* mode, using frequent mini-injections to replace electrons lost to collisions. Top-up mode allows you to get beam on your sample without interruption.

- *First-generation* SLS feature a storage ring originally intended for another purpose, generally particle physics.
- *Second-generation* SLS were designed and built from the start to be x-ray sources.
- *Third-generation* SLS are designed and built to have insertion devices in the straight sections. In addition, they feature electron beams with significantly better collimation (Altarelli and Salam 2004).
- *Fourth-generation* SLS feature order of magnitude or better reductions of *emittance* (a measure of the size and dispersion of the electron beam in the storage ring). This is usually accomplished by using a greater number of smaller magnets in what is known as a *multibend achromat* (MBA) lattice (Eriksson 2016).

2.1.2 Energy-Recovery Linacs

The process of emitting synchrotron radiation inherently increases the emittance of the electron beam, which is then partially countered by the lattice of magnets used to steer and shape the beam. An electron beam produced by a linear accelerator, in contrast, does not produce synchrotron radiation until it passes through an insertion device such as an undulator or FEL, and can thus have very low emittance.

The problem, of course, is that in a linear accelerator, a given pulse of electrons would only pass through the insertion device once. Since the energy required to boost an electron pulse to relativistic speeds is considerable, using a linear accelerator as a source of synchrotron radiation is very energy-inefficient.

Energy-recovery linacs (Gruner and Bilderback 2003) solve this problem by routing the electrons back to the linac, where they are decelerated and most of the energy is recovered. That energy is then used to accelerate a new pulse of electrons. The net effect is to produce radiation similar to that created by fourth-generation storage-ring light sources, with reasonable energy efficiency.

Currently, there are dozens of light sources scattered across five continents. In Table 2.1, we list some of the more prominent facilities.

The number of XAFS endstations in this table is approximate as of 2023 and can change from year to year.

Recovery linacs, along with storage rings the feature MBA lattices, are sometimes referred to as 'fourth-generation synchrotron light sources.'

2.1.3 Tabletop Sources

In recent years, a number of 'tabletop' systems have been introduced (Zimmermann et al. 2020), allowing XAFS data to be collected in a local laboratory, or even in the

Table 2.1   Some Major Light Sources

Facility	Initials	Location	Number of XAFS Endstations	Comments
Advanced Light Source	ALS	Berkeley, CA	9	UV and soft x-ray
Advanced Photon Source	APS	Near Chicago, IL	13	
ALBA		Barcelona, Spain	1	
ANKA		Karlsruhe, Germany	4	
Australian Synchrotron	AS	Near Melbourne, Australia	3	
Beijing Synchrotron Radiation Facility	BSRF	Beijing, China	4	
Berliner Elektronenspeicherring-Gesellschaft für Synchrotronstrahlung	BESSY II	Berlin, Germany	15	
Canadian Light Source	CLS	Saskatoon, Canada	11	
DAFNE		Near Rome, Italy	1	
Diamond Light Source		Near Oxford, England	9	
Elettra		Trieste, Italy	12	
European Synchrotron Radiation Facility – Extremely Brilliant Source	ESRF-EBS	Grenoble, France	13	
Hiroshima Synchrotron Radiation Center	HiSOR	Hiroshima, Japan	2	
Indus 2		Indore, India	2	
Kurchatov Synchrotron Radiation Source	SIBR	Moscow, Russia	3	
Linac Coherent Light Source	LCLS	Near San Jose, CA	2	FEL facility
Microtron Accelerator for X-rays IV	MAX IV	Lund, Sweden	3	
National Synchrotron Light Source II	NSLS II	Near New York, NY	5	
PETRA III		Hamburg, Germany	3	
Photon Factory	PF	Tsukuba, Japan	13	
Pohang Light Source	PLS-II	Pohang, Korea	3	
Ritsumeikan University Synchrotron Radiation Center		Near Kyoto, Japan	7	
Saga Light Source	SAGA-LS	Saga, Japan	3	
Shanghai Synchrotron Radiation Facility	SSRF	Shanghai, China	4	
Singapore Synchrotron Light Source	SSLS	Singapore	1	
Sirius		Campinas, Brazil	3	
Solaris		Krakow, Poland	4	
Soleil		Near Paris, France	7	
Stanford Synchrotron Radiation Lightsource	SSRL	Near San Jose, CA	17	
Super Photon Ring-8 GeV	SPring-8	Hyogo, Japan	12	
Swiss Light Source	SLS	Aargau, Switzerland	5	
SwissFEL		Aargau, Switzerland	3	FEL facility
Synchrotron Light for Experimental Science and Applications in the Middle East	SESAME	Allan, Jordan	1	
Synchrotron Light Research Institute	SLRI	Nakhon Ratchasima, Thailand	4	
Taiwan Light Source	TLS	Hsinchu, Taiwan	5	
Taiwan Photon Source	TPS	Hsinchu, Taiwan	4	Some XAFS beamlines still under construction at time of this writing
Ultraviolet Synchrotron Orbital Radiation Facility	USORV	Okazaki, Japan	2	Soft x-ray

field. While these devices necessarily produce many fewer photons than traditional multiuser facilities, they have a number of important use cases:

- Experiments which would not qualify for scarce beamtime at a multiuser facility, such as 'routine' quality assurance testing of manufactured samples
- Student training (Zimmermann et al. 2020; Bleiner 2021)
- Experiments where it is difficult or inadvisable to transport the objects being studied long distances, such as measurements on valuable works of art or cultural artifacts (Luiten 2016)
- Experiments on substances which present unusual safety hazards for multiuser facilities (Zimmermann et al. 2020)
- Experiments on proprietary materials (Bleiner 2021)
- Long-duration *in situ* or *operando* experiments, where it is desirable to make measurements of a single sample over a period of days, weeks, or months (Zimmermann et al. 2020)
- Preliminary experiments to lay the groundwork for a future visit to a multiuser facility (Zimmermann et al. 2020)

In cases such as these, the advantages of having a local system under the control of a single experimental team may outweigh the relative paucity of photons the system produces, particularly if the samples being studied or the experimental questions being asked don't require the high x-ray flux available at modern beamlines.

2.2 Bending Magnets and Insertion Devices

While the capabilities of synchrotrons are important, it is the combination of synchrotron characteristics and the insertion device (or bending magnet) that constrains the capabilities of a particular beamline.

2.2.1 Brilliance

Though brilliance is the figure of merit for a source, the quantity that actually matters in bulk XAFS experiments is *intensity* in some bandwidth—that's brilliance without the 'per square milliradian' part. For microprobe experiments (see Section 3.2 of Chapter 3), however, brilliance is critical.

Optical components can do a lot to modify the x-rays provided by a bending magnet or insertion device. The x-rays can be focused by mirrors, for example. As another example, a monochromator (Section 2.3) can be used to discard photons not in the desired energy range, leaving the beam more nearly monochromatic.

But the principle of 'conservation of étendue' (Chaves 2008) assures us that the number of photons per second per square millimeter of area per square milliradian of solid angle within a bandwidth of 0.1% of any given energy (a quantity known as *brilliance*) cannot be increased by optical means (Altarelli and Salam 2004).

The brilliance of several sources in the United States is shown, as a function of energy, in Figure 2.1.

As can be seen from the figure, bending magnets on a second-generation storage ring can achieve a brilliance figure of about 10^{13} (usual units) at around 10 keV (a typical energy for XAFS measurements).

2.2.2 Wigglers

The complicated-seeming definition of brilliance allows for a number of trade-offs—for example, a beam can be focused to have more photons per area, but at the expense of radial divergence.

By placing a series of magnets of opposite polarity in a row, electrons can be made to wiggle back and forth, giving off more x-ray radiation than would be the case for the single turn caused by a bending magnet. This array of magnets is called a *wiggler*. The x-rays from wigglers are used in much the same way as the x-rays from bending

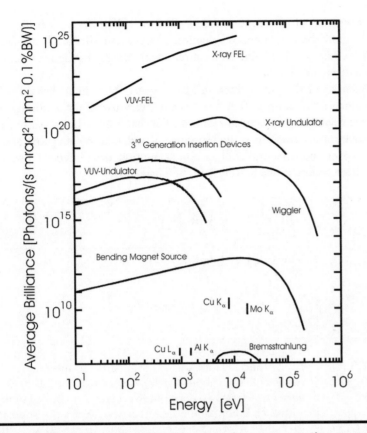

Figure 2.1 Brilliance as a function of energy for representative types of x-ray source. (Reprinted from U. Pietsch, V. Holý, and T. Baumbach, *High-Resolution X-Ray Scattering: From Thin Films to Lateral Nanostructures*, 2nd edition, 2004. Copyright 2004 by Springer-Verlag.)

magnets, but with a significant enhancement of brilliance (often more than two orders of magnitude, depending on the number of poles and the field strength).

2.2.3 Undulators

Like wigglers, *undulators* are insertion devices composed of magnets of opposite polarity. But the deflections caused by an undulator are small enough that the x-rays emitted from each turn overlap, and thus interfere. At some energies, this interference is constructive. This causes dramatic increases in the intensity (and thus the brilliance) at the energies for which the undulator is tuned, and almost complete cancellation at other energies.

The peak brilliance at the tuned energies can easily exceed that of a bending magnet by more than four orders of magnitude (but more brilliance is not always better—see Box 2.1).

2.2.4 Free Electron Lasers

So-called *free electron lasers* (FEL) provide extremely intense x-rays in extremely short pulses (Emma et al. 2010). In an FEL, electrons traveling through an undulator produce radiation which in turn spatially separates the electrons into microbunches. As the microbunches continue through the undulator they trigger self-amplified spontaneous emission (SASE), producing highly coherent x-ray radiation. It is this process that justifies the 'laser' terminology, even though in many respects, the process is quite

If you'd like to read more about insertion devices, the works of Bunker (2010) and Kim (2009) are both good places to start.

You'll sometimes see FELs referred to as a type of fourth-generation light source, but I think that's confusing! To me, the generation refers to the storage ring or energy-recovery linac system. Upgrading to the next generation should require building a new ring or engineering a major retrofit to an existing one. Some FELs, on the other hand, are built as beamlines on an existing third-generation ring.

different from that of conventional lasers (Bleiner 2021). The peak brilliance of an FEL can be more than ten orders of magnitude higher than that of a bending magnet.

An FEL could be a beamline on a conventional storage-ring light source or could be fed by a linear accelerator.

The intensity of the pulses from an FEL is generally so high that a single pulse severely damages the sample. They are, however, very useful for obtaining subpico-second time resolution of chemical reactions. For that kind of investigation, an automated system starts a reaction, waits a specified period, takes a data point, and repeats the process over and over until full spectra at multiple time intervals after reaction start have been collected (see Section 3.1.6 of Chapter 3).

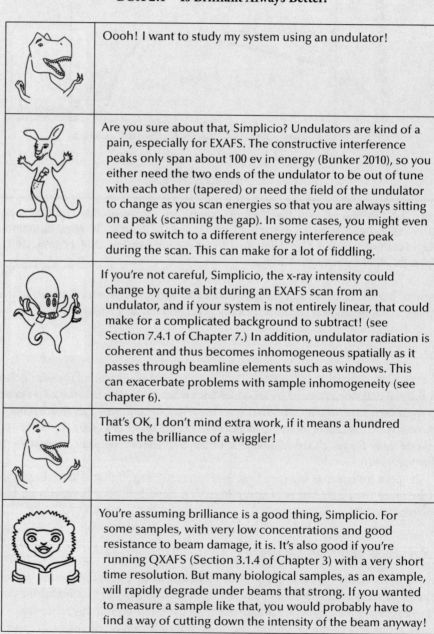

BOX 2.1   Is Brilliant Always Better?

	Oooh! I want to study my system using an undulator!
	Are you sure about that, Simplicio? Undulators are kind of a pain, especially for EXAFS. The constructive interference peaks only span about 100 ev in energy (Bunker 2010), so you either need the two ends of the undulator to be out of tune with each other (tapered) or need the field of the undulator to change as you scan energies so that you are always sitting on a peak (scanning the gap). In some cases, you might even need to switch to a different energy interference peak during the scan. This can make for a lot of fiddling.
	If you're not careful, Simplicio, the x-ray intensity could change by quite a bit during an EXAFS scan from an undulator, and if your system is not entirely linear, that could make for a complicated background to subtract! (see Section 7.4.1 of Chapter 7.) In addition, undulator radiation is coherent and thus becomes inhomogeneous spatially as it passes through beamline elements such as windows. This can exacerbate problems with sample inhomogeneity (see chapter 6).
	That's OK, I don't mind extra work, if it means a hundred times the brilliance of a wiggler!
	You're assuming brilliance is a good thing, Simplicio. For some samples, with very low concentrations and good resistance to beam damage, it is. It's also good if you're running QXAFS (Section 3.1.4 of Chapter 3) with a very short time resolution. But many biological samples, as an example, will rapidly degrade under beams that strong. If you wanted to measure a sample like that, you would probably have to find a way of cutting down the intensity of the beam anyway!

Even if your sample can take it, at some point more flux doesn't bring you all that much. Suppose a wiggler gives you enough intensity so that you get decent signal to noise by measuring for 1 second at each energy (we'll talk more about scan times in Section 8.9 of Chapter 8). An undulator might let you get the same signal-to-noise ratio in 0.01 seconds. But if it takes a third of a second to change energy from one point to the next, then you're not really going a hundred times faster—you're only going three times faster. Add in the time to go from the end of one scan to the start of the next, the time for changing samples, and the like, and it's not as much help as it sounds.

On the other hand, if you do have a sample that really needs an intense beam, either because the concentration of the absorbing element is very low or because you're going to be slewing the monochromator very quickly (QXAFS), then an undulator might make sense. We don't want to scare you away from undulators—just understand that brighter is not *always* better!

2.3 Monochromators

To collect the XAFS spectra, you need a way to select x-rays of particular energies (actually narrow ranges of energy). The most common way to do this is a *monochromator*, a device that uses diffraction from crystals (usually a pair) to ensure that only x-rays near the desired energy travel toward the sample. We'll discuss monochromators in more detail in Chapter 8.

The combination of the crystals used for the monochromator, their orientation, and geometrical considerations will limit any beamline to a range of usable energies. This will generally be specified on the Web page dedicated to that beamline; for example, the DIFFABS beamline at Soleil can provide x-ray energies from 3 to 23 keV, while the SAMBA beamline at Soleil can provide x-ray energies from 4.8 to 40 keV. Thus, the barium K edge (37.4 keV) could be measured on SAMBA but not on DIFFABS, although the barium L_3 edge at 5.2 keV could be measured on either. The energy range that a beamline can provide is one of the most important characteristics when deciding whether to apply for time there!

2.4 Measurement Modes and Detectors

2.4.1 Measurement Modes

All XAFS experiments require the determination of the fraction of x-ray photons absorbed as a function of energy. There are three common methods by which the amount of absorption is deduced:

1. The measured intensity of the x-rays downstream of the sample can be compared to the measured intensity of the x-rays upstream of the sample; the difference must be due to absorption. This is called the *transmission geometry.*
2. When an atom absorbs an x-ray, it is left in an excited state, with a vacancy in one of its core orbitals. This orbital can be filled by an electron from a higher level, resulting in the emission of a lower energy x-ray in a random direction

The Soleil synchrotron has a particularly clear set of Web pages describing its beamlines at www.synchrotron-soleil.fr/Recherche/LignesLumiere. Each page provides a description, energy range, energy resolution, source (i.e., bending magnet or insertion device), intensity measurements, lists of detectors, examples of science in multiple fields, and more! If you're new to thinking about beamlines, the Soleil pages are a good place to start. Even if you don't plan to go to Soleil, they can get you thinking about the kinds of information you'd like to know about beamlines at the facilities that you are considering.

Most XAFS experiments are done by having a monochromator select a particular energy and remain there for a few seconds for measurement before moving on to the next energy, but there are alternatives, some of which we'll explore in Chapters 3 and 4.

Many experiments will use more than one of these modes at once!

Figure 8.2 in Chapter 8 shows us the transmission and fluorescence geometries schematically. Electron yield measurements are made directly at the location of the sample.

(i.e., the probability is isotropic). A detector placed perpendicular to the direction of the beam can measure some of these photons, and the intensity compared to the intensity upstream of the sample. This is called a *fluorescence* measurement.

3. Instead of emitting a photon, the excited atom can return to the ground state by releasing a high-energy electron (called an *Auger-Meitner* electron). The number of electrons emitted can be counted and compared to the intensity of x-rays upstream of the sample. This is called an *electron yield* measurement.

2.4.2 Ion Chambers

Transmission mode uses the x-ray version of the Beer–Lambert Law (Section 6.3.1 of Chapter 6).

The most common type of detector is the *ion chamber*. X-rays passing through these gas-filled detectors ionize some of the gas; the resulting cascades of charged particles are collected and amplified to form a measurable current. These detectors, and this process, will be described in detail in Section 8.6 of Chapter 8.

Ion chambers are simple, reliable, exhibit good linearity, and can handle high count rates. They are by far the most common detector used for the upstream intensity measurement needed for all measurement modes, and are also very common for the downstream measurement in transmission. A special design of ion chamber, known as a Lytle detector (Stern and Heald 1979), is sometimes used to measure fluorescence.

2.4.3 Energy-Discriminating Fluorescence Detectors

An ion chamber counts x-rays of every energy. But in the fluorescence geometry, we are only interested in photons of a few specific energies: the ones that correspond to fluorescence from the atom the edge of which we are measuring. Other x-rays, such as those scattered from the sample, constitute undesirable background. *Energy-discriminating detectors* solve this problem by counting only photons within a chosen energy range (subject to the resolution limits of the detector). The most common form of energy-discriminating detector is a *multielement semiconductor detector*, using either germanium (Oyanagi et al. 1998) or silicon (Lechner et al. 1996).

One popular type of multielement semiconductor detector is the *silicon drift detector*, described in Lechner et al. (1996).

In these kinds of energy-discriminating detectors, the energy discrimination takes place *after* detection; that is, x-ray photons of all energies are counted, the energy of each determined, and then those that fall within the desired range reported to a measurement computer. Because these detectors count and analyze each photon separately, they can sometimes fail to count a photon while they are still analyzing the previous one (i.e., there is some *dead time* after a measurement), or they can count two photons as if they were one of higher energy (*pileup*). The total count rate (i.e., including background counts) of a single element of a detector of this kind is, therefore, limited to a few hundred thousand photons per second at most (Bunker 2010; Thompson 2009). To achieve higher count rates, multiple elements are used.

2.4.4 Current-Mode Semiconductor Detectors

'Multielement' sounds like it's referring to the ability to discriminate between chemical elements, but actually means that the detector is made of multiple identical *elements*, each of which acts as a separate detector.

There are also semiconductor detectors designed to be used in *current mode*, where the total number of photons is counted without discriminating in energy. These detectors therefore are employed somewhat like ion chambers. One disadvantage of semiconductor detectors operated in this mode is that they are sensitive to visual light as well as x-rays, and thus, the sample and detector must be kept in the dark while measurements are being made (Bunker 2010).

2.4.5 Wavelength-Dispersive Fluorescence Detectors

The energy-dispersive detectors discussed in Section 2.4.3 analyze all photons incident on the detector, but only count those near a specified energy. As we discussed, this imposes limits on the count rate.

In contrast, *wavelength-dispersive detectors* use diffraction to select photons of the desired energy before they enter the detector proper. Thus, energy is selected by the geometry of the detector relative to the sample, rather than by electronic analysis, and there is effectively no limit on count rates. Energy resolution in this kind of detector is extremely good. Because these detectors use crystal diffraction off of a crystal to select the desired energy, they are also referred to as *crystal analyzers* or *diffractive analyzers*.

Ironically, the energy resolution of the simplest forms of wavelength-dispersive detectors are 'too good' for efficient fluorescence measurements. The problem is that lifetime broadening and other factors make the width of the fluorescent peaks broader than the crystal acceptance of the analyzer, leading to a large fraction of the fluorescent photons being rejected. The acceptance can be improved by introducing strain into the crystals, but this in turn reduces the diffraction efficiency (Heald 2015).

In addition, since the wavelength discrimination of these detectors is determined by geometry, they will only accept fluorescent photons off of a small spot on the sample. In order to avoid substantial reductions in detection efficiency, in practice this means the beamline must provide a focused beam (Bunker 2010). Depending on the detector geometry, it may be desirable to have a beam smaller than 100 microns across, although some detector geometries allow for spot sizes up to 1 mm (Heald 2015). It is important to realize that this geometrical limit applies not just to dimensions parallel to the surface of the sample but perpendicular to it as well; if, as is usually the case, the geometry is optimized for detection of fluorescent photons arising from the surface of the sample, photons arising from more than a few tens of microns depth may not be counted.

2.4.6 Electron Yield

Detectors for electron yield experiments must, of course, detect electrons rather than x-rays. As with fluorescence detectors, electron yield can be measured in *total yield* (Kemner et al. 1994) or *partial yield*, that is, limited to electrons of certain energies (Nakanishi and Ohta 2011).

2.4.7 Other Detectors

While the x-ray detectors we've discussed so far are the most common in general use, many other types of detectors have been used, especially for specialized purposes such as spatially resolved detection. Thompson (2009) and Bunker (2010) provide many examples of additional types.

2.4.8 Beamline Equipment

Beamlines will always provide you a basic set of detectors such as ion chambers, and may also have available more specialized options. In some cases, the light source may also have a pool of detectors that can be reserved or borrowed for your beamtime.

But it is *not* the case that all beamlines have, for example, energy-discriminating detectors. The detectors available when using a beamline are an important consideration when deciding where to apply for time.

Passivated implanted planar silicon detectors are current-mode semiconductor devices that have become a popular substitute for Lytle detectors.

Of course, fluorescence photons originating from deep within a sample are less likely to make it to *any* kind of detector because of absorption within the sample. The additional loss with wavelength-dispersive detector due to geometric effects is only significant if the sample does not absorb the fluorescent photons strongly.

Most beamlines will also let you bring your own detectors! Total electron yield detectors, for example, are not all that expensive and can even be made in-house without too much trouble. If it's something you're going to use over and over at different beamlines or if you'd like a detector customized for your application, it's worth considering the possibility. But if you do, make sure you contact the beamline scientists well in advance! Getting it to work with the electronics of a particular beamline may be much more complicated than just plugging in a cable.

2.4.9 Simultaneous Probes

It is generally possible to make other measurements simultaneously with XAFS. Complementary probes could include almost any technique that you could use in your home laboratory: cyclic voltammetry, differential scanning calorimetry, infrared spectroscopy, x-ray diffraction—the list of possibilities is very long!

In some cases, beamlines are set up to facilitate specific complementary techniques. In others, you may have to bring your own measurement equipment with you, adapting it to make sure it doesn't interfere excessively with the path of x-rays to the sample and from there to the x-ray (or electron yield) detector.

2.4.10 Microprobe

On *microprobe* beamlines, the incident x-rays can be focused down to a spot on the scale of microns for hard x-rays (even smaller for soft x-rays). *Spot size* is therefore an important figure of merit for these lines, although of course not the only one. These beamlines include a motorized sample stage capable of positioning the sample precisely relative to the beam and frequently provide visual or electron microscopy to image a sample (Bertsch and Hunter 2001). We will discuss microprobe experiments further in Section 3.2 of Chapter 3.

2.5 Getting Beamtime

There are many ways to get beamtime at a light source, but the most common is through peer-reviewed general user programs. Under this kind of system, you submit a proposal explaining how you want to use the beamtime, which other scientists then review and rate. The strongest proposals are then awarded beamtime.

While the proposal requirements for every synchrotron are different, here are some general pieces of advice when applying for general user time:

- Consider your choice of beamlines carefully. At some light sources, you apply for time on a specific beamline, while at others you apply to the facility and let them choose the line. Either way, it helps to know what beamlines will work best for your experiment and which won't work at all.
- Contact beamline scientists for lines you're considering before you apply to discuss the gist of your idea. They may have good suggestions for how to make the most of their line. On the other hand, don't send them a full proposal to preview—they're busy people! Just a few paragraphs in an e-mail or a few minutes via videoconference are sufficient to get the conversation started.
- Your proposal should be specific about how you plan to use your time. Indicate information about the number of samples and how they differ, the number of standards, the edges that you will measure, and rough time estimates.
- Be realistic. Build in some time for sample changes, unexpected problems, and the like. If the experiment is straightforward and you intend on running 24 hours a day (hopefully there's more than one of you!), then adding about 50% to the actual scan time should account for setup, sample changes, and challenges. Of course, if the experiment is more involved (perhaps an *operando* experiment involving wet chemistry), then your cushion should be larger. If your samples have a standard form, and the beamline features a high degree of automation, a smaller cushion would work—but you should always leave *some* room for the unexpected!
- Realize your reviewers may not be specialists in your field. They probably know XAFS better than you do, but you know your system better. Describe the

scientific questions you hope the experiment will answer clearly, briefly, and without jargon. Describe the importance of your investigation in a way that would make sense to someone outside your field. But don't go on at length about how wonderful XAFS is or how it works; the reviewers know that already!

■ Indicate why synchrotron radiation is important for your experiment. If the reviewers think your scientific questions can be answered in some other way, they are less likely to recommend that you be awarded beamtime.

■ Describe your expertise for designing and running an XAFS experiment. If this will be your first time to a synchrotron, say so—light sources like new users! But also show that you've done your homework. If you've taken workshops on XAFS, or if your advisor has used the technique before, mention that. If you have nothing else to fall back on, you could even mention that you've familiarized yourself with the principles of XAFS through this book.

■ Say where you will get the expertise to analyze your data after you collect it. If you're going to work with more experienced colleagues, say so and name them.

■ Give a sense of what approaches you plan to use for analyzing the data: fingerprinting, modeling, and so on.

■ Try to stay brief. Reviewers are volunteers; provide them the key points, but don't drown them in excessive detail or verbiage.

WHAT I'VE LEARNED IN CHAPTER 2, BY SIMPLICIO

■ A beamline can get x-rays (in order of increasing brilliance) from a *bending magnet*, a *wiggler*, an *undulator*, or a *free electron laser*.

■ Undulators can produce more intense x-rays than wigglers and bending magnets, but there are additional complexities associated with using them for EXAFS, and not all experiments benefit from more intense x-rays.

■ Measurements may be performed in *transmission*, *fluorescence*, or by using *electron yield*.

■ Common detectors include *ion chambers* (able to handle high count rates), *multielement semiconductor detectors* (able to discriminate the energy of x-ray photons), and *wavelength-dispersive detectors* (able to handle high count rates and discriminate the energy of x-ray photons, but require small spot size).

■ Applications for beamtime should be clear, specific, and brief.

References

Altarelli, M. and A. Salam. 2004. The quest for brilliance: Light sources from the third to the fourth generation. *Europhysics News*. 35:47–50. DOI:10.1051/epn:2004204.

Bertsch, P. M. and D. B. Hunter. 2001. Applications of synchrotron-based x-ray microprobes. *Chem. Rev.* 101:1809–1842. DOI:10.1021/cr990070s.

Bleiner, D. 2021. Tabletop beams for short wavelength spectroscopy. *Spectrochim Acta B*. 181:105978. DOI:10.1016/j.sab.2020.105978.

Bunker, G. 2010. *Introduction to XAFS*. New York: Cambridge University Press.

Chaves, J. 2008. *Introduction to Nonimaging Optics*. Boca Raton, FL: CRC Press.

Emma, P. et al. 2010. First lasing and operation of an Ångstrom-wavelength free-electron laser. *Nat. Photonics*. 4:641–647. DOI:10.1038/nphoton.2010.176.

Eriksson, M. 2016. The multi-bend achromat storage rings. *AIP Conf. Proc.* 1741:020001. DOI:10.1063/1.4952780.

Gruner, S. M. and D. H. Bilderback. 2003. Energy recovery linacs as synchrotron light sources. *Nucl. Instrum. Methods Phys. Res. A.* 500:25–32. DOI:10.1016/S0168-9002(03)00738-1.

Heald, S. M. 2015. Strategies and limitations for fluorescence detection of XAFS at high flux beamlines. *J. Synch. Rad.* 22:436–445. DOI:10.1107/S1600577515001320.

Kemner, K. M., J. Kropf, and B. A. Bunker. 1994. A low-temperature total electron yield detector for x-ray absorption fine structure spectra. *Rev. Sci. Instrum.* 65:3667–3669. DOI:10.1063/1.1144489.

Kim, K.-J. 2009. Characteristics of synchrotron radiation. In *X-Ray Data Booklet*, eds. A. C. Thompson and D. Vaughan, 2–1 to 2–16. Berkeley, CA: Lawrence Berkeley National Laboratory.

Lechner, P., S. Eckbauer, R. Hartmann, S. Krisch, D. Hauff, R. Richeter, H. Soltau, L. Strüder, C. Fiorini, E. Gatti, A. Longoni, and M. Sampietro. 1996. Silicon drift detectors for high resolution room temperature X-ray spectroscopy. *Nucl. Instrum. Methods Phys. Res. A.* 377:346–351. DOI:10.1016/0168-9002(96)00210-0.

Luiten, O. J. 2016. *KNAW-Agenda Grootschalige Onderzoeksfaciliteiten : Smart*Light: A Dutch Table-Top Synchrotron Light Source.* Koninklijke Nederlandse Akademie van Wetenschappen (KNAW).

Nakanishi, K. and T. Ohta. 2011. Improvement of the detection system in the soft X-ray absorption spectroscopy. *Surf. Interface Anal.* 44:784–788. DOI:10.1002/sia.3870.

Oyanagi, H., M. Martini, and M. Saito. 1998. Nineteen-element high-purity Ge solid-state detector array for fluorescence X-ray absorption fine structure studies. *Nucl. Instrum. Methods Phys. Res. A.* 403:58–64. DOI:10.1016/S0168-9002(97)00927-3.

Robinson, A. L. 2009. History of synchrotron radiation. In *X-Ray Data Booklet*, eds. A. C. Thompson, 2–21 to 2–28. Berkeley, CA: Lawrence Berkeley National Laboratory.

Stern, E. A. and S. M. Heald. 1979. X-ray filter assembly for fluorescence measurements of x-ray absorption fine structure. *Rev. Sci. Instrum.* 50:1579–1582. DOI:10.1063/1.1135763.

Thompson, A. C. 2009. X-ray detectors. In *X-Ray Data Booklet*, eds. A. C. Thompson and D. Vaughan, 4–32 to 4–39. Berkeley, CA: Lawrence Berkeley National Laboratory.

Zimmermann, P., S. Peredkov, P. M. Abdala, S. DeBeer, M. Tromp, C. Müller, and J. A. van Bokhoven. 2020. Modern x-ray spectroscopy: XAS and XES in the laboratory. *Coordin. Chem. Rev.* 423:213466. DOI:10.1016/j.ccr.2020.213466.

Chapter 3

Spatial and Temporal Resolution

There's a Time and Place for Some Things

I thought there was a time and place for everything!	
In everyday life, is there a time and place for the air you breathe?	
The air has characteristics that have a time and place. Next week, in the town I work in, it's supposed to be smoggy.	
So because of what you're interested in ('will it be smoggy where I work on Tuesday?') you care about time and place for that question. But for other questions ('what percentage of normal air is oxygen?'), you don't. Whether time and place are important in an experiment depends on the question you are asking.	

DOI: 10.1201/9780429329555-5

3.1 Time-Resolved Experiments

In many experiments, one of your goals is to understand how the system evolves with time. Regardless of the time scale of that evolution, it's possible to investigate that time evolution using x-ray absorption fine structure (XAFS), but the method and instrumentation you use will depend on the time scale.

3.1.1 Evolution on a Scale of Months or Years: Just Wait!

One option for systems that evolve on a timescale of months or years is to simply measure them months or years apart. This depends, of course, on the XAFS measurement not interfering with the sample or its evolution. It won't work on samples that would be vulnerable to radiation damage in an x-ray beam, including most biological samples. It also wouldn't be suitable for samples that need to have their function destroyed in order to prepare them for measurement: if a battery cell needs to be destructively disassembled for a measurement, for example, then a given cell could only be measured at one point in its evolution.

For certain studies on the aging of materials, however, this might be a viable approach. It still has the disadvantage of having to get multiple sets of beam time in order to do the measurements.

The issue of confounding the effect of time evolution with inherent differences between the samples of a set can be mitigated to a large extent by careful monitoring of each item of the set during its time evolution. Voltage curves of battery cells, for example, can be measured for each cycle and compared; plants can be photographed after each day of growth. This requires having more members of the set than are expected to be measured at the end of the experiment, so that any members that show deviation from the cohort during time evolution can be removed from the experiment.

3.1.2 Freezing Time: Quenched Samples

An alternative is to prepare systems that are as identical as possible, allow them to start evolving in time, and then stop the time evolution at different points for each sample. A set of battery cells, for example, could be prepared so that one is pristine, one has been cycled once, another ten times, and another a hundred times. Or samples could be prepared from seedlings at various amounts of time after sprouting. In chemistry, stopping a reaction while it is in progress can be referred to as *quenching*; this extends that idea to all measured systems that undergo time evolution.

The advantages to using a set of quenched samples are that they can all be measured during one trip to a beamline, and they can be prepared for measurement in whatever manner is most convenient. The primary disadvantage is that differences between the items in the set at the start of their time evolution can be misinterpreted as the effect of time evolution. If, for example, the battery cell which had been cycled the most times also happened to have a subtle manufacturing defect, the effect of the defect and the effect of the cycling would be difficult to disentangle.

3.1.3 Evolution on a Scale of Tens of Minutes

If a system evolves on a scale of 15 minutes to a few hours, it's often possible to measure it *in situ* (see Section 5.5 of Chapter 5) using conventional XAFS techniques. As we'll learn in Chapter 8, a conventional extended x-ray absorption fine structure (EXAFS) scan usually takes about 20 minutes, but that can be shortened to as little as 10 minutes at the cost of some signal-to-noise ratio (see Section 8.9.3 of Chapter 8). X-ray absorption near-edge spectroscopy (XANES) scans can be even faster, with a scan taking as little as 2 minutes. If a system takes, for example, 3 hours to reach its endpoint, it might not change much during the course of a single 10-minute EXAFS scan.

Of course, this depends on the system being able to 'age' appropriately at the beamline. If the sample requires special environmental conditions, or if it is vulnerable to beam damage (e.g. it is a living system), then quenched samples can work for this time scale too.

In many systems that evolve on this time scale, such as some catalysis studies, the time evolution may be most rapid at the start of the process. Thus, the first few scans might be difficult to interpret, as significant changes are occurring within a single scan. As the rate of change slows, later scans may still provide valuable information.

3.1.4 Quick XAFS

Traditionally, XAFS experiments used the monochromator to select a particular energy and then collected a measurement there for a period of time (usually measured in seconds; see Section 8.9 of Chapter 8) before moving on to the next energy. For time-resolved measurements on a scale of seconds, however, this means that too much time is lost while the monochromator is moving from one energy to another without measurement being done. In addition, there is a significant delay between scans while the monochromator slews from the last energy of one scan back down to the first energy of the next one.

To address these issues, a technique known as quick XAFS (QXAFS) was developed, in which the monochromator is continually rotated through the energy range of the scan, with data collected on the fly (Frahm 1989). In addition, scans alternate between starting at a low energy and rotating to a high one, and starting at a high energy and rotating to low, thus nearly eliminating the time gap between scans. While some beamlines can operate in either QXAFS or traditional mode, an increasing number are optimized for QXAFS measurements (Dent 2002; Ressler 2003).

There are three primary factors that control the time resolution of a QXAFS scan:

1. **How quickly the monochromator can rotate through an energy range.** This is thus a characteristic of the endstation at a beamline and can often be looked up in publications or on the beamline's web page. (Or, of course, just ask the beamline scientists!)
2. **Detector response time.** Usually, a QXAFS beamline will provide detectors capable of making discrete measurements fast enough to keep up with the monochromator. But if you're bringing your own detector for a specialized experiment, then the response time of the detector needs to be considered as well.
3. **The desired signal-to-noise ratio.** This depends on characteristics of the beamline, characteristics of the sample, and the kind of information being sought in the experiment. (See Chapter 8.)
4. **The energy range of the measurement.** As always, the time resolution of a XANES-only measurement will be better than one that collects XANES and EXAFS data, simply because the energy range is shorter.

Some QXAFS beamlines can now achieve sub-millisecond time resolution for full EXAFS scans (Sekizawa et al. 2013), although time resolutions of tens of milliseconds are more typical.

3.1.5 Energy-Dispersive XAFS

For even better time resolution, a *polychromator* (also known as an *energy-dispersive monochromator*) can be used. This device, typically built around a curved silicon crystal, sends different energies of x-rays through the sample at different angles (see Figure 3.1). The emergent x-rays are then measured by a position-sensitive detector (e.g., a charge-coupled device), with the position of the measurement corresponding to the energy of the x-rays (Dent 2002; Pascarelli et al. 2006).

This geometry, known as *energy-dispersive EXAFS*, has the advantage that it measures an entire spectrum simultaneously, and thus can be very fast, with a time resolution that does not depend on the energy range of the scan. It has some notable limitations, however (Ressler 2003). For one thing, measurements can only be made in the transmission mode; this effectively limits samples to those that are uniform, fairly concentrated, and with thickness on the order of micrometers (see Chapter 6).

You'll sometimes see the speed of the mono- chromator given as a frequency measured in Hz: '50 Hz for a 1.5 keV scan at the copper edge.' 50 Hz would mean 20 ms per complete cycle, but a complete cycle includes *two* scans: one of increasing energy and one decreasing. So a monochromator operating at 50 Hz is taking 10 ms per scan.

Signal to noise can also be helped by repeating the experiment multiple times, combining the data from each set of scans.

Yes, but now we're back to the issue we had with quenched samples: is each trial of the experiment really the same? If, for instance, there are temperature fluctuations from one trial to the next, the time evolution might occur at a different rate, muddying the ability to combine data from different trials.

No single acronym for energy-dispersive EXAFS has taken hold. Just in the references cited in this section, it's referred to as ED-XAS, EDE, DEXAFS, DXAFS, and DXAS!

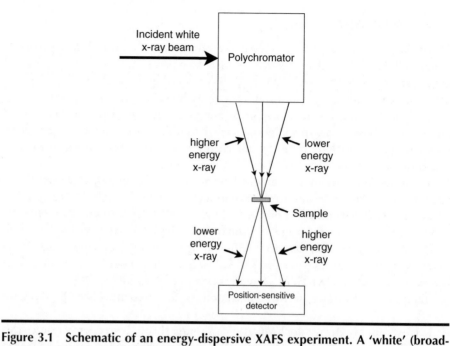

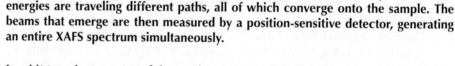

Figure 3.1   Schematic of an energy-dispersive XAFS experiment. A 'white' (broad-spectrum) x-ray beam is dispersed by the polychromator so that x-rays of different energies are traveling different paths, all of which converge onto the sample. The beams that emerge are then measured by a position-sensitive detector, generating an entire XAFS spectrum simultaneously.

Sometimes a polychromator is used with a downstream slit which is rapidly scanned across the beam, mimicking a traditional XAFS scan, with one energy being measured at a time. This is called turbo-XAFS (Pascarelli et al. 1999a). Turbo-XAFS restores the ability to measure incident and transmitted intensities simultaneously, as well as measuring fluorescence. It's pretty much a quick x-ray absorption fine structure (QXAFS) experiment using energy-dispersive geometry.

In addition, the intensity of the incident x-rays is difficult to measure simultaneously with the transmitted intensity, making the signal more sensitive to instabilities in the beam (Pascarelli et al. 1999b).

Full EXAFS scans can be achieved in 100 microseconds (Sekizawa et al. 2013). Since the entire spectrum is collected simultaneously, the time resolution is generally limited by detector response time.

3.1.6 Pump-Probe

Taking a continuous series of x-ray absorption measurements with a time resolution of picoseconds or femtoseconds runs into practical difficulties which are likely insurmountable; for one thing, x-ray detectors with such a short time response do not exist.

There is an alternative, however; rather than have the detectors provide the time resolution, as is the case in energy-dispersive XAFS, have the x-ray beam itself deliver pulses of very short duration. If there is enough downtime between pulses for the detectors to associate a measurement with each specific pulse, the time of the measurement can be known with great precision.

Here is an analogy, dear friends. Suppose there is a very reliable bus (the x-ray pulse) that comes to a certain stop every hour on the hour. A person (the detector) knows they arrived at the stop sometime between 1 and 2 pm, but is not sure exactly when. When the bus comes, the person knows it must be exactly 2 pm. In order to know the time of measurement (boarding the bus) to the nearest minute, the person only has to be able to determine times of events in general to the nearest hour.

In order to translate these measurements into a series of spectra that can be used to track the time evolution of a process, a *pump-probe* technique is employed. In this scheme, a 'pump' (typically a laser) is used to trigger the process to be studied, with an ultra-short duration x-ray pulse (the 'probe') used to make the measurement after a short, well-controlled delay. The system is then returned to its initial state (for reversible processes, this simply means waiting until the system relaxes), the delay is adjusted to a new value, and the pump-probe sequence is repeated. By scanning the delays through a range of values, the time evolution of the system may be measured. This is sometimes referred to as a *stroboscopic* technique.

Typically, the electron bunches in a third-generation storage ring have a width of 50 to 100 picoseconds (Cherugi 2022), limiting the time resolution of this technique if the pulses in the storage ring are used as the probes. (Continuing the analogy from Mandelbrant's sidebar, this would be like having a bus that stopped for several minutes to let passengers on.)

It is, however, possible to produce pulses of shorter duration.

One technique (*slicing*) uses an optical laser to modulate the energy of a brief portion of the electron bunch. The higher energy electrons are then spatially separated from the rest of the bunch and used to produce the ultra-short duration pulse (Schoenlein et al. 2000). This technique can produce pulses as short as 100 femtoseconds (Cherugi 2022). Even shorter pulses can be produced by free-electron lasers, with sub-femtosecond durations now demonstrated (Duris et al. 2020).

While there are many variations, both in beamlines and in designs for particular experiments, we will now describe a typical pump-probe XAFS experiment:

1. A laser is used as a pump to trigger the process which is to be measured.
2. The pump is linked, with a chosen time delay, to an ultra-short x-ray probe produced either by slicing an electron bunch from a storage ring or by a free-electron laser (FEL).
3. The probe is used as incident radiation to produce a single 'shot' measurement. Typically, this represents a measurement at a single energy at a single moment in the time evolution, although it is also possible to use energy-dispersive measurements to collect measurements for multiple energies in one shot.
4. Conditions are set for the next shot. Many shots may be taken under identical conditions to improve signal-to-noise. For measurement modes which select a single energy per shot, sometimes that energy must be changed between shots in order to produce a new spectrum. And in order to find a picture of time evolution, sometimes the time delay between pump and probe must be changed. The order in which these things are done will depend on details of the beamline and the experiment.
5. The system is reset. For reversible processes, this may happen by allowing sufficient time to pass for the system to relax back to its initial state. For irreversible processes, another technique must be found, so that a pristine sample (or section of sample) is measured for each shot.
6. Steps 1 through 5 are repeated until the desired energy range, time range, and signal-to-noise ratio are all achieved. The number of shots required is the product of the number of different energy values, the number of different time values, and the number of shots per combination of time and energy. It is not unusual for an experiment to require millions of shots.

Even though the duration of each probe may be measured in attoseconds or picoseconds, the time between shots is often on the order of one hundred microseconds. This gap between probe pulses is necessary for detectors to be able to discriminate between pulses and for the sample to relax or be refreshed. (Which of those two factors determines the length of the gap depends on the particular beamline and experiment.) An experiment requiring millions of shots thus may require several minutes to complete. If the time between shots is on the order of a hundred milliseconds (perhaps because of the need to continually expose pristine sample to the beam), the time for an experiment can easily stretch into many hours.

3.2 Microprobes

If you're a material scientist studying a homogeneous material, then it doesn't matter what part of your sample you measure. But many times, XAFS is used to study objects where the scientific questions relate to how they differ from place to place: where in a

FEL endstations dedicated to XAFS are a relatively recent phenomenon, and so improvements to configurations and techniques are happening rapidly. The gap between shots is an area with a clear potential for improvement, as is the number of shots needed to collect a good spectrum. If a third edition of this book is ever produced, I expect we'll be updating this section!

plant do toxins concentrate? How does material building up near the cathode degrade an electrochemical cell after multiple cycles? What pigments were used in a damaged section of a piece of artwork, and how can the damage be mitigated? In those cases, it is desirable to collect spatially resolved XAFS data (*μ-XAFS*). Beamline endstations equipped for those kinds of studies are often called *x-ray microprobes* (or occasionally *x-ray nanoprobes*, if the resolution warrants it).

Almost all microprobe endstations will include the following features:

■ A 'small' spot size for the x-ray beam. (How small is small varies greatly depending on the beamline and the desired application!) A minimum spot size about 1 micron in diameter is typical, although some beamlines feature resolutions better than 100 nm, with continued improvement possible (Sung et al. 2015).
■ The ability to position the beam at desired coordinates on the sample, with a precision better than the spot size.
■ Some way of aligning and positioning the sample so that the coordinates of the sample stage correspond in a known way to a position on the sample.
■ The ability to collect x-ray absorption data. This could employ any of the geometries, techniques, and equipment described elsewhere in this book.
■ X-ray fluorescence mapping of the sample.

Most microprobe endstations will also include the following features:

■ The ability to view the sample in sufficient resolution while positioning it to see where the beam will hit. If the spot size is relatively large (e.g. 1 mm in diameter), then this may be a simple camera image; for small spot sizes, the visual image may itself be produced by some kind of visual microscope.

Some microprobe endstations feature the following:

■ The ability to do other kinds of measurements (e.g. x-ray diffraction) at the same locations chosen for XAFS measurements.
■ The ability to do tomography by rotating samples.
■ Ability to control sample environment (e.g. gases, temperature).
■ A high degree of automation.

When you bring a sample to a beamline, it is usually because you have some combination of two kinds of questions: where in your sample is something of interest located, and what are the characteristics of the sample at particular spots you are interested in? For example, for a plant exposed to high levels of arsenic from its environment, you might want to know what part of the plant that arsenic concentrates in and what is the chemical form of the arsenic stored in each location.

To locate areas of interest, the most common procedure is to make an x-ray fluorescence (XRF) map of the sample, using the fluorescence lines of one or more elements of interest. Most microprobe beamlines have the ability to make maps of this kind, taking measurements at each spot on the sample (or representative subsets).

In some cases, it is possible to use XANES features to, for example, calculate the valence of an element. Much as an XRF map can show where an element is present in a sample, μ-XANES can sometimes be used to create a map of valence states across the sample (Nuyts et al. 2015).

Once an area of particular interest has been found, the beam can be moved to that spot and a full XAFS spectrum collected, allowing for detailed analysis of the composition and local environment.

WHAT I'VE LEARNED IN CHAPTER 3, BY SIMPLICIO

- Systems that age on a scale of months or years can be repeatedly measured at long intervals, but this requires getting beam time more than once.
- An alternative is to prepare quenched samples: start with a set of samples that are as identical as possible, allow them to start evolving in time, and then stop the time evolution at different points for each sample.
- Systems that evolve on a scale of tens of minutes can be measured using conventional XAFS, or can be prepared as quenched samples.
- For systems that evolve on a scale of milliseconds to minutes, QXAFS endstations can be used.
- Energy-dispersive XAFS, in which an entire spectrum is collected simultaneously, can be even faster than QXAFS, but can only be used in transmission mode.
- Pump-probe methods can work on time scales as short as femtoseconds, but must have a method to reset or replace the sample between each shot.
- Microprobe beamlines allow spectra to be collected from one spot on a sample. Spot size is typically measured in microns, but can be smaller.
- Microprobe beamlines are usually also capable of making x-ray fluorescence maps showing elemental abundance. These maps can be used to decide which spots to study by XAFS.

References

Cherugi, M. 2022. Time-resolved optical pump/X-ray absorption spectroscopy probe. *Int. Tables Crystallogr.* 1. Early view chapter. DOI:10.1107/S1574870720004760.

Dent, A. J. 2002. Development of time-resolved XAFS instrumentation for quick EXAFS and energy-dispersive EXAFS measurements on catalyst systems. *Top. Catal.* 18:27–35. DOI:10.1023/A:1013826015970.

Duris, J., Li, S., Driver, T. et al. 2020. Tunable isolated attosecond X-ray pulses with gigawatt peak power from a free-electron laser. *Nat. Photonics.* 14:30–36. DOI:10.1038/s41566-019-0549-5.

Frahm, R. 1989. New method for time-dependent x-ray absorption studies. *Rev. Sci. Instrum.* 60:2515–2518. DOI:10.1063/1.1140716.

Nuyts, G., S. Cagno, S. Bugani, and K. Janssens. 2015. Micro-XANES study on Mn browning: Use of quantitative valence state maps. *J. Anal. At. Spectrom.* 30:642–650. DOI:10.1039/c4ja00386a.

Pascarelli, S., O. Mathon, M. Muñoz, T. Mairs, and J. Susini. 2006. Energy-dispersive absorption spectroscopy for hard-X-ray micro-XAFS applications. *J. Synchrotron Radiat.* 13:351–358. DOI:10.1107/S0909049599000096.

Pascarelli, S., T. Neisius, and S. De Panfilis. 1999a. Turbo-XAS: Dispersive XAS using sequential acquisition. *J. Synchrotron Radiat.* 6:1044–1050. DOI:10.1107/S0909049599004513.

Pascarelli, S., T. Neisius, S. De Panfilis, M. Bonfim, S. Pizzini, K. Mackay, S. David, A. Fontaine, A. San Miguel, J. P. Itié, M. Gauthier, and A. Polian. 1999b. Dispersive XAFS at third-generation sources: Strengths and limitations. *J. Synchrotron Radiat.* 6:146–148. DOI:10.1107/S0909049506026938.

Ressler, T. 2003. Application of time-resolved in-situ X-ray absorption spectroscopy in solid-state-chemistry. *Anal. Bioanal. Chem.* 376:584–593. DOI:10.1007/s00216-003-1987-x.

Schoenlein, R. W., S. Chattopadhyay, H. H. W. Chong, T. E. Glover, P. A. Heimann, W. P. Leemans, C. V. Shank, A. Zholents, and M. Zolotorev. 2000. Generation of femtosecond

X-ray pulses via laser–electron beam interaction. *Appl. Phys. B.* 71:1–10. DOI:10.1007/s003400000372.

Sekizawa, O., T. Uruga, M. Tada, K. Nitta, K. Kato, H. Tanida, K. Takeshita, S. Takahashi, M. Sano, H. Aoyagi, A. Watanabe, N. Nariyama, H. Ohashi, H. Yumoto, T. Koyama, Y. Senba, T. Takeuchi, Y. Furukawa, T. Ohata, T. Matsushita, Y. Ishizawa, T. Kudo, H. Kimura, H. Yamazaki, T. Tanaka, T. Bizen, T. Seike, S. Goto, H. Ohno, M. Takata, H. Kitamura, T. Ishikawa, T. Yokoyama, and Y. Iwasawa. 2013. New XAFS beamline for structural and electronic dynamics of nanoparticle catalysts in fuel cells under operating conditions. *J. Phys. Conf. Ser.* 430:012020. DOI:10.1088/1742-6596/430/1/012020.

Sung, N.-E., I.-J. Lee, K.-S. Lee, S.-H. Jeong, S.-W. Kang, and Y.-B. Shin. 2015. A submicrometer resolution hard X-ray microprobe system h. of BL8C at Pohang light source. *J. Synch. Rad.* 22:1306–1311. DOI:10.1107/S1600577515014071.

Chapter 4

High-Energy-Resolution Techniques

A Photon Identity Crisis

In an XAFS fluorescence experiment, a photon is absorbed, leaving a core hole. Then the hole is filled by an electron in a higher energy level, producing a fluorescent photon.	
In an XAFS fluorescent experiment, a photon is scattered inelastically. Some of its energy goes to the atom it scattered from.	
While Dysnomia and Mr. Handy seem to be disagreeing, they're both giving accurate descriptions of the same experiment.	
Friends! We can discuss this later in the chapter. But let us begin by discussing energy resolution, which is in the title of this chapter.	

DOI: 10.1201/9780429329555-6

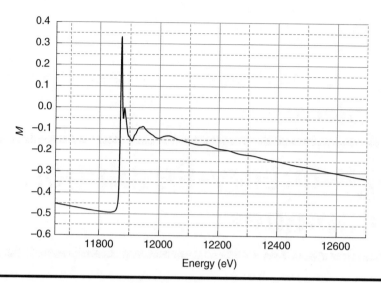

Figure 4.1 **X-ray absorption arsenic K-edge spectrum for an oxidized gallium arsenide sample.**

4.1 Energy Resolution

While EXAFS spectra vary fairly gradually with energy, XANES features can be quite sharp. Consider, for example, the spectrum shown in Figure 4.1. EXAFS is the gentle wiggles visible above 11,900 eV, changing on a scale of tens of eV. The XANES (just below 11,900 eV in this spectrum) changes much more rapidly, showing a sharply peaked white line and then, just a few eV higher in energy, a smaller satellite peak.

It should be clear that the energy resolution of this spectrum is sufficient for EXAFS analysis; improving the energy resolution further wouldn't make those gentle curves any sharper.

But what about the XANES? Is it possible that the white line is, in actuality, even sharper than what the spectrum shows, and thus should show even stronger absorption at its peak? We can see one satellite peak just above the white line; is it possible that the satellite itself has structure, perhaps consisting of multiple closely-spaced peaks?

In many cases, the energy resolution of a standard x-ray absorption experiment limits or distorts the information that can be deduced from XANES.

While energy resolution can be expressed in eV, another common metric is *resolving power*, which is the ratio of the measured energy to the energy resolution. For hard x-ray beamlines making measurements in transmission mode, resolving power is typically on the order of a few thousand.

The energy resolution of a XANES experiment is affected by several factors (see Section 8.5.2 of Chapter 8), but the limiting factor is *core-hole lifetime broadening*.

The XAFS process involves the formation of a core hole with a finite lifetime; eventually, an electron in a higher energy level will fill the core hole and the atom will either fluoresce or emit an Auger-Meitner electron. According to the Heisenberg uncertainty principle, this finite lifetime corresponds to an uncertainty in energy, thus limiting the energy resolution.

Core-hole lifetime broadening has been tabulated (Krause and Oliver 1979) and is shown in terms of the fractional energy resolution in Figure 4.2.

Over most of the range commonly used for XAFS, the lifetime broadening is, therefore, on the order of a few electron volts. It is also worth noting that for the *K* and L_3 edges, the fractional energy resolution owing to the core-hole lifetime is only weakly dependent on energy.

'Energy resolution' is usually defined by the full-width at half maximum (FWHM) of the signal measured in the theoretical case of a process with definite energy.

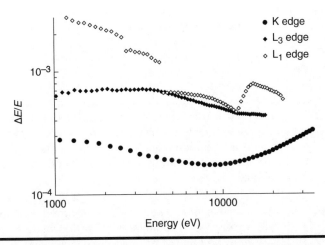

Figure 4.2 Fractional energy resolution due to core-hole lifetime (Source: From Krause, M. O. and J. H. Oliver. *J. Phys. Chem. Reference Data.* **8, 329–338, 1979).**

Figure 4.2 shows that *L* edges generally have worse energy resolution than *K* edges at the same energy; for example, the platinum L_3 edge has a poorer energy resolution than the arsenic *K* edge, since both are at around 12 keV. While it's also true that the *fractional* energy resolution of a given element's L_3 edge is worse than its *K* edge, the energy resolution as measured in eV is better, since its L_3 edge will be at a much lower energy than its *K* edge. For instance, the L_3 edge of iodine has a lifetime broadening of 3.1 eV, while its *K* edge has a lifetime broadening of 10.6 eV!

BOX 4.1   Determining Absorption by Measuring Fluorescence Is a Scattering Experiment

	How can absorption also be scattering? A photon is either absorbed or scattered; it can't be both at the same time.
	Can't it? Please describe the process of using fluorescence to measure absorption, Simplicio.
	An x-ray photon is incident on the sample, and is absorbed by an atom in the sample. It's absorbed! It's gone. It doesn't scatter. The energy from the photon is used to excite one of the atom's core electrons into an unoccupied state, or to knock it out of the atom altogether. Either way, that leaves a core hole—an unoccupied orbital in the first or second shell. After a little bit, one of the electrons in the atom falls into the core hole. The energy released creates a fluorescent photon, which is then detected. Every time we detect one of those fluorescent photons, we know that one of the incident photons was absorbed.
	So what you're telling me is that a photon goes into the sample, and then a photon comes out. That sounds like scattering to me! The scattering is inelastic, because it lost some energy to the electron, but it's still scattering.

	But it's not the same photon!
	How do you know that? Photons aren't labeled. They have energy, momentum, and spin, but otherwise they're identical. If a photon goes in with one energy and comes out with another, how do we know it's not the same photon?
	The process could be described as absorption and then fluorescence, or it could be described as inelastic scattering, but that's just to give us a mental model. Those are different ways of describing the same physical phenomenon.

Since absorption and then fluorescence can be thought of as inelastic scattering, it can be considered a single process with a lifetime longer than that of a core hole in a transmission experiment. The lifetime broadening Γ is given by combining in quadrature the lifetime broadening Γ_a of the 'absorption' core hole with the lifetime broadening Γ_f of the hole left at the end of the process (Proux et al. 2017):

$$\Gamma = \frac{1}{\sqrt{\dfrac{1}{\Gamma_a^2} + \dfrac{1}{\Gamma_f^2}}} \tag{4.1}$$

Since the hole left at the end of the process is in a higher energy state than the initial core hole, Γ_f is smaller than Γ_a, and thus, Γ is smaller than Γ_a—perhaps quite a bit smaller. Since the effective lifetime is longer, the uncertainty in energy is smaller, and the energy resolution better. This allows for energy resolutions well below 1 eV (Hayama et al. 2021).

4.2 High-Energy-Resolution Fluorescence-Detected X-Ray Absorption Spectroscopy

XAFS experiments which take advantage of the improved energy resolution of fluorescence experiments relative to transmission are referred to as high-energy-resolution fluorescence-detected x-ray absorption spectroscopy (HERFD-XAS).

It should be stressed that the high energy resolution employed in these experiments refers to the energy of the incident photons, not the measured intensity of the fluorescent photons. It would be possible to measure HERFD-XAS using a total-yield fluorescent detector with no energy discrimination at all, although in practice an energy- or wavelength-dispersive detector is often used to improve signal to noise.

Current HERFD-XAS beamlines can achieve hard x-ray energy resolutions of 0.2 eV or better, allowing the examination of XANES features not visible in conventional measurements (Hayama et al. 2021).

Beamlines optimized for HERFD-XAS require the energy selected by the monochromator to be very stable and reproducible. This has benefits aside from the

HERFD-XAS doesn't help if you're interested in EXAFS instead of XANES.

Figure 4.3   Detector arm for beamline I21 at the Diamond Light Source (cc-by-sa/2.0 - © Bill Nicholls - geograph.org .uk/p/6074853).

measurement of HERFD-XAS, and such beamlines often allow for a number of different modes of operation.

4.3 Resonant Inelastic X-Ray Scattering

While HERFD-XAS is used in the same way as a standard XAFS experiment, but with better energy resolution, the information from resonant inelastic x-ray scattering (RIXS) is quite different.

For RIXS experiments, not only is the energy of the incident photon precisely controlled, but the energy of the fluorescent photon is also measured with high precision. In practice, this often results in a two-dimensional plot with the energy of the fluorescent photon on one axis (or, quite often, the difference between the energy of the fluorescent photon and the energy of the incident photon), and the energy of the incident photon on the other. This makes these experiments a hybrid between x-ray absorption experiments of the kind described throughout this book, and x-ray emission experiments which are outside this book's scope.

Because the energy of the fluorescent photon is measured, these experiments are often framed as 'inelastic scattering' rather than fluorescence, with the measured photon referred to as the 'scattered' photon (see Box 4.1). Depending on the perspective of the researcher, RIXS can also be referred to as resonant x-ray emission spectroscopy (RXES), resonant x-ray fluorescence spectroscopy (RXFS), or resonant Raman spectroscopy (Ament et al. 2011).

To make measurements matching the energy resolution of the incident photons, the energy of the scattered photons must also be measured with a precision of a fraction of an eV. In practice, as described in Section 2.4.5, this requires wavelength-dispersive detectors. To achieve sub-eV energy resolution in those detectors requires very long detector arms, as can be seen in Figure 4.3. Current beamlines are designed for resolving powers as high as 70,000 for both the monochromator and the detector (Jarrige et al. 2018).

The energy lost by the scattered photon can be anywhere from a fraction of an eV to several hundred eVs, allowing a wide range of phenomena to be studied. The ability to probe transfers of relatively small amounts of energy allow studies of band structure and various kinds of solid-state excitations that are normally inaccessible to XAFS.

To learn more about RIXS, the review article by Ament et al. (2011) is a good place to start.

WHAT I'VE LEARNED IN CHAPTER 4, BY SIMPLICIO

■ Fluorescence can be thought of as one type of inelastic scattering of a photon.
■ The average lifetime of the fluorescence process is longer than the average lifetime of absorption processes, meaning that the energy resolution can be better. High-energy-resolution fluorescence-detected x-ray absorption spectroscopy (HERFD-XAS) uses this to improve the energy resolution of the incident x-rays.
■ If a wavelength-dispersive detector is used to also measure the energy of the fluorescent x-rays with high-energy-resolution, then the experiment is often referred to as resonant inelastic x-ray scattering (RIXS).

References

Ament, L. J. P., M. van Veenendaal, T. P. Devereaux, J. P. Hill, and J. van den Brink. 2011. Resonant inelastic x-ray scattering studies of elementary excitations. *Rev. Mod. Phys.* 83:705. DOI:10.1103/RevModPhys.83.705.

Hayama, S., R. Boada, J. Chaboy, A. Birt, G. Duller, L. Cahill, A. Freeman, M. Amboage, L. Keenan, and S. Diaz-Moreno. 2021. Photon-in/photon-out spectroscopy at the I20-scanning beamline at diamond light source. *J. Phys.–Condens. Mat.* 33:284003. DOI:10.1088/1361-648X/abfe93.

Jarrige, I., V. Bisogni, Y. Zhu, W. Leonhardt, and J. Dvorak. 2018. Paving the way to ultra-high-resolution resonant inelastic x-ray scattering with the SIX beamline at NSLS-II. *Synch. Rad. News.* 31:7–13. DOI:10.1080/08940886.2018.1435949.

Krause, M. O. and J. H. Oliver. 1979. Natural widths of atomic *K* and *L* levels, *Kα* x-ray lines and several *KLL* Auger lines. *J. Phys. Chem. Reference Data.* 8:307–327. DOI:10.1063/1.555595.

Proux, O., E. Lahera, W. Del Net, I. Kieffer, M. Rovezzi, D. Testemale, M. Irar, S. Thomas, A. Aguilar-Tapia, E. F. Bazarkina, A. Prat, M. Tella, M. Auffan, J. Rose, and J.-L. Hazemann. 2017. High-energy resolution fluorescence detected x-ray absorption spectroscopy: A powerful new structural tool in environmental biogeochemistry sciences. *J. Environ. Qual.* 46:1146–1157. DOI:10.2134/jeq2017.01.0023.

Chapter 5

Experimental Design

The End Justifies the Means

That subtitle sounds a little harsh, but in science it's often true. We go to light sources and spend hours collecting data and weeks analyzing it not for the sake of having data, but because we want to learn something about our samples.

It's easy to forget that. You can start to think there are things you're 'supposed to do' in an XAFS experiment, and do a lot of stuff that doesn't make sense for your goals.

A corollary, my friends, is that newer or fancier or more expensive isn't always better. Sometimes a simple experiment is the best at answering a question.

5.1 Identifying Your Questions

The first step to planning an x-ray absorption fine structure (XAFS) experiment is to identify the scientific questions you want to answer.

Most questions you might want XAFS to answer can be put into one of two categories: either you want to know what materials are in your sample (*speciation*), or you'd like to know specific characteristics of the substances in your sample (*characterization*). Table 5.1 gives examples of speciation questions, and Table 5.2 gives examples of characterization questions.

DOI: 10.1201/9780429329555-7

Table 5.1   Speciation Questions

Question	F	L	P	C
Is the sample pure X?	✓			
Is constituent X present?	✓		✓	
Which of several possible substances are present?		✓	✓	
What is the proportion of each constituent of a mixed sample?		✓		✓
How do qualitative features of a sample change with conditions?	✓	✓		✓

Source: Questions in this table are based in part on the National Synchrotron Light Source Online Orientation, which is licensed under a Creative Commons Attribution-Share Alike 3.0 United States License.

Check marks indicate whether the question is well suited to fingerprinting (F), linear combination analysis (L), principal component analysis (P), and curve fitting to a theoretical standard (C).

Table 5.2   Characterization Questions

Question	F	L	P	C
How does a modified version of X differ from pure X?				✓
How does the local structure differ from the known average structure?				✓
What is the immediate coordination environment of a particular element?				✓
What is the oxidation state of a particular element?	✓			
What is the environment of a particular dopant element in a material?	✓			✓
How do subtle features of a sample change with conditions?	✓			✓

Source: Questions in this table are based in part on the National Synchrotron Light Source Online Orientation, which is licensed under a Creative Commons Attribution-Share Alike 3.0 United States License.

Check marks indicate whether the question is well suited to fingerprinting (F), linear combination analysis (L), principal component analysis (P), and curve fitting to a theoretical standard (C).

5.2 Automation

Some beamlines feature varying degrees of automation. In many cases, multiple samples can be mounted at once, with the beamline rotating one sample at a time into the beam for measurement according to a script provided by the user.

Figure 5.1 shows an example of a sample wheel from Beamline for Materials Measurement (6-BM) at the National Synchrotron Light Source II. Blank wheels can be provided to users in advance of an experiment so that they can come to the beamline with their samples already mounted. The beamline also has wheels pre-mounted with a variety of reference foils and compounds.

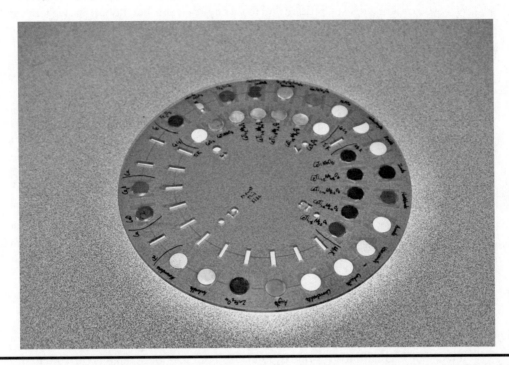

Figure 5.1 Double sample wheel for the Beamline for Materials Measurement at the National Synchrotron Light Source II.

There is also the possibility of using software to identify points of interest in a microprobe XRF map that fit a user-specified condition and then collecting XAFS at those points. In the future, artificial intelligence may play an increasing role in identifying and evaluating spectra, allowing for adjustments such as automatically collecting additional scans if the signal-to-noise ratio is poor—or 'giving up' on a sample or point if no absorption edge is observed.

Some beamlines allow fully remote experiments, with the samples shipped to the beamline.

BOX 5.1   Will Automation Eliminate In-Person Experiments?

	Maybe in the future, users won't ever have to go to a synchrotron—it will all be automated!
	There are already some experiments where users ship their samples to the beamline for measurement by beamline scientists, and don't come in person.
	I myself have participated in such experiments when my measurement plan is straightforward, my dear Mr. Handy. But I do not favor them in general. While I like to have a plan, I know that the best plans have the flexibility to change as data is collected and preliminary analysis is done. I prefer to be at the beamline myself, in full control of the experiment, able to make changes as conditions and results warrant. In fact, I generally bring many more samples and standards with me than I expect to have time to measure, so that I have options should the results of the measurements from early in the run provide surprises.

 What you describe doesn't require you to be physically present, Robert. As we learned during the COVID-19 pandemic, many things can be done remotely without being fully automated.

 Sure, Carvaka, but that doesn't extend to working with my custom *operando* rigs! They're fidgety, and I don't want anyone else mucking around with them. I've got one device I have to adjust using a rubber mallet. No offense, Mr. Handy, but I'm not asking a beamline scientist to whack my pride and joy with a mallet.

 No offense taken, Dysnomia! I'll support users in performing remote experiments that use standard equipment and routine protocols, but as soon as there's anything custom involved they'll need to come to the beamline themselves.

 Oh. I sometimes work with radioactive samples. I've been trained on the safety requirements, but I suppose that's also the kind of thing where I should be physically present.
Automation is great, and for some experiments I can see how it might save a trip to the beamline. But a lot of science and engineering is always going to be on the cutting edge one way or another, so I guess there will always be good reasons for users to go in person to light sources.

5.3 Experimental Plan

XAFS is best at comparing things: different model structures for a compound, different samples to each other, and different conditions for a single sample. Think about your experiment accordingly—don't just plan to measure your most important sample. Measure that important sample and other substances that it can be compared to.

For many studies, you will spend more than half your time measuring standards; this is particularly true if you're planning to do linear combination analysis.

For other studies, you want a series of samples that are related in some way: soil samples as a function of depth, thin films as a function of composition, and so on.

Sometimes you will only measure one or two samples, but under a number of different conditions or as a function of time. Cyclic voltammetry, *operando* catalysis, and time-resolved chemical reactions are all examples where the number of distinct samples may be small, but the number of spectra is not.

In addition to samples, you should think about people. Assuming it's not a remote experiment, who will go to the light source? Ideally, you'd like a team of two to four. If the experiment will not be highly automated, or if safety issues require constant supervision, you'll want a team on the larger end of that range, so that you can measure both night and day. Regardless, the group should ideally include people with experience and those just learning the ropes. Synchrotron experience and familiarity with the system are both important; often, the expert in synchrotron measurements is a novice to the system being studied and vice versa. Beamtime is a great time to learn from each other!

You should also consider the question of who will analyze the data. It is important that there are open lines of communications between the people most familiar with the system, the person analyzing the XAFS, and the people conducting other kinds of complementary measurements.

5.4 *Ex Situ* Experiments

The 'traditional' XAFS experiment is *ex situ*: that is, samples are prepared in a convenient form for measurement at the beamline. The sample preparation techniques described in Chapter 6 are largely meant for *ex situ* experiments.

5.5 *In Situ* Experiments

Ex situ experiments often entail the destruction and homogenization of the thing being measured: typically, it might be ground to a fine powder (see Chapter 6). While this facilitates measurement, it might be undesirable for a number of reasons:

Every discipline has its own definition of *in situ* outside the context of XAFS measurements (Wikipedia 2023). This can mean that there's some 'gray area' as to what constitutes an *in situ* XAFS experiment. Although fields such as archaeology, botany, zoology, and cultural studies often use the term to refer to measurements outside of a laboratory, in the context of XAFS it likely means bringing an item to a synchrotron, but leaving it at least partially intact. Soil samples are another good example: generally, a geochemist might use *in situ* to refer to measurements done at the site the soil is collected, but in the context of XAFS, it often refers to measurements where the soil structure was left intact (i.e., not pulverized or mixed).

■ The samples might be a precious artifact, such as a historically significant painting.

■ The experimental questions depend on how speciation or structure varies from point to point within a sample: for example, are the arsenic compounds in the leaves of a plant different from those in its roots?

■ The experimental question has to do with the time evolution of a sample in some environment of interest.

■ The material being measured is very short-lived. While the experimental question might involve only the initial state, and not the subsequent time evolution, it might not be practical to extract the material of interest and prepare a sample before it decays.

■ The material being measured is fragile or readily suffers environmental degradation. If a material readily oxidizes in response to air, for example, it might be easiest to measure it in the context in which it was formed.

■ The experimental questions have to do with changes in the sample after it is exposed to different conditions or use patterns. This is often the case for electrochemical cells, for example.

Measuring a sample *in situ* means more difficult data collection and analysis: the signal to noise of the spectrum may be poor, or there may be distortions introduced by issues such as self-absorption (see Section 6.4 of Chapter 6).

In situ experiments address these issues by measuring the sample in its original context: an electrochemical coin cell, a whole plant, a painting, a biological cell, a chemical solution, etc. Notice that microprobe experiments are almost always *in situ*, and time-resolved experiments are very likely to be (depending in part on how the phrase is defined).

5.6 *Operando* Experiments

Consider an *in situ* experiment on an electrochemical coin cell. In the experiment, the cell might be charged to a certain voltage, removed from the charging equipment, and then subjected to an XAFS measurement. It could then be removed from the beam, reattached to charging equipment and charged to a different voltage and then remeasured. This might be repeated several times. The XAFS measurements are *in situ* because the function of the cell is not destroyed during the process. But the cell is not actually charging or discharging *while* being measured.

In contrast, imagine measuring the cell while it is hooked up to a circuit, simultaneously measuring current and voltage. That would be an *operando* experiment, since

That is true, my dear Mr. Handy, but I am willing to tolerate some degradation in the quality of my spectra if the alternative is grinding up a masterpiece by Velasco.

In biology, the analogous term to *operando* is *in vivo*, in which measurements are conducted on a living organism.

the XAFS measurements would be taken while the device was serving its intended function.

Usually, the term *operando* is reserved for experiments in which a measure of the operating function is also being collected simultaneously with the XAFS. In addition to the electrochemical example given above, common cases include measuring a catalyst via XAFS at the same time as its catalytic activity is being monitored.

5.7 Multimodal Measurements

In *operando* experiments, data on the functioning of a sample is conducted simultaneously with XAFS data. That's one example of a *multidimensional* experiment, in which more than one kind of data is collected, either simultaneously (as in *operando* experiments) or in succession.

If the different measurements involve different x-ray instruments, geometries, or detectors, the experiment is referred to as *multimodal*. For example, a microprobe experiment which uses XRF to create a map of a sample and then XAFS to measure individual points can be considered multimodal. Even an experiment which collects both transmission and fluorescence XAFS data from the same samples could be considered multimodal.

Techniques featured at multimodal beamlines can include (Chu 2020) conventional XAFS, x-ray diffraction, x-ray fluorescence mapping, transmission x-ray microscopy, x-ray emission spectroscopy, RIXS, and more. Since each of these techniques is sensitive to different kinds of information, combining them can allow for a powerful scientific case to be made during a single synchrotron visit.

WHAT I'VE LEARNED IN CHAPTER 5, BY SIMPLICIO

- The first thing we should do is to identify the scientific questions we would like to be answered about our system.
- XAFS experiments are best at comparing things, so we should be measuring more than just our best sample. Usually, we'll also be measuring either standards or a series of samples or samples under a series of conditions.
- Many experiments use *ex situ* samples that have been prepared to optimize measurement. Experiments can also be performed *in situ*, in which a manufactured, biological, or environmental sample is measured in its original context, or *operando*, in which a device is measured while performing its intended function. (*In vivo* is the analogous term to *operando* for biological samples.)
- If more than one kind of data is collected on a sample, the experiment is *multidimensional*.
- If an experiment uses more than one kind of x-ray instrument, detector, or geometry, then it is *multimodal*.

References

Chu, Y. S., W.-K. Lee, R. Tappero, M. Ge, X. Huang, and X. Xiao. 2020. Multimodal, multidimensional, and multiscale x-ray imaging at the national synchrotron light source II. *Synch. Rad. News.* 33:29–36. DOI:1080/08940886.2020.1751520.

Wikimedia Foundation. In situ. Accessed September 9, 2023. *Wikipedia.*

Chapter 6

Sample Preparation

Cut Twice, Measure Once

Wait .... Is that right?	
Dear Simplicio, you don't want to have to take the same sample to a synchrotron to be measured more than once, do you?	
No, but...	
Would it make more sense, my friend, if we said 'do your physical sample preparation carefully, so that you don't waste beam time?'	
Oh, OK. But that wouldn't make a very good chapter title. I guess we should leave it the way it is.	

DOI: 10.1201/9780429329555-8

6.1 XAFS Samples

It's possible to perform XAFS analysis on nearly any sample of material. But the physical characteristics of the sample and the conditions under which you wish to measure it will affect how you prepare your sample and which measurement techniques you choose to use.

Some examples of sample type:

- Bulk solids
- Natural heterogenous materials, such as soil or tissue
- Gases
- Liquids, solutions, suspensions, and emulsions
- Frozen materials
- Unoriented powders
- Oriented crystals
- Thin films
- Manufactured objects, such as electrochemical cells and cultural artifacts

Special environmental conditions under which samples can be measured include:

- High or low temperature
- High or low pressure
- Electromagnetic fields
- Wet cells of hydrated environments
- *In situ* and *operando* environments
- Alternative atmospheres

Although XAFS can work with almost any type of sample, that doesn't mean that all types work equally well or are equally easy to deal with. Some samples, for instance, will not be able to be prepared with an 'ideal' thickness or concentration, resulting in a degradation of data quality or a trickier analysis.

Likewise, some environments, such as corrosive atmospheres or extreme temperatures, will complicate your choice of sample holder and window. Finally, some materials (particularly organics) can be damaged by x-rays, and this may require modification of the experiment. But chances are someone has attempted something similar before, and checking the literature and online resources for similar experiments should be your first step in designing an XAFS experiment for an 'unusual' sample.

6.2 Absorption

6.2.1 Undesirable Photons

In theory, an XAFS spectrum is supposed to represent the probability of photon absorption as a function of energy. Depending on the measurement technique employed, this will be some function of the ratio of the intensity of x-rays or electrons coming out of or through the sample compared with the intensity of x-rays heading into it. But a real experiment is not so simple. Frequently a detector will measure some photons that you aren't interested in. Likewise, photons that you would like to have counted might not be, particularly in fluorescence measurements.

Here are the most common examples of photons that we don't want to count in a transmission experiment:

- **Scattered photons.** Some of the photons that don't pass through the sample may nevertheless be deflected into the detector downstream of the sample (the I_t detector).

- **Harmonics**. Monochromators usually allow through some harmonics: photons with energies that are integer multiples of the desired energy (although not necessarily all of the integer multiples, depending on the monochromator crystal and its orientation).
- **Distribution tails**. Monochromators, despite the name, do not produce perfectly monochromatic beams. In addition to harmonics, there will be a spread of energies around the desired energy; some photons with an energy electron volt or two higher or lower than the nominal value are often present.

The XAFS community uses the expression *thickness effects* to describe the impact of these photons (Stern and Kim 1981). This term is a little misleading—with a perfect beam and no scattering, even thick samples would not exhibit these 'thickness effects!' In this book, we'll instead usually refer to the effect of *undesirable* photons.

Although these real-life complications are always present to some extent, for some samples the effect on the spectrum is negligible, whereas in others the data will be severely distorted. Why the difference? Often, it comes down to sample thickness and uniformity. For instance, a transmission sample that is too thick may leave very few of the desired photons downstream of the sample to be counted, and most of the counts might then be due to the undesirable photons. On the other hand, a transmission sample that is too thin will absorb very few photons, and statistical noise could overwhelm the signal.

6.2.2 The Absorption Coefficient

Our next step is to make this quantitative. Table 6.1 lists the variables we'll be working with.

The absorption coefficient μ is a quantity that expresses the fraction of photons that are absorbed when an x-ray beam is passed through a given thickness of a particular material:

$$\mu(E) \equiv -\frac{dI}{dx} \tag{6.1}$$

When XAFS scientists calculate 'absorption,' they are actually computing *attenuation*, that is, the reduction in intensity of a narrow beam. This is due mostly to actual absorption of photons (called *photoelectric absorption*), but also includes the reduction due to photons that scatter out of the beam.

Table 6.1   Symbol List for This Chapter

Symbols in This Book	Alternate Symbols	Units	Meaning
μ_m	μ	cm²/g	Mass absorption coefficient
μ		cm⁻¹	Absorption coefficient
Absorption	μ	—	Absorption
x	t	cm	Sample thickness
A		cm²	Sample area
V		cm³	Sample volume
m		g	Sample mass
ρ		g/cm³	Sample density
I_0		counts, V, A	Intensity of incident x-rays
I_t	I, I_1	counts, V, A	Intensity of x-rays transmitted through sample
I_f		counts, V, A	Intensity of fluorescent x-rays

BOX 6.1   Which μ?

	According to the table, it seems that μ could mean three different things. How are we supposed to know which one someone means?
	You can tell by the units! The mass absorption coefficient has units like cm²/g, the absorption coefficient has units like cm⁻¹, and absorption has no units.
	You really should be able to tell by the context of the equation—who cares what symbol someone uses?
	I care, dear Carvaka! It is confusing, and we should all try to make the world a less confusing place. Just because other people sometimes use the symbols in sloppy ways does not mean we should. In this book, we will always use the symbols in the first column.

The absorption coefficient is dependent on the energy of the photons absorbed. Every chemical element absorbs progressively less as energy is increased (with the exception of the sharp jumps near its own edge energies). Also notice that it depends on the density of a material; a more densely packed material will have more atoms in a given thickness, and thus a higher absorption coefficient. Therefore, it is often useful to work with the mass absorption coefficient:

$$\mu_m(E) \equiv \frac{\mu(E)}{\rho} \tag{6.2}$$

μ_m is directly proportional to the probability that a given atom in the material will absorb a photon and can generally be calculated from tabulated data. For example, the μ_m for elemental arsenic, just above its K edge, is about 179 cm²/g (Elam et al. 2002).

6.3 Sample Characteristics for Transmission

6.3.1 Thickness, Mass, and Absorption Lengths

The definition of absorption coefficient can be easily integrated to give the decrease in intensity found in transmission, a result known as *Bouguer's Law*:

$$I_t = I_o e^{-\mu_m \rho x} \tag{6.3}$$

or, equivalently,

$$\mu_m \rho x = \mu x = \ln\left(\frac{I_o}{I_t}\right) \tag{6.4}$$

X-ray mass absorption/ attenuation coefficients can be found online. Free software such as HEPHAESTUS (Ravel and Newville 2005) can even calculate the effective coefficient for a material given its chemical formula, saving you the necessity of calculating the average over the atoms present by hand.

Table 6.2 Transmitted Photons as a Function of Number of Absorption Lengths

Number of Absorption Lengths	Percent of Photons Transmitted
0.001	99.9
0.01	99
0.1	90
1	37
2	14
3	5
4	2
6	0.2
10	0.005

Colloquially, people often refer to μx for a particular sample as the number of *absorption lengths* it has.

This allows you to find the thickness of a sample that corresponds to a desired number of absorption lengths. For example, consider a thin film made of zinc oxide (ZnO). Just above the zinc K edge, the mass absorption coefficient of zinc is 254 cm²/g and for oxygen it's 7 cm²/g (Elam et al. 2002). The percent by mass of zinc in ZnO is 80.3, with the rest being oxygen. So just above the zinc edge, $\mu_m = 0.803 \times 254 + 0.197 \times 7 = 205$ cm²/g. Because ZnO has a density of 5.6 g/cm³, one absorption length is 8.7 μm (i.e., $x = 8.7$ μm makes $\mu_m \rho x = 1$).

Bouguer's Law tells us that the fraction of the photon intensity that makes it through the sample is exponentially dependent on the number of absorption lengths present. Although it's a simple calculation to do yourself, Table 6.2 is provided to impress upon you the implications of this exponential dependence.

Perhaps you have noticed that we have started writing μ_m instead of $\mu_m(E)$, for the purpose of keeping the equations from becoming too cluttered. I encourage you not to forget that absorption coefficients are functions of energy.

Therefore, if you have a 'thin film' of ZnO 87 μm (10 absorption lengths) thick and shine x-rays on it with energies just above the zinc K edge, virtually none of the x-ray photons will get through! There's little point in attempting a transmission measurement on such a sample. Fluorescence would be better, but that has its own perils, which are discussed later in this chapter. Better yet would be using a thinner sample, but that's not always possible or convenient.

I have heard Bouguer's Law called Beer's Law, the Beer-Lambert Law, and other variations. A purist might insist the laws actually refer to different aspects of exponential absorption, but in my opinion, common usage has made the terms essentially interchangeable.

With samples that are thin films, solutions, gases, or sectioned solids, you generally know (or can choose) how thick they are. But what about powders? It is difficult to directly measure the thickness of a thin layer of powder, and there may be some air spaces between grains. Fortunately, by measuring the area over which a layer of powder is spread, it is easy to convert the thickness to a mass:

$$\mu_m \rho x = \mu_m \frac{m}{V} x = \mu_m \frac{m}{Ax} x = \frac{\mu_m m}{A} \tag{6.5}$$

As you can see, you don't even need to know the density of the material! So if, for example, you want a layer of ZnO powder that corresponds to one absorption length just above the zinc K edge, and you want the sample to be 2.0 cm², then you need 9.8 mg of powder. If you use 100 mg of powder, you're barely going to get any signal through at all.

6.3.2 Uniformity

It is extremely important that the sample be as uniform as possible over the width of the beam. This is because the log of sums is not the same as the sum of logs, and so unevenness results in distortion of the data (see Box 6.2). It is therefore imperative that you do whatever you can to your particular sample to make it as evenly distributed as you can. *Pinholes*, the usual term for spots where the beam can reach the I_t detector without traversing any sample, are an obvious sign of nonuniformity. Less obvious, but potentially just as important, is the effect of particle shape and size. If the individual particles comprising your sample are spheres with diameters greater than about 0.2 absorption lengths, then the sample will be inherently uneven and substantial distortion will result. For a detailed analysis of these effects, see Scarrow (2002).

If we consider a sample made of a single layer of spherical particles, then a photon coming in along the path shown by the arrow on the left would see a much thicker sample than one coming in along the arrow to the right.

BOX 6.2   Quantifying the Effect of Sample Nonuniformity

 Friends! Let us illuminate this idea further by exploring it quantitatively. For our example, consider a sample with two different regions within the beam, one of thickness x_1 and area A_1, the other of thickness x_2 and area A_2. The intensity in each region obeys Bouguer's Law, as we have learned

$$I_{t1} = I_0 e^{-\mu_m \rho x_1} \quad \text{and} \quad I_{t2} = I_0 e^{-\mu_m \rho x_2} \tag{6.6}$$

This gives us a total transmitted intensity of

$$I_t = \frac{I_{t1}A_1 + I_{t2}A_2}{A_1 + A_2} = \frac{I_0}{A_1 + A_2} \left(A_1 e^{-\mu x_1} + A_2 e^{-\mu x_2} \right) \tag{6.7}$$

If the usual experimental procedure is used, we would then calculate absorption to be

$$Absorption_{meas} = \ln \frac{I_0}{I_t} = \ln \left(\frac{I_0}{\frac{I_0}{A_1 + A_2} \left(A_1 e^{-\mu x_1} + A_2 e^{-\mu x_2} \right)} \right) \tag{6.8}$$

$$= \ln \left(\frac{A_1 + A_2}{A_1 e^{-\mu x_1} + A_2 e^{-\mu x_2}} \right)$$

But in XAFS, our analyses are generally based on the energy dependence of absorption. So let us consider the ratio of the absorptions at two different energies:

$$\frac{Absorption_{meas}\left(E_b \right)}{Absorption_{meas}\left(E_a \right)} =$$

$$\frac{\ln \left(\dfrac{A_1 + A_2}{A_1 e^{-\mu(E_b)x_1} + A_2 e^{-\mu(E_b)x_2}} \right)}{\ln \left(\dfrac{A_1 + A_2}{A_1 e^{-\mu(E_a)x_1} + A_2 e^{-\mu(E_a)x_2}} \right)} \tag{6.9}$$

Unless $x_1 = x_2$, this ratio depends explicitly on A_1, A_2, x_1, and x_2, meaning that the shape of our XAFS spectrum no longer depends only on the absorption coefficient μ, but also on the details of the sample preparation in a way that is probably not known by us when we conduct the experiment. This will therefore distort our analysis.

How large is this effect? To see, let us consider the case where half the sample is twice as thick as the other half, so that $x_2 = 2x_1$ and $A_1 = A_2 = 0.5A$.

$$Absorption_{meas} = \ln\left(\frac{A}{0.5Ae^{-\mu x_1} + 0.5Ae^{-2\mu x_1}}\right) =$$

$$\ln\left(\frac{2e^{\mu x_1}}{1 + e^{-\mu x_1}}\right) = \ln 2 + \mu x_1 - \ln\left(1 + e^{-\mu x_1}\right) \tag{6.10}$$

If the sample were of uniform thickness x_1, the absorption would be μx_1; if it were of uniform thickness $x_2 = 2x_1$, then the absorption would be $\ln 2 + \mu x_1$. So we see that the $-\ln\left(1 + e^{-\mu x_1}\right)$ is trying to provide an absorption somewhere between the two; the problem is that it is itself dependent on μ in a nonlinear way. To further explore this effect, let us suppose that at energy E_a, μx_1 is 1.000, and at energy E_b, μx_1 is 1.100. The ratio of absorptions measured with our hypothetical sample would not be 1.100/1.000 = 1.100, as desired, but 1.091.

That $-\ln\left(1 + e^{-\mu x_1}\right)$ gets very small if the sample were very thick, right? So I don't have to worry as long as the thinnest part of my sample is very thick.

Yeah, but in practice, that's not the way it works. It's really hard to make a sample thicker without making the variation in the thickness bigger too. Putting lots of uneven layers on top of each other makes the variation in thickness increase in just the right way to cancel out the improvement from having a thicker sample. And there are other problems with thick samples, as we'll see shortly.

There are a couple of things you can do to assure that you are measuring a uniform region of your sample. One is to make the sample as uniform as possible in the first place; techniques for doing this will be covered later under the specific sample types. Another is to try to find a uniform section of a sample after it is made. If aberrations are noticeable visually, it will help to mount the sample in such a way that the uniform part is placed in the beam. You can also scan as a function of position of the sample to find its most uniform part (See section 8.8 of Chapter 8).

6.3.3 A Short Digression on Noise

As you probably know, *noise* refers to random, unbiased variations in a measurement; that is, it will average out if sufficient data are taken. But that's a broad definition. Let's start by discussing *shot noise*, also known as *Poisson noise* or noise due to *counting statistics*.

This noise arises from concepts of probability: if on average 100 photons are absorbed in a chamber per second, then during a particular second, it might be 93 and during another second, it might be 104 or any other number fairly close to 100.

But what's 'fairly close?' It turns out that such measurements tend to vary as the square root of the count (Bevington and Robinson 2003); in the earlier example, we'd expect a variation of about ± 10 photons in any given measurement.

This means that if we had more photons, say, an average of 1,000,000 per second, then the typical variation in the *number* of photons we count would be much greater (1,000, in this case), but the typical *percentage* variation would be smaller (1,000/1,000,000 = 0.1%, as compared to 10/100 = 10% earlier). In fact, it's easy to convince yourself that the percentage variation goes inversely as the square root of the number of counts, so that if we increase the number of counts by two orders of magnitude the percentage variation goes down by one order of magnitude.

XAFS beamlines at modern light sources can easily produce beams strong enough to produce 10^{10} photons per second in a detector, yielding shot noise on the order of 0.001% of the signal.

But counting statistics isn't the only source of noise. Another prominent source, for instance, is the electronics that amplify and process the signals from the detectors. With modern synchrotrons and concentrated, well-prepared samples, electronic noise is likely to be much more significant than shot noise.

There are also other sources of noise. A number of effects show up as systematic errors if they change slowly (or not at all) relative to the data collection time, but as noise if they fluctuate rapidly. In this category are effects such as changes in the detectors themselves (perhaps the temperature or pressure of the gas in an ion chamber) and effects tied to the orbit of the electrons in the storage ring.

EXAFS features are usually less than one tenth the amplitude of the major XANES features. So shot noise of 0.001% of the overall signal would be more than 0.01% of the EXAFS signal.

Even with modern light sources, there are still times that shot noise is important. For instance, if you are taking time-resolved data or using a microprobe with a sample vulnerable to beam damage, the number of photons per data point measured in your transmission chamber could be fewer than a thousand. And for fluorescence measurements, shot noise is often the dominant source of noise.

6.3.4 Optimum Thickness for Transmission

It's clear that 10 absorption lengths is too thick for a transmission sample, because very few photons will make it through to be counted. There's a similar problem for very thin samples: too few photons are then absorbed. But that's not the only problem with samples that are too thin or too thick.

If a powder transmission sample is **too thin**, then,

■ Very few photons are absorbed by the sample. Why is this a problem? Consider the signal-to-noise ratio due to noise in I_o and I_t. Using standard methods for the propagation of uncertainty, we find the signal-to-noise ratio is

$$
\frac{\mu x}{\Delta(\mu x)} = \frac{\mu x}{\Delta\left(\ln\frac{I_o}{I_t}\right)}
$$

$$
= \frac{\mu x}{\frac{I_t}{I_o}\sqrt{\frac{I_t^2(\Delta I_o)^2 + I_o^2(\Delta I_t)^2}{I_t^4}}} = \frac{\mu x}{\sqrt{\frac{(\Delta I_o)^2}{I_o^2} + \frac{(\Delta I_t)^2}{I_t^2}}}
$$

(6.11)

As we take the limit of thinner and thinner samples, μx (the signal) approaches zero, but the expression in the denominator (the noise) levels off at some nonzero value. The result is that the signal-to-noise ratio becomes very poor.

■ The samples will have a small edge jump (see Section 6.3.5). This also depresses signal relative to noise.

■ With extremely thin samples, it may be difficult to make the sample uniform.

If, on the other hand, a powder transmission sample is **too thick**, then,

- Very few photons make it through the sample. Noise in I_t can swamp the signal. (Comparatively, the effect of noise in I_o is negligible for thick samples.)
- The absorption of harmonics will be much less than for photons of the desired energy. A beam including 1% harmonics before traveling through a thick sample could easily be 70% harmonics after. The same is true of photons scattered within the hutch.
- Limited monochromator resolution also affects thick samples more. Just above the edge of a thick, concentrated sample, nearly all the desired photons may be absorbed, but those just a few eV lower may make it through. Although this is similar to the effect of harmonics and scattered photons, monochromator resolution only has a noticeable effect on features that vary rapidly in energy, such as the edge and the white line.
- It is difficult to judge uniformity by inspection, either with the naked eye or with a microscope.

Over the years, XAFS experts have analyzed thickness effects and come up with recommendations for the 'ideal' thickness of a transmission sample. A few of these recommendations are summarized in Table 6.3.

This table is presented here primarily to make the point that the *precise* thickness of a transmission sample is not crucial.

It's worth noting that the recommendations for total absorption include *all* the material between the I_o and I_t detectors, not just whatever you deign to call the 'sample.' Windows, substrates, and air (particularly at low energies) also contribute to this absorption.

Don't worry too much about exactly what thickness is ideal! Think instead about what the issues are as you get thinner or thicker. For example, if you have a sample that is particularly thin—perhaps only 0.2 absorption lengths— you should do what you can to improve the signal to noise. If, on the other hand, you have a sample that has a thickness of 3 absorption lengths, you must take special care to eliminate harmonics from your beam!

On the other hand, if it is convenient to do so, one should certainly strive to make transmission samples roughly 1–2.5 absorption lengths in thickness. With powders, that is not difficult to do, and is a good compromise between thin and thick.

Table 6.3   Recommendations for Sample Thickness

Reference	Recommended Number of Absorption Lengths	Comments
Scarrow 2002	2–2.5	Shot noise dominant
Scarrow 2002	1–1.3	Electronic noise dominant
Scarrow 2002	1–1.5	General recommendation
Kelly et al. 2008	< 2.5	
Lee et al. 1981	2.6	Based only on shot noise
Bunker 2010	< 3	
We recommend	~1	

6.3.5 Edge Jump

The difference in absorption between energies before the edge and energies just after the edge is referred to as the *edge jump* (or *edge step*) and is often written $\Delta\mu x$. In theory, the edge jump in transmission is directly proportional to the thickness of the sample and also to the concentration of the target element in the absorbing material. Because of undesirable photons, this won't be exactly true, particularly for thicker samples.

Because fine structure is proportional to the edge jump, a tiny edge jump means tiny fine structure, and thus a poor signal-to-noise ratio. But a really large edge jump is sensitive to undesirable photons.

In this section, we haven't defined *edge jump* very rigorously, but for our current discussion, we don't need to. We will be more careful about it in Section 7.3.4 of Chapter 7.

Most XAFS experts recommend aiming for an edge jump of around 1 (but see Box 6.3). Beyond 1.5, it is important to take great care that thickness effects due to undesirable photons aren't distorting your spectrum.

BOX 6.3   Edge Jump or Total Absorption?

 You've given me two different criteria to worry about: the total absorption should be around 2 and the edge jump should be around 1. But often I can't do both! If my sample has a small concentration of the target element, then if I make the total absorption right, I'll get a small edge jump, but if I make the edge jump right, I'll get a big total absorption. What should I do?

 The total absorption is more important than the edge jump, Kitsune, because if the total absorption is too high, undesirable photons are almost certain to distort your spectrum. If there's not much of your target element in your sample, then you can make the sample a bit thicker than normal—maybe 2.5 absorption lengths—if you're really careful to keep harmonics out of the beam. If that means the edge jump is only 0.2 or even 0.02, that's still OK. And if your sample is even more dilute than that, you should probably just bite the bullet and use fluorescence.

 But it's so much easier to check if I've got the edge jump right and not worry about total absorption—all I have to do is look at the spectrum!! Measuring the total absorption is hard.

 No it's not! Just look at I_t before you put the sample in and then look again after. If it drops by a factor of 20 or more, then your sample is at least 3 absorption lengths thick, which is probably thicker than you want.

BOX 6.4   Scary Graphs

 It's kind of traditional to show you graphs of 'bad' data caused by 'bad' sample preparation, maybe with the idea that it will scare you into making sure you always make a perfectly ideal sample. But your sample may not be the kind of thing you can make textbook perfect. And let's be honest; sedimentation, sieving, grinding, and weighing each take time. Those of us who don't have a big research group to fall back on often face a bottom-line decision: Is it better to get beautiful data on 5 samples or decent data on 20?

 That depends. If I were a graduate student and those five samples were the heart of my dissertation, and I were making my argument based on finding coordination numbers to within ± 10%, then I would do everything I could to get the best possible data. If instead I were doing some preliminary work to try to understand the speciation in a range of samples, then I might be happier measuring a larger number of samples.

Whatever you decide, make sure you give a sense of what you did in any eventual publication. If you sieved the samples, for instance, say so. That way your audience can judge for themselves how far to trust the data.

So we're not going to show you a comparison of 'perfect' data to 'bad' data. Instead, we're going to compare 'typically good' data to data taken from samples that are extremely thin, thick, or uneven. That way you can clearly see what kinds of problems arise in each case.

To illustrate the comparison, we are going to use graphs of $\chi(k)$ collected for at the zinc K edge of zinc ferrite. Graphs of this type were introduced in Section 1.4.1 of Chapter 1, and will be discussed in more depth in Section 7.4 of Chapter 7.

For our 'typically good' data, we're using a single iron K-edge scan of zinc ferrite. The edge jump of the sample was 1.0, and the total absorption, counting the tape, was 2.9 absorption lengths. That's a little on the thick side, so we tried extra hard to reduce the harmonic content of the beam. The sample was ground finely and spread thinly on layers of tape, but not sieved or separated by sedimentation.
Our first comparison is with a thin sample of the same material. For the thin sample, the edge jump was 0.09 and the total absorption was 0.5 absorption lengths.

Why is the edge jump of the thin sample less than 10% of the edge jump of the 'typically good' sample while the total absorption is more than 15%? Because there's some 'overhead' associated with the layers of tape used to seal the sample.
You can tell from the label on the y-axis of Figure 6.1 that these data have been weighted by k^3. As you'll learn in Section 7.4.5 of Chapter 7, multiplying by some power of k is typically done because EXAFS signals get smaller at higher values of k. But while k weighting is used to keep the signal about the same amplitude at high k as at low, it also increases the noise. That's why the high-k portion of a weighted spectrum almost always looks noisier than the low-k part.
In this case, we can see that the thin sample is much noisier than the typically good sample (look at the jagged features visible in the thin sample's high-k spectrum). But despite the typically good sample being on the thick side, there's no evidence of trouble from harmonics, which would show up at low k as well as high.

Although most of us can recognize noisy data from a single scan because of the point-to-point variation at the high-k end of a k-weighted spectrum, taking additional scans should show that the noise varies from scan to scan in an unpredictable way. One should always collect more than one scan on any sample one is interested in, if only to have a way of judging the amount of noise that is present.

Next up is a really thick sample (Figure 6.2). This sample has a whopping edge jump of 6.1 and a total absorption of 8.4 absorption lengths. Only about 1 photon in 5000 makes it through the sample, and you can bet that a lot of those are undesirable harmonics.

There's not much evidence of noise this time, despite the small number of photons making it through—just a hint above 11 Å⁻¹. But see how the amplitude is suppressed? That's due to harmonics. Because the photons constituting the harmonics have a much higher energy than those in the fundamental, they have an easy time passing through the sample and are not affected by the edge or the fine structure. At energies with a lot of absorption, such as near the white line, many of the photons belong to the harmonics, and so the feature is suppressed. Preedge features, on the other hand, would not be suppressed nearly as much, because many more of the photons belonging to the fundamental make it through at those energies. This pattern is particularly easy to see if we look at the XANES (Figure 6.3).

If you didn't realize the thick spectrum was distorted, an attempt to analyze the XANES would likely lead to fundamental errors even in qualitative tasks like phase identification!

Our final comparison is with a sample made to be uneven, with multiple thick and thin regions, using a total amount of material comparable to the 'typically good' sample. The edge jump is 0.6, which is fine. But the lack of uniformity gets to the spectrum, as can be seen in Figure 6.4.

You can see a distortion that's a bit like what we had for the thick sample, although this time a lot of it's due to the phenomenon Mandelbrant walked us through in Box 6.2. But the high *k* part of the spectrum is even more messed up than it was for either the thick or thin sample. As we'll get to in Chapter 8, uneven samples are particularly vulnerable to problems in I_o or nonlinearity in the detectors, and that kind of thing is contributing to the problems we see here at high *k*.

Honestly, uneven samples are the worst of all worlds; spots that are particularly thin or thick degrade the signal-to-noise ratio, thick spots exacerbate the effect of harmonics, and the lack of uniformity introduces additional distortions.

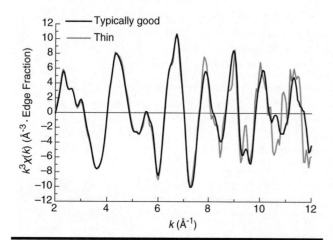

Figure 6.1 k^3-Weighted $\chi(k)$ for two samples of zinc ferrite; one prepared fairly well and the other prepared to be thinner than ideal.

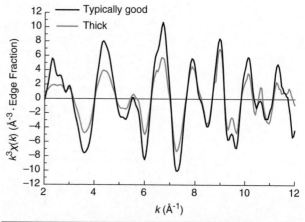

Figure 6.2 k^3-Weighted $\chi(k)$ for two samples of zinc ferrite; one prepared fairly well and the other prepared to be thicker than ideal.

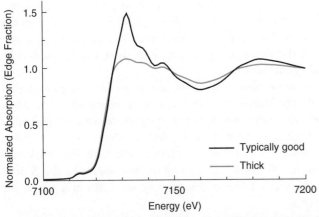

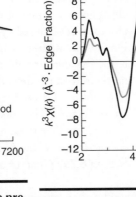

Figure 6.3 **XANES for two samples of zinc ferrite; one prepared fairly well and the other prepared to be thicker than ideal.**

Figure. 6.4 k^3-**Weighted** $\chi(k)$ **for two samples of zinc ferrite; one prepared fairly well and the other prepared to be uneven.**

6.4 Sample Characteristics for Fluorescence

6.4.1 From Incidence to Detection

In a transmission experiment, the direct measurement is of photons that made it through the sample unscathed; absorption by the sample is deduced by comparing the intensity of photons downstream of the sample to those upstream.

In contrast, fluorescence experiments measure photons emitted by atoms that have absorbed an incident x-ray. This introduces several subtleties into the measurement. To understand them, let's look at an overview of the process that brings a signal photon to the detector:

1. As in transmission, some of the incident photons of the energy E selected by the monochromator are absorbed in the I_o chamber. There are also a few undesirable photons (harmonics, etc.) in the beam.
2. Also as in transmission, on the way from the I_o detector to the sample, photons may be absorbed or scattered by windows, air, and the like.
3. Once the photons penetrate the sample, many things can happen to them. They may be scattered elastically, scattered inelastically, absorbed by the element of interest, or absorbed by another element. Or, if the sample is fairly thin, they may make it out the other side. The only one of these processes that we actually want to measure is absorption by the element of interest.
4. The atoms of the element of interest that do absorb a photon are now missing an electron in one of their lower energy states; they have a *core hole*. This hole may be filled in various ways; the case that we're interested in here is when an electron from a higher shell falls into the hole, with the difference in energy between the two energy levels appearing as a photon. Note that this *fluorescent* photon is at a lower energy (called E_f) than the incident photon was, because there's still a hole—it's just that the hole is now in a higher level.
5. Fluorescent photons are emitted isotropically; that is, the direction of any individual fluorescent photon is random. Most will not head toward the I_f detector.
6. Those that do head toward the detector stand a chance of being absorbed or scattered again before emerging from the sample. They do have an advantage over the incident photons, though, because E_f is below the absorption edge.

As we discussed in Box 4.1 of Chapter 4, the fluorescence process we are interested in measuring can also be described as a particular kind of inelastic scattering of the incident photon.

7. After emerging from the sample, the fluorescent photons still have the usual assortment of windows and air to penetrate. But in addition, some fluorescent setups include a filter. The filter is designed to screen out some of the undesirable photons. Some of the fluorescent photons we *do* want will themselves be absorbed or scattered by the filter.

8. Finally, the desired photons, along with a lot of undesirable *background*, will reach the detector. If the fluorescence detector is an energy-discriminating detector, then the detector itself is able to screen out photons with energies more than about 200 eV higher or lower than the desired energy, which is good. Unfortunately, detectors of this type have a limit on how quickly they can count photons, so that there are difficulties if the number of photons in the desired energy range is too *high*. For more information on detectors and filters, see Section 8.7 of Chapter 8.

Compared to transmission, then, there's a lot going on!

One of the things to notice is that there aren't going to be nearly as many photons going into the fluorescent detector as would go into the I_t detector in a well-designed transmission experiment. After all, not all of the absorption generates fluorescent photons and then only some of those head in the right direction. We will examine the ramifications of this in Chapter 8.

To understand how sample characteristics influence fluorescence measurements, let's consider four extreme cases.

6.4.1.1 Limiting case: Thin and concentrated

This case works a lot like transmission and in fact can often be measured simultaneously in that mode. Most of the photons make it through the sample, but those that do not interact primarily with the target element. The probability of interaction depends on the absorption coefficient $\mu_*(E)$ of the target element at each energy, but the fluorescent photons don't have to worry much about getting absorbed on the way out to the detector. Simple!

We can use this logic to work out how $\mu_*(E)$ depends on I_f. Suppose a fraction a of the incident photons that aren't transmitted through the sample end up generating fluorescent photons in our fluorescence detector. In other words:

$$I_f = a\left(I_o - I_t\right) \tag{6.12}$$

or, solving for I_t,

$$I_t = I_o - \frac{1}{a}I_f \tag{6.13}$$

In the limit of a thin, concentrated sample, a is only weakly dependent on the energy of the incident photon, and will not itself contribute to XAFS, which includes only relatively rapid variations in $\mu_*(E)$.

Substituting into Equation 6.4 gives us:

$$\mu_*(E)x = \ln\left(\frac{I_o}{I_t}\right) = \ln\left(\frac{I_o}{I_o - \frac{1}{a}I_f}\right) \tag{6.14}$$

$$= -\ln\left(1 - \frac{I_f}{aI_o}\right)$$

Because the sample is thin, $I_f << aI_o$, and we can use a Taylor expansion to write:

$$\mu_*(E)x = -\ln\left(1 - \frac{I_f}{aI_o}\right) \approx -\left(-\frac{I_f}{aI_o}\right) \tag{6.15}$$

This leads directly to the simple result:

$$\mu_*(E) \propto \frac{I_f}{I_o} \tag{6.16}$$

6.4.1.2 Limiting case: Thin and dilute

This is similar to the thin concentrated case, except that in step 3 in Section 6.4.1, we'll have relatively less absorption by the element of interest relative to other events. This leads to a weaker signal relative to the background, and thus the signal-to-noise ratio will be worse than in the thin concentrated case.

6.4.1.3 Limiting case: Thick and concentrated

Initially, this sounds as if it should be even better than the thin concentrated case. A thicker sample means more signal, right?

Unfortunately, it depends on what is meant by 'signal.' Because the sample is thick and concentrated, most of the incident photons will interact with the target element. Some will still be elastically scattered, but at least most of the absorption will come from the desired element. Above the edge, that does mean a lot of fluorescent photons arriving at your detector from the element of interest.

But here's the rub: Above the edge, incident photons are likely to be absorbed by the element of interest regardless of what $\mu_*(E)$ is. And because the fluorescent photons are necessarily lower in energy than the absorption edge being studied, they will be able to make it out to the fluorescent detector with relative ease regardless of what $\mu_*(E)$ is.

The result is that your 'signal' will show a jump at the edge, followed by a featureless horizontal line—no EXAFS!

This is admittedly an oversimplification. The processes that are described in Section 6.4.1.4 for the thick dilute case apply weakly to this case as well, producing some EXAFS in the signal, but the features will be greatly suppressed compared to the other limiting cases.

For historical reasons, the suppression of signal in thick concentrated samples is referred to as *self-absorption* or sometimes *overabsorption*.

6.4.1.4 Limiting case: Thick and dilute

Suppose we have a sample that is very thick, but that has only a small fraction of atoms of the element of interest.

As in the thin dilute case, absorption and scattering by elements other than the element of interest contribute a substantial background. But in a surprising twist, the other elements help alleviate the self-absorption problem!

One way in which they help is by absorbing many of the incident photons. Unlike in the thick concentrated case, where the element of interest is likely to absorb the photons whether $\mu_*(E)$ is high or low, the element of interest has competition from the other elements. If at a particular energy it does not absorb well, the other elements are likely to get the photon first. This helps make the fluorescence from the element of interest lower at energies at which $\mu_*(E)$ is low, which of course is the behavior we would like.

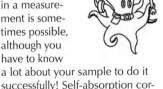

Correcting for self-absorption in a measurement is sometimes possible, although you have to know a lot about your sample to do it successfully! Self-absorption correction schemes will be addressed in Section 7.3.5 of Chapter 7.

But that's not all. At energies at which $\mu_*(E)$ is low, photons absorbed by that element are likely to be absorbed at a greater depth. In turn, fluorescent photons from those absorption events have to travel through more of the sample to get back out. Those fluorescent photons are lower in energy than the edge energy of the target element, and thus, they pass through other atoms of the target element easily on their way back out. But the other elements in the sample generally absorb *more* strongly at the energy of the fluorescent photons than they do at the incident energy, simply because, in the absence of edges, absorption is greater at lower energies. Fluorescent photons absorbed at a relatively great depth therefore have less chance of making it back out of the sample than those absorbed at more shallow depths. Thus, this effect contributes to suppressing fluorescence at energies for which $\mu_*(E)$ is low, which is once again what we'd like to see to compensate for the self-absorption problem described in Section 6.4.1.3.

The physics of the thick, dilute case is really tricky to understand. Most people just memorize that self-absorption is a problem with thick, concentrated samples measured in fluorescence, and that for both the thin concentrated and thick dilute cases, $\mu_*(E) \propto \dfrac{I_f}{I_o}$.

Although these effects should make $\dfrac{I_f}{I_o}$ sensitive to variations in $\mu_*(E)$, you might expect the functional dependence to be somewhat complicated. In fact, the derivation of the correct dependence is somewhat involved (Pfalzer et al. 1999). But through something of a mathematical miracle, it turns out that in the thick, dilute limit that

$$\mu_*(E) + B_e(E) \propto \frac{I_f}{I_o} \tag{6.17}$$

You and your shortcuts, Dysnomia! Some of us *like* thinking through the physics...

where $B_e(E)$ is a slowly varying function of energy due to effects such as absorption by other elements in the sample. Once that background is subtracted out, Equation 6.17 is the same as Equation 6.16!

It is important to realize that there is no physical reason that the proportionality for the cases in Sections 6.4.1.1 and 6.4.1.4 should have turned out to be the same—it's just one of those lucky happenstances of nature that it did.

6.4.2 Understanding Your Sample

The assumption that $B_e(E)$ is a slowly varying function of energy, and thus can be subtracted out in a data processing step, is better for an energy-discriminating fluorescence detector than for something like a Lytle detector.

Thick dilute and thin concentrated samples are the least problematic when it comes to fluorescence, but as with transmission, sometimes you don't have control over the form of your sample. If you are measuring a new thin film technology, you may be stuck with something that is thin and dilute, for instance. And if you are measuring, say, a priceless ancient relic that is thick and concentrated, it is not an option to try to cut it down to size! In this section and in Chapter 8, we provide some methods for coping with self-absorption, background photons in fluorescence measurements, and low count rates. But it's important to realize ahead of time which issues are most likely to be major ones for your samples, so that you can be prepared to take the actions necessary to get usable data.

There are many different circumstances that can arise when considering fluorescence experiments on samples, but they all face the issues we've already described. To show you how these issues can interact, let's take a few examples.

It may be *better* for an energy-discriminating detector, but in most cases, it's a good assumption for a Lytle detector too.

6.4.2.1 A thin film on a thick substrate, such as some common solar cell designs

If the film is thin compared to one absorption length of the material, then self-absorption will not be much of an issue. (We'll be quantitative about what 'not much of an issue' means in Section 6.4.3.) But the photons that penetrate the film will hit the substrate, and it can fluoresce or scatter the photons. In that circumstance, the biggest problem will be background photons.

If the film is several absorption lengths thick, then self-absorption will be the main issue. The person who made it may refer to it as a 'thin film,' and it may be less than half a millimeter thick, but it might as well be a boulder as far as the fluorescent signal is concerned.

6.4.2.2 A heterogenous soil sample with the element of interest concentrated in grains sparsely scattered through the material

What matters here is the grain size. If the individual grains are several absorption lengths thick, then those parts of the sample are thick concentrated and will suffer from self-absorption. It doesn't matter that only a small part of the cross section of the beam is hitting a grain; the part that is, and thus the part that is measuring $\mu(E)$ for your target element, is seeing a concentrated sample. (To put it another way, there is little correlation between μ and the depth of the grain that does the absorbing.) In fact, this is potentially the worst of both worlds; the sample is concentrated in the sense that it suffers from self-absorption, but dilute in that the signal-to-noise ratio will be poor.

There are a couple of methods for dealing with such a sample. One is to use a microprobe beamline capable of focusing the beam on to just one grain. You'll still have self-absorption effects to deal with, but at least the number of counts will improve.

Another solution would be to grind up the sample, breaking up the grains into pieces much smaller than an absorption length in size. This makes the sample truly thin dilute and is probably the best route if it's feasible to do and if you don't need spatial information.

6.4.2.3 A material in solution

Usually solutes qualify as dilute, and you are also usually able to choose how deep to make the sample cell. Therefore, the thick dilute limit should be achievable. Care must be taken, however, with the sample cell that is used to hold the solution. If part of the beam intersects the sample cell, it will contribute to the background. The author of this book learned that lesson the hard way, by spending several hours collecting spectra that turned out to be due to the stainless-steel screws used to hold a solution cell together.

6.4.2.4 A gas

If the thickness of the cell is chosen well, it may be possible to measure the gas as a thin concentrated sample. But in that case, the beam will penetrate to the back wall of the sample cell. Therefore, care must be taken to make the back wall a material that will not fluoresce extensively at the energies of interest. Even better would be a window at the back wall that allowed most of photons that made it that far through, so that scattering is also minimized.

6.4.3 The Mathematics of Self-Absorption

The dependence of the measured intensity at the fluorescence detector on several key factors is given in this proportion (Pfalzer et al. 1999):

$$
I_f \propto I_o \left[\frac{\mu_*(E)}{\frac{\mu(E)}{\sin\phi} + \frac{\mu(E_f)}{\sin\theta}} \left(1 - e^{-\left(\frac{\mu(E)}{\sin\phi} + \frac{\mu(E_f)}{\sin\theta}\right)x} \right) + B_e(E) \right] \tag{6.18}
$$

The meaning of the symbols in this proportion is summarized in Table 6.4.

Table 6.4   Symbols Used in Equation 6.18

Symbol	Meaning
I_f	Photon intensity at fluorescence detector
I_o	Photon intensity at incident detector
E	Energy of incident photons
E_f	Energy of fluorescent photons
$\mu_*(E)$	Absorption coefficient of target element associated with measured edge
μ	Total absorption coefficient of sample
π	Incident angle relative to the sample surface
θ	Angle to fluorescence detector relative to the sample surface
x	Sample thickness
$B_e(E)$	A slowly varying background function due to scattered photons, absorption by windows, etc.

Even this rather complicated-looking expression is considerably simplified. For example, depending on the detection scheme, there may be more than one E_f, as different electrons can fill the core hole. Likewise, there is actually a range of angles to the fluorescent detector; to maximize counts, fluorescent detectors are usually given a large area and placed near the sample. But this expression is still helpful for estimating the size of self-absorption effects and for understanding how to minimize them.

For example, consider a thick sample. In that case, the exponential term approaches 0, simplifying the expression.

Now consider a dilute sample. If the sample is dilute, then μ will be primarily due to atoms other than the target element. As long as those elements don't have edges near that of the target element, μ will be a slowly varying function of E, and thus I_f will roughly be proportional to $\mu_*(E) I_o$. This gives the happy result that, assuming the slowly varying $B_e(E)$ can be subtracted off in some way (we'll see how this is done in Section 7.4.1 of Chapter 7), the energy dependence of the absorption of the target is now easy to measure:

Proportionalities 6.19 and 6.20 could be written to look exactly the same, since x is a constant for a given sample and could thus be made part of the proportionality constant. But we left it this way to emphasize that the agreement of the two proportionalities is a mathematical coincidence. The physics of the two cases is completely different. In fact, doubling the thickness of a thin concentrated sample doubles the signal (as long as the thin approximation isn't violated!), whereas doubling the thickness of a thick dilute sample has no effect.

$$\mu_*\left(E\right) \propto \frac{I_f}{I_o} \tag{6.19}$$

What about a thin concentrated sample? In that case, the exponential can be expanded in a Taylor series, with the result that the troublesome denominator cancels out, leading to

$$\mu_*\left(E\right)x \propto \frac{I_f}{I_o} \tag{6.20}$$

which also directly yields the energy dependence of the absorption of the target.

6.4.4 Geometrical Factors

Because fluorescence is isotropic, it might seem that the fluorescence detector could be placed at any angle relative to the incident beam and be equally effective. But although fluorescence is isotropic, scattering is not. In the elastic limit, the probability of scattering in a given direction depends on the angle α relative to the incident direction and the angle β relative to the plane of polarization (Heitler 1954):

$$P(\alpha,\beta) \propto \left(1 - \sin^2\alpha\cos^2\beta\right) d\alpha\, d\beta \tag{6.21}$$

Because synchrotron radiation is generally horizontally polarized, there should therefore be no elastic scattering if the fluorescent detector is mounted on the same table used for the sample ($\beta = 0$) and the I_o and I_t detectors, so that the line from it to the sample is perpendicular to the beam ($\alpha = 90°$), as shown in Figure 6.5.

For inelastic scattering, such as Compton scattering, the angular dependence is more complicated and the minimum shifts a bit toward higher α (Heitler 1954). The minimum is quite broad, so 90° is still an appropriate choice. Of course, fluorescent detectors are designed and placed so as to intercept a large solid angle, so that they can count a reasonable fraction of the fluorescent photons. Even though there may not be many photons scattered toward the center of the detector, the large solid angle intercepted by the detector prevents scattering from being completely eliminated.

The desire to minimize scattering tells us where to put the detector, but it doesn't tell us how to orient the sample. Suppose we use ϕ to represent the angle the incident beam makes with the surface of the sample, and θ to represent the angle the path from sample to fluorescence detector makes with the surface of the sample (Figures 6.6 and 6.7). The position of the fluorescence detector tells us that,

$$\phi + \theta = 90° \tag{6.22}$$

but doesn't tell us which combination of ϕ and θ to pick.

Frequently, the geometry is chosen so that $\phi = \theta = 45°$—in fact, sample holders are often made that way. But there's nothing particularly special about that choice, and sometimes you'll want to use a different geometry.

Consider, for instance, one of our problematic limiting cases: a thick concentrated sample (Section 6.4.1.3). We can now see how self-absorption comes out of Equation 6.18; the problem is that the $\mu(E)$ in the denominator has a large contribution from $\mu_s(E)$ and thus will partially cancel the $\mu_s(E)$ in the numerator. But $\mu(E_f)$ is also in

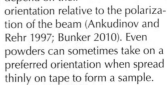

This may be a good time to mention that the spectra of samples such as single crystals depend on their orientation relative to the polarization of the beam (Ankudinov and Rehr 1997; Bunker 2010). Even powders can sometimes take on a preferred orientation when spread thinly on tape to form a sample.

Figure 6.5 Detector geometry.

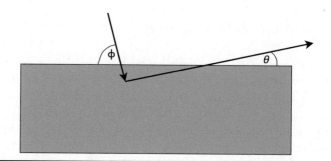

Figure 6.6   Grazing exit.

Figure 6.7   Grazing entry.

If you think about the discussion of thick concentrated samples earlier in Section 6.4.1.3, you'll see why a grazing exit works to minimize self-absorption effects. Because the problem was that fluorescent photons could make it out of the sample too easily, forcing the fluorescent photons to travel through more sample will alleviate the effect. And *that*, Dysnomia, is why I prefer to understand the physics. If I had tried to get by on memorization, I never would have understood grazing exit.

the denominator, and it does not contribute to canceling the energy dependence that we want to measure. After all, $\mu(E_f)$ is measured at a single energy—the fluorescent energy—and is thus just a number. (A more realistic treatment would acknowledge that there are several different fluorescent energies for the target element, but that doesn't change the argument: When measuring a given edge, the energies of the fluorescent photons do not depend on the energy of the incident photons.) So if there was just a way of emphasizing the $\mu(E_f)$ term in the denominator at the expense of the $\mu(E)$ term, we could minimize the self-absorption effect.

And, fortunately, there is! If we choose to make θ small, $\dfrac{\mu(E_f)}{\sin\theta}$ will be large, $\mu(E_f)$ will dominate the denominator, and self-absorption effects will be minimized. Since θ is small, Equation 6.13 tells us that ϕ will be near 90°, with the beam hitting the sample nearly face on, and then exiting at a shallow angle. This geometry is often referred to as grazing exit and is shown in Figure 6.6.

What about the other problematic case, a thin dilute sample? In that case, the best geometry is the opposite: grazing entry, shown in Figure 6.7.

In grazing entry, the beam effectively sees a thick dilute sample (or at least thicker), which will improve the signal-to-noise characteristics.

6.4.5   How Thick Is 'Thick'?

Point taken, Carvaka. Along the same lines, people should realize that grazing exit will only help if the surface of the sample is smooth on the scale of an absorption length.

How thick does a concentrated sample have to be before self-absorption starts to become a problem?

To answer that question, let's consider the ultimate in concentrated samples: a film composed of a single metal. In that case, $\mu \approx \mu_*$. Let's also assume that you're using a sample holder where $\phi = \theta = 45°$.

The Taylor series expansion of the exponential in Equation 6.18 would then be

Grazing entry can also reduce problematic scattering and fluorescence from the substrate.

$$e^{-\left(\frac{\mu(E)}{\sin\phi} + \frac{\mu(E_f)}{\sin\theta}\right)x} \approx 1 - \left(\frac{\mu_*(E)}{\sin 45°} + \frac{\mu_*(E_f)}{\sin 45°}\right)x$$

$$+ \frac{1}{2}\left[\left(\frac{\mu_*(E)}{\sin 45°} + \frac{\mu_*(E_f)}{\sin 45°}\right)x\right]^2 + \dots$$

(6.23)

Because we're just trying to see roughly how strongly the self-absorption effect depends on the sample thickness, let's make some more simplifications: ignore $\mu_*(E_f)$, because the target absorbs much more poorly below the edge, and assume $B_e(E)$ can be removed. With those simplifications, and using the fact that $\sin 45° = \frac{1}{\sqrt{2}}$, substituting the first three terms of the expansion into Equation 6.18 gives

$$I_f \propto I_o\sqrt{2}\,\frac{\mu_*(E)}{\mu_*(E)}\left(1-1+\sqrt{2}\mu_*(E)x-\frac{1}{2}\sqrt{2}\mu_*(E)^2 x^2\right)$$

$$\propto I_0\mu_*(E)x\left(1-\frac{1}{2}\mu_*(E)x\right)$$

(6.24)

Thus, if a concentrated sample is 0.1 absorption lengths thick for photons at the incident energy, self-absorption will reduce the signal by about 5%, enough to introduce noticeable errors into the analysis. This means that for many common edges, a concentrated sample less than a micron thick may still be significantly affected by self-absorption!

6.4.6 *How Concentrated Is 'Concentrated'?*

Now let's see how concentrated a thick sample can be before it starts to exhibit significant self-absorption. As before, let's assume $\phi = \theta = 45°$. This time, we'll imagine a sample thick enough that the exponential term goes to zero. As usual, we'll assume $B_e(E)$ can be removed during data reduction. Finally, let's break $\mu(E)$ into two parts: $\mu_*(E)$, the absorption due to the target element, and $\mu_{other}(E)$, the absorption due to everything else. That yields:

$$I_f \propto \sqrt{2}I_o\,\frac{\mu_*(E)}{\mu_{other}(E)+\mu_*(E)+\mu(E_f)}$$

$$\approx \sqrt{2}I_o\,\frac{\mu_*(E)}{\mu_{other}(E)+\mu(E_f)}$$

(6.25)

$$\left(1-\frac{\mu_*(E)}{\mu_{other}(E)+\mu(E_f)}+\ldots\right)$$

The factor in parentheses is acting to damp out fine structure. You can see that if the target element is responsible for 1% of the total absorption at some energy, then it will reduce the signal by about 1% at that energy—acceptable for most experiments.

6.5 Which Technique Should You Choose?

In *XAFS for Everyone*, we have intentionally refrained from providing hard and fast 'rules' for when fluorescence, transmission, or electron yield should be used. Instead, we've provided you with the tools for understanding what issues will crop up as you use samples that stray from the 'ideal' sample for each technique.

For example, suppose you have a powder that is dilute enough so that a sample one absorption length thick would only have an edge jump of 0.03. Measuring in transmission is not out of the question—pushing the thickness to 2.5 absorption lengths would give an edge jump of 0.08, which probably will be OK in terms of signal-to-noise. But with a sample that thick in absorption, you'll have to be careful about harmonics. You

Great! So this soil sample I have that's 1% chromium shouldn't be a problem.

Not so fast!! The mass absorption coefficient for chromium just above its edge is about 500 cm²/g. 'Soil' can mean a lot of things, but typical soils are mostly oxygen (~30 cm²/g above the chromium edge), silicon (~150 cm²/g), aluminum (~120 cm²/g), and iron (~80 cm²/g). Depending on the proportions present in your sample, the 1% chromium by mass could be responsible for 10% of the absorption above its edge!

This chapter has a boatload of information about sample effects, how to estimate their size, and how to minimize them. But I'll level with you; sometimes I just dump some powder on a tape and measure in transmission or take a manufactured object and measure it in fluorescence. I'm still going to get a spectrum, and it will probably still tell me something about the material. If my question is not particularly subtle (e.g., mostly elemental metal or mostly oxide?), then I'm OK. When I need a precision measurement, though, I think all this through.

I can see your point, my dear Dysnomia, but I *always* think everything through. Better safe than sorry, I say.

could also choose to make a thick sample and measure in fluorescence, but it's a little on the concentrated side for that, so you'd have to be careful about self-absorption, perhaps using a shallow angle of incidence if the surface can be made very smooth. Another good approach would be to measure the sample both ways; since harmonics and self-absorption both tend to reduce the size of EXAFS features, you could then pick the spectrum where they were suppressed less, or feel confident in the data quality if they both match!

6.6 Preparing Samples

One of the most important things you can do, especially when working with reactive chemicals, is to test run everything you can before you get to the synchrotron. Set up as much of your system as possible at your home institution. This will ensure that your Teflon containers aren't going to dissolve unexpectedly; or that the adhesive on your Kapton tape won't melt when you try to measure a temperature series; or that a tube with glacial acetic acid pumping through it won't come loose and spray everywhere, causing every scientist in the facility to stop by at some point and say, 'Oh, so you're the one responsible for that awful smell.' The experimental floors of light sources are usually very open places, and the ramifications of a catastrophic failure are often much worse than they would be in the familiar environment of a dedicated academic or industrial laboratory. At the very least, failures of this sort make a mess and jeopardize the experiment and at the worst could compromise the safety of people and the environment.

Dissolving Teflon? Melting Kapton adhesive? An acetic acid stink bomb? Those are curiously specific examples. You don't suppose our esteemed author is speaking from experience, do you?

While we're on the subject, we should stress that one thing we will not try to do in *XAFS for Everyone* is to provide safety guidelines for hazardous samples. Hazards vary in type and severity, and precautions that are adequate for a suspected carcinogen might not be appropriate for a strong oxidizer. And although safety personnel at synchrotrons can help you plan your experiment, no one there is likely to be as familiar with your materials as you are. In the open environment on a synchrotron experimental floor, precautions may be necessary that aren't required in your home laboratory and safety personnel can help guide you through those issues. But when it comes to questions like the chemical compatibility of your sample holder with your sample, you shouldn't trust that someone else is going to foresee hazards that might occur.

6.6.1 Powder

Powder is a common sample form in the XAFS world and can produce good data if you prepare it properly. If you are planning to run the sample in transmission, the hardest part is making sure the individual grains are all considerably smaller than one absorption length—if they aren't, you will not be able to make a uniform sample and the spectra will be distorted.

There are several approaches that can help make sure the grains are small. Depending on how much sample you have, you may find it helpful to use a sieve to select for the smaller particles first. However, 635 mesh is the finest sieve readily available, and it still allows through particles up to 20 μm. For many samples that is more than one absorption length.

Be careful while minimizing your powder size that you do not inadvertently expose it to something it will react with!

Using a mortar and pestle (or a grinding mill such as a Wig-L-Bug) is sometimes an effective way to reduce the powder to a fine enough grain. If your substance will not grind, or you are concerned it has not been ground finely enough, you may want to try sedimentation, described in Section 6.6.1.2.

Another common technique for selecting only the smallest particles is to spread them *thinly* on adhesive tape. Larger particles tend to brush off the adhesive, but the smallest ones stick.

Finally, uniformity can be measured at the beamline, and the most uniform section of the sample chosen. This won't work unless most of the particles are small, but if an occasional 'boulder' found its way in, you may be able to avoid shooting through it.

6.6.1.1 Fillers

Some people find it inconvenient to work with the small amounts of sample called for—a 10 or 20 μm layer isn't very much!

So many people mix the sample with a filler. This can also reduce the orientation problem Kitsune mentioned in Section 6.4.4.

If you do decide to dilute your sample with another substance, it is important to avoid contaminating your sample or causing it to react. You also need to choose a filler composed of elements with low atomic numbers (much lower than your sample!), so that the filler doesn't absorb too much. Typically, carbon black, boron nitride, or sugar are used (we've also heard of cellulose or mineral oil being effective). Make sure the material is not contaminated with an element you are trying to measure! Always know how much of the mixture is the target element, so that you can still calculate the appropriate thickness.

6.6.1.2 Procedure for sedimentation

1. Grind the material as finely as possible.
2. Place the powder into the bottom of a test tube or centrifuge tube.
3. Fill with a liquid such as high-purity acetone or isopropyl alcohol. In theory, the liquid could be anything that doesn't react with your sample (which is one reason water is often a bad idea!), but the density and viscosity of the liquid will affect how long the process takes.
4. Assess how long it takes for the bigger particles to settle. This will depend on the density of the sample and liquid, the viscosity of the liquid, and the length of the tube.
5. Use a pipet to collect the liquid above the settled sample, and either put it in a liquid sample holder and seal it or allow the liquid to evaporate and use the powder that is left behind. Depending on what material you are using, it can be difficult to get very many particles this way. If you only have a small amount of sample, you may have to recycle the remaining mixture in the tube by letting the liquid evaporate. Start again at step 1 when the powder is dry.

Contamination can be a particular concern with carbon black, which often contains some iron.

BOX 6.5   Fillers and Uniformity

Fillers are also a method for achieving uniformity, right? It's a lot easier to make something a uniform thickness if you've got more of it.

Yes and no. It's true that fillers can help you make the thickness uniform, and that's important. But if the individual grains of sample are large, you've still got a problem. In some places, the beam will pass through a big chunk of sample and not much filler; in others, mostly filler and not much sample. By having a nice smooth surface, you might actually be less likely to catch that kind of nonuniformity before getting to the beamline.

Some researchers use white filler (e.g., sugar) with dark samples and black filler (e.g., carbon black) with light samples to make it easier to see the kind of nonuniformity Professor Carvaka just described. Some people even use epoxy; it is clear and relatively easy to visually inspect for uniformity, but it can be difficult to cut to the desired thickness.

As a biologist, I have ready access to a microtome, which is a tool for slicing thin layers. So epoxy works well for me.

6.6.1.3 Procedure for spreading on tape

1. Perform whatever initial steps you are using to select small particles (sedimentation, sieving, and grinding).
2. Many different kinds of tape are acceptable: Kapton, Teflon, Mylar, and Scotch all work. If it's the first time you've used a given type of tape at a given edge, collect a spectrum on a blank piece of the tape to make sure that it doesn't include any elements in the energy range you are using. Just getting chemical compositions from a spec sheet is not enough, because those specifications don't generally include the adhesive. Likewise, while Kapton, for instance, is produced by DuPont, different companies process and resell it, and there's no guarantee that they chose the same adhesive when they turned it into a roll of tape. Once you have the tape, measure and cut strips of a convenient size. The size you choose will depend to some extent on how much sample you're willing to waste and personal preference; 2.5 × 1.0 cm is not atypical.
3. Weigh each strip before spreading sample on it. Do not assume all the strips you cut are the same weight.
4. A rubber policeman or some other soft spatula-like tool works the best for spreading. (You may want to practice without sample on a piece of tape to get your technique down—if you spread too firmly you can rub off the adhesive.) The finely ground powder should spread like paint on the tape (see Figure 6.8). Different powders spread with varying ease, but for the most part, you should be able to 'paint' pretty easily. Don't be disappointed that most of the powder doesn't stick to the tape; that's the whole idea! Make sure to spread as evenly as possible and as vigorously as possible without taking off adhesive. This should ensure any remaining large particles will be brushed off. Keep an eye out for uneven spots, pinholes, and cracks. If you can't fill them, you will want to remember where they are so as to avoid them when mounting the sample (step 6).
5. Weigh the finished tape and subtract its initial mass to find the mass of the sample you've added. You will probably need to repeat steps 3 and 4 a few more times to reach your target mass. But if you have too many layers of tape, then the total absorption will become high.

   How many is too many? Unfortunately, it's hard to look up the absorption of a layer of tape in a table or computer program. In part, it's the question of the adhesive again; it's easy to find the absorption coefficient of Kapton or Mylar, for instance, but often very difficult to ascertain what adhesive a given manufacturer used or how much of it they applied. It's not a bad idea, therefore, to

Rather than use a spatula or a brush, sometimes I just use my gloved finger to spread a sample. Of course, I never do this with highly toxic samples! Be especially cautious with nanoparticulate samples; in some cases, they can work their way through gloves.

Figure 6.8 **This picture shows the use of a brush rather than a spatula to spread the sample. That can work well as long as enough pressure is put on the brush. Notice the visible grains of material off to the sides of the tape—those are probably large grains that didn't stick well, and that we therefore are happy did not end up on the sample. (Photo courtesy of National Synchrotron Light Source. This photo is licensed under a Creative Commons Attribution-Share Alike 3.0 United States License.)**

measure the absorption of some blank strips of the tape you're using the first time you go to collect data.

With very thin tapes (e.g., 0.3 mil Kapton) at energies above 6 keV or so, even 20 layers can work. Off-the-shelf Scotch Magic Tape is a bit thicker and limiting to half that number is reasonable. But absorption is strongly energy dependent— if you're working at 15 keV, you could shoot through several millimeters of tape and barely notice; at 4 keV (the calcium *K* edge), you won't want to go much above three or four layers. When you have roughly the total sample mass you're shooting for (or the maximum number of tapes you're comfortable with, whichever comes first), continue to step 6.

6. Sandwich your tape pieces between larger pieces of tape, so that they are sealed together. Yes, this does mean two additional layers of tape that the x-rays have to pass through, but synchrotrons generally frown on exposed powders, even if the material is something completely harmless like table salt. If some of your powder gets on beamline equipment and no one notices until later, the synchrotron staff will have no way of knowing the substance is not very dangerous and will have to regard it as such.

   If you're at a low enough energy so that an additional layer makes a difference, you might be able to fold the top cover tape around the edges of the sample instead of having a full piece on the bottom.

   If you run out of sample before reaching your target mass, you may have to fold your tape stack a few times as well. This can be tricky, but tweezers and your spatula should help.

7. (Optional) When you have your sample as you need it, it's helpful to find something to use as a frame, as this will aid alignment. Ordinary razor blades work well because they have a little window in the middle, although it's a bit small for some beams. (Just cut off the sharp edge with shears before using it, making sure to wear safety goggles as you do so.)

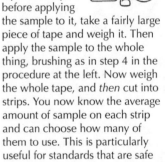

If you have plenty of sample, here's a handy shortcut. Instead of cutting the tape into strips before applying the sample to it, take a fairly large piece of tape and weigh it. Then apply the sample to the whole thing, brushing as in step 4 in the procedure at the left. Now weigh the whole tape, and *then* cut into strips. You now know the average amount of sample on each strip and can choose how many of them to use. This is particularly useful for standards that are safe and cheap, so that you don't mind throwing away half of the strips if you don't need them.

Some researchers like to turn their sample into a paint and do away with adhesive tape altogether. To do this, one can dissolve the sample in Duco cement thinned with acetone. It can then be painted directly on to a thin surface like a sheet of Kapton or Mylar. This does not, however, relieve one of the necessities of making the particles small first, and one still has to be careful that the procedure does not cause any undesired chemical reactions.

Another way to deal with powder samples is to use a frame of a known depth. A typical razor blade, for instance, is about 300 µm thick away from the cutting edge. Using about one absorption length's worth of sample diluted with enough filler to fill the window is pretty easy, as neither brushing nor pressing is involved. The downside is that the total thickness is set by the frame depth, and so the filler may absorb a lot more than you'd like, particularly at low energies.

It is also possible to put powder samples, diluted if appropriate, into capillaries. For transmission, it is important that the front and back of the capillary be flat, as ordinary round capillaries will lead to the kind of distortion Mandelbrant discussed in Box 6.2. You can try using 'square capillaries' or 'rectangle capillaries' in Web searches to find suppliers. Be sure to calculate how much absorption you'll get, including from the capillary walls, before deciding what dimensions of capillary to purchase.

Polyethylene glycol can be a good choice for a filler, because when it is pressed, it polymerizes, forming a solid pellet.

Be particularly careful for fluorescence measurements that you are measuring your sample and not the frame! It's a good idea to take a scan with just a blank frame in place so you can understand how it interacts with the beam.

Finally, do your best to put the most even part of the sample in the window and tape the sample to the frame along the edges, being careful not to partially overlap the window with the additional tape. Now you are ready to collect XAFS data.

6.6.1.4 Powders without all that tape: Pressing into a pellet

At low energies, even three or four layers of tape may introduce considerable additional absorption. And no matter how carefully you spread, nonuniformity in the adhesive may introduce nonuniformity into your layers. For highly dilute samples, trace elements present in the tape or adhesive might contribute to the spectrum, if too many layers are used. And, as Kitsune mentioned in Section 6.4.4, brushing thin layers on tape can sometimes cause microcrystalline samples to take on a nonrandom orientation. For all of these reasons, people sometimes prefer to press their sample into a pellet:

1. In most cases, pressing into a pellet will involve mixing with a filler; pellets a few microns thick don't tend to hold together very well! Decide on how many absorption lengths of filler and of sample you want; 0.5 of each might be typical. Use the method of Equation 6.5 to turn this into a ratio of masses. Combine the sample and filler in the proper ratio, mix, and grind. Make sure the result is a completely homogenous, very fine powder. If you have plenty of your sample and the toxicity is low, it is easiest to work with much more material than you need and discard the surplus; thoroughly mixing 0.5 g of sample with 10 g of filler is much easier than mixing 5 mg of sample with 100 mg of filler, even though you'll be wasting 99% of the mixture.

2. Using the area of your particular pellet presser's pellet chamber and the number of absorption lengths of filler you desire, find the required mass of filler from Equation 6.5. Repeat with the sample, and add the two to find the desired mass of the pellet. Pour that amount of the mixture into the pellet chamber.

3. Press the pellet with force for 10 seconds or so. Be careful while removing it from the chamber, as it may easily crumble when handled. Sandwich the pellet in tape to make sure it does not break apart.

6.6.1.5 Air-sensitive powders

Materials that are even mildly oxygen or water vapor sensitive become more so when ground into a fine powder, as the surface-to-volume ratio can be dramatically increased relative to a bulk sample. In cases where this is a concern, samples can be prepared in a glove box, glove bag, or anaerobic chamber using one of the methods described earlier. Static electricity builds up easily in ultradry environments, so a static brush or other static control system is quite helpful. Some light sources have glove boxes available onsite; talk with beamline staff in advance to see what the requirements are for access.

BOX 6.6   Pick Three of Five

We've talked about five methods for getting particles small enough to make a uniform sample: grinding, sieving, sedimentation, spreading thinly on tape, and choosing the most uniform part of the sample at the beamline. At energies from 5–15 keV, choosing three of these techniques is usually good enough, unless you're doing ultraprecise work.

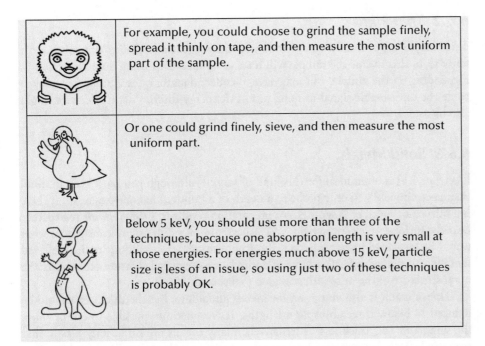

	For example, you could choose to grind the sample finely, spread it thinly on tape, and then measure the most uniform part of the sample.
	Or one could grind finely, sieve, and then measure the most uniform part.
	Below 5 keV, you should use more than three of the techniques, because one absorption length is very small at those energies. For energies much above 15 keV, particle size is less of an issue, so using just two of these techniques is probably OK.

Once your sample is prepared, how do you keep it protected during transport and measurement? Sandwiching the sample in Kapton tape itself provides some protection—Mylar is even better. Heat-sealing the resulting sample in a low-density polyethylene (LDPE) bag (ordinary supermarket food storage bags are LDPE) can sometimes protect sensitive samples for a day or two.

The protection afforded by aluminized Mylar bags is orders of magnitude better than LDPE. Depending on the energy at which you are measuring, however, these bags may be several absorption lengths thick, rendering them unsuitable for transmission experiments. If that is the case, you can heat-seal your sample inside an LDPE bag, and heat-seal that inside an aluminized Mylar bag for transport. Once you are ready to measure, the Mylar bag could then be cut open and the LDPE bag left intact for the duration of the measurement. A similar strategy can be used with sealed thermos-like transport containers.

Another option for the measurement itself is to measure under vacuum or inert gas. Many beamlines will have a cryostat available for measurements below room temperature. Crysotats are generally set up to measure under vacuum, nitrogen, or helium, because otherwise condensation of ice on the sample would be a major issue. The cryostat can therefore be used to maintain a nonreactive atmosphere, even if you plan to measure at room temperature. The quality of the vacuum or gas environment may not be good by the standards of ultrahigh vacuum or a glove box, but is often sufficient to extend the lifetime of air-sensitive samples sealed in Kapton or Mylar for long enough to collect good data. And for many samples, measuring at the lower temperatures available with a cryostat can also dramatically slow down any reactions if they do begin to occur.

One size does not fit all—the energy at which you are planning to measure and the sensitivity of your particular material will determine the optimum strategy. So the bottom line with air-sensitive samples is to avoid guesswork! Test out your protection systems in your home laboratory well before you leave for the synchrotron, and calculate the absorption length of the materials you are using. If you do, you will very likely get the data you want. If you don't, then you are rolling the dice with your valuable beam time.

When shipping air-sensitive samples, try not to forget that they could undergo external pressure changes. Even driving a sample over a range of mountains could result in substantial pressure changes—putting one in the hold of an airplane is that much more of a test. Under those circumstances, sealed bags may spring leaks that then allow outside air in. To avoid problems of that kind, talk to your shipping company about what you are trying to do. It's a good idea anyway, as it will keep you from running afoul of safety, environmental, and customs regulations.

6.6.2 Thin Films

This sample type usually comes mounted on a substrate. Unless the substrate is fairly thin, this means the sample will have to be measured using fluorescence, as is. Depending on the sample, you may have to collect data for quite a bit longer to overcome the unfavorable signal-to-noise ratio. An energy-discriminating detector might also be useful.

6.6.3 Solid Metals

Electron yield is often ideal for this kind of sample, although you must keep in mind that electron yield will only probe to a depth of a hundred nanometers or so. If there is a thin oxide layer, or if synthesis or preparation results in a different microstructure near the surface, an electron-yield XAFS spectrum may differ significantly from what would be measured in transmission. Electropolishing or ion milling are the preferred methods for removing surface layers if desired, as mechanical abrasion can modify the microstructure of the near-surface region (Saksl et al. 2006).

Transmission is also an option for metals and alloys, but the thickness should be reduced to below three absorption lengths. It is sometimes possible to cut or shave the sample to this thickness. Alternatively, many metals are sufficiently plastic that they can be rolled or pounded into thin sheets. Because this is very likely to introduce defects into the microstructure, it is a good idea to anneal your sample after it has been made into a foil. Regardless of the method you use to produce a suitably thin transmission sample, if you will be measuring edges with markedly different atomic numbers, a separate foil of appropriate thickness can be made for each element that is to be measured.

Finally, you could measure in fluorescence, but may then have to guard against self-absorption or try to correct for it after the measurement.

6.6.4 Solutions and Liquids

Solutions and liquids make excellent XAFS samples, because they are generally quite uniform. Commercially available x-ray fluorescence (XRF) cups work very well for fluorescence or for transmission at high energies. At low energies, however, XRF cups are too deep for transmission measurements, but flat capillaries such as those Kistune discussed in Section 6.6.1 may work. In addition, making your own sample holder is not difficult (Scarrow 2002):

1. Nonhomogenous liquid samples must be mixed as well as possible. If the separation time is long, you may be able to mix the sample just before scanning. But if the mixture separates on a shorter time period, a method will have to be devised to keep it mixed as the scans take place.

   If you have a choice of solvent for a solution, organic solvents tend to be more transmissive and less reactive than water, and so they are often a good choice. For samples that need to stay at a particular pH, buffer salts generally don't change the total absorption much and thus are fine to use (but when in doubt, calculate the absorption!). If you are planning to freeze the sample, adding a drop of glycerol will inhibit crystal nucleation and thus produce a more uniform sample.

2. Your sample holder need be no more complicated than the hole in a razor blade, although you should of course choose (or make) something with a depth appropriate to your experiment. Cover both windows with tape (if you're at a particularly low energy, you can use 5 μm polyethylene). Make sure you verify that your sample will not dissolve the tape or its adhesive!

3. Puncture a hole in one end of one of the windows with a needle-tipped syringe. In the other end of the same window, puncture a hole, slant the holder, and carefully inject sample into the chamber. If it is a high surface tension liquid like water, inject into the lower end. Otherwise, inject into the high end. Do this slowly to avoid air bubbles.

4. If you want to freeze this sample (for example, to quench a reaction or prevent a mixture from separating), put it in a freezer, holes side up. If it is an organic or something you are worried about cracking, put the cold source at one end of the sample, so that it doesn't freeze all at once.

5. Seal the holes with small pieces of tape.

6.6.5 Gases

Gases work great in transmission, because their thickness will only be dependent on the container, and they are by nature uniform mixtures.

The only thing you have to be concerned with is choosing a proper container and windows. Make sure you know how your gas interacts with the materials of the holder; even when you are sure it is properly contained, leave it in its container for an extended period of time before bringing it to the synchrotron. You may be surprised by what happens, and it will be a dilemma best dealt with before your scheduled beam time. If your gas starts to deposit solids on your windows, you will soon be suffering from serious inhomogeneity; see the advice in Section 6.6.7.

6.6.6 Environmental

If your sample is soil or some other heterogeneous mixture, fluorescence is usually the method of choice, provided it is dilute and there are no large grains rich in the element being measured. An XRF tub makes a good sample holder, but almost anything will serve. While preparing your sample, be sure to pull out any pebbles, twigs, and other solid objects that might significantly interfere with the beam. Sieving may even be appropriate, depending on the sample. Though uniformity is not the issue that would be in transmission, you still need to get rid of any concentrated regions of the element of interest on a scale comparable to or larger than an absorption length. Keep in mind while arranging your sample that you can mask off the area of the sample you are most interested in to make sure it is hit by the beam. Some samples can be prepared almost as if they were powder samples; in that case, both fluorescence and transmission can be collected at the same time to see which gives you better quality spectra.

6.6.7 In Situ *and* Operando

For these categories of experiment, you will probably find that it is a challenge to achieve ideal thickness and uniformity. These experiments are therefore often measured in fluorescence.

Make sure you test as much of your system as possible before your trip to the synchrotron. This will ensure that you can prepare for whatever otherwise unexpected circumstances you will encounter on your run. If you are trying to observe a chemical reaction in action, or an unstable sample, there are a variety of uniformity issues you may run into. For instance, if gases are evolving in solution, bubbles can form. Even in fluorescence, bubbles are disruptive—for one thing, they change the path length from the sample to the detector. In transmission, of course, the distortion they introduce is severe. Similarly, solutions or gases can deposit solids on your windows. Even if a precipitate is what you are interested in measuring, if the solid remains stuck on the window, you will lose the ability to observe the subsequent progress of the reaction.

If either bubbles or deposition is a severe problem, it may be necessary to consider an *ex situ* version of the experiment, in which aliquots are periodically removed from an ongoing reaction and measured after the reaction has been quenched.

Finally, for *operando* measurements don't become so concerned with optimizing the XAFS spectra that you lose sight of the process you are trying to study. If you are investigating the interaction of a powdered catalyst with a flowing gas, for instance, the gas must be able to flow freely through the powder, perhaps necessitating the intentional introduction of air gaps by sieving the sample and only using particles larger than some size. Such choices will necessarily introduce nonuniformity into the sample, but that is preferable to collecting perfect spectra on a material that never undergoes the reaction you wish to study.

BOX 6.7 All Is Not Lost

	I don't think I paid enough attention to all of the things in this chapter when I made these samples, but I'm already at the beam line! I guess I'll just give up and go home…
	Hold on! A bad sample doesn't have to make or break you. After all, sometimes there is nothing you can do to perfect a sample! Optimizing the beam by proper detuning and by minimizing slit size can make all the difference. So can the quality of your detectors (what combination of gases you have in them, how well they are lined up, and so on). If you read Chapter 8 to learn how to hone your beams and detectors, you can probably salvage your experiment. As long as two out of three are good (beam, detectors, and sample), your results are likely to be usable.
	Also, the question you are trying to answer will partially determine how much a bad sample will hinder your data. For instance, phase identification is more likely to remain possible in the event of a thick or uneven sample than figuring out coordination numbers. For more detail on this, see Chapter 13.
	And hey, the best way to handle bad samples is as a learning experience. Have you ever noticed how XAFS experts can take one glance at your spectrum and tell you that your sample is too thick or too thin or too uneven or whatever? They can do that because they've run tons of bad samples, either accidentally or intentionally. If you discover you've got an iffy sample and you're unable to correct it on the fly, you can at least see what the spectrum looks like, and even, if you have time, play around with how that interacts with changes you make to the beam.
	Yay! All is not lost. Thanks, guys!

WHAT I'VE LEARNED IN CHAPTER 6, BY SIMPLICIO

- For transmission, it is very important that samples be of uniform thickness on the scale of an absorption length. This means that for powders, the grain size must be very small.
- To make the particles small, it is usually OK to choose three of the following five methods when working at energies of 5–15 keV: grinding, sieving, sedimentation, spreading thinly on tape, and finding the most uniform part of the sample at the beamline. At lower energies, more care must be taken.
- For transmission, samples should be *roughly* one absorption length thick above the edge we're going to measure.
- For transmission, an edge jump of *roughly* one is ideal, but it can be quite a bit less and still be OK.
- Transmission samples that are too thin suffer from poor signal to noise, those that are too thick from harmonics, and those that are uneven from multiple issues.
- Fluorescence samples work best when they are thin (compared to an absorption length) and concentrated, or thick and dilute. Thin, dilute samples exhibit poor signal to noise; thick, concentrated samples suffer from self-absorption.
- *Grazing entrance* can help with thin dilute samples and *grazing exit* with thick concentrated ones.
- Many sample problems can be anticipated and corrected for in advance, if we're willing to do the calculations.

References

Ankudinov, A. L. and J. J. Rehr. 1997. Relativistic calculations of spin-dependent x-ray absorption spectra. *Phys. Rev. B*. 56:R1712–R1715. DOI:0.1103/PhysRevB.56.R1712.

Bevington, P. R. and D. K. Robinson. 2003. *Data Reduction and Error Analysis for the Physical Sciences*, 3rd ed. Boston, MA: McGraw-Hill.

Bunker, G. 2010. *Introduction to XAFS*. New York: Cambridge University Press.

Elam, W. T., B. D. Ravel, and J. R. Sieber. 2002. A new atomic database for X-ray spectroscopic calculations. *Rad. Phys. Chem.* 63:121–128. DOI:10.1016/S0969-806X(01)00227-4.

Heitler, W. 1954. *The Quantum Theory of Radiation*, 3rd ed. Oxford: Oxford University Press.

Kelly, S. D., D. Hesterberg, and B. Ravel. 2008. Analysis of soils and minerals using X-ray absorption spectroscopy. In *Methods of Soil Analysis Part 5: Minerological Methods*, ed. A. Klute, 387–463. Madison, WI: American Society of Agronomy.

Lee, P. A., P. H. Citrin, P. Eisenberger, and B. M. Kincaid. 1981. Extended x-ray absorption fine structure—Its strengths and limitations as a structural tool. *Rev. Mod. Phys.* 53:769–806. DOI:10.1103/RevModPhys.53.769.

Pfalzer, P., J.-P. Urbach, M. Klemm, S. Horn, M. L. denBoer, A. I. Frenkel, and J. P. Kirkland. 1999. Elimination of self-absorption in fluorescence hard x-ray absorption spectra. *Phys. Rev. B*. 60:9335–9339. DOI:0.1103/PhysRevB.60.9335.

Ravel, B. and M. Newville. 2005. ATHENA, ARTEMIS, HEPHAESTUS: Data analysis for X-ray absorption spectroscopy using IFEFFIT. *J. Synchrotron Radiat.* 12:537–541. DOI:10.1107 /S0909049505012719.

Saksl, K., D. Zajac, H. Franz, Š. Michalík, and J. Z. Jiang. 2006. Comparisons between XAFS signals measured in transmission and total electron yield mode. Hasylab Annual Report.

Scarrow, R. 2002. Sample preparation for EXAFS spectroscopy. www.xafs.org/Workshops/ NSLS2002.

Stern, E. A. and K. Kim. 1981. Thickness effect on the extended-x-ray-absorption-fine-structure amplitude. *Phys. Rev. B*. 23:3781–3787. DOI:10.1103/PhysRevB.23.3781.

Chapter 7

Data Reduction

Synchrotron Driver's Training

But we haven't done data collection yet. How can we talk about data reduction when we don't know how to collect it yet?	
Well, which did you learn to do first: how to drive a car, or how to buy one? Even though you've got to buy a car before you drive it, you really should know *how* to drive before you buy. It's the same thing with data. If you know something about how it's going to be processed before you get to the beamline, you'll make better decisions when it comes time to collect it.	
There's plenty of raw data that you can practice with online before you ever step foot in a synchrotron. For instance, check out the links to databases at xafs.xrayabsorption.org.	

7.1 Preprocessing

Depending on the data source, there are a few specialized tasks you might do right at the start.

7.1.1 Rebinning

In this chapter, we'll go through data reduction in pretty much the order you'll do it in. But that order's not set in stone. For example, should you merge individual scans before or after normalization? With good data, either order is OK, but in other cases, one might work better than the other.

Some data collection modes produce EXAFS data that are very finely spaced in energy, but contain relatively few counts per energy point. For example, quick-XAFS is collected by slewing the monochromator continuously through the energy range of the scan. As we will see later in this chapter, it is not necessary to have very fine resolution in energy in the EXAFS region (the XANES region can be another story). Therefore, it is sometimes a good idea to *rebin* the data on to a coarser grid. This improves signal to noise, and thus makes the subsequent steps of data reduction more accurate.

DOI: 10.1201/9780429329555-9

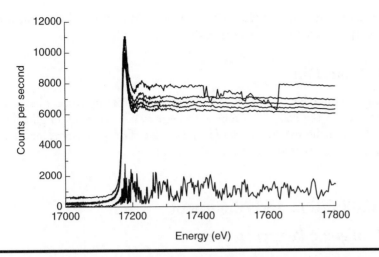

Figure 7.1   Spectra collected by a multielement detector showing bad channels.

7.1.2 Selecting Channels and Scans

Often, not all the data you collect is usable. Consider, for example, Figure 7.1, showing data from six different channels of a multielement detector.

The vertical translation between scans is not a concern; we'd expect some channels to record more counts than others, just from geometrical considerations. As long as they're all recording fluorescence from the sample, they should all have the same shape.

But there are two channels that do not have the same shape as the other four, and, in fact, do not appear to be proper XAFS spectra. There's the one at the bottom, which looks normal enough in the preedge and shows the beginning of an edge, but then doesn't show the rest of the edge and looks extremely noisy. There's also the channel at the top, which looks OK, if a bit noisier than the others, until about 17,400 eV. From there until about 17,630 eV, it deviates sharply from the slow variations shown by the other scans, only to return to a more typical pattern at the end.

Something clearly went wrong with those two aberrant channels during data collection, so those data should be rejected from the analysis.

While we've given an example in terms of a multielement detector, similar screening may apply in other cases. For instance, a time-resolved *operando* experiment may be plagued by occasional air bubbles which disrupt individual scans. Even a conventional *ex situ* transmission measurement may suffer from occasional bad scans due, for example, to a failure of the synchrotron beam.

7.1.3 Calculating Unnormalized Absorption

This is a 'specialized task' only in that it depends on how the measurement was performed. For transmission, the absorption coefficient at each energy is proportional to the natural log of the ratio of the I_o measurement channel to the I_t measurement channel, while for an ideal fluorescence sample (or an electron yield sample) it's proportional to the ratio of the I_f measurement channel to the I_o measurement channel. Because other factors, such as the thickness of the sample and detector characteristics, are buried in the proportionality constant, it's technically incorrect to claim that these quantities are the absorption. Oddly, there is no standard nomenclature for what quantities such as the natural log of the ratio of the I_t measurement channel to the I_o measurement channel should be called—although 'raw data' is an informal term you will probably hear on the synchrotron floor. In this book, we will call these quantities *unnormalized absorption*, and symbolize them by $M(E)$.

Smoothing is a fancier alternative to rebinning. A simple smoothing routine, for instance, could replace each *y*-value with a three-point moving average, and then repeat enough times to get the balance between signal to noise and resolution you want. But a moving average degrades peaks, so a better smoothing method is a Savitsky–Golay filter (Savitzky and Golay 1964).

The effect of smoothing is not as clear-cut as rebinning! Many in the field recommend against it!

The data shown in Figure 7.1 are from the L_3 uranium edge of a soil sample. In actuality, there were 25 channels of good data; only four good channels have been shown in order to avoid cluttering the graph.

Actually, some differences in shape are OK sometimes, because steps like background subtraction will take them out anyway. If you read this chapter to the end and think it through, you'll get the idea for what can be tolerated and what's a problem.

In casual usage, almost no one distinguishes between the intensity of x-rays at some point and the measurement of that intensity. It's quite possible for the I_1 channel to read higher than the I_o channel, if, for instance, the gain is set higher. Novices are often surprised: 'How can I_1 be 3.78 volts when I_o is 2.34 volts? Shouldn't the sample absorb photons?' Their confusion is understandable, as it comes from our playing fast and loose with the difference between the x-ray intensity and our measurement of it!

Personally, my preference is to avoid throwing away data early in the process. One can always perform subsequent steps, such as normalization and background subtraction, in such a way as to exclude problem regions.

When people compare 'modern' beamlines to those from decades ago, they often think about aspects such as flux, time resolution, and spatial resolution. But great strides have also been made in energy stability and reproducibility. You should still measure references simultaneously with the sample if possible, but on many beamlines it's less of a concern than it used to be.

I do not disagree, Kitsune, but it is still important to calibrate on each visit to a beamline. While some beamlines may remain stable for a day or two, that doesn't mean that they will remain stable for months, or that other users won't have changed the calibration.

You may wonder why we suggest you use the reference to align, rather than the data from the sample. That's because materials used for references tend to be very stable and not susceptible to beam damage. Your sample, on the other hand, may change gradually under the beam, confusing attempts at alignment!

If you measure a reference simultaneously, you'll want to calculate the unnormalized absorption due to the reference at this time as well.

7.1.4 Truncation

Data may be unusable above or below some energy. This might occur because another edge appears in the data, for instance, or because of a severe set of glitches. In such cases, it's reasonable to truncate the data just inside of the problem region so that it will not interfere with subsequent steps.

7.2 Calibration and Alignment

7.2.1 Aligning Reference Scans

Some monochromators don't retain perfect energy calibration over the course of multiple measurements. In some cases, the drift in energy is steady, for example, 0.5 eV per scan. In other cases, the monochromator maintains calibration for sustained stretches, but occasionally jumps by an electron volt or two. There are even reports of beamlines where the energy calibration seems to depend weakly on the time of day, perhaps due to changes in thermal stresses from sunlight warming the building or the ground.

If you measured the reference data simultaneously with your sample, then you should align the reference scans with each other. Unless the reference data are very noisy, alignment is usually most precise if you look at the derivative spectrum near the edge. It is better to look at more than one peak in the derivative spectrum when you align.

Align the scans of data by using the same shifts that were necessary to bring the references into alignment.

Occasionally, you may not have a simultaneous or near-simultaneous reference measurement. This may occur, for instance, if a fluorescence sample is too thick to allow any photons to be transmitted to a reference channel, and if a long series of uninterrupted measurements need to be performed on the sample—perhaps it's a battery being cycled or an *in situ* chemical reaction. In such circumstances, alignment is more difficult to assure. At a minimum, a reference should be measured before and after the sample, so that the total shift can be ascertained. It's also a good idea to take several scans of references and standards before beginning your main experiment, so that you can ascertain the behavior of the monochromator. If there is a steady energy drift, then the measurements before and after the data on the sample are collected can be used to estimate shifts in between. If the energy is prone to jumps, then, frankly, you have a problem. Discuss the issue with beamline personnel to see if it can be resolved.

7.2.2 Merging

After alignment, scans on a sample measured under identical conditions should be merged. Usually this is simply a matter of averaging the scans together.

Occasionally, however, the amount of noise present may differ significantly between the scans. At times, the amount of signal may also differ. For instance, you might measure a fluorescence scan with a Lytle detector, and then decide that the gain on the detector should be increased by a factor of 10, after which you collect two more scans. The scans can still be averaged together, but how should they be weighted for that average?

It turns out there is not a simple answer to that question. Suppose, for instance, that one scan has a noise level of a and a signal of c, while a second scan has a noise level of b and a signal of d. Let's consider what happens when we average the two scans together, but with the second scan having a weight of w relative to the first.

Noise adds in quadrature, so the noise level of the first scan added to w times the second scan would be

$$Noise = \sqrt{a^2 + (wb)^2} \tag{7.1}$$

Signal just adds, so it becomes

$$Signal = c + wd \tag{7.2}$$

This gives

$$\frac{Signal}{Noise} = \frac{c + wd}{\sqrt{a^2 + (wb)^2}} \tag{7.3}$$

We want to find the value of the weight w that maximizes that ratio. A little calculus leads to the result

$$w = \frac{da^2}{cb^2} \tag{7.4}$$

This result can be illuminated by considering some special cases. First, suppose that all noise is strictly shot noise from measuring the signal. In that case, $a \propto \sqrt{c}$ and $b \propto \sqrt{d}$, leading to

$$w = 1 \quad \text{(noise due to signal counting statistics only)} \tag{7.5}$$

For the next case, consider a constant signal to noise ratio, but with different amounts of signal. In some circumstances, changing the gain on an amplifier attached to a detector might have roughly this effect (see Section 8.6.4 in Chapter 8). In that case, $\frac{a}{c} = \frac{b}{d}$, leading to

$$w = \frac{c}{d} \quad \text{(constant signal to noise)} \tag{7.6}$$

This weighting implies that the scans should be normalized first (see the following text), and *then* merged.

Finally, consider the case where the noise level is independent of the signal. This might be the case if the signal is dominated by background due to other processes, such as scattering and fluorescence from elements other than the target. In that case, $a = b$ and

$$w = \frac{d}{c} \quad \text{(constant noise)} \tag{7.7}$$

While in theory the proper weighting could be chosen by using Equation 7.4, there is a practical difficulty: how does one know signal and noise for an individual scan?

This section is written as if it is talking about a series of scans taken on a sample under static conditions. But it also applies to the different channels of data collected by a multielement detector.

Wow, this is tedious! Depending on the kind of noise we've got, we're either supposed to weight scans by their edge jumps, or by one over their edge jumps, or by something in between? In any real situation, 'something in between' will probably be the right answer. And you know what's in between one over the edge jump and the edge jump? Weighting 'em all equally!

So here's what I do. I just average all the scans with equal weighting. If a scan looks extra noisy, I see if including it makes the average look noisier or not. If it makes it noisier, I leave it out. That's much easier than all this fooling around trying to find the perfect weighting. Maybe I end up with a signal to noise that's not quite the best it could be, but it saves me a lot of time that I could be spending on more important things.

BOX 7.1 Estimating Noise

 Friends! The Standards and Criteria Committee of the International X-ray Absorption Society has identified four methods for estimating the noise in an EXAFS measurement (IXS Standards and Criteria Committee 2000). While their recommendations are aimed at estimating the noise in the EXAFS, we can use them as a basis for making our own list of four different methods that could be used to help estimate the noise present in individual absorption measurements:

1. Compare the data to a smoothed version. Glitches and sharp features, such as the edge and the white line, should be avoided when making the comparison. The best region to choose would probably be a smooth region well past the edge.
2. Perform a rough data reduction on the scan, and look at the amplitude of the Fourier transform well above the value of R where features are expected to be seen
3. Estimate the noise on the basis of theory and beamline characteristics; for instance, if the number of counts is small, shot noise is likely to be dominant
4. Compare an individual scan to the average of all scans

Mandlebrant, using suggestions from the IXS, covers *noise* in Box 7.1. The relative amount of *signal* present in different scans can be estimated by determining the edge jump for each scan.

7.2.3 Calibrating

If you have a reference scan, or, better yet, a merged set of reference scans, you can now calibrate. The most common method is to find the first peak of the first derivative, and assign it a tabulated value for that material.

There is a problem with this procedure, however. The same material may be assigned slightly different values by different tables. That's because, while monochromators are pretty good at maintaining energy calibration over the range of an EXAFS scan, they do not generally measure larger energy intervals to high accuracy, such as the 30 keV difference from the sulfur K edge to the iodine K edge. In fact, those edges would not generally even be measured at the same beamline!

Thus, we should expect tabulated values to be a bit inaccurate— experience shows they sometimes differ from each other by an electron volt or two at any given edge. Of course, the scientific community gradually learns how to do that kind of measurement better, with the result that accuracy should improve over time. This would argue that we should use the most recent table that has been published, perhaps (Kraft et al. 1996).

The problem with using the most recent table, however, is that it makes comparison between an older and a newer paper confusing. If the old paper placed the K edge of cadmium metal at 26,711 eV for calibration purposes and the new one places it at 26,713.3 eV, then we have to mentally shift all the figures from the old paper by 2.3 eV when comparing side by side.

So many XAFS researchers stick with the tables that were current at the time modern XAFS research began, namely those found in the work by Bearden and Burr 1967. The Bearden and Burr values are also built into many XAFS software packages.

7.3 Finding Normalized Absorption

7.3.1 Deadtime Correction

As described in Chapters 2 and 5, energy-discriminating detectors have a limited dynamic range, and as they approach the top of that range, they become nonlinear. This nonlinearity can be corrected for, although the details depend on the detector in use. If using an energy-discriminating detector, ask your beamline scientist about deadtime correction.

Wait—are you telling me to intentionally use older, less accurate tables to calibrate my data? That sounds crazy!

7.3.2 Deglitching

As discussed in Chapter 8, *glitches* are sharp deviations in the measured absorption that are consistent from scan to scan. Small glitches can be removed in a later step by Fourier filtering, but large glitches can distort procedures such as normalization and background subtraction and, if present, should be removed before those steps.

Usually, when we talk about glitches being 'small' or 'large,' we are talking about the maximum amount of deviation they cause from the signal. But more important is the width of the glitch. Even a 'small' glitch can be extremely disruptive if it extends over a range of, say, 0.25 Å^{-1}.

If you follow the procedures we describe in Section 8.9.1 of Chapter 8, you will have somewhat oversampled your data in energy space, and thus the loss of one point will not significantly affect your analysis. One method for dealing with a narrow glitch, therefore, is to simply remove that point (or points) from the data. Another method would be to interpolate a value from the surrounding data. Since the data are oversampled, the exact method used is not important.

If a glitch is broad, however, there is nothing you can do: the data under the glitch are irrevocably damaged. If the glitch is both broad and also sufficiently deep to dominate your signal, then the game is up: you cannot use that part of the signal. Fortunately, beamline scientists generally know the energies of broad, deep glitches on their beamlines, and avoid scheduling users whose experiments would be severely curtailed.

Absolute energy calibration, my dear Simplicio, is not important in XAFS; what is important is the shifts between spectra. If we all agree to use the same values for our elemental metals, it will facilitate observing those shifts, even with papers by different authors or from different eras.

7.3.3 Choosing E$_o$

Once you have merged and calibrated your data, you can choose a point to represent the zero energy for the photoelectron.

In informal conversation, the width of a glitch is often given in terms of the number of separate points in energy space that it affects; to wit, 'a three-point glitch.'

BOX 7.2   'Choose' or 'Find'?

'Choose a point?' That's a strange way of putting it. Don't you mean 'find the point'? I mean, the zero energy for the photoelectron is a concrete idea, and I should be able to find the energy that corresponds to it, right?

You might think so, Simplicio, but as we'll see in Section 13.1.5 in Chapter 13, the zero energy for the photoelectron is not all that well defined. That makes the definition somewhat arbitrary, and thus leaves it up to us to choose.

This is one case where smoothing your data is justified! Since the primary goal in choosing E_o is consistency, smoothing out noise is very helpful, even if it distorts the data in some way, because the distortion will at least be consistent! So it is typical, for instance, to use a smoothed derivative spectrum when choosing E_o.

While the choice of E_o is somewhat arbitrary, it is helpful to choose it in a consistent and easily described way. That way, equivalent choices can be made on different spectra, facilitating their comparison.

Following are some common ways in which E_o is chosen:

■ At the first peak in the first derivative; in other words, at the first inflection point in the spectrum
■ At the largest peak of the first derivative. If you are going to be comparing spectra which are quite different from each other, such as a metal and its oxides, this is not a good choice, as which peak is largest may differ from sample to sample.
■ At some other well-defined peak of the first derivative
■ At the top of the white line
■ Halfway up the edge. This choice generally requires some kind of curve to be fit either through the edge or through the pre- and postedge, so that 'halfway up' has a reproducible definition.
■ At the first peak in the first derivative of the reference spectrum
■ At the same energy that was chosen for another related spectrum
■ Wherever the algorithm embedded in your analysis software decides to put it. While this is OK for a quick look at the data, *we recommend against* blindly trusting the analysis software when you are performing a careful analysis.
■ The 'onset of the edge,' that is, the energy at which the rising absorption attributable to the edge can first be discerned. *We recommend against this choice.* Such a definition is strongly dependent on the size of the edge relative to the background, as well as the amount of noise in your measurement.

7.3.4 Normalization

If one has noisy XANES data, using the same E_o for all spectra in a series is a good procedure, as opposed to trying to identify inflection points that are being pulled to and fro by noise.

The unnormalized absorption $M(E)$ has the same shape as $\mu(E)$, but can also be dependent on a bewildering variety of other factors, including sample thickness, the solid angle subtended by a fluorescence detector, gains on amplifiers, gases used in the detectors, the use of filters and collimators, and more.

As we saw in Section 1.6 in Chapter 1, all forms of XAFS analysis depend on comparing spectra, whether they are the spectra of different samples, the spectra of a sample and empirical standards, or the spectra of a sample and theoretical standards. So we have to somehow divide out all the other factors—that is, we have to normalize the spectra.

But how do we know the proper normalization to use? After all, we are likely comparing spectra that are not identical, and in some cases, such as linear combination fitting, we may be comparing spectra that are quite dissimilar in fine structure.

To wrap our minds around this issue, we'll consider $M(E)$ to be due to a combination of the following effects:

The notation we are using in this section is, as far as we know, unique to this book. Unfortunately, the common usage tends to blur the distinctions between measured quantities like $M(E)$ and the underlying quantities associated with the sample, such as $\mu(E)$. For instance, many sources will refer to 'measuring' the absorption coefficient $\mu(E)$ and determining a 'smooth background' $\mu_o(E)$. But in fact, it is rare to determine the absorption coefficient $\mu(E)$ during an EXAFS experiment, especially when measuring thick dilute samples in fluorescence. What is actually measured is $\mu(E)$ modified by some scaling factor; that is, what we have here called the unnormalized absorption $M(E)$.

■ A coarse structure due to absorption by the element of interest, consisting of a featureless preedge curve, followed by a sharp edge, and by a slowly varying postedge curve. We'll refer to this as the *intrinsic background* and use the symbol $B_i(E)$.
■ A slowly varying background function due to the response of the detectors and the presence in the beam path of elements other than the element of interest. We'll refer to this as the *extrinsic background* and use the symbol $B_e(E)$.
■ A scaling factor that incorporates the concentration of the element of interest, the thickness of the sample, and some detector effects such as the gain on a Lytle detector. For reasons which will become clear shortly, we'll call this $\Delta M_o(E_o)$.

■ The fine structure we are interested in, expressed relative to $B_i(E)$, which we will call $\chi(E)$.

From the above definitions, we can write down the following equation:

$$M(E) = B_e(E) + B_i(E)\left[1 + \chi(E)\right] \tag{7.8}$$

We'd like to have the normalized version of that equation, though, so that we can quantitatively compare different spectra:

$$\frac{M(E)}{\Delta M_o(E_o)} = \frac{B_e(E)}{\Delta M_o(E_o)} + \frac{B_i(E)}{\Delta M_o(E_o)}\left[1 + \chi(E)\right] \tag{7.9}$$

Fortunately, we know what $\dfrac{B_i(E)}{\Delta M_o(E_o)}$ looks like when a given element is present in a sample—there should be an edge. Figure 7.2 shows $\dfrac{B_i(E)}{\Delta M_o(E_o)}$ for an arsenic K edge.

Since all the information about the atomic environment is in the fine structure, Figure 7.2 now applies to the K edge of any arsenic-containing sample. (We have plotted relative to E_o because the value of E_o will vary by a few electron volts from sample to sample, in part because it depends on oxidation state.)

Our goal is to compare spectra, and $\dfrac{B_i(E)}{\Delta M_o(E_o)}$ is the same for all spectra at that edge, so it provides the key to making a consistent comparison. But there's still a bit of ambiguity in this procedure; while $\dfrac{B_i(E)}{\Delta M_o(E_o)}$ must be the same for all spectra at that edge, we're allowed to pick how it should be scaled. So we make the simplest choice, and decide that the edge jump (the height of the vertical line in the graph) will be 1.

Thus, we have a plan for normalizing our data. First, remove the fine structure from our data, leaving us a plot that looks something like Figure 7.2. We'll call this plot $M_o(E)$. Comparison to Equation 7.8 shows us that

$$M_o(E) = B_e(E) + B_i(E) \tag{7.10}$$

We still don't know much about $B_e(E)$—it depends on details of the detectors and the sample. But we do know one thing about it—it varies slowly. The big jump in $M_o(E)$

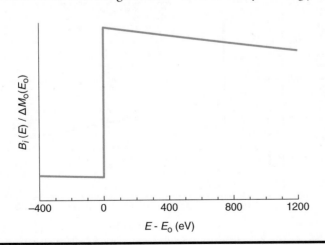

Figure 7.2   Schematic of normalized intrinsic background for an arsenic *K* edge.

Equation 7.8 is fine for transmission. But for fluorescence or electron yield measurements, detector response, which is part of $B_e(E)$, is a multiplicative factor, even though Equation 7.8 shows it as additive. (The effect of other elements is still additive). If the energy dependence of the incident and fluorescence detectors are the same, that's no big deal; the response will divide out when we normalize. But to the extent the energy dependence of the detector responses is not the same (or corrected for), Equation 7.8 will fail, distorting the $\chi(E)$ found by the procedure in this chapter.

This is one good reason why detectors with relatively flat energy response are valued. It is unlikely that the energy response of a solid-state detector used to measure fluorescence will be identical to that of an ion chamber used to measure the incident beam. But if the energy response in both cases is nearly constant over the energy range of the measured spectrum, the distortion introduced will be small.

Since we have plotted Figure 7.2 relative to E_o, one may wonder why we specify that it is for the arsenic K edge. The reason is that the decay seen above the edge in Figure 7.2 would be slightly different for other elements.

We used the term *edge jump* in Section 6.3.5 in Chapter 6 for $\Delta\mu x$, but here we are using it for $\Delta M_o(E_o)$. In the case of an ideal transmission experiment with identical I_o and I_t detectors, the two quantities would be the same, but, in general, they are not. If you want to be fastidious, you could refer to $\Delta\mu x$ as the *absorption edge jump* and $\Delta M_o(E_o)$ as the *measured edge jump*. But, alas, no one in the field is that meticulous, not even I.

must be entirely due to $B_i(E)$. And since we know that we have chosen to have an edge jump of 1, we know that $\Delta M_o(E_o)$ must be equal to the edge jump of $B_i(E)$, which is equal to the edge jump of $M_o(E)$, which is something we can measure! So now you can see why we decided to call the scaling factor $\Delta M_o(E_o)$: it is the change in $M_o(E)$ at the edge energy E_o.

Normalization, therefore, comes down to making a good estimate of the edge jump $\Delta M_o(E_o)$. It is usually easy to make a rough estimate of the edge jump. For example, consider Figure 7.3, a spectrum of the arsenic K edge of a sample of oxidized gallium arsenide.

If we imagine Figure 7.3 without the fine structure, we might estimate that M_o rises from about –0.50 just below the edge to about –0.09 just above it, giving an edge jump of 0.41. We could therefore normalize the spectrum by dividing all the y-values by 0.41, giving us Figure 7.4.

To complete the process of putting this spectrum into a standard form, we should place it in a standard position vertically; conventionally, it is shifted so that the bottom of the edge lies at 0. We have to do this because $B_e(E)$ will be different for different spectra. In this case, we'll shift the spectrum up by 1.2 units, giving us Figure 7.5.

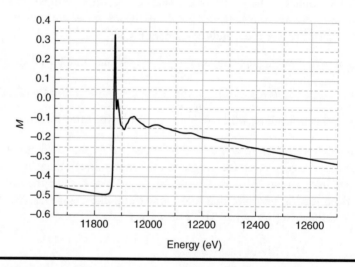

Figure 7.3 **Arsenic *K*-edge spectrum of oxidized gallium arsenide.**

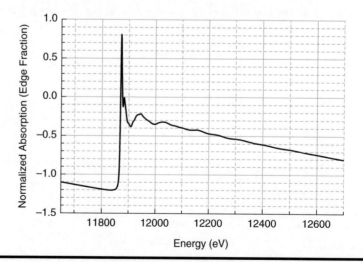

Figure 7.4 **The spectrum from Figure 7.3 normalized by the edge jump.**

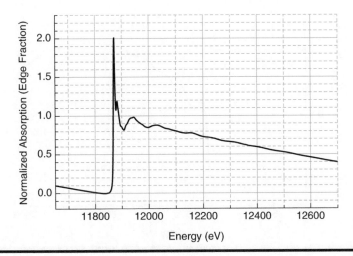

Figure 7.5 **The normalized spectrum from Figure 7.4 shifted up so as to have the edge begin near zero.**

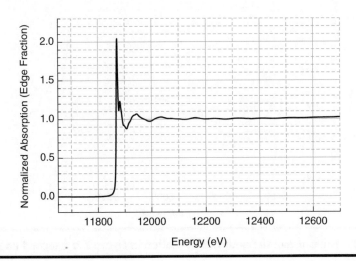

Figure 7.6 **The normalized spectrum from Figure 7.5 with the slow variation before and after the edge subtracted out.**

We have now successfully normalized our spectrum. Some researchers (and some software) like to also take a stab at subtracting out the slow variation before and after the edge is present in $M_o(E)$, as shown in Figure 7.6.

In this particular case, the normalization procedure was done in an ad hoc way: an edge jump was chosen that seemed about right, followed by a shift that seemed to bring the data up to around zero just below the edge, followed by subtraction of best fit lines above and below the edge. The result is just as good a normalization as any analysis software will give you.

There are two problems, though, with this seat-of-the-pants procedure: it is time consuming and subjective.

Normalization routines used in analysis software automate the process, and in a perfect world they alleviate both problems. Alas, the world is not perfect. Figure 7.7, for instance, shows the result of a default normalization of the manganese K edge of a manganese zinc ferrite sample performed by ATHENA, a popular software package for data reduction.

The result is obviously poor; the edge jump of the 'normalized' data is around 0.5, not 1.0. (The flattening routine also failed badly, but that's a less serious concern.)

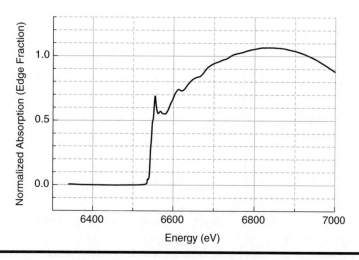

Figure 7.7   A poor attempt at normalization by data processing software.

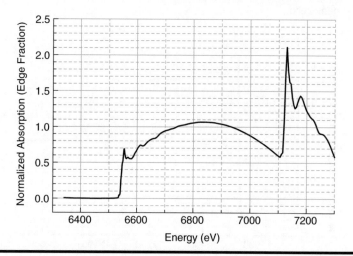

Figure 7.8   The cause of the poor normalization in Figure 7.7: a second edge.

You may have noticed that we aren't giving details on things like the algorithm used by ATHENA or how it was tweaked to yield Figure 7.9 rather than Figure 7.7. That's because different analysis software uses different algorithms for normalization. Each software package comes with documentation (hopefully!) describing its procedure and how to intervene when things go awry. Our point is that these packages are all in essence time-saving devices and methods for enforcing consistency from spectrum to spectrum; your eye is still the best judge of proper normalization!

In this case, this happened because there was another edge in the data, as can be seen in Figure 7.8, where we plot the data over a somewhat larger energy range.

While in this case the culprit was an additional edge, many different anomalies can cause automated normalization routines to perform poorly, including, for instance, a large white line in conjunction with a relatively short XANES scan.

Allowing software to normalize data in a completely automated fashion does remove subjectivity, but it also reduces accuracy—sometimes, as in this example, to an absurd degree. There is no avoiding some subjective evaluation at some stage of the process; you need to decide for yourself how to strike a balance between accuracy, objectivity, and ease of processing.

Figure 7.9 uses the same data as in Figure 7.7 and is also normalized by ATHENA, but with just a little user intervention to keep it from being distracted by the second edge.

When choosing normalization for a set of spectra, one important goal is to try to be as consistent as possible. If the estimates of the edge jumps for all the spectra are 10% smaller than they 'should' be, for instance, that has much less of an impact on analysis than if one spectrum is 3% too small and another is 3% too large.

In any case, you should use the normalization process to estimate the uncertainty in your edge jump. Consider, for instance, Figure 7.10, which shows a sulfur *K*-edge XANES spectrum of L-cysteic acid.

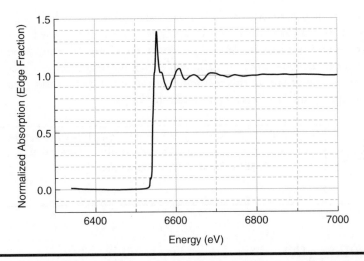

Figure 7.9 **The data from Figure 7.7 with a more appropriate normalization.**

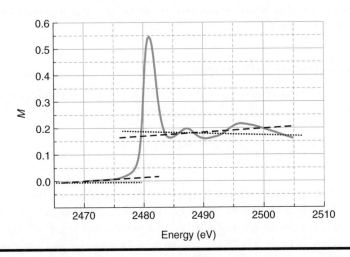

Figure 7.10 **A difficult normalization. The dotted and dashed lines show two different plausible attempts to determine the edge jump. (Source: Spectral data from Dhez, M., ID21 sulfur XANES spectra database, 2009, available at http://www.esrf .eu/UsersAndScience/Experiments/Imaging/ID21/php.)**

It's not entirely clear what the edge jump should be. Does M_0 trend up after the white line, like the dashed line shows, or slightly down, like the dotted line? Likewise, there isn't quite enough of the preedge to decide how M_0 behaves there. The result is that we could reasonably argue for an edge jump of anywhere from about 0.15 to about 0.19. If we split the difference and selected an edge jump of 0.17, that would still imply an uncertainty of ± 0.02/0.17, or ± 12%. The uncertainty you determine in this way should be carried through subsequent analysis, a point we'll reinforce in the following chapters.

7.3.5 Self-Absorption Correction

In Section 6.4 of Chapter 6, we discussed self-absorption, a source of distortion in fluorescence measurements. An examination of Equation 6.18 suggests that it should be possible to correct for self-absorption, if you know the geometry of the measurement and $\mu(E) - \mu_*(E)$, that is, the absorption due to other elements in the sample and elsewhere in the beam path (and other edges of the same element as well).

The tough normalization choices in Figure 7.10 could have been avoided if the data had been collected over a bigger energy range. Whenever you can, collect at least a few points well below and above the edge, even if you're only interested in XANES.

Self-absorption corrections are only approximate! Iida and Noma (1993) includes an excellent discussion of the way uncertainties propagate through the process. The conclusion is that any correction is likely to introduce an uncertainty on the order of 5% for the amplitude of features such as the white line. While that's tolerable for most analysis, it's better to design your experiment so that self-absorption isn't a problem in the first place!

As long as the other elements do not have edges near the energy range you have measured, $\mu(E) - \mu_*(E)$ will vary fairly slowly over the range and will not be sensitive to structural information. Thus, if you know the elemental composition of your sample, you can estimate $\mu(E) - \mu_*(E)$, and, therefore, correct for self-absorption in your spectrum. Variations on this method, first published by Iida and Noma (1993), have been implemented in many data analysis packages.

For XANES analysis, data reduction is complete once we've normalized the spectrum. The rest of this chapter will be devoted to procedures related to EXAFS analysis.

7.4 Finding χ(*K*)

Normalization is enough to allow XANES to be compared, but for EXAFS we want to go further and find $\chi(k)$, so that we can think in terms of the description introduced in Section 1.2 of Chapter 1.

7.4.1 Background Subtraction

Equation 7.9 guides us in extracting the EXAFS signal $\chi(E)$. Rearranging and using the definition given by Equation 7.10, we see that

$$\frac{B_i(E)}{\Delta B_i(E_o)}\chi(E) = \frac{M(E) - M_o(E)}{\Delta M_o(E_o)} \tag{7.11}$$

The right side of the equation looks a little like what we did during the normalization process, particularly if we do something similar to what was done in Figure 7.6, where we subtracted off a slowly varying function from our measured $M(E)$ and also normalized by the edge jump $\Delta M_o(E_o)$. But what we subtracted off was really not all that much like $M_o(E)$— most notably, $M_o(E)$ is supposed to include the general shape of the edge, but we made no attempt to remove the edge from our data! Therefore, some kind of additional background subtraction is necessary.

The first thing to realize about background subtraction is that it will necessarily involve fitting some kind of smooth curve to the data. While an approximation to the intrinsic background $B_i(E)$ is available in tables, the extrinsic background $B_e(E)$ certainly is not—it's due to too many disparate experimental effects. So we can't calculate $M_o(E)$; we have to extract it from our data.

Next, you should realize that there are lots of good software packages that perform background subtraction for you, using a variety of schemes that people have found to be effective. While it's a good idea to read about how the particular package you are using performs its background subtraction in its documentation, the process of background subtraction will almost certainly be semiautomated for you.

But semiautomated is not the same thing as automated. Your analysis program will still provide you with methods to influence the background, and, just as we saw with normalization, it will sometimes be necessary to override a program's default choice.

For example, consider Figure 7.11, an attempt by the IFEFFIT software package to find the background function $\dfrac{M_o(E)}{\Delta M_o(E_o)}$ for the oxidized gallium arsenide spectrum we normalized earlier in this chapter.

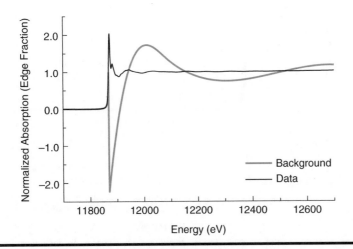

Figure 7.11   A poor background function that doesn't follow the general trend of the data.

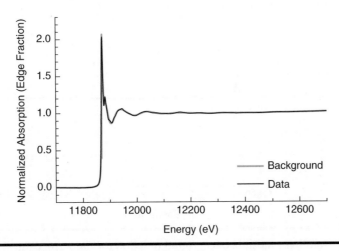

Figure 7.12   Another poor background function that follows the fine structure of the data too well.

It is evident that something has gone terribly awry—the background function doesn't look much like the data at all.

Figure 7.12 shows another attempt at a background for the data.

This time we've got the opposite problem. The background follows the wiggles in the data too faithfully; it's supposed to only be the 'slowly varying' part.

Figure 7.13 shows one more try.

We finally have what looks like our Goldilocks solution: not too wiggly, but wiggly enough to follow the general trend through the EXAFS.

But Goldilocks only had three beds to choose from, and we have many. Figure 7.14 shows yet another.

It is not immediately clear whether the background shown in Figures 7.13 or 7.14 is preferable. The difference is primarily in the XANES, but the background is only used for EXAFS. Are they different in the EXAFS region? One way to tell is to zoom in on that part of the graph.

Figure 7.15 shows that the backgrounds *are* slightly different in the EXAFS region. If you look carefully, you'll see that the gray line is slightly above the dashed line around 12,150 eV, then the dashed one is above around 12,350 eV, then the gray around 12,500 eV, and finally, the dashed line is once again on top toward the end.

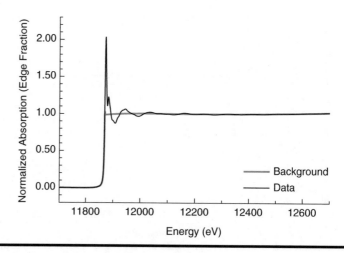

Figure 7.13   A plausible background function.

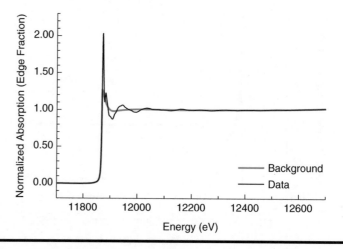

Figure 7.14   Another plausible background function for the same data.

The EXAFS region of the two backgrounds is indeed slightly different, but it's hard to tell which is better just by looking. Also, if you're planning to Fourier transform the data, it's not clear if the two backgrounds would produce different Fourier transforms in the range that you will use for the analysis.

So let's jump ahead and use Figure 7.16 to look at the magnitudes of the Fourier transforms of the data extracted using the four different backgrounds.

The absurd-looking background from Figure 7.11 produces an absurd-looking Fourier transform graph, without anything like the distinct shells usually seen in EXAFS. But the other three backgrounds all produce Fourier transforms which are effectively identical above 2.5 Å. The background from Figure 7.12, however, which we rejected as following the data too closely, shows no peak around 1.4 Å in its Fourier transform.

At this point we have to review our expectations. Our sample is oxidized gallium arsenide. We do not know the exact structure of our sample—'oxidized' is a bit vague. But we do expect to see gallium–oxygen bonds. Those typically have bond lengths of about 1.7 Å (Belsky et al. 2002), which, after undergoing the usual 0.3 to 0.5 Å shift (see Chapters 1 and 13), could correspond to the large peak seen when the backgrounds from Figure 7.13 or Figure 7.14 are used. As we suspected, the background subtraction from Figure 7.12 is actually removing part of our signal, namely, the large peak around R = 1.4 Å.

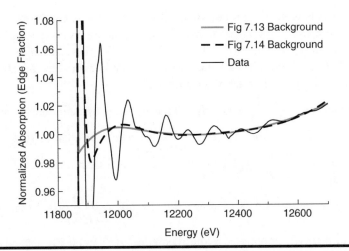

Figure 7.15   Comparison of the backgrounds from Figures 7.13 and 7.14.

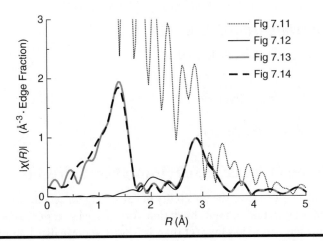

Figure 7.16   Fourier transforms of the oxidized gallium arsenide data using the backgrounds shown in Figures 7.11 through 7.14.

Figures 7.13 and 7.14 are very similar above 1.5 Å. The height of the main peak at 1.4 Å is slightly different, and the structure below 1.0 Å is quite different. It is not clear-cut which background is better, but that's OK if we're using the Fourier transform: we won't use the data below 1.0 Å anyway, since it does not correspond to the EXAFS scattering process we described in Section 1.2 of Chapter 1. The small difference in height of the first peak we will treat in much the same way we did the alternate normalizations in Section 7.3.4: we will try each in our ultimate fits, and if both yield acceptable results, we will include differences between them as part of our reported uncertainties.

7.4.2 χ(E)

Now that we have one (or more!) acceptable background functions, we can use Equation 7.11 to find $\frac{B_i(E)}{\Delta B_i(E_o)}\chi(E)$. From there, in theory, we can do a pretty good job of getting $\chi(E)$, since $B_i(E)$ is supposed to be nearly independent of chemical environment and can thus be tabulated (see Rehr et al. 1994 for why we say 'nearly' independent).

Of course, you should always be prepared to discover that your sample is not what you thought it was! If the sample had been reduced back to pure gallium arsenide somewhere along the line, then there would be no peak around 1.3 Å even with a good background.

You might be wondering what the structure below 1.0 Å is if it's not signal. Part of it may be due to the intrinsic background—see, for instance Rehr et al. (1994). Part may be extrinsic background; in this case, there is a gallium edge only about 1500 eV below the arsenic edge, and the gallium EXAFS may show up weakly as a stretched-out signal low in the Fourier transform. Part may be truncation effects (see Sections 7.5.2 through 7.5.4).

I wouldn't put it that way, Kitsune. Intrinsic and extrinsic backgrounds are both part of M_o, which we have tried to subtract from our data. Any remaining 'structure,' except for the truncation effects, represents a failure of our background subtraction procedure, and while we might be able to attribute a feature which is hard to remove as having some physical cause, it is risky to discuss physical causes for features we have tried to remove!

Whatever, guys. We're not going to analyze that part of the Fourier transform anyway, so I don't particularly care why it looks the way it does.

In practice, however, that is not usually done at this stage. For linear combination analysis or principal component analysis, a factor that affects all spectra in identical ways is irrelevant to the analysis, and can be left out. For curve fitting to a theoretical standard, $B_i(E)$ is often addressed in a subsequent step known as the *McMaster correction*, described in Section 13.1.3.3 in Chapter 13.

Also note that $\dfrac{B_i(E)}{\Delta B_i(E_o)}$ is exactly 1.00 at the edge by definition. It then gradually decreases above the edge. Therefore, $\dfrac{B_i(E)}{\Delta B_i(E_o)}\chi(E)$ is similar to $\chi(E)$, except that the oscillations of $\dfrac{B_i(E)}{\Delta B_i(E_o)}\chi(E)$ are somewhat smaller than those in $\chi(E)$ when well above the edge.

For these reasons, EXAFS users occasionally call $\dfrac{B_i(E)}{\Delta B_i(E_o)}\chi(E)$ the 'experimentally determined $\chi(E)$' as opposed to the 'theoretical $\chi(E)$' implicitly defined in Section 7.3.4. Both are generally shortened to just $\chi(E)$ in practice, with context hopefully making it clear which is meant.

7.4.3 Converting from E to k

This part is easy. Basic physics tells us that

$$k = \frac{1}{h}\sqrt{2m_e\left(E - E_o\right)} \tag{7.12}$$

where m_e is the mass of the electron.

While your data analysis software will convert from a function of E to a function of k at the click of a button, it's often useful to be able to make a rough estimate in your head. For that purpose, it's good to know that $k = 10$ Å^{-1} is roughly 400 eV above the edge. Since energy above the edge is proportional to k^2, that means 20 Å^{-1} is about 1600 eV above the edge, 5 Å^{-1} is about 100 eV above the edge, and so on.

7.4.4 A Second Chance at Self-Absorption Correction

In Section 7.3.5, we discussed the possibility of correcting for self-absorption. Because EXAFS is a small variation in total absorption, it is somewhat easier to derive an accurate self-absorption correction for $\chi(k)$ than it is for the normalized absorption. One well-known correction of this type is given in Booth and Bridges 2005. A good brief review of self-absorption correction methods is given by Klementiev (2010).

7.4.5 Weighting χ(k)

$\chi(k)$ typically decreases in amplitude at high k. For that reason, the data are usually weighted by k, k^2, or k^3 for purposes of both presentation and analysis. In other words, rather than $\chi(k)$ being used directly, $k\chi(k)$, $k^2\chi(k)$, or $k^3\chi(k)$ is used. This is colloquially referred to as using a *k-weight* of 1, 2, or 3, respectively.

There is no single guideline for which k-weight to use in a given circumstance. For a discussion of how the members of our panel decide which k-weight to use, see Box 15.2 in Chapter 15. For a way of using k-weight to help distinguish between scattering species, see Section 13.3 in Chapter 13.

7.5 Finding the Fourier Transform

7.5.1 About Fourier Transforms

As discussed in Section 1.4.2 in Chapter 1, a Fourier transform is a way of picking out periodic signals from data. Mathematically, let us preliminarily define the EXAFS Fourier transform with k-weight w as

$$\tilde{\chi}(R) \equiv \frac{1}{\sqrt{2\pi}} \int_{-\infty}^{+\infty} k^w \chi(k) e^{i2kR} dk \qquad (7.13)$$

Readers familiar with Fourier transforms in other contexts will notice the usual definition has been tweaked for ease of use with EXAFS. In particular, the kernel is e^{i2kR} rather than e^{ikR}, since $2kD$ is the argument of the sine function in the motivational derivation we used in Section 1.2 in Chapter 1. Equation 7.13 is not the only definition we could have used, and in fact several variations have been used by members of the EXAFS community. But the one given in Equation 7.13 is particularly common.

7.5.2 Data Ranges Are Finite

While we can imagine $\chi(k)$ extending over all real k (or at least all positive k), in practice we only have a finite amount of usable data. So we would like to replace Equation 7.13 with a finite Fourier transform:

$$\tilde{\chi}(R) \equiv \frac{1}{\sqrt{2\pi}} \int_{k_{max}}^{k_{min}} k^w \chi(k) e^{i2kR} dk \qquad (7.14)$$

Can we make the $\tilde{\chi}(R)$ defined by Equation 7.14 compatible with Equation 7.13? Certainly. One way to do it is to define the value of $\chi(k)$ as zero when outside of the interval from k_{min} to k_{max}. If that is done, Equations 7.13 and 7.14 clearly give the same $\tilde{\chi}(R)$. This is a very common method of envisioning what is happening when we take a finite Fourier transform. It is not, however, particularly intuitive, at least at first. What 'frequencies' are present in a function which is zero everywhere except over a small range? Answering that leads to discussions of convolutions of functions, 'boxcars,' and 'sinc' functions.

So we're going to use a different method for visualizing the effect of evaluating a Fourier transform over a finite interval.

Consider, first, what you would think if you saw Figure 7.17, which shows a function over a finite interval from 3 to 10 without being able to see the function below 3 or above 10.

If you had to guess, what would you expect the function looks like below 3 and above 10? Probably you'd come up with Figure 7.18.

In other words, you might expect the function to 'keep doing what it was doing.'

The '~' on top of $\tilde{\chi}(R)$ is there as a reminder that $\tilde{\chi}(R)$ is a complex function, with a real and an imaginary part. We will often drop this reminder, and use $\chi(R)$ and $\tilde{\chi}(R)$ interchangeably.

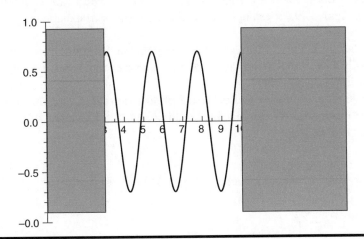

Figure 7.17   What's behind the gray boxes?

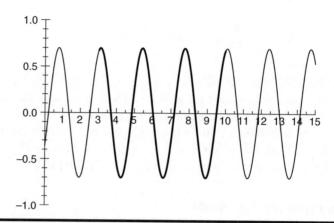

Figure 7.18   A reasonable guess as to what was hidden in Figure 7.17: a periodic extension of the function.

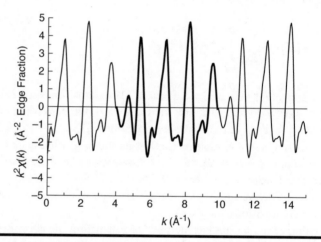

Figure 7.19   A periodic extension of $\chi(k)$ for a metallic iron spectrum.

We were inspired to use a Fourier transform in the first place because we expected $\chi(k)$ to be a sum of quasi-periodic signals. Yes, we know they are not exactly periodic—for one thing, there are a number of factors which decrease the amplitude with k. On the other hand, if we k-weight the data, we may be roughly compensating for that decrease.

Since we have a quasi-periodic function over a finite range, a very crude way to extrapolate it out to lower and higher k is to just have it repeat (this is called *periodic extension*). Figure 7.19, for example, shows the same crude extrapolation applied to $\chi(k)$ data for an iron foil.

Superficially, it looks like a reasonable extrapolation. It's almost certainly not what $\chi(k)$ actually does in the extrapolated region, but on the other hand, it surely wouldn't suddenly drop to zero, either.

Please note that the y-axis tells us that the data are plotted using a k-weight of 2.

How does the expression in Equation 7.13 relate to that in Equation 7.14 for this kind of extrapolation? We can imagine breaking Equation 7.13 into a sum of integrals of the repeated section:

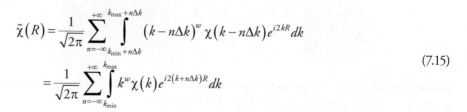

$$\tilde{\chi}(R) = \frac{1}{\sqrt{2\pi}} \sum_{n=-\infty}^{+\infty} \int_{k_{\min}+n\Delta k}^{k_{\max}+n\Delta k} (k-n\Delta k)^w \chi(k-n\Delta k) e^{i2kR} dk$$

$$= \frac{1}{\sqrt{2\pi}} \sum_{n=-\infty}^{+\infty} \int_{k_{\min}}^{k_{\max}} k^w \chi(k) e^{i2(k+n\Delta k)R} dk$$

(7.15)

where $\Delta k = k_{max} - k_{min}$. The integrals aren't identical, because while we can construct the function so that the $k^w \chi(k)$ part repeats, the phase of the kernel, $e^{i2(k+n\Delta k)R}$, will, in general, be different for different integrals in the series. In fact, over an infinite sum, that means the integrals will usually cancel out.

But there are some exceptions. If $2R\Delta k$ is a multiple of 2π, then the phase term will be the same for all integrals, and they will not cancel. In fact, if there were N repeats, the result would be just N times the integral in Equation 7.14. This is just a scaling factor—for those select values of $2R\Delta k$, the shape of the Fourier transform is identical whether we consider $\chi(k)$ to be zero outside the interval we're using or whether we consider it to repeat.

Let's take the condition that $2R\Delta k$ is a multiple of 2π and solve it for R:

$$R = \frac{\pi}{\Delta k} j \quad \left(j \, an \, \text{integer}\right) \tag{7.16}$$

The fact that the shape of the Fourier transform at those values of R is the same whether we assume $\chi(k)$ repeats or goes to zero suggests that we just don't have enough information to know what happens on a scale finer than that in R. For instance, if we Fourier transform a $\chi(k)$ from 3 to 10 Å^{-1}, it is unreasonable to think we could resolve peaks which were much closer than $\dfrac{\pi}{(10-3)} \approx 0.4$ Å apart.

Figure 7.20 plots the amplitude of the Fourier transform of the data from Figure 7.19 in a way which emphasizes this limited knowledge.

Compared with the way the Fourier transforms of EXAFS are usually drawn, that's not much to go on. But it, along with the similarly sparse plot of the phase of the Fourier transform, is really all we have. The smooth plots of peaks and shoulders we usually see are a great aid to the eye (for instance, it's hard for us to take in the significance of the point just below 5 Å being a bit higher than the point just below 4 Å), but they don't actually contain more physically meaningful information!

You may have noticed that for pedagogical reasons we chose the k range in Figures 7.18 and 7.19 so that the repeating function didn't have a sharp discontinuity. But is that actually useful to analysis as well? As a first attempt at understanding that issue, Figure 7.21 takes the simple sine wave from Figure 7.18, but chooses a different subset of the data to repeat.

Figure 7.21 has a funny sharp double peak in it around 3 and 10.5. How does that affect the Fourier transform? Figure 7.22 compares the Fourier transform of the functions shown in Figures 7.21 and 7.18.

What's 'much closer than 0.4 Å apart'? There's a degree of arbitrariness in all peak resolution criteria, but many describe the ability of EXAFS to resolve peaks to be half the spacing found by stepping Equation 7.16 by one integer, in this case giving 0.2 Å. This factor of one half is in the spirit of the criterion for peak resolution first proposed by Lord Rayleigh, although he himself simply wrote that the limit he specified was 'a state of things inconsistent with good definition' (Rayleigh 1874).

Keep in mind that resolution is *not* the same as precision. We could tell if a peak position shifted by 0.05 Å (say, from one sample to the next) even if the resolution was 0.2 Å, because the relative heights of all the points around the peak would change to reflect the shift. But a resolution of 0.2 Å does mean there's no hope of catching a dip in the Fourier transform that was less than 0.2 Å or so across, because the dip would just get averaged into the surrounding peaks.

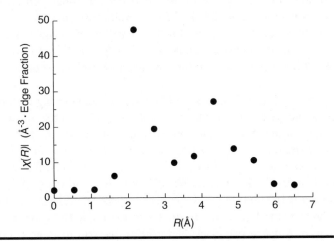

Figure 7.20 The Fourier transform of the spectrum shown in Figure 7.19.

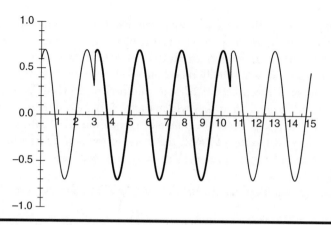

Figure 7.21 **Periodic extension of a different segment of sine wave from that shown in Figure 7.18.**

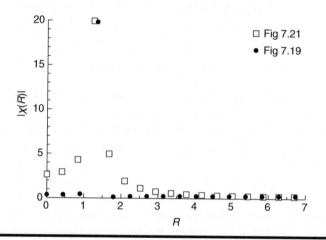

Figure 7.22 **Effect of choice of endpoints on Fourier transform of a sine wave.**

The transform of Figure 7.18 shows only one value of R for which the magnitude of the Fourier transform is significantly different from zero; that corresponds to the single frequency sine wave in the original function. The transform of Figure 7.21 shows the same peak, but with a substantial broadening; the Fourier transform is trying to handle the sharp features corresponding to the double peak.

It is important to realize that this is *not* an artifact of using a repeating function, because we already know we would get the same result by assuming the function was zero outside of the selected interval. Rather, it's an artifact of the interval we chose—it doesn't contain a whole number of oscillations of our periodic function.

Let's change the interval we use for our iron foil, and see if something similar happens.

The mismatch in Figure 7.23 is actually pretty severe, with big discontinuities at the repeats, but our eye doesn't pick it up as easily as it did with the sine wave. That, as we'll see, is a clue to the effect on the Fourier transform, shown in Figure 7.24.

This time, it's hard to summarize what happened. The peak around 4.4 Å actually appears to have become *sharper*, since it falls further at 4 and 5 Å. But, there seems to be a small peak around 5.6 Å that wasn't evident before, and the peak around 2.2 Å seems to be roughly the same width. Of course, inspection of the graph is made more difficult by the fact that Equation 7.16 directs us to sample the Fourier transform at different values of R in the two cases. Nevertheless, we can imagine a single smooth curve going roughly through both sets of data for the peak near 2.2 Å, but that doesn't

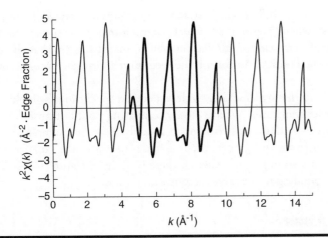

Figure 7.23 A slightly different choice of endpoints for the periodic extension shown in Figure 7.19.

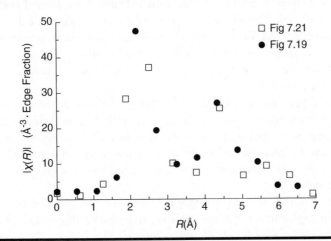

Figure 7.24 Effect of choice of endpoints on Fourier transform of a spectrum of iron metal.

seem to be the case for the peak(s) above 4 Å. So, does the peak around 5.6 Å belong there, or not? And why didn't everything just get broader, the way it did with the sine curve?

To understand the answer to these questions, go back to the EXAFS equation (on the inside front cover) or to the initial discussion in Section 1.2 in Chapter 1. $\chi(k)$ is a sum over paths. While each path is quasi-periodic, the periods of the paths in k-space are different. When we take the Fourier transform, we are doing it in part so that we can separate the contributions of different paths.

Each path, then, does behave a bit like the sine wave did, giving a narrower Fourier transform when the interval chosen contains a whole number of its periods. But, if we satisfy this condition for one path, we are unlikely to have satisfied it for another, as different paths have different periods. That's why our eye doesn't particularly object to Figure 7.23 the way it does to Figure 7.21, even though there's a big discontinuity—it may very well be that Figure 7.23 shows a more continuous pattern at *some* frequencies than Figure 7.19, and our eye (really our brain, of course), does have some intuitive ability to do that kind of pattern analysis.

So that may be why the peak near 4.4 Å got narrower—we may actually be giving that particular frequency a smoother repeat. But in doing so, other peaks could get broader, and in general it's very difficult to judge whether we came out ahead.

The considerable differences between the magnitude graphs of the two Fourier transforms shown in Figure 7.24, which stem entirely from a very modest change in the range of data selected for the transform, might suggest that the game is up: what good is the Fourier transform if it depends on arbitrary choices we make during processing?

The answer is that we should *never* draw detailed conclusions by looking at a single EXAFS Fourier transform. We are *always* comparing Fourier transforms, whether it is a sample to a standard, or two samples to each other, or a sample to a theoretical calculation. Since we will process each of those spectra in the same way, we will be all right. It is not so important whether there is 'really' a feature at 5.6 Å, so long as spectra corresponding to the same structure all produce it.

7.5.3 Windows

Makers of keyboard instruments such as harpsichords and pianos faced a similar problem hundreds of years ago. The periods of different notes can't be made to line up, just as the periods of different paths can't. Their eventual solution was to distort all notes a bit, so that no musical key is perfect for a given melody, but none is awful—a solution known as *equal temperament*.

Although we will process all spectra we wish to compare in the same way, we are still facing a somewhat unsatisfactory state of affairs. What if, for instance, we are comparing the Fourier transform of data from a sample to the Fourier transform of a theoretical calculation, but we have not quite aligned the energies of the two spectra properly. (In Section 13.1.5 of Chapter 13, we'll learn that this is, in fact, a nontrivial problem.) That relatively modest misalignment might cause us to inadvertently select modestly different intervals in $\chi(k)$, which in turn could lead to substantial differences in the Fourier transforms!

In addition, broadening is at odds with one of the reasons we wanted to take a Fourier transform in the first place, causing paths to 'leak' into each other.

The solution to this problem is to *smoothly* reduce all paths to zero, or at least to a small value, at the endpoints of the interval chosen. That will distort *all* paths a bit, but will avoid sharp discontinuities. It will also avoid having a sharp dependence on the interval chosen.

To accomplish this, we multiply $\chi(k)$ by a *window* function that has the desired properties of being small at the endpoints and being smooth throughout. Two popular window functions for EXAFS are shown in Figure 7.25.

Both functions have adjustable parameters so that they can be made more or less sharply peaked; the versions shown are typical.

Figure 7.25 raises another point. Windows work by suppressing data near the endpoints of the region selected for analysis. But it is of course easier for a given number of free parameters to yield a good fit when the data set is smaller. An incautious

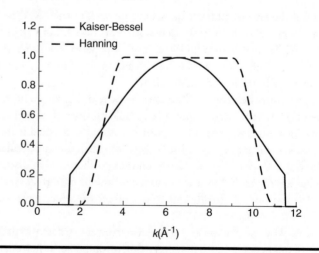

Figure 7.25   Two examples of window functions.

application of windows, by decreasing the importance of data near the endpoints, could therefore create misleading effects in fitting statistics. A solution is to consider windowed data to effectively cover a smaller range than unwindowed data; for example, both windows shown in Figure 7.25 have about the same amount of data for statistical purposes as unwindowed data ranging from 3 to 10 $\mathring{A}^{-1}$. Taking this a step further, let us agree to describe the windows in Figure 7.25 as both being 'applied to a data range' of 3 to 10 $\mathring{A}^{-1}$. Under this convention (nearly universally used in the literature), windows not only suppress data near the endpoints but also 'borrow' some data from just outside the endpoints.

Analysis software usually incorporates this convention! That means you have to be careful when you choose the range of data to Fourier transform: you may think you are cutting off the data just before a big glitch or just after a problem with the background, for instance, but you are actually including a little of the problem area! Be sure to check where the window actually begins and ends!

7.5.4 Zero Padding

Some computer algorithms for computing *fast Fourier transforms* require data sets with a number of equally spaced points that is an integral power of 2: 256 points, perhaps, or 512. The 'equally spaced' part is no problem for EXAFS data, as we can just rebin and/or interpolate data to fall on an equally spaced grid in k-space (0.05 $\mathring{A}^{-1}$ is the spacing typically used). In fact, most EXAFS analysis software does this rebinning automatically when converting from E to k.

Getting an integral power of 2 for the number of points is a bit trickier, though. One option would be to use a grid spacing that depends on the size of the range chosen. For instance, the unwindowed data from 3 to 10 $\mathring{A}^{-1}$ could be placed on a grid $\frac{10-3}{127} \approx 0.0551$ $\mathring{A}^{-1}$ apart, so that there were 128 points in the data set and the data spacing was about 0.05 $\mathring{A}^{-1}$. But that's a bit unwieldy, as then the data would have to be rebinned and interpolated every time the range or the window was changed.

Why 0.05 $\mathring{A}^{-1}$ for the spacing? I'll answer in Section 7.5.8, once we've talked a bit about back-transforms.

Instead, most software packages opt to *zero pad* the data; that is, to add enough zeroes to the end of the $\chi(k)$ data to bring the total number of points to some convenient power of 2.

In addition to the computational advantages, zero padding also results in a finer grid in R, in accord with Equation 7.16. For instance, zero padding to a total of 2048 points spaced 0.05 $\mathring{A}^{-1}$ apart results in a grid with a spacing of $\frac{\pi}{(2048)(0.05)} \approx 0.03068$

$\mathring{A}$. This, of course, does not result in an actual improvement in resolution, but can make it easier to visually interpret the resulting graph. For example, Figure 7.26 shows the two unwindowed Fourier transforms from Figure 7.24, but with zero padding to 2048 points.

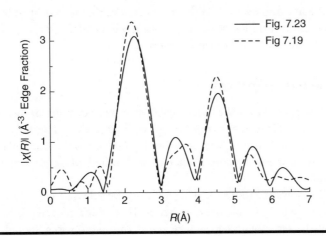

Figure 7.26 **The transforms shown in Figure 7.24, but with $\chi(k)$ zero padded to 2048 points.**

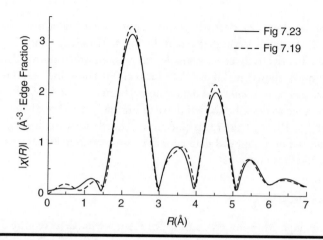

Figure 7.27   The transforms from Figure 7.26, redone using a Hanning window.

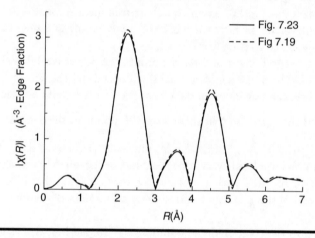

Figure 7.28   The transforms from Figure 7.26, redone using a Kaiser–Bessel window.

A careful comparison of Figure 7.26 with Figure 7.24 will show that zero padding has had other effects aside from filling in smooth features between the points. That is to be expected; rather than the repeating function shown in Figure 7.19 or Figure 7.23, we are now transforming a repeating function that has a large flat section in it, and thus the discontinuities at the terminations will be different from before.

It is, however, much easier to visually take in the versions that have been zero padded. There are some disconcerting differences between the two zero-padded traces: the peak near 2.2 Å, for instance, seems to show a modest shift, and the peak between 6 and 7 Å is only present in one of the transforms.

7.5.5 Choice of Windows

Windowing is supposed to make us less sensitive to small changes in the range of data we select. Let's see if it works on our zero-padded data. First, Figure 7.27 shows the result of applying the Hanning window from Figure 7.25 to the iron foil data.

The two are now much more similar to each other! Mark up a victory for the Hanning window.

Figure 7.28 shows the result of trying a Kaiser–Bessel window.

In this case, Kaiser–Bessel does even better! That's not a general result. Sometimes Hanning does better. But it *is* a general result that windowing should make your transform less sensitive to small changes in the endpoints of the data range being transformed.

BOX 7.3 'Which Window Should I Choose?'

	OK. So which window should I choose? My analysis software gives me lots of choices: Kaiser–Bessel, Parzen, Welch, Gaussian, Hanning, Hamming, Norton–Beer, and more. And then even once I choose a window type, there are parameters that change how quickly the window drops near the endpoints.
	One could follow a procedure like the one used above, my dear Simplicio. Simply try different windows on the data, vary the endpoints of the range being transformed slightly, and see which windows show the least sensitivity to those endpoints. As for the parameters that affect the sharpness of the drop, those should be chosen so that the drop is as sharp as possible consistent with removing sensitivity to the endpoints.
	That's an interesting approach, Robert. But it could confuse people looking at your data, say in an article or presentation. If you change the window you're using from data set to data set, then the Fourier transforms of those sets will be more difficult to compare visually. In fact, that's probably the best justification for Simplicio's software having so many choices. If you're trying to compare your data to a figure published in the literature, it's best to use the same window they did, as well as the same *k*-weight and endpoints.
	Meh. Why do you have to make things so complicated, Robert? If your results depend on what window you used, you don't have very robust results anyway. I just use a Hanning window with 1.0 Å sills most of the time. If the data are pretty noisy, but extend up far in *k*, then I might use a Kaiser–Bessel window with sharpness parameter 3 so that I can emphasize the cleanest part of the data most while still capturing the noisy high *k* part.
	Why stop there, friend Dysnomia? We could always use the same window—perhaps your Hanning window with 1.0 Å sills. That way all of our data could be compared with ease.
	YAWN. This is perhaps the least interesting discussion we've had yet. Can we move on so that we can get to some actual science?

7.5.6 Fourier Transforms Are Complex

An examination of Equation 7.14 reveals that a Fourier transform is a complex function, with a real part and an imaginary part or, equivalently, a magnitude and a phase. So far, all the plots we've shown in this chapter have been of the magnitude of the function. In fact, observant readers may have noted that we labeled the *y*-axis $|\chi(R)|$—the *R* tells us that we're plotting the Fourier transform, and the bars designate that we're plotting the magnitude of the function.

But what does the phase look like? Figure 7.29 shows the phase corresponding to Figure 7.23, with the same windowing and zero padding as was used to generate Figure 7.28.

The primary trend is a line of roughly constant negative slope, but with several discontinuities. This suggests that plotting the derivative of the phase might be χ(*k*), as we originally constructed it, is real. But the definition of a Fourier transform suggests that we should be operating in a complex space, and that χ(*k*) itself should have a phase. One's choice of that phase is arbitrary, as it does not vary with *k*; that is to say, it is constant. One common convention is to choose the phase so that the imaginary part of χ(*k*) is zero, which is what we have done in constructing Figure 7.29. Another convention in use is to choose the phase so that χ(*k*) is imaginary. The only effect of this choice on Figure 7.29 is to translate the plot up or down.

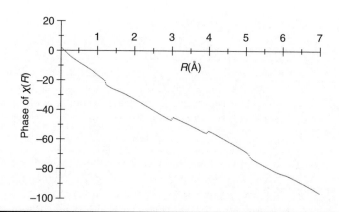

Figure 7.29 The phase corresponding to the magnitude labeled 'Figure 7.23' in Figure 7.28.

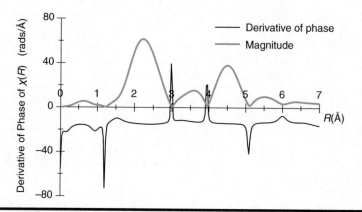

Figure 7.30 The derivative of the phase plot from Figure 7.29 along with the corresponding magnitude.

Note that the derivative of the phase of the Fourier transform is independent of this initial arbitrary choice, which is another good reason for looking at it!

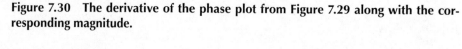

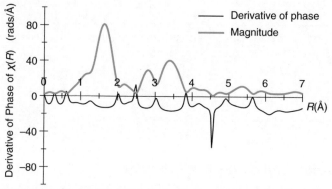

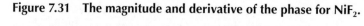

Figure 7.31 The magnitude and derivative of the phase for NiF$_2$.

Magnitude and phase do not have the same units (Section 18.3.2 in Chapter 18 for a discussion of the units of magnitude). An arbitrary scaling factor has, therefore, been applied to the magnitude in Figures 7.30 and 7.31 to facilitate visual comparison to the phase.

interesting. Figure 7.30 shows a plot of the derivative of the phase along with the magnitude (which was already plotted in Figure 7.28).

It is evident that the discontinuities in phase correspond to minima in the magnitude, with a roughly constant negative derivative in between.

Figure 7.31 gives another example, this time for the nickel *K* edge of nickel (II) fluoride.

This time, the segments of negative derivative between the spikes show more structure, but the general pattern is similar.

There is another way to represent Fourier transforms, however, rather than plotting the phase and the magnitude, and that is to plot the real part and the imaginary part. The real part is the magnitude multiplied by the cosine of the phase; the imaginary part is the magnitude multiplied by the sine of the phase. Plotting either will provide us with information that is due to both phase and amplitude. Figure 7.32 shows the real part of the iron foil spectrum from Figures 7.28 and 7.29.

With a little practice, you can imagine the magnitude as a kind of envelope defining the amplitude of the oscillations in Figure 7.32: there's a big peak around 2.3 Å, a smaller one at 3.5 Å, and another bigger one around 4.5 Å. In fact, you'll sometimes see the magnitude and the real part of the Fourier transform plotted together in the literature, like in Figure 7.33.

As for the phase, the relatively flat sections of the derivative are responsible for the sinusoidal behavior of the real part within each magnitude peak. The discontinuities in phase, however, are clearly visible between magnitude peaks; the oscillations in the real part appear to 'stutter' there.

The imaginary part is qualitatively similar to the real part, but with a phase shift, as seen in Figure 7.34.

The real part and the imaginary part still provide independent information; one cannot be used to predict the other precisely. But, since the phase only changes rapidly when the magnitude is small, it *is* possible to know *roughly* what one will look like

Plots of Fourier transform phase are not seen much in the EXAFS literature, and I have never seen a plot of its derivative. Figures 7.30 and 7.31 are provocative to me. There is clearly some qualitative information that can be extracted from the magnitude (even though it's not a radial distribution function, many of the peaks do correlate to scattering shells, and in later chapters we'll show you some tricks for learning about your sample by manipulating them). Can the phase plot also be used to directly extract qualitative clues about the sample's structure? To my knowledge, little investigation has been done in this area.

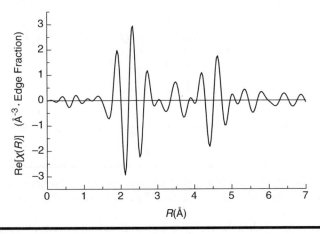

Figure 7.32   The real part of the iron foil Fourier transform from Figures 7.28 and 7.29.

For the rest of this chapter, we're going to continue to use both the NiF₂ example *and* the iron foil example. The NiF₂ spectrum is a nice example, because it is so complicated, while the iron foil is a nice example because it is relatively simple. Pay attention to the captions so you know which one you're looking at!

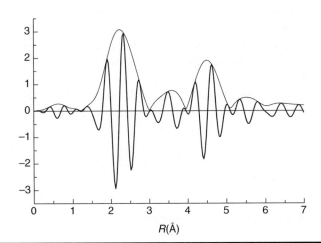

Figure 7.33   A plot showing the magnitude and the real part of the iron foil Fourier transform on the same scale.

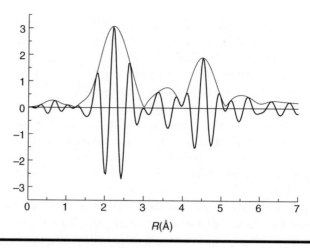

Figure 7.34 A plot showing the magnitude and the imaginary part of the iron foil Fourier transform on the same scale.

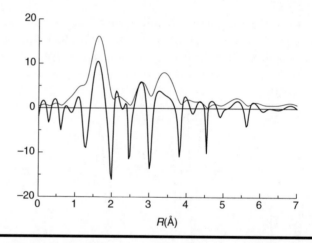

Figure 7.35 The darker line that goes below the *x*-axis is the second derivative of the magnitude for the NiF$_2$ data, multiplied by a negative scaling factor.

given the other. This makes the real part (or the imaginary part) a very useful way of presenting the Fourier transform, because it encodes most of the magnitude information and most of the phase information in one plot.

By the way, as long as we're talking about interesting ways to present Fourier transform data, here's one borrowed from infrared spectroscopy. Compute the second derivative of the magnitude of the Fourier transform, and then multiply by a negative scaling factor. Figure 7.35 shows what it looks like for the nickel fluoride data shown from Figure 7.31.

This process makes peaks appear sharper. It is important to realize that it doesn't actually improve resolution or add information; it's a visual trick. But the same can be said of most of the processing we've explored in this chapter. In this case, it clearly demonstrates that the peak between 3 and 4 Å is not symmetric, but rather has some characteristics of a poorly resolved double peak. This property is very difficult to see by inspection of the original transform!

7.5.7 'Corrected' Fourier Transforms

Since the absorbing atom is known, it's possible to mathematically remove the part of $\delta(k)$ that is due to the central atom (see Section 13.3 of Chapter 13). If there is only a single scattering species, and it is known, then all of the effect of $\delta(k)$ can be removed.

BOX 7.4   In Which We Agree Not to Use Corrected Fourier Transforms

	Great! So by correcting the Fourier transform, we can get a true radial distribution function, right?
	No! Multiple scattering and the mean free path still play a role, as do the effects of taking a finite interval discussed in Sections 7.5.2 through 7.5.5.
	In fact, I think 'correcting' the transform makes things worse. By making the Fourier transform look a little more like a radial distribution function, without it actually being a radial distribution function, you're just going to confuse people.
	As long as one explains carefully what one has done, it is scientifically valid to present a corrected Fourier transform. But I must admit, I agree with dear Dysnomia: I do not see the value in it. Do any of the rest of you have anything to say in favor of the practice?
	Not us!
	Then let us agree. We will not use corrected Fourier transforms to present our results.

7.5.8 Back-Transforms

By definition, the inverse Fourier transform of the Fourier transform of a function results in the original function, if both are performed from −∞ to ∞. If we use Equation 7.14 to define the Fourier transform, then the inverse Fourier transform is

$$k^w \chi(k) = \frac{1}{\sqrt{2\pi}} \int_{-\infty}^{+\infty} \tilde{\chi}(R) e^{-i2kR} dR \qquad (7.17)$$

The argument in this section is loosely based on Section 6.8 in the work by Teo (1986).

One will sometimes encounter $\chi(q)$ as the symbol for the back-transform. While fairly common, this notation does not emphasize that $\tilde{\chi}'(k)$ and $\chi(k)$ can and should be compared on the same graph.

An inverse transform *per se*, though, is not very useful in EXAFS analysis—since we get $\chi(R)$ from $\chi(k)$, there is little value in perfectly reversing the process.

It turns out, however, that reversals that are modified in some way, so that they do not reproduce the original $\chi(k)$ with complete fidelity, can be useful. The resulting modified $\chi(k)$ is called a *back-transform* of the data, and is symbolized in this text as $\tilde{\chi}'(k)$.

One way to motivate the idea of a back-transform is to go back to the way we developed the EXAFS equation in Section 1.2 of Chapter 1. On the basis of the way we developed the equation, $\chi(k)$ can be thought of as

$$\chi(k) = \sum A(k,R)\sin\left[\Phi(k,R)\right] \tag{7.18}$$

where the sum is over scattering paths (both single and multiple), $A(k, R)$ incorporates all the amplitude effects, and $\Phi(k, R)$ incorporates all the phase effects. But phase and amplitude are difficult to pin down for the real-valued function shown in Equation 7.18. If it exhibits a shoulder after a peak, for instance, is it because the phase advanced more slowly in that region, or because the amplitude increased there?

Equation 7.18 can be rewritten in terms of complex exponentials, though

$$\chi(k) = \sum \frac{1}{2i} A(k,R) e^{i\Phi(k,R)} - \sum \frac{1}{2i} A(k,R) e^{-i\Phi(k,R)} \tag{7.19}$$

$\chi(k)$ is still real-valued, but we're now thinking of it as a complex function that happens to have zero for its imaginary part; a conceptual shift laying the groundwork for what is to follow. (This is a common convention, but not the only one; see Robert's comment at the start of Section 7.5.6.)

Next, take a Fourier transform of the function in the usual way. The first term will yield nonzero values for positive R, while the second will yield nonzero values for negative R. In fact, since $\chi(k)$ is real-valued, $\chi(R) = \chi^*(-R)$.

Now for the key step: *throw away* the values for negative R before taking the back-transform. Neglecting the practical effects discussed in Sections 7.5.2 through 7.5.4, we'll recover *only the first term*, and thus we have a function that is now complex, with a magnitude of $\sum \frac{1}{2} A(k,R)$ and a phase of $-\frac{\pi}{2} + \sum \Phi(k,R)$. Since it's easy to extract the magnitude and phase of a complex function, we now have a way of getting $\sum A(k, R)$ and $\sum \Phi(k, R)$ from $\chi(k)$. By using the EXAFS equation, we even have an approach that can be used to estimate $f(k)$ and $\delta(k)$ from experimental data. Before the development of theoretical methods such as FEFF, this approach was used to extract $f(k)$ and $\delta(k)$ from known, 'model' compounds for analysis of unknowns. Many clever variations on this approach were developed—see Teo (1986) and Bunker (2010) for specifics. While the details of such techniques are beyond the scope of this book, they are still in use by some researchers, and, for some purposes, have advantages over more widely used approaches.

The real part of the back-transform (again, ignoring the effects discussed in Sections 7.5.2 through 7.5.4 for the moment) is given by

$$Re\left[\sum \frac{1}{2i} A(k,R) e^{i\Phi(k,R)}\right] = \sum \frac{1}{2} A(k,R)\sin\left[\Phi(k,R)\right]$$
$$= \frac{1}{2}\chi(k) \tag{7.20}$$

To cancel out the factor of 1/2, $\tilde{\chi}'(k)$ is usually defined with an extra factor of 2 out front:

$$\tilde{\chi}'(k) \equiv \frac{2}{\sqrt{2\pi}} \int\limits_{R_{min}}^{R_{max}} \tilde{\chi}(R)e^{-i2kR} dR \qquad (7.21)$$

Ignoring the effects discussed in Sections 7.5.2 through 7.5.4, if $R_{min} = 0$, and $R_{max} = \infty$, then $\text{Re}\left[\tilde{\chi}'(k)\right] = k^w\chi(k)$, which is thought to be a pleasing way of arranging things.

More important for our purposes is what happens if we choose R_{min} and R_{max} so as to select just part of the Fourier transform. For instance, slowly varying parts of $\chi(k)$ aren't EXAFS signal; they're part of the background $M_o(E)$. In fact, many background subtraction algorithms essentially use this strategy to determine $M_o(E)$. And it's not just low-frequency oscillations that you might want to filter out. Because of effects discussed in Chapters 1 and 10, the EXAFS signal decreases at large R, so rapid oscillations are probably noise rather than signal. Finally, you might want to focus on scattering paths of only a certain length; for instance, you might want to know what the signal due to just the near-neighbors looks like. Figure 7.36 shows the back-transform of the iron foil spectrum shown in Figure 7.23 using different filtering ranges.

The 0.0–10.0 Å plot is essentially unfiltered (compare to Figure 7.23). But the k-weighting and forward-transform window have now been incorporated into the function, so that it goes smoothly to zero at the left and right edges of the plot. The 1.2–3.0 Å plot corresponds to the 'nearest-neighbor' peak shown in Figure 7.28, and indeed appears to include just a single frequency. (The structure of metallic iron includes atoms at two different distances in that peak, but the swath of $\chi(k)$ we chose for the forward transform was too small to resolve those peaks.)

The 1.2–4.0 Å plot is perhaps the most interesting. It clearly corresponds to two magnitude peaks in the forward transform, but the effect of the second frequency on $\tilde{\chi}'(k)$ is somewhat subtle. While the information content is the same, it is easier for our brains to take in the information in Figure 7.28 than the trace shown here.

So why use $\tilde{\chi}'(k)$ in this way at all? Because it can be quite helpful for understanding how features in the raw data relate to the forward transform $\chi(R)$. In Figure 7.36, for instance, the small peaks near 6.0, 7.6, and 8.9 Å^{-1} don't appear in $\tilde{\chi}'(k)$ when it is filtered to include only frequencies from 1.2 to 4.0 Å. Those small bumps, then, must be key to the structure of $\chi(R)$ *above* 4.0 Å. So, if we believe those small peaks in $\chi(k)$ are legitimate, and not some kind of noise or measurement artifact, then we also gain trust in the high-R structure. If, on the other hand, we believed them to be caused by

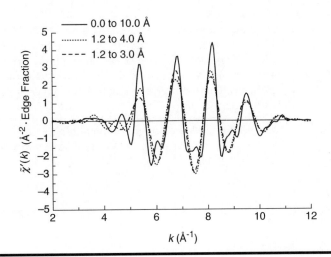

Figure 7.36   Back-transforms of the iron foil spectrum from Figure 7.23.

some glitches in the measurement (perhaps because I_o showed large glitches at those positions), then we would also have to distrust the high-R structure.

Similarly, if you are comparing two spectra (for example, data you have collected with a calculated model spectrum), $\tilde{\chi}'(k)$ allows differences to be studied as a function of k. Are the differences only at very low k? Perhaps comparable backgrounds were not used in the same cases. Only at very high k? Perhaps signal to noise at high k is an issue. Somewhat throughout the spectrum, but more severely for higher k in $\tilde{\chi}'(k)$? Perhaps the nearest-neighbor light elements match, but more distant shells of metal atoms are different (see Section 13.3 in Chapter 13). And so on … back-transforms can be quite useful as diagnostic tools!

7.6 Wavelet Transforms

Now, at last, we can see why collecting or binning $\chi(k)$ data at 0.05 Å⁻¹ intervals is oversampling. If we back-transform $\chi(R)$ data from 0 to 10 Å, the logic of Section 7.5.2 tells us that we cannot resolve $\tilde{\chi}'(k)$ peaks closer than $\frac{1}{2}\frac{\pi}{(10-0)} \approx 0.16\text{Å}^{-1}$ apart.

Since all the physical information about the system is likely contained within the 0–10 Å range, it must not be necessary to sample $\chi(k)$ on a finer grid than that. Therefore, 0.05 Å⁻¹ represents a mild degree of oversampling, and the claim in Section 7.3.2 that a one-point glitch can be removed without disrupting the analysis is verified.

BOX 7.5 How Does the Fourier Transform Depend on the Type of Scattering Atom?

	You've convinced me that $\chi(R)$ is not the same as the radial distribution function around the absorbing atom. But they're related—peaks in the radial distribution function will usually correspond to peaks in $\chi(R)$, for example, although they'll be shifted by a few tenths of an angstrom, and there might be some peaks in $\chi(R)$ that come from multiple scattering or are artifacts of data reduction.
	Right! We can't read bond lengths straight off of a Fourier transform, but we usually have a general idea which peaks come from which scattering shells.
	OK…but sulfur is going to scatter electrons a lot differently than oxygen. What part of the Fourier transform is showing us the difference between different kinds of scattering atoms?
	The atomic number of the scattering atom affects the amplitude of the Fourier transform, so that heavier atoms will tend to create stronger peaks. In addition, and perhaps more importantly for distinguishing scatterers, different species will have different impacts on the phase of the Fourier transform.
	Are there things I can look for in the phase to tell me what kind of scatterer I might have? Without having to actually use a model, I mean.

Not reliably. I usually go back to $\chi(k)$ or maybe $\tilde{\chi}'(k)$. We'll talk more about this in Section 13.3.

I think wavelet transforms might be the kind of thing Simplicio is looking for. They do a better job than Fourier transforms at visually separating scatterers not just by the absorber–scatterer distance but also by atomic number. Let's conclude this chapter by explaining how they work and how to use them.

7.6.1 Comparison of Fourier and Wavelet Transforms

Before we dive into the intricacies of wavelet transforms, let's take one more look at Fourier transforms so that we can understand how wavelet transforms are similar, and how they differ.

When we perform a Fourier transform, we multiply our data function, $k^w\chi(k)$, by a *basis function*, e^{i2kR}, integrating over all values of k. This then yields a complex-valued function of R. For Fourier transforms, these basis functions are perfectly periodic, and are thus well-chosen to pick out periodic patterns in our data function.

Since different scattering elements will tend to contribute most to $\chi(k)$ in different ranges of k (see Section 13.3), we can try to achieve Simplicio's goal by choosing basis functions that select out a single region of k-space at a time, while still having enough of a quasi-periodic character to be sensitive to EXAFS-type signals. This kind of basis function is called a *wavelet*. For EXAFS analysis, the most common function chosen is based on the *Morlet wavelet* (see, for example, Funke et al. 2005), given by the equation:

$$\Psi(k) = \frac{1}{\sqrt{2\pi}s}\left(e^{i\eta k} - e^{-\eta^2 s^2/2}\right)e^{-k^2/\left(2s^2\right)}$$ (7.22)

Most references use the symbol σ in the equation for the Morlet equation, rather than s. We have chosen to use s to avoid confusion with the mean square radial displacement found in the EXAFS equation (Equation 1.17).

The Morlet wavelet has a periodic term, $e^{i\eta k}$, which looks much like the basis function for the Fourier transform, with the parameter η standing in for $2R$. But this function is then multiplied by a Gaussian factor, $e^{-k^2/(2s^2)}$, localizing it to a specific region of k-space with a width related to the parameter s.

$e^{-\eta^2 s^2/2}$ ensures the wavelet transform fulfills the *admissibility criterion*, (Funke et al. 2005). For typical values of η and s, it's small enough to be negligible.

The real part of the Morlet wavelet is shown in Figure 7.37, for four different combinations of the parameters η and s. As expected, η controls how rapidly the wavelet oscillates, while s controls how much k-space is effectively sampled.

As given in equation 7.22, the Morlet wavelet would focus on the low-k part of $\chi(k)$; in that sense, it would tend to run counter to k-weighting (Section 7.4.5). It also would tend to pick out frequencies corresponding to $R = \eta/2$ in $\chi(R)$. In order to fulfill Simplicio's goal, we need to be able to center it on different regions of $\chi(k)$, and make it sensitive to different frequencies.

For a review of some alternatives to Morlet wavelets, see Xia et al. (2018).

To accomplish both these goals, we translate and dilate the original wavelet (the *parent function*) to create new wavelets $\psi(k')$ (*child functions*):

$$\psi(k') = \frac{1}{\sqrt{a}}\Psi\left(\frac{k'-k}{a}\right)$$ (7.23)

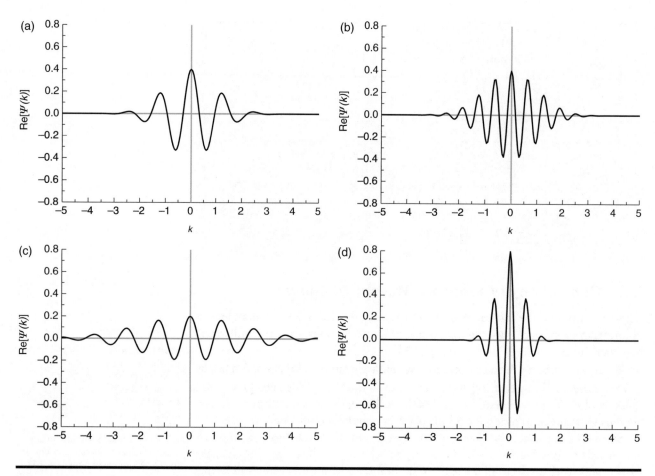

Figure 7.37 **Real part of the Morlet wavelet. (a)** $\eta = 5$ **and** $s = 1$. **(b)** $\eta = 10$ **and** $s = 1$. **(c)** $\eta = 5$ **and** $s = 2$. **(d)** $\eta = 10$ **and** $s = 0.5$.

The factor of $\frac{1}{\sqrt{a}}$ is used to normalize the child function.

k, of course, is the center of the child wavelet, while a measures how much the parent function has been stretched along the k-axis. Since the parent function picks out frequencies of $R_m = \eta / 2$, the child function will pick out frequencies of $R = \frac{R_m}{a} = \frac{\eta}{2a}$. Rearranging, $a = \frac{\eta}{2R}$, so that

$$\psi(k') = \sqrt{\frac{2R}{\eta}} \Psi\left(\frac{2R(k'-k)}{\eta}\right) \tag{7.24}$$

This then leads to the *wavelet transform* used in EXAFS analysis:

$$W(k,R) = \sqrt{\frac{2R}{\eta}} \int_{k_{min}}^{k_{max}} k'^{w} \chi(k') \Psi^*\left(\frac{2R(k'-k)}{\eta}\right) dk' \tag{7.25}$$

Note the following features of the wavelet transform:

■ Like the Fourier transform, the wavelet transform results in a complex-valued function.
■ Unlike the Fourier transform, the wavelet transform results in a function of *two* variables, k and R.

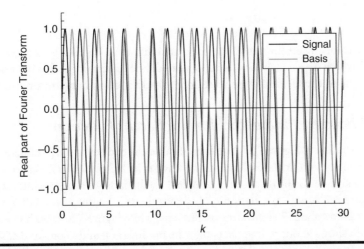

Figure 7.38 **Comparison of a simple sinusoidal 'signal' with the real part of a basis function used for a Fourier transform. The basis function used here has a slightly different frequency from the signal, causing it to come in and out of alignment.**

- The wavelet transform and the Fourier transform depend on R in similar ways, yielding similar interpretations and caveats.
- The dependence of the wavelet transform on k relates to the atomic number of the scattering atom. While this information is implicitly present in the complex Fourier transform, the wavelet transform makes it a function of a separate variable, aiding in visual representations and analysis.

7.6.2 The Uncertainty Principle

For a strictly periodic function of k' such as $\cos(\omega k')$, a Fourier transform taken over all values of k' will yield a single positive spike corresponding to the frequency ω. That's because only a basis function with the same frequency as the signal will consistently align with it over all k'. Any other function will sometimes align, contributing positively to the integral, and sometimes be opposite, contributing negatively. Over all k', the effects cancel out, yielding zero for the value of the integral. Figure 7.38 illustrates this effect: around k = 8, for example, the basis function and signal align pretty well, contributing positively, but around k = 16 they are nearly perfectly out of alignment, contributing negatively, only to come back into alignment again around k = 24, and so on.

But when, as is the case in EXAFS analysis, the Fourier transform is taken over a limited range of k', basis functions with frequencies *close* to that of the signal will still align pretty well over the range being sampled. So if, for example, the k range for the functions shown in Figure 7.38 was chosen to be 5 to 11, the product of the signal and the basis function would be contributing positively much more often than they were contributing negatively. Thus, as the frequencies of the basis function were varied, the magnitude of the Fourier transform would no longer be a spike, but rather a rounded peak with positive contributions from basis functions with frequencies *close* to that of the signal. The narrower we make our k' range, the greater the range of frequencies that will keep the alignment similar over most of the range, and the wider the peaks we get in R-space. The same will certainly be true of a wavelet transform: the more narrowly we sample the spectrum around a particular value k, the wider the peaks that we get in R-space.

If this sounds familiar, it's because we came to a similar conclusion using a somewhat different argument in Section 7.5.2. Based on Equation 7.16, we can reasonably write

You'll get a spike at $-\omega$ as well, but for EXAFS analysis that's not physically meaningful.

Recall that the basis functions for Fourier transforms are actually $e^{-i\omega t}$. Thus for a signal of the form $\chi(k) = \cos(\omega t)$ the real part of the basis function aligns, while for $\chi(k) = \sin(\omega t)$ the imaginary part is in alignment.

$$\sigma_R \sigma_k = \alpha \tag{7.26}$$

Where σ_R is the uncertainty in R, σ_k is the uncertainty in k, and α is a constant on the order of 1 which depends on how we choose to define uncertainty. This is a mathematical analogue to the Heisenberg uncertainty principle found in quantum mechanics.

7.6.3 Choosing Parameters for the Parent Function

Take another look at Figure 7.37. You'll notice that the choices of the parameters η and s affect how many significant oscillations the wavelet includes. Figure 7.37a, for instance, with $\eta = 5$ and $s = 1$, shows about 3 complete oscillations with significant amplitude, while Figure 7.37b, with $\eta = 10$ and $s = 1$, shows about 7. Figure 7.37c, with $\eta = 5$ and $s = 2$, again shows about 7 significant oscillations, while Figure 7.37d, with $\eta = 10$ and $s = 0.5$, again shows about 3. It appears that the product of η and s control the 'shape' of the parent function; this can also be seen by careful consideration of Eq. 7.22.

Since higher values of ηs have more oscillations, they will be better at picking out one frequency over another; that is, they will be better at selecting for a given value of R. On the other hand, since s is a measure of the width of the parent function, smaller values of s correspond to smaller uncertainties in the k being sampled.

It should be clear that σ_k is directly proportional to s, with the proportionality constant depending only on the definition of uncertainty chosen, but on the order of 1.

You might have a different opinion about how many of those oscillations are 'significant.' That's OK! The point is that two parent wavelets with the same ηs show the same number of significant oscillations, no matter what reasonable criterion you use.

But what about σ_R?

Suppose we imagine a sampling window stretching from $k-s$ to $k+s$. In that interval, which is $2s$ wide, a periodic signal with frequency (in k-space) of $2R_o$ will undergo a number of oscillations equal to $\dfrac{2s}{2\pi \big/ 2R_o} = \dfrac{2sR_o}{\pi}$. If a wavelet transform is calculated for a slightly different value of R we'll call R', then as long as the number of oscillations in that same interval of the basis function with frequency $2R'$ is within about $\pm\frac{1}{4}$ of the number of oscillations of the signal with frequency $2R_o$, the wavelet transform will still yield a result with significant magnitude. A change of $\frac{1}{4}$ of an oscillation over the interval would lead to the equation:

$$1/4 = \frac{2sR'}{\pi} - \frac{2sR_o}{\pi}. \tag{7.27}$$

σ_R (the uncertainty in R) is directly proportional to $R' - R_o$, with the proportionality constant as usual depending on our definition of uncertainty. Rearranging Eq. 7.27 yields

$$\sigma_R \sim R' - R_o = \frac{\pi}{8s} \tag{7.28}$$

Since for the parent function σ_k is proportional to s, and σ_R is proportional to $1/s$, we once again have Eq. 7.26 as we would expect.

But what about for the child functions?

The translation just recenters the function, and thus doesn't have any effect on these uncertainty relationships.

Stretching the parent function by a factor of a, also increases the uncertainty σ_k by a factor of a. But following the same argument as before, that means the uncertainty σ_R is *decreased* by the same factor of a. This is true regardless of our precise definition of uncertainty.

What values, then, should we choose for s and η when performing wavelet transforms for EXAFS analysis?

We suggest you start by thinking about the product ηs. As we saw in Figure 7.37, when $\eta s = 5$ we only get roughly three significant oscillations. Choosing ηs much less than that would mean we pretty much only get the k-dependence back, with very little information on R at all, in which case there's little point in doing the transform. As we increase ηs our resolution in R-space improves, but we lose resolution in k-space; in the extreme case, we'd be back to just taking a Fourier transform. In practice, we might compute wavelet transforms for several different values of ηs, with the lower values yielding transforms that provide better resolution in k, and the higher values yielding transforms that provide better resolution in R.

Once we've chosen a value of ηs, what is the effect of a particular choice of η?

To see that, let's combine Equations 7.24 and 7.22:

$$\psi(k') = \sqrt{\frac{2R}{\eta}}\,\Psi\left(\frac{2R(k'-k)}{\eta}\right) = \sqrt{\frac{2R}{\eta}}\,\frac{1}{\sqrt{2\pi}s}$$

$$\left(e^{i\eta\left(\frac{2R(k'-k)}{\eta}\right)} - e^{-\frac{\eta^2 s^2}{2}}\right)e^{-\frac{\left(\frac{2R(k'-k)}{\eta}\right)^2}{2s^2}}$$

$$= \sqrt{\frac{1}{s}}\sqrt{\frac{R}{\pi \eta s}}\left(e^{i2R(k'-k)} - e^{-\frac{\eta^2 s^2}{2}}\right)e^{-\frac{(2R(k'-k))^2}{2\eta^2 s^2}} \tag{7.29}$$

BOX 7.6 Does The Parent Function for Morlet Wavelets have Two Independent Parameters, or only One?

	Clearly, the parent function for Morlet wavelets has two independent parameters, η and s.
	In a formal sense, sure. But the shape of the parent function is only dependent on the product ηs, which is a single parameter. And since we're going to dilate the parent function anyway, the child functions only depend on a single independent parameter in the parent (the product ηs). If $\eta = 100$ and $s = 0.05$, the result is the same as if $\eta = 5$ and $s = 1$.
	That is not quite true, my dear Carvaka. The normalization of the child function includes an extra factor of $\sqrt{\frac{1}{s}}$.

That factor just affects normalization, though. Some papers don't even give values for the wavelet transforms on their graphs—in those cases, that normalization factor literally makes no difference at all!

They provide graphs without a scale? I am appalled!

Part of the issue is that different authors may use different normalization schemes for their wavelets. We've chosen the most common in Eq. 7.25, but it's not the only possibility.

As long as the same normalization is used for whatever we're comparing (a sample to a standard, two samples to each other, or a sample to a theoretical calculation), the normalization we choose has no effect on our analysis. That's why some authors leave off the values entirely.

This is another discussion that is mostly about semantics. The product ηs affects our analysis, but the *individual* values for η and s have no effect whatsoever. Can we move on now?

Since only the product ηs matters to the analysis, let's identify some useful ηs values:

- $\eta s = 1$ is a useful lower limit. Using that value for parameterizing the parent wavelet will sample roughly one oscillation in $\chi(k)$, providing good discrimination as a function of k, at the expense of providing poor resolution as a function of R. Using lower values of the product ηs increasingly leads to reproducing $\chi(k)$ —that is, not transforming the data at all—and is thus not useful.

- $\eta s = 200$ is a useful upper limit. For EXAFS data, this parameterization will generally yield a result very similar to a Fourier transform, with no discrimination as a function of k and the best possible resolution as a function of R. Wavelet transforms using high values of ηs (perhaps 15 or above) are sometimes called *overview wavelet transforms*.

- $\eta s = 5$ samples roughly three oscillations of $\chi(k)$, making it a good compromise when you want pretty good discrimination as a function of k while also getting some information as a function of R. In contrast to an overview wavelet transform, a transform generated using $\eta s = 5$ could be called a *detail wavelet transform*.

- $\eta s = 10$ is useful if you'd like good discrimination as a function of R, while also getting some information as a function of k.

There's nothing magic about those particular values of ηs. If you'd like to use $\eta s = 7$ at some point in your analysis, for example, that's just fine! Just don't get carried away trying to fine tune these parameters to some kind of 'ideal' values.

WHAT I'VE LEARNED IN CHAPTER 7, BY SIMPLICIO

- Scans should be aligned in energy before being merged, preferably using a reference.
- There are many valid ways to pick an E_o for a spectrum. The key is to be consistent and to be able to easily describe how the choice was made.
- *Normalizing* a spectrum means dividing by its edge jump and shifting it so that the edge rises from zero to one (ignoring the fine structure). It is important to estimate the uncertainty in normalization, and to use a consistent method on spectra that are going to be compared.
- It may be possible to partially correct for self-absorption mathematically, either before finding $\chi(k)$ or after, but it is better to collect data that doesn't show much self-absorption in the first place.
- To find $\chi(k)$, an appropriate *background* function must first be found. The background function should not vary so slowly that it can't follow the general trend of the spectrum, nor so rapidly that it follows the EXAFS wiggles. More than one background function may appear acceptable; the effect of different choices of background functions can be compared at later stages in the process.
- 100 eV is roughly 5 Å^{-1}, 400 eV is roughly 10 Å^{-1}, 1600 eV is roughly 20 Å^{-1}, and so on.
- Taking the Fourier transform of a finite function introduces truncation effects. Windows can be used to reduce the impact of those effects on analysis. The precise choice of window is not all that important, as long as the choice is consistent for data being compared.
- Fourier transforms can be broken into a magnitude and a phase, or a real part and an imaginary part. The real part and the imaginary part contain similar information.
- While 'correcting' a Fourier transform for phase shifts is scientifically legitimate, it probably creates as many problems as it solves, and all of my friends recommend against it.
- Back-transforms turn the real-valued $\chi(k)$ into a complex $\tilde{\chi}'(k)$. They are also useful for seeing how different ranges of R show up as features in $\chi(k)$.
- Wavelet transforms allow us to get a rough idea of the dependence of an EXAFS signal on k and R simultaneously.
- Wavelet transforms using Morlet wavelets as parent functions depend on the product of two parameters ηs. Large values (e.g. 200) provide good resolution in R and poor resolution in k, while small values (e.g. 1) provide good resolution in k and poor resolution in R, with intermediate values (often 5 to 10) yielding a compromise between resolution in k and R.

REFERENCES

Bearden, J. A. and A. F. Burr. 1967. Reevaluation of x-ray atomic energy levels. *Rev. Mod. Phys.* 39:125–142. DOI:10.1103/RevModPhys.39.125.

Belsky, A., M. Hellenbrandt, V. L. Karen, and P. Lukch. 2002. New developments in the Inorganic Crystal Structure Database (ICSD): Accessibility in support of materials research and design. *Acta Crystallogr. B.* 58:364–369. DOI:10.1107/S0108768102006948.

Booth, C. H. and F. Bridges. 2005. Improved self-absorption correction for fluorescence measurements of extended x-ray absorption fine-structure. Physica. Scripta. T115:202–204. DOI:10.1238/Physica.Topical.115a00202.

Bunker, G. 2010. *Introduction to XAFS.* New York: Cambridge University Press.

Dhez, M. 2009. ID21 sulfur XANES spectra database. http://www.esrf.eu/UsersAndScience/ Experiments/Imaging/ID21/php.

Funke, H., A. C. Scheinost, and M. Chukalina. 2005. Wavelet analysis of extended x-ray absorption fine structure data. *Phys. Rev. B.* 71: 094110. DOI:10.1103/PhysRevB.71.094110.

Iida, A. and T. Noma. 1993. Correction of the self-absorption effect in fluorescence x-ray absorption fine structure. *Jpn. J. Appl. Phys.* 32:2899–2902. DOI:10.1143/JJAP.32.2899.

IXS Standards and Criteria Committee. 2000. Error reporting recommendations: A report of the standards and criteria committee. http://www.ixasportal.net/ixas/images/ixas_mat/ Error_Reports_2000.pdf.

Klementiev, K. 2010. XANES dactyloscope: A program for quick and rigorous XANES analysis for Windows. User manual and tutorial.

Kraft S., J. Stümpel, P. Becker, and U. Kuetgens. 1996. High resolution x-ray absorption spectroscopy with absolute energy calibration for the determination of absorption edge energies. *Rev. Sci. Instrum.* 67:681–687. DOI:10.1063/1.1146657.

Rayleigh, Lord 1874. On the manufacture and theory of diffraction-gratings. *Philos. Mag.* 47:81–93.

Rehr, J. J., C. H. Booth, F. Bridges, and S. I. Zabinsky. 1994. X-ray absorption fine structure in embedded atoms. *Phys. Rev. B.* 49:12347–12350. DOI:10.1103/PhysRevB.49.12347.

Savitzky, A. and M. J. E. Golay. 1964. Smoothing and differentiation of data by simplified least squares procedures. *Anal. Chem.* 36:1627–1639. DOI:10.1021/ac60214a047.

Teo, B. K. 1986. *EXAFS: Basic Principles and Data Analysis.* New York: Springer-Verlag.

Xia, Z., H. Zhang, K. Shen, Y. Qu, and Z. Jiang. 2018. *Physica B.* 542:12–10. DOI:10.1016/j. physb.2018.04.039.

Chapter 8
Data Collection

Finding Truth and Beauty by Closing Time

There's nothing I'd like better than coming home from my beamtime with a beautiful spectrum!	
When one places beauty over all else, Simplicio, that is what one gets.	
If you prioritize beauty over truth, you might end up getting lied to.	
And if you spend all your time chasing after beauty, you may wind up with very little to show for it!	
Um, we are talking about *spectra*, right?	
Whether or not our friends were referring to the trade-offs inherent in collecting XAFS data, those trade-offs are crucial to understand. Let us keep them in mind while we read this chapter.	

DOI: 10.1201/9780429329555-10

8.1 Noise, Distortion, and Time

Researchers often describe a spectrum as 'beautiful' when it is free of evident noise and the features appear sharp. But as our panel just discussed, a beautiful spectrum is not necessarily an accurate one. In fact, conditions that minimize noise often introduce systematic error, distorting the spectrum. As an example, consider the 'thin' and 'thick' spectra from Box 6.4, shown together in Figure 8.1.

The thin-sample spectrum looks like it has somewhat more noise than the thick-sample one. For example, it shows a couple of little bumps around 8 Å that look as if they might represent random variation, and the peak near 10 Å is particularly irregular.

But, as was discussed in Box 6.4, the thick-sample spectrum is significantly compromised by harmonics. Although it is less noisy than the thin-sample spectrum, it is much more distorted. When faced with a choice between the two, the thin-sample spectrum is preferable.

When collecting data, there is a natural tendency to continually tweak the collection conditions so as to maximize signal-to-noise ratio, since it is easy to see the improvement. Systematic distortion, on the other hand, is less evident. It is, therefore, crucial to keep sources of systematic error in mind, so that you do not inadvertently distort your data in the pursuit of noise reduction.

Another aspect that must be balanced against noise and distortion is time. It is always possible to improve the signal-to-noise ratio by collecting more data; that is inherent in the definition of noise! But is it better to collect beautiful data on two samples or usable data on ten? That depends on the particular study being conducted and the scientific questions being asked.

One spectrum 'looks' noisier than the other? That is a vague statement. It would be better to examine multiple scans of each sample and look for variation between scans. It would be better yet to quantify that variation, yielding a standard deviation as a function of k.

Sure, Robert. But an experienced eye can do pretty well recognizing noise—or at any rate things that are not EXAFS—even in a merged spectrum (see Section 8.10 for examples of features that are neither noise nor signal). And anyway, the point of this section is to warn against optimizing data collection to produce the best-looking spectrum.

8.2 Detector Choice

There are many special detectors that have been designed or adapted for XAFS, some of which were discussed in Section 2.4 of Chapter 2. If you're using, say, a wavelength-dispersive geometry or an electron-yield detector, this is likely to have been part of your experimental design, decided on long before arriving at the light source. But if detector choice is not an integral part of your experimental design, and you are using a typical beamline, then you are most likely faced with several choices: transmission using an ion chamber; fluorescence using a Lytle, passivated implanted planar silicon (PIPS), or similar total-yield detector; or fluorescence using an energy-discriminating detector. Even on many specialized lines, such as those dedicated to microprobe or quick-XAFS, you may be faced with a choice of this nature.

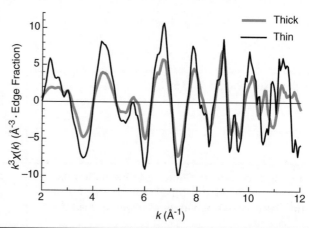

Figure 8.1 $k^3\chi(k)$ **for the thin and thick zinc ferrite samples discussed in Box 6.4.**

As a rule of thumb, signal-to-noise ratio generally improves as you go from transmission to total-yield fluorescence to energy-discriminating fluorescence, but the possibilities for systematic error also increase.

Transmission measurements using an ion chamber have the advantage that, with a proper choice of gas (see Section 8.6.3), there is often no practical limit on the intensity of x-rays that can be measured. Self-absorption is also not an issue, although inhomogeneous samples and harmonics can be. Of course, some samples are unsuitable for transmission because they are inherently thick or are mounted on a thick substrate.

Transmission, however, exhibits poor signal-to-noise ratio when the concentration of the absorbing element is low or when the sample is particularly thin or thick.

When using a total-yield fluorescence detector, on the other hand, self-absorption must be considered when the concentration of the absorbing element is high (see Section 6.4 of Chapter 6). Scattering and fluorescence from elements other than the absorbing element contribute to the background, and this can degrade the signal-to-noise ratio, particularly when the concentration of the absorbing element is very low.

Energy-discriminating fluorescence detectors reduce the background from scattering and other elements, yielding the best signal-to-noise ratio for low-concentration samples. But linearity of these detectors can be an issue, particularly when the number of counts is high.

8.2.1 Predicting Signal-to-Noise Ratio in Transmission

Many researchers treat signal-to-noise ratio as a matter of trial and error, to be done only after arrival at the beamline. Although a certain amount of trial and error is often part of the data collection experience, it is helpful to be able to compute an estimated signal-to-noise ratio in advance.

The first step in this kind of calculation is to estimate the flux of photons incident on your sample. This depends on many factors that are out of your control once you select a beamline, such as characteristics of the light source, the bending magnet or insertion device used by the beamline, the resolution of the monochromator, and gases and windows in the beam path before it enters the hutch. Other factors may be under your control, such as the positions of slits used to control the size of the beam and, of course, the duration of time that data are collected at each energy (the *integration time*). Beamline scientists are usually happy to provide you an order-of-magnitude estimate of the flux you can expect for your experiment. For modern light sources with the slits positioned so as to provide the maximum flux consistent with good energy resolution, flux may range from about 10^{10} photons per second for some bending magnet lines to perhaps 10^{14} photons per second for some undulators.

As an example, suppose you are using a wiggler that produces a flux of 10^{12} photons per second on the sample. If your I_o detector collects 10% of that flux, it will measure about 10^{11} photons per second. Poisson statistics (Section 6.3.3 of Chapter 6) tells us that the noise in the number of I_o counts should be about the square root of the number of photons measured or 3×10^5 for a 1-second measurement. The signal-to-noise ratio in I_o is impressive: about 300,000 to 1.

Now suppose the sample is a pure copper reference foil, 2 absorption lengths thick above the edge. As shown in Table 6.2, only 14% of the photons will emerge from the sample to be detected by the I_t detector. If the I_t detector is set up to absorb most of those photons, then we will have roughly 10^{11} photons per second measured in I_t. The signal-to-noise ratio in I_t will, therefore, be about 300,000 to 1, just as it is in I_o.

(Under these circumstances, it is doubtful that signal-to-noise ratio will be quite as good as these calculations indicate, as electronic noise will probably dominate over the shot noise we have calculated.)

When wavelength-dispersive detectors (also known as crystal analyzers) are an option, then for a given experiment signal-to-noise tends to be worse than for an energy-discriminating detector. Energy resolution will be much better, however, and for experiments which have strong signals anyway the maximum count rates can be higher.

The top value of around 10^{14} photons per second is unlikely to rise much, since it is limited by considerations such as the heat load on beamline components, including the monochromator, not to mention beam damage to the sample.

In pulse-probe experiments with free electron lasers, the peak flux can be much higher than 10^{14} photons per second, my dear Carvaka.

That's true, Robert, but in pump-probe experiments most of the time there's no beam on the sample at all. Still, you're right that even average flux on those beamlines can exceed that on standard undulators, since sample damage is not a concern in a pump-probe experiment.

What we've done so far, however, is estimate the signal-to-noise ratio in the total signal measured in the detectors. If all we were interested in was an edge jump, that would be good enough. But we're interested in fine structure, so the variations in signal we are looking at are much less than the total signal. For XANES, we're often looking at features about 10% the size of the total signal; for EXAFS, the features may be nearly that large at the start of the EXAFS range, but drop rapidly in amplitude with increasing energy. Although the rate of this drop varies greatly from material to material and also depends on conditions such as temperature, for EXAFS dominated by low-*Z* scatterers (e.g., oxides), the signal may drop by an order of magnitude every couple of inverse angstroms in *k*-space (see Section 13.3 of Chapter 13). For reasons discussed in Section 13.3 of Chapter 13, our pure copper foil will not see the sharp drop begin until perhaps 7 Å^{-1} above the edge, but from that point on suffers a similar rate of decline.

Although these rules of thumb are very rough, let's use it to try to make a back-of-the-envelope calculation of the value of *k* at which the signal drops to the level of the noise in the I_t detector for a 1-second measurement in our example. Since the total signal is about 10^{11} photons, the size of the EXAFS features in the low-*k* part of the range might amount to differences of about 10^{10} photons, giving a signal-to-noise ratio of less than 30,000 to 1. If the size of these features was to drop by an order of magnitude for every 2 Å^{-1} above 7 Å^{-1}, it would take about 8 Å^{-1} more for the signal to drop to a level comparable to the noise; that is, we could still make out the signal up to about 15 Å^{-1} if we took a single scan with an integration time of 1 second per point. This is comparable to what is actually observed, validating our rough calculation.

BOX 8.1 Thickness Effects on Signal-to-Noise Ratio

	Friends, why are we treating I_t as if it were the signal? The measured signal is $\ln(I_0/I_t)$, and from there we normalize the data before analyzing it. For EXAFS, we also subtract a background and convert to $\chi(k)$. Do these steps change this analysis?
	The first-order Taylor expansion of $\ln[a/(b - x)]$ is $x + a$, a constant, so small changes in I_t translate proportionally into small changes in the signal. Normalization and transformation to the unweighted $\chi(k)$ leave that proportionality in place. If, say, $\chi(k)$ at 13 Å^{-1} is 1% of its value at 9 Å^{-1}, then the change in I_t relative to the background at 13 Å^{-1} is also 1% of its value at 9 Å^{-1}.
	But, friend Dysnomia, we are not looking only at small changes relative to the background. In the text above, we have calculated the noise on the entire measurement, including the background. Then we compare with the signal, which has been normalized by the edge jump. The edge jump is not small; for an ideal sample, it is likely around 1, as we discussed in Section 6.3.5 of Chapter 6. Let us attempt to work this out in detail. Suppose there is a feature that is of amplitude x in the unweighted, experimentally determined $\chi(k)$. I agree that it is also of amplitude x in the experimentally determined $\chi(E)$. It corresponds to a difference relative to the background in the normalized spectrum: $$x = \frac{M(E) - M_o(E)}{\Delta M_o(E_o)} \qquad (8.1)$$

Thus,

$$M(E) = x\Delta M_o(E_o) + M_o(E) \tag{8.2}$$

To estimate the signal-to-noise ratio more accurately, we want to know what fraction y of $I_t(E)$ the feature represents. To do this, we introduce a smoothly varying background function $I_{tb}(E)$, which is analogous to $M_o(E)$:

$$\ln\left(\frac{I_0(E)}{yI_{tb}(E) + I_{tb}(E)}\right) = x\Delta M_0(E_0) + \ln\left(\frac{I_0(E)}{I_{tb}(E)}\right) \tag{8.3}$$

Assuming y is small compared to 1, as should be the case for an EXAFS feature, we can use the rules of logarithms to write

$$-y + \ln\left(\frac{I_0(E)}{I_{tb}(E)}\right) \approx x\Delta M_0(E_0) + \ln\left(\frac{I_0(E)}{I_{tb}(E)}\right) \tag{8.4}$$

$$y \approx x\Delta M_0(E_0)$$

Right. For a reasonably normal transmission sample, with an edge jump around one, the fractional effect of a feature on I_t is more or less the same as the fractional effect on $\chi(k)$. We have been doing very rough, back-of-the-envelope calculations to estimate noise, which could easily be off by a couple of orders of magnitude for other reasons. I don't think we have to worry about your correction much.

Not for samples close to the ideal thickness, no. But Equation 8.4 does show us that signal-to-noise ratio degrades for very thin samples, in accord with experience. For very thick samples, on the other hand, although Equation 8.4 at first seems to suggest that signal-to-noise ratio improves, I_t will also be dropping, which will suppress signal-to-noise ratio in our calculation simply because the total I_t is lower.

Next, suppose that instead of a metal foil, we have a finely ground soil sample with a concentration of copper such that 1% of the absorption just above the copper edge is due to copper and 99% to other elements. (As Section 6.4.6 of Chapter 6 discusses, this suggests the sample is considerably *less* than 1% copper by mass). If the total absorption is still 2 absorption lengths, it means that the size of our signal is reduced by a factor of hundred relative to the foil, without affecting the noise at all. For 1-second scans in the low-k part of the EXAFS region, the signal-to-noise ratio is now perhaps 300 to 1. In addition, unless the copper is present as tiny nodules in the soil, it is likely that the copper EXAFS is dominated by low-Z scatterers, meaning that the signal will start dropping sooner than it did for the metal. If we suppose the signal begins to drop rapidly after about 4 Å^{-1} (typical for an oxide), then the signal will be comparable to the noise roughly at 8 Å^{-1} above the edge. At this point, we could say that we have reasonably good XANES data, but any EXAFS analysis would be significantly compromised.

If we imagine the concentration of copper being such that it is responsible for only 0.1% of the absorption, even XANES features yield signal-to-noise values of only around 30 to 1, and we can see features only in the very beginning of the EXAFS region.

One could imagine addressing this problem with longer integration times and maximum scans. But the nature of Poisson statistics means that we need to increase the duration by a factor of 100 for every factor of 10 by which we'd like to improve the signal-to-noise ratio. To measure a transmission EXAFS spectrum of our 0.1% soil sample with a signal-to-noise ratio comparable to the 1-second integration time, copper foil spectrum would require an integration time of about 10^9 seconds per point—that's roughly 30 years! Although increased integration times, or equivalently, multiple scans (see Section 8.9.2) can improve signal-to-noise ratio, the improvement will only be modest and cannot make up for problems such as very low concentration.

8.2.2 Signal-to-Noise Ratio in Fluorescence

Signal-to-noise ratios are poor in transmission when the absorbing element is dilute, the sample is much thinner than ideal, or the sample is much thicker than ideal. Let us consider those cases one at a time.

8.2.2.1 Very thin samples

For very thin samples, Equation 8.4 tells us that the signal in I_t, expressed as a fraction of I_t, drops in proportion to the edge jump. Although this was derived in Box 8.1, we can also understand it conceptually: for very thin samples, the transmission is nearly 100% at all energies, making the shot noise nearly the same for, say, a 0.01 absorption-length sample as for a 0.1 absorption-length sample. The small variations that represent our signal are proportional to the edge jump, however, and thus so is the signal-to-noise ratio.

For a thin, concentrated sample, however, the *total* signal in the fluorescence detector is primarily due to fluorescence from the absorbing element, and thus the signal-to-noise ratio is proportional to the *square root* of the edge jump, as would be expected from Poisson statistics. While the geometry and physics of fluorescence contributes to significantly lower total counts than transmission (for one thing, the fluorescence detector is usually positioned so as to capture only a modest fraction of the fluorescent photons), the weaker dependence on the edge jump makes fluorescence better for thin samples.

This discussion ignores scattering. We will discuss scattering, and how to minimize it, in Section 8.7.1.

For example, suppose a fluorescence detector is capturing 1% as many photons as an I_t detector when the edge jump is 1. The signal-to-noise ratio would then be three times worse for fluorescence than transmission. (In addition, this sample would suffer from substantial self-absorption!) But now suppose the concentrated sample was so thin that the edge jump was only 0.01. The signal-to-noise ratio in transmission would have become 100 times worse than for the thicker sample, but in fluorescence it would only be 10 times worse than before. Since the fluorescence started three times worse, the fluorescence would now have a signal-to-noise ratio that is about three times better than the transmission. The difference would become even greater for thinner samples.

8.2.2.2 Very thick samples

This case is straightforward. In transmission, signal-to-noise ratio degrades because the total number of photons detected in the I_t detector decreases. For example, at 9 absorption lengths, transmission has dropped to about 0.01%, or roughly 10,000 times less than for a sample of ideal thickness for transmission. This causes the signal-to-noise ratio to degrade by a factor of 100, in accord with Poisson statistics.

In fluorescence, on the other hand, the fluorescence signal plateaus for thick samples and does not drop.

8.2.2.3 Dilute samples

In practice, signal-to-noise ratio is the least of your worries for a very thick, concentrated sample. In transmission, harmonics will be a problem. In fluorescence, self-absorption is an issue. One possible solution, if you can't make the sample thinner, might be to measure in electron yield.

At first consideration, it might seem that there should be no special advantage for total-yield fluorescence with dilute samples. After all, if elements other than the one for which XAFS is being sought are doing most of the absorbing, then they should be doing most of the fluorescing as well, right?

Not necessarily. It turns out that low-Z atoms are much more likely to fill a core-hole by an Auger-Meitner process than by fluorescence. Oxygen, for example, fills a K-shell vacancy through fluorescence only 0.8% of the time and silicon 5% of the time, while copper does so 44% of the time (Krause 1979).

More importantly, most of the fluorescence from low-Z elements never makes it to the detector. The most prominent fluorescence lines from silicon, for instance, are at about 1700 eV. At that energy, the absorption length of *air* is less than a centimeter. The combination of gases and windows in the path from the sample to fluorescence detector will filter out almost all of the fluorescence from these low-Z elements.

The fact that most absorption by low-Z elements does not result in counts in the fluorescence detector can result in a dramatic improvement in the signal-to-noise ratio. To see how large this effect can be, let's return to the example from the end of Section 8.2.1, in which copper is responsible for only 0.1% of the absorption of a sample. Let us assume the rest of the absorption is from low-Z elements and that no fluorescence from those low-Z elements reaches the fluorescence detector. Let's also assume that geometric and physical factors combine so that only 1% of the absorption by copper results in fluorescence is counted in the detector.

With approximately 10^{12} photons per second getting absorbed by the sample, about 10^9 of those are due to copper, leading to about 10^7 photons per second being counted in the fluorescence detector. Shot noise in 1 second will be the square root of that value or about 3×10^3 photons. If at the start of the EXAFS range the features have an amplitude around 10% of the total signal, the signal-to-noise ratio for those features will be on the order of $10^6/(3 \times 10^3) \sim 300$ to 1. This is a factor of ten improvement over our estimate for transmission. The relative improvement would be greater for even more dilute samples.

We've referred to fluorescence detectors such as Lytle detectors as 'total-yield detectors.' But that's only true for photons that penetrate the detector! Materials in the beam path act as a high-pass filter; this effect can be intentionally enhanced, as we will discuss in Section 8.7.2. In the same section, we learn that it is also possible to use low-pass filters. By choosing an appropriate combination, we can cause our experimental set up to have some degree of energy discrimination, even though the detector *per se* does not.

8.2.3 Energy-Discriminating Fluorescence Detectors

In Section 8.2.2, we neglected the effect of photons scattered into our fluorescent detector. We also focused on the case where the element of interest is much higher Z than the bulk of the sample. This is because, in part, these effects can be mitigated by shielding, filters, and geometry (Section 8.7). The calculations in Section 8.2.2 compute best-case scenarios, in which the mitigation is entirely successful.

However, as Section 8.7 shows, there is a limit to how well such procedures work with total-yield fluorescence detectors. If a sample is very dilute, or if it includes a substantial amount of elements with atomic number similar to or higher than the absorbing element, then the fluorescent background from scattering and fluorescence from nontarget elements may be the dominant source of noise.

This problem can often be alleviated by the use of an energy-discriminating detector (see Section 2.4.3 of Chapter 2), which will only count photons near the energy of the desired fluorescence. This can improve signal-to-noise ratios to levels closer to the results of the kinds of calculations done in Section 8.2.2.

This does not mean, however, that the signal-to-noise ratio achieved by energy-discriminating detectors is always better than total-yield fluorescence detectors. That is because energy-discriminating detectors saturate at high flux, and so in some circumstances, the intensity of the x-rays must be lowered (e.g., by narrowing the pre-I_0 slits).

For example, suppose an energy-discriminating multielement detector becomes nonlinear at count rates above $\sim 3 \times 10^6$ total photons per second. Although the detector can discriminate photons within about 200 eV of the target fluorescence from the background, it does so *after* detection, so that it is the total number of counts, including the background, that must be kept below 3×10^6 per second. If only 1% of the photons counted by the detector are within the chosen energy range, which is not atypical for circumstances in which energy-discriminating detectors are used, we would thus be limited to perhaps 3×10^4 photons per second of signal. Poisson statistics tells us that the shot noise is about 200 photons per second. For XANES or low-k EXAFS features with amplitudes about 10% of the total signal, which in this case would be about 3000 photons per second, this implies a signal-to-noise ratio of only about 15 to 1! Even if there was no fluorescent background at all, the signal-to-noise ratio for such a feature would be only about 150 to 1.

Notice that these calculations, unlike those for transmission or total-yield fluorescence, do not depend on the flux of the incoming beam—once your detector has reached its maximum counting rate, you wouldn't do any better by upgrading from a bending magnet to a wiggler, for example—you'd just have to reduce the flux of the beam back to the same level anyway!

Also notice that for the dilute sample described in Section 8.2.2 (0.1% absorption by copper), the flux of fluorescence photons from the absorbing element alone is higher than the maximum count rate of most energy-discriminating detectors.

In short, energy-discriminating detectors are useful primarily for extremely dilute samples or ones where the fluorescent background is very high for other reasons, such as the presence of substantial amounts of elements with atomic number comparable to or greater than the absorber.

8.2.4 Wavelength-Dispersive Fluorescence Detectors

In general, wavelength-dispersive detectors (see Section 2.4.5 of Chapter 2) do not have the issues of nonlinearity at high count rates experienced by energy-discriminating fluorescence detectors. This can make them a good choice if the signal is relatively strong. If the signal-to-noise for an energy-discriminating detector is already poor due to *low* count rates, however, it may be even worse in a wavelength-dispersive detector, since a smaller proportion of the desired photons are likely to be counted in the detector.

8.3 Mode Choice

On some beamlines, you have a choice between operating in quick-XAFS mode (see Section 3.1.4 of Chapter 3), or a traditional step-scan mode. If you need the improved time resolution provided by quick-XAFS, of course you'll want to select that mode. But what if you don't? Quick-XAFS will still achieve your desired signal-to-noise ratio more quickly, because data is collected continuously, without pauses between steps and between scans. Also, a greater proportion of the time will be spent collecting data, reducing beam damage to sensitive samples.

On the other hand, quick-XAFS is a more sophisticated experiment, requiring additional steps in data reduction. Depending on the preprocessing software available at the beamline, quick-XAFS experiment can generate immense amounts of data, which can be unwieldy to store and process. In addition, quick-XAFS provides much less control over scan parameters (Section 8.9) than traditional step scans.

For experiments in which time resolution is not needed and beam damage to the sample is not a concern, it may therefore often be preferable to use a step scan even if quick-XAFS is available.

8.4 Before You Begin

8.4.1 Plan Your Beamtime

Before you begin to collect data, take stock of what you hope to accomplish. Decide which of your samples are of high priority and which are lower. Know which edges have to be measured on each. Have an estimate of which ones will take longer, perhaps because they are particularly low concentration. Also keep in mind any that will be particularly challenging, perhaps because they will be measured *operando*, are at an edge that is rarely used on the beamline, or have a tricky mix of elements or substrates.

At that point, you have some choices to make. If you plan to measure multiple edges on some of your samples, will you measure all the edges on one sample before switching to the next or all the samples at one edge before changing energies? That will depend on how difficult it is to align your samples, how crucial it is to know that you're hitting the same spot, and how much of a hassle it is to change between the edges you'd like to measure.

Beamtime is often precious, and many experimental runs are performed with little sleep, and/or with some team members who have less experience—everyone has to have a first time to the synchrotron and it's a lot easier if it's with someone who knows the ropes! Try to plan the challenging stuff for when the experienced people are up *and alert*—and for when beamline staff are around. Three in the morning on the third night of a run is a good time to have your new undergrad measuring powder standards, but it is not such a good time to make your first attempt at setting up your *in situ* chemical reaction quick-XAFS experiment.

On the other hand, you would also like to measure samples you know quite a bit about before you measure the ones that are more unfamiliar, so that you can make sure everything is working properly. Therefore, whatever edge you start on, you will probably start by measuring a reference foil—these are often available to borrow from the beamline or perhaps from a central repository in the synchrotron. (If this is your first time coming to a particular synchrotron, however, ask before you come!) After the reference foil, you might want to measure a standard or two, so that you can again work out any kinks that might be present. You can even do some quick fitting at the beamline, further confirming that everything is going well. At that point, you can get into the trickier stuff. Of course, on a very short run, you may have to get to your high-priority samples more quickly than that.

The best laid plans, however, often go off track. Leaving high-priority samples to the last few hours of the run is a bad idea: the synchrotron beam may go down, a beamline amplifier might burn out, or your sample might turn out to have some surprises in store for you. In addition, information you learn from earlier samples may influence your priorities for later ones—perhaps the sample you thought you were going to use as a baseline has an interesting story of its own to tell, and you decide the other baseline samples you brought along, which you had thought of as low-priority, may now be the subject of your next paper.

8.4.2 Get to Know Your Beamline

Every XAFS beamline is different. Some are optimized for microprobe work, others for quick-XAFS. They may have mirrors to reduce the spot size of the beam, or mirrors to eliminate harmonics, or both. Monochromators come in a variety of designs and use a variety of feedback systems to maintain alignment. Special sample environments may be available and motors may be enabled that allow the sample to be moved and rotated in various ways. Parts of the beam path may be through vacuum, or helium, or just open to the air (inside a hutch from which people are excluded when beam is present, of course!).

Figure 8.2 shows a schematic of a 'typical' XAFS beamline. Actual beamlines may differ in several ways, including the addition, substitution, omission, or reordering of some of the components shown.

Airplane pilots take a walk around the outside of their plane before taking off. You should take a walk along your beamline before taking data! It's better to get an idea of the layout when you are getting started, rather than having to search for some component at two in the morning when something has gone wrong!

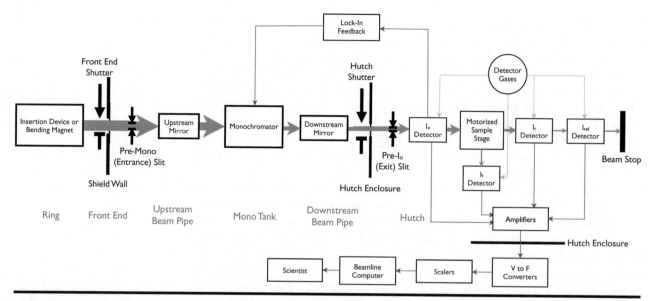

Figure 8.2 **(See color insert.) Schematic of a typical beamline. X-rays are shown in magenta, signals in blue, and gases in green. Brown labels refer to regions used to identify the location of components; e.g., a beamline scientist might refer to some particular control as being 'at the front end,' or 'in the hutch.'**

8.5 Optimizing the Beam

Although it is the responsibility of light source and beamline staff to provide you a stable and well-characterized beam, there may be a few aspects of the beam that are under your control. What is the responsibility of users and what is the responsibility of staff varies from facility to facility, beamline to beamline, and by level of experience of the user. Beamline scientists are usually happy to discuss these matters with you, train you for the aspects that you can control, and provide you information about the aspects that you cannot. *Because every beamline is different, their advice should supersede anything we say in this section.*

8.5.1 Aligning the Beam

The beam leaving the monochromator is not uniform. At angles above and below its center, it will drop off in intensity, the energy of the fundamental will change as dictated by Bragg's Law, and the energy content (e.g., the relative amount of harmonics) will change as well.

In addition, depending on the design of the beamline optics, the position of the beam may change gradually with energy to the extent that its center may be a few millimeters higher or lower at, for example, the silver K edge (25.5 keV) than at the iron K edge (7.1 keV).

For these reasons, it is a good idea to align the beam at the beginning of any experimental run and when moving from one edge to another at a significantly different energy. On some beamlines, the beamline scientist may also recommend that you realign after every refill of the storage ring.

Typically, beam alignment will involve opening up any pre-I_o slits very wide. On some beamlines, you will have control over a premono slit as well; if so, it could be adjusted up or down to maximize the signal in I_o. On the other hand, moving the premono slit will change the heat load on the monochromator, which may take a while to re-equilibrate in response. Whether or not adjusting the premono slit is a good idea is another of those things the beamline scientist will know!

You may be wondering what constitutes a 'significantly different' energy. That will depend on the beamline and is something that your beamline scientist will know. Some lines may need to be realigned even for a move of a couple of keV, while some may be stable across their whole energy range.

The alignment itself often involves moving the table on which the detectors are located up and down, although it's also possible it will be done by moving the monochromator crystals. In either case, effect on I_o is monitored, and the beam position is chosen to be in the middle of the broad maximum for I_o.

8.5.2 Choosing Pre-I_o Vertical Slit Width

The beam emerging from the monochromator will have an energy that varies vertically. Therefore, narrowing that beam will allow through a smaller range of energies. It will also, of course, cut down the flux incident on the sample.

It is worth presenting a rough argument as to how the energy resolution of the beam should depend on the vertical width of the slit. Although any actual case will be more complicated, depending on the design of a particular beamline's optics, it will allow us to get a sense of the dependence.

Monochromators can be thought of as taking advantage of the first-order peak given by Bragg's Law:

$$2d \sin \theta = \lambda = \frac{hc}{E} \tag{8.5}$$

Differentiating, we get

$$2d \left(\cos \theta \right) \delta \theta = -\frac{hc}{E^2} \delta E, so \left| \frac{\delta E}{E} \right| = \frac{2Ed}{hc} \left(\cos \theta \right) \delta \theta$$

$$= \frac{2Ed}{hc} \delta \theta \sqrt{1 - \sin^2 \theta} \tag{8.6}$$

$$= \frac{2Ed}{hc} \delta \theta \sqrt{1 - \left(\frac{hc}{2Ed} \right)^2} = \delta \theta \sqrt{\left(\frac{2Ed}{hc} \right)^2 - 1}$$

For typical beamline optics, geometries, and energies, the fractional broadening $\left| \frac{\delta E}{E} \right|$ is on the order of 10^{-4} when the pre-I_o slit is narrowed vertically to 1 mm or so. Equation 8.6 tells us that for a given geometry, this fractional broadening should increase with energy.

Vertical slit width is not the only factor affecting energy resolution, however. As discussed in Section 4.1 of Chapter 4, core-hole lifetime broadening is also a significant contributor.

Finally, diffraction from a monochromator has an inherent angular spread known as the Darwin width, after the pioneering work of Charles Galton Darwin (1914). For diffraction from a single crystal, this width is comparable to the broadening from the core-hole lifetime. In a typical monochromator, however, the x-rays diffract twice. By slightly misaligning the two diffraction conditions, it is possible to dramatically reduce the monochromator broadening below that specified by the Darwin width (DuMond 1937).

If it weren't for lifetime broadening, and, to a lesser extent, monochromator broadening, we would face a difficult trade-off: the more we narrow vertically the pre-I_o slits, the better the energy resolution. This would also result in a lower flux, however, and thus worse signal-to-noise ratio. But the lifetime and monochromator broadenings establish a limit on how good an energy resolution we can achieve, and it usually turns out that we can get close to that limit with only a modest reduction in flux.

The vertical width of the pre-I_o slit should, therefore, be chosen by beginning with the slit fairly wide open and collecting the XANES spectrum of a standard, such as

On some beam-lines, the pre-I_o slit has little effect on resolution, because upstream optical components already limit the beam to a resolution comparable to that dictated by the monochromator and lifetime broadenings. If that's the case, the pre-I_o slit is used to provide sharp definition to the beam, but not to adjust resolution.

a metal foil. The slit should then be narrowed and the standard measured again; presumably, the resolution will visibly improve, as seen in Figure 8.3. This process should be repeated until the resolution of the spectra ceases to improve.

The effects of the lower resolution of the 2.00 mm spectrum in Figure 8.3 are complicated by alignment and normalization, but that has been done because it is always what happens in practice: alignment is done using reference foils, and reference foils are as affected by changes in energy resolution as the sample!

The most straightforward difference of energy resolution on the 2.00 mm slit spectrum is the suppression of the small shoulder around 6545 eV. The peak near 6557 eV is also suppressed, as can be seen most easily in the derivative plot.

Feature positions are also shifted somewhat—in the case of the 6545-eV shoulder, the derivative spectrum of the 2.00 mm spectrum shows the inflection point roughly half an electron volt higher than is the case for the other two spectra. The inflection point around 6549 eV, in contrast, appears shifted somewhat to lower energy. While the shifts are small, they are on the order of shifts used for analysis in fingerprinting (Chapter 9), linear combination analysis (Chapter 10), and principal component analysis (Chapter 11).

Since fractional broadening due to vertical slit width increases with energy (for a given width), while fractional broadening due to the core-hole lifetime is roughly energy-independent (at least for K and L_3 edges), the vertical slit width needs to be

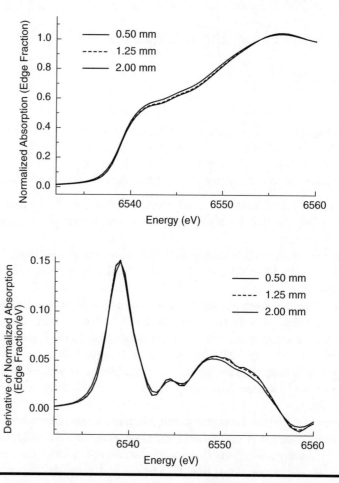

Figure 8.3 Comparison of (a) the derivative of XANES and (b) XANES for the *K* edge of manganese foil as a function of pre-I_o vertical slit width. Scans have been normalized and aligned. The 0.50 and 1.25 mm spectra are effectively indistinguishable, but the 2.00 mm shows several subtle differences.

narrowed for higher energy edges to maintain maximum resolution. Therefore, this procedure should be repeated whenever energy is changed by more than a few keV.

Alternatively, if your experiment is not highly flux-sensitive (e.g., your sample is concentrated, smaller than the vertical slit width with no postslit focusing, or being used with an energy-discriminating detector that is saturating), this procedure can be done once at the highest energy you will be using for your experiments. This will assure the best possible energy resolution at all edges at the cost of some flux at the lower energies.

8.5.3 Reducing Harmonics

As described in Chapter 6, harmonics in the beam will distort transmission spectra, especially for thick samples. For spectra measured in fluorescence, harmonics do not cause distortions, but they still degrade signal-to-noise ratio by increasing the fluorescent background. Since Bragg's Law allows harmonics to pass through the monochromator at the same angle as the fundamental, it is necessary to find a way to suppress them.

There are two common methods of harmonic suppression: harmonic rejection mirrors and detuning.

8.5.3.1 Harmonic rejection mirror

Mirrors designed for x-rays work only at very shallow angles; at steeper angles, nearly all the x-rays are absorbed or transmitted rather than reflected. This effect is energy-dependent; the higher the energy of the x-rays, the smaller the angle with the surface necessary for efficient reflection (Kirkpatrick and Baez 1948). Mirrors can, therefore, be used as low-pass filters for x-rays; by adjusting the orientation of the mirror, the cut-off energy can be chosen as desired. Since harmonics are at least twice the energy of the fundamental (see Section 6.2.1 of Chapter 6), it is not difficult to choose a cut-off energy that suppresses harmonics without affecting the fundamental.

It must be noted, however, that while the reflection efficiency of x-ray mirrors decreases rapidly with energy above the cut-off, it does not go to zero (Bilderback and Hubbard 1982). Thus, while harmonic rejection mirrors are very effective with samples of typical thickness (see Section 6.3.4 of Chapter 6), extremely thick samples measured in transmission may still reveal evidence of harmonics.

In addition, if the coating of a harmonic rejection mirror includes an element with an edge in the energy range being measured, the XAFS spectrum of the coating will modulate the reflectivity of the mirror, with the result that its spectrum will show up in I_o. If the detector chain is not completely linear, then this will prevent accurate measurements of this edge in a sample. In such cases, the harmonic rejection mirror may sometimes be moved out of the beam path and detuning (see below) used instead.

8.5.3.2 Detuning

A typical monochromator employs two crystals, although sometimes either one crystal in two sections (a *channel cut* monochromator) or four crystals are used instead. Figure 8.4 shows a detail of a double-crystal monochromator.

The first crystal diffracts x-rays of the desired energy, along with harmonics, on to the second crystal, which then diffracts them back along the original direction, albeit displaced vertically.

What if the two crystals are slightly misaligned? Diffraction peaks are not infinitely sharp, so some photons of the desired energy will still travel down the beam pipe, through the slits, and into the detectors and sample. The intensity of diffracted x-rays as a function of angle around a diffraction peak is called the *rocking curve*.

Unfortunately, the flux of most beamlines falls off at high energy! This means that for high-energy edges, the trade-off between flux and resolution may be a more difficult choice to make, particularly for dilute samples. Remember that energy resolution is more important for XANES analyses than for EXAFS, because EXAFS features are broader in energy.

Some crystal orientations, such as Si(111) and Si(311), do not pass the second harmonic, making the job of harmonic rejection somewhat easier.

A detailed calculation of Bragg diffraction using the dynamical theory of x-ray diffraction (Hart and Berman 1998) is shown in Figure 8.5.

This calculation presents several important features:

- The peaks are shifted slightly relative to the angle predicted by the Bragg equation.
- The shift is smaller for harmonics than for the fundamental.
- The peaks for harmonics are significantly narrower than the peak for the fundamental.

This, then, suggests a method of reducing the harmonic content of the beam: intentionally misalign one crystal relative to the other. This is called *detuning* the monochromator.

For the geometry shown in Figure 8.4, detuning is equivalent to using two offset reflectivity curves, as shown in Figure 8.6, with the combined result given by their product.

The reflectivity curves for the fundamental of the two crystals have significant overlap, resulting in a flux roughly half that the crystals would have if perfectly aligned. But the curves for the second harmonics hardly overlap at all. Thus, the harmonics have been suppressed, as desired.

The implementation of detuning is highly idiosyncratic, varying from beamline to beamline. It is usually not as simple as just using a small motor to deflect one crystal relative to the other, because the alignment must be maintained and the beam kept at the same vertical position (or the experimental table moved to match changes in the vertical position of the beam) as the energy selected by the monochromator is changed. Frequently some kind of electronic lock-in system is employed, which uses small adjustments to keep the desired detuning as the energy is scanned. It is the job of the beamline scientist to let you know how to maintain detuning on their line. They can also let

Depending on the line and the beamline scientist, you may be given training ranging anywhere from 'this is the lock-in system—don't touch it!' to detailed instructions requiring you to monitor an oscilloscope. Beamline scientists know their own feedback systems and know what users have to do (or not do) to keep them working right!

Figure 8.4 Detail of double-crystal monochromator. Arrows show path of x-ray beam of selected energy.

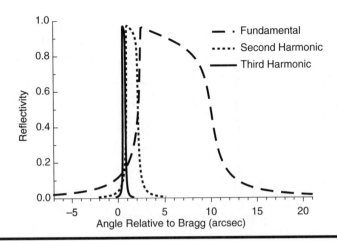

Figure 8.5 **Calculated reflectivity for fundamental and first two harmonics of a perfect silicon crystal, using the (220) orientation with a fundamental energy of 8 keV. The x-axis is relative to the prediction of the Bragg equation and the y-axis is the fraction of incoming x-rays at the given energy that are reflected. A plot of intensities would show the same shapes, but would also depend on the intensity of the incident beam as a function of energy. Typically, synchrotron sources produce beams that are more intense at 8 keV than higher energies and thus the peak intensity of the fundamental would be higher than the peak intensity of the harmonics. Calculations performed using software** χ_{oh}**-ON-THE-WEB (Source: From Lugovskaya, O. M. and S. A. Stepanov. *Sov. Phys. Crystallogr.* 36, 478–481, 1991).**

you know how much detuning is normal for their line—for example, they might advise you to 'detune 30%' (meaning misalign the crystals so that the reading in I_o drops by 30%). Using the principles you have learned in this book, you can adjust their advice up or down, as appropriate. For example, it might be wise to detune more than the typical amount if you are measuring an unusually thick sample in transmission.

It is also important to realize that there is often a limit to how far you can suppress harmonics. Depending on the arrangement of the optics, surface imperfections on the monochromator crystals, and the structure of the beam incident on the monochromator, there may be some reflection of harmonics even far from their diffraction peaks. (See Parratt et al. [1957] for an early experimental measurement of these 'remote tails'; Hart and Rodrigues (1978) includes a more recent comparison of experiment to the predictions of the dynamical theory of x-ray diffraction.) If, say, the off-peak effective reflectivity of each crystal to the second harmonic is 1%, then one part in 10,000 of the intensity of that harmonic will reach the sample regardless of detuning. At some point, further detuning does no good; it is unlikely that detuning 99% will give any benefit over detuning 90% and will result in a degradation of signal-to-noise ratio.

8.5.3.3 Testing for harmonics

If you are concerned about harmonics, it is simple to determine at what level they are present:

1. Set the monochromator to the energy you are most interested in.
2. Measure I_o and I_t with no sample in place. Compute $M = \ln(I_o/I_t)$.
3. Place a piece of aluminum foil between the I_o and I_t detectors (i.e., where a transmission sample would go) and calculate the new value of M.
4. Continue to add aluminum foil, one or more layers at a time, measuring M each time. (At high energies, you will want to add multiple layers of aluminum foil at a time, as adding one will make very little difference).
5. Graph the result, as shown in Figure 8.7.

Most beamline scientists will tell you to detune their monochromator in a particular direction from the fully tuned state. One reason for this is practical: the motors or geometry may not be set up to go very far in one of the directions—a channel cut crystal might even be at risk of breaking if stressed too far in the wrong direction. Another reason is that monochromator glitches may be worse in one direction than the other. In any case, the instruction to detune in a particular direction is *not* connected to the asymmetries evident in Figure 8.5; a little consideration will reveal that whether you rotate one of the crystals so as to move it left or right along the x-axis of Figure 8.5, the result ends up looking like Figure 8.6.

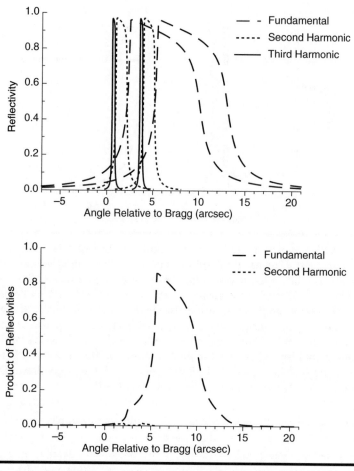

Figure 8.6 (a) **Reflectivity curves for two Si (220) crystals with 3 arcsecond detuning. (b) Product of reflectivities for the two crystals. The third harmonic would not be visible on this scale.**

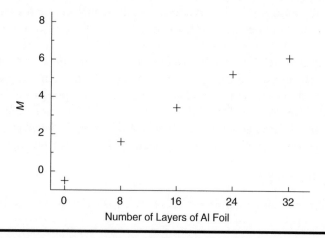

Figure 8.7 **Results of a test for harmonics conducted near the nickel edge.**

Dark current from the I_t amplifier will produce an effect similar (but not identical—see Tran et al. 2003) to that seen in Figure 8.7. Make sure you check the offsets (Section 8.6.4) before using this method to check for harmonics!

6. Assume the I_t reading is made up of two parts, I_{tf} and I_{th}, the first of which is due to photons of the desired energy and the second of which is due to harmonics. Since harmonics are absorbed by aluminum much weaklier than the fundamental, make the rough approximation that I_{th} is the same for all measurements (i.e., no harmonics are absorbed by the aluminum). Likewise, neglect the

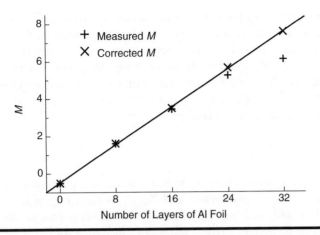

Figure 8.8 Results of correcting data from Figure 8.7 by assuming the presence of harmonics. In this case, the best linear fit was achieved by assuming that 0.18% of the incident counts were due to harmonics.

contribution of harmonics to the I_o signal. Choose a trial value for I_{th}, calculate $\ln(I_o/(I_t - I_{th}))$ for each measurement, and plot the result.

7. Adjust the value of I_{th}, either manually or using software, until the resulting data are linear, as in Figure 8.8.

8. The fraction of harmonics present can be estimated by taking the ratio of the value of I_{th} necessary to achieve a linear fit to the value of I_t measured with no aluminum foil. In addition, the thickness of sample at which harmonics start to become a concern can be directly determined by examining when the measured and corrected values of M begin to differ significantly on the chart. In Figure 8.8, harmonics become significant when the sample is thicker than 4 absorption lengths.

It is possible to make a rough correction for harmonics in your data using this method— this could be useful if you have a sample that is concentrated and intrinsically thick and if you have detuned to the point that you can't suppress the harmonics any further. But keep in mind that this correction is very rough, as it ignores the energy-dependent absorption of the harmonics by the aluminum, as well as the presence of harmonics in the I_o signal! A somewhat more careful analysis and correction is discussed by Tran et al. (2004).

8.6 Ion Chambers

For transmission, the most common detectors are ion chambers (also known as ionization chambers). The Lytle detectors (Stern and Heald 1979) often used for fluorescence are also a kind of ion chamber. Even when other detectors are used for fluorescence or electron yield, an ion chamber is usually still used to measure I_o.

8.6.1 Physics of Ion Chambers

In an ion chamber, some fraction of x-ray photons entering the chamber get absorbed by atoms in the fill gas, ionizing them and forming electron–ion pairs. Each electron is likely to have nearly the same kinetic energy (less whatever energy it took to ionize it and the recoil energy of the ion) as the original x-ray photon, but a much shorter mean free path, meaning it is likely to undergo a collision with another molecule of gas, ionizing it as well. These collisions will continue in a cascade. In the meantime, the excited ions will relax to their ground states, emitting fluorescent photons and Auger-Meitner electrons in the process, many of which will cause further ionizations, enhancing the cascade. Eventually, most of the energy of the original x-ray will go into ionizing the gas, although some will show up as heat in the detector or fluorescence out of it. The resulting ions and electrons are pulled out of the detector by an electric field placed across the chamber, typically on the order of 10^4 V/m for a bending magnet beamline.

The different forms of signal in the detector chain can cause some muddiness of nomenclature. It is not unusual to hear someone say 'Don't let the current in I_o exceed 10 volts,' for instance. The final readout on the computer is sometimes referred to as 'current,' sometimes 'voltage,' and sometimes 'counts,' as the signal has been expressed in all three forms during its journey from detector to monitor.

This creates a current, which is then sent to a *current amplifier*, which converts the current to a voltage. The voltage is then typically passed to a *voltage to frequency converter* and from there to a *scaler*, which at last passes the value to a computer. The readout on the computer is typically scaled to match the voltage output of the current amplifier.

It's possible to estimate the number of photons corresponding to a given reading on a current amplifier. If we ignore energy lost to heat and fluorescence, then the current I coming out of the detector is given by

$$I \approx 2e \frac{N}{t} \frac{E_{photon}}{E_{ionization}} \qquad (8.7)$$

where e is the charge on an electron, N/t is the number of photons absorbed per unit time, E_{photon} is the energy of a photon, and $E_{ionization}$ is the energy of an average ionization event. (The factor of 2 appears because each ionization even results in an ion of charge +1 and an electron of charge –1, both of which contribute to the current.)

$E_{ionization}$ is 41 eV for helium, 36 eV for nitrogen, 26 eV for argon, and 24 eV for krypton (Thompson 2009).

We can, therefore, estimate the typical current output from a detector. Suppose, for instance, that 10^{11} photons per second are absorbed in an ion chamber, as in the first example in Section 8.2.1. Suppose the gas is nitrogen and the energy is 6000 eV, which is just above the chromium edge. Equation 8.7 predicts a current of 5×10^{-6} A.

8.6.2 Limitations

Consideration of the physical process described in the previous section tells us that ion chambers cannot respond instantaneously to changes in photon flux—unless they are specially designed for rapid response, they are likely to have response times measured in milliseconds. Although fast enough for traditional XAFS measurements, this may be an issue in some quick-XAFS experiments.

In addition, while ion chambers have the ability to handle higher flux than many other kinds of detectors, the ability is not unlimited. Above $\sim 10^{11}$ photons/s/cm³ (Thompson 2009), the density of ions and electrons becomes large enough that significant recombination begins to occur, causing the detector to become nonlinear.

8.6.3 Choosing Fill Gases

Ordinary air is not generally used for ion chambers, in part because the moist air can itself support a current between the plates comparable to the current we would like to measure. In addition, this current is strongly dependent on small changes in conditions and can thus show large fluctuations and even hysteresis-like effects (Carlon 1988).

Ideally, a gas should be chosen for I_o that absorbs enough photons to provide good statistics for the incident beam, but not so many as to substantially impact the statistics in the transmitted beam (or fluorescent or Auger-Meitner counts, for experiments using those modes)—a target of 0.1 absorption lengths is a good rule of thumb. The absorption can be calculated using the same methods as are used for samples in Section 6.3.1 of Chapter 6.

For example, consider a measurement at the arsenic K edge (11,867 eV) using a 15-cm long ion chamber to measure I_o. If we are measuring EXAFS, we might be most concerned about signal-to-noise ratio well above the edge, so let's calculate the absorption at 12,500 eV. Helium (15 cm) at that energy is only 0.0003 absorption lengths, which is much too little. Nitrogen gives 0.04 absorption lengths, which is in the right ballpark. And argon gives 0.90 absorption lengths, which is a bit more than ideal. Nitrogen, therefore, would be the most appropriate gas for I_o in this case.

For I_t or I_f, we would prefer to absorb most of the photons—we will improve our counting statistics, and we don't need them further down the line.

BOX 8.2   Don't Get Carried Away!

	But I do need photons after I_t—my beamline has a reference there and then the reference detector!
	Sure, but the reference is primarily there just so we can check the energy calibration. We don't need a lot of photons for that.
	There's a more important reason we don't want to absorb too many photons in I_t, though. In most detector designs, the collecting plates don't extend all the way to the entrance window—there are guard electrodes there instead (Bunker 2010). Suppose you choose a fill gas for the I_t detector so that it gives 50 absorption lengths. That means most of your x-rays will be absorbed in the first 5% of your chamber. Do you really trust your detector to behave linearly when you're forcing the first few millimeters to do all the work? These detectors were designed to have absorption occurring throughout their length, so it's best to use them that way. A bit less than 2 absorption lengths is a good target for I_t, I_f, and I_{ref}.

Some beamlines have the ability to mix gases, so that you could use mostly nitrogen and a little bit of argon for this example. But just picking the best of the standard gases usually works just fine; I_o is not generally the dominant source of your noise!

Continuing our arsenic K edge example, suppose the I_t chamber were 30 cm long (it is not unusual to have the I_t chamber longer than the I_o chamber, since we want more absorption there). Nitrogen would give 0.08 absorption lengths, which is a bit low for I_t—we would be throwing a lot of photons away. Argon, on the other hand, gives 1.8 absorption lengths, which is ideal. So we use nitrogen in I_o and argon in I_t.

8.6.4 Amplifiers

Current amplifiers generally give users the ability to control their *gain*. A gain of 1 means that 1 V is output for every 1 A of current input (the gain may also be described by specifying the units; e.g., 200 μA/V). They also have a maximum voltage they can output without distortion, typically 5 or 10 V, depending on the model of amplifier. If the gain is set too high, the amplifiers will saturate, distorting the readings received by the computer.

However, overly low gains also present a problem. In the typical detector chain, the output of the amplifier is sent to a voltage to frequency converter, which turns it into a series of pulses with a frequency dependent on the input voltage. The pulses are then counted by a scaler, which sends the result to a computer. Each of those components, along with the cabling connecting them, is subject to various kinds of error: dropped counts, rounding errors, electronic noise, and so on. In a well-functioning system, an

It can be quite useful to use Equation 8.7 to estimate the number of photons being detected and thus the shot noise. For example, suppose I_t reads 3.5 V with a gain of 1×10^8, at 7500 eV, using nitrogen gas. The current is thus 3.5×10^{-8} A, and Equation 8.7 tells us there are about 5×10^8 photons per second. If we were to use five scans, with an integration time of 3 seconds, that would give a total of about 8×10^9 photons per data point. We could then use calculations like those in Section 8.2 to estimate the amount of shot noise expected.

Note that saturating an amplifier is different from saturating a detector, a particular concern with energy-discriminating detectors.

There is no point in getting too fussy about gains, as long as you don't saturate the amplifier. With decent electronics, an I_t that ranges from 0.1 to 0.7 V over the course of a scan won't be a lot noisier than one that ranges from 1 to 7 V. If it were 0.01–0.07 V, though, I might start to worry.

Keep an eye on the detector readouts on the beamline computer each time you close the hutch shutter. If the values differ significantly from zero, you need to measure the offsets again!

A *fluorescence spectrum*, such as Figure 8.9, is a spectrum showing the relative amounts of photons coming off of a sample as a function of the energies of those photons; the energy of the incident photons is held constant. An *absorption spectrum measured in fluorescence*, on the other hand, is an x-ray absorption spectrum showing the relative amounts of absorption as a function of the energy of the incident photons, as determined by the amount of fluorescence coming off of the sample.

EXAFS feature represented by the difference between the amplifier outputting 4.720 and 4.755 V will be much larger than these sources of noise. But the same signal could be given by the difference between 0.004720 and 0.004755 V if the amplifier is set at a lower gain. We are now asking the components in the signal chain to distinguish differences of a few parts per million of their full dynamic range. This is a much more challenging task, and our spectrum may appear noisy or distorted.

Therefore, we should choose the gain on each amplifier to be as high as it can be without risking saturation. While I_o is likely to vary only modestly across the course of a spectrum, I_t (and I_{ref}, if it is also measured in transmission) are most likely to saturate the amplifiers either just below the edge or at the highest energy of the scan and should be checked at those energies. I_f, on the other hand, is most likely to saturate an amplifier just above the edge, particularly if there is a strong white line feature.

On some beamlines, you also need to be concerned with *dark current*, the output of a current amplifier when there is no input signal. Why is there dark current? Some detector chains are not as linear at very low voltages. Also if zero signal was to correspond to zero output voltage, random variations could sometimes result in negative voltages—an eventuality that some detector chains are not designed to handle. So some current amplifiers provide the ability to choose a level of dark current (also known as an *offset*). Your beamline scientists will know what setting works best for the electronics on your beamline.

If there is dark current, however, it is imperative that it be subtracted back off before calculating a spectrum! This is easily done by software on the beamline computers, but to do it, the computer has to know how big the offsets are. Beamline software will typically have a feature that allows the output of the detector chains to be counted for a few seconds *when the hutch shutter is closed*. The measured offsets are then automatically subtracted from the detector readings displayed by the software.

8.7 Suppressing Fluorescent Background

If you are using a Lytle, PIPS, or other total-yield fluorescent detector, the fluorescent background should be suppressed because it contributes significantly—often dominantly—to noise. If you are using an energy-discriminating detector, the fluorescent background contributes significantly to the total counts and can thus force you to reduce the flux to avoid dead-time nonlinearity and saturation; in other words, it may force you to reduce your signal. Either way, suppressing fluorescent background is key to optimizing your signal.

Figure 8.9 shows a typical fluorescence spectrum obtained from an energy-discriminating detector.

Notice that the peaks appear about 500 eV wide at their base. This may surprise you if you've heard that such detectors can achieve energy resolutions of better than 200 eV. There are several reasons for this:

■ Energy resolutions for detectors are usually given as full width at half maximum (FWHM). That's a useful measure if you are interested in knowing how precisely you can pin down the energy of a peak. But that's not generally the question when using an energy-discriminating detector for XAFS—you know where the fluorescent lines are, but want to be able to choose energies so that one is not 'contaminated' by counts belonging to another. For that purpose, it's more important to know the broadening of the peak at the point it rises above the background at its base. This can easily be two or more times the FWHM.

■ Some energy-discriminating detectors have the ability to trade-off dead time with energy resolution—the better the energy resolution, the greater the dead

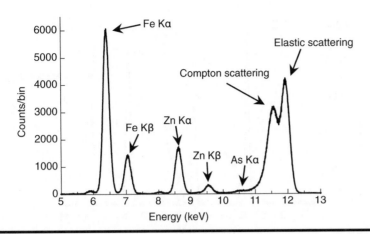

Figure 8.9 **Fluorescence spectrum, measured by a silicon drift detector for a soil sample exposed to x-rays with energy 11,967 eV (100 eV above the arsenic *K* edge). The bins are 4 eV wide.**

time, and the lower the rate at which counts can be collected without distortion. Thus, energy-discriminating detectors are often not operated at their best energy resolution.

■ Fluorescent lines usually consist of several closely related transitions that are too close in energy to be resolved, but far enough apart in energy to broaden the peaks. The copper *K*α peak, for example, comprises two prominent transitions at 8027 and 8046 eV.

The fluorescence spectrum in Figure 8.9 barely shows the arsenic peak we're interested in; it's the low feature around 10,500 eV. The peaks at lower energy are due to other elements in the soil, notably iron and zinc. Above the arsenic peak are the photons that were scattered from the sample. The highest energy peak is due to *elastic scattering*, that is, photons that scatter off of the sample losing very little energy in the process. The lower of the two scattering peaks is due to *Compton scattering*, in which the photon scatters off of a single electron (Compton 1923). Unlike fluorescent lines, both peaks will shift upward as the monochromator energy is increased.

It is useful to know the energy at which the Compton peak will appear, relative to the incident x-ray energy. Assuming the scattering is at 90° (see Section 6.4.4 of Chapter 6), the formula is

$$\Delta E = \Delta \left(\frac{hc}{\lambda} \right) \approx -\frac{hc}{\lambda^2} \Delta \lambda = -\frac{E^2}{hc} \frac{h}{m_e c} = -\frac{E^2}{m_e c^2} = -\frac{E^2}{511\,keV} \qquad (8.8)$$

At 11,967 eV, Equation 8.8 tells us to expect the Compton peak to be about 300 eV less than the elastic peak, in agreement with Figure 8.9.

8.7.1 Suppressing Scatter Peaks

To illustrate most clearly the effect of filters and, in the next section, shielding, suppose we are interested in measuring the gold L_3 edge of a sample using fluorescence and that the fluorescence spectrum looks like the one in Figure 8.10.

The spectrum clearly shows the gold *L*α line we're interested in measuring, but there are also a substantial number of counts coming from calcium, iron, and scatter peaks. (There's also a gold *L*β line hidden in the scatter peaks.)

To suppress the scatter peaks relative to the gold *L*α, we can insert a filter between the sample and the detector. The filter needs to be something that absorbs more at the

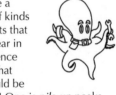

There are a couple of kinds of artifacts that can appear in fluorescence spectra that you should be aware of! One is *pile up* peaks, caused by two photons arriving very closely together in time and being counted as one. They show up as small peaks at an energy corresponding to twice the energy of a large peak or the sum of the energies of two large peaks. Another possible artifact is *escape* peaks, in which a fluorescent photon from the detector substrate (usually silicon or germanium) escapes the detection region, lowering the detected energy by an amount equal to its energy. It is important to realize that in both cases, the recorded events do not correspond to the energy of actual photons! For example, if there is a very large amount of calcium in your sample, there may be a pile up peak at twice the energy of the *K*α line of calcium, that is, at 7.4 keV, which is close to the *K*α line of nickel. But the pile up calcium peak is actually made of two *K*α calcium photons, each with energy 3.7 keV, which can be suppressed accordingly (e.g., by the technique described in Section 8.7.2).

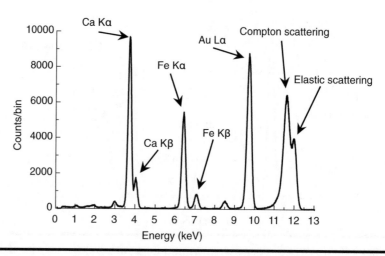

Figure 8.10 **Fluorescence spectrum, measured by a silicon drift detector for a sample exposed to x-rays with energy 12,000 eV (just above the gold L_3 edge). The sample consists of gold/iron nanoparticles embedded in a matrix of calcium acetate.**

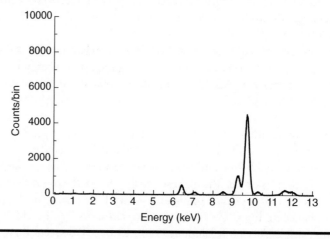

Figure 8.11 **Fluorescence spectrum for the sample from Figure 8.10, with a 3 absorption-length gallium filter placed between sample and detector.**

Make sure to understand the part written in italics, my friends! The difference between the absorption edge of an element and the fluorescent line resulting from it is fundamental to the idea of a filter.

Many people learn the 'Z-1' rule for choosing filters. Under that rule, they just choose a filter with an atomic number one less than the edge that they are measuring. That works pretty well for *K* edges, but not so well for *L* edges, so you do need to understand the principles involved.

energy of the scatter peaks than at the energy of the gold *L*α line. Since that requires more absorption at higher energy, it means *we need something with an edge between the 9.7 keV energy of the gold* Lα *fluorescence line and the 11.9 keV energy of the gold* L₃ *absorption edge.* Gallium, with an edge at 10.4 keV, is a good choice.

Filters do not need to be pure; they are often partially or entirely oxidized and may include low-*Z* binders or matrices to provide structural integrity. The thickness of filters is usually specified by giving the number of absorption lengths just above their edge.

Figure 8.11 shows a fluorescence spectrum for the same sample as in Figure 8.10, taken under the same conditions, except that a gallium filter, 3 absorption lengths thick above its edge, was inserted between sample and detector.

Three absorption lengths of gallium at its edge is about 23 μm. Twenty-three μm of gallium at 12 keV, where the scatter peak is located, is roughly 2 absorption lengths, meaning the scatter peak should be decreased by nearly 90%. In contrast, it is only about half of an absorption length at 9.7 keV, where the gold *L*α line is located, meaning that the intensity of that line is only decreased by about 40%. In addition, the usual rise of absorption with lower energies means that 23 μm of gallium is about 7 absorption lengths at the calcium *K*α line, meaning that 99.9% of that fluorescence is suppressed!

While that sounds good—we've removed most of the scatter and the low-*Z* fluorescence at the price of less than half of the signal—it may not always represent an improvement. For example, consider a total-yield fluorescent detector (Lytle, PIPS, etc.). Suppose that, without a filter, there are 10,000 counts per second in the line we want to measure and 2,000 counts per second in the scatter peaks and low-*Z* fluorescence. (Such a spectrum would look very different from Figure 8.10.) Poisson statistics predict that the noise in one second without a filter would be about $\sqrt{12,000}=110$ counts. Reducing the scatter by 90% and the signal line by 40% would reduce the total number of counts per second to 6,200, and the number of noise counts to 79 per second. But the signal-to-noise ratio, rather than being 10,000:110 = 91, would *drop* to 6,000:79 = 76. The filter in that case would be counterproductive.

In addition, filters have another, undesired, effect. Look at Figure 8.11 again and note the new peak around 9.3 keV. That is the *K*α line for the gallium, which has now been inserted into the beam path. However, it is much smaller than the amount of scatter and low-*Z* fluorescence it has replaced for two reasons. First of all, some of the gallium excitations resulted in Auger-Meitner electrons or lower-energy gallium fluorescent lines. Second, since fluorescence is isotropic, not all of the gallium *K*α fluorescence travels into the detector.

With energy-discriminating detectors, fluorescence by the filter can be particularly annoying, as it often ends up overlapping the fluorescent line you are trying to measure!

One solution to suppressing fluorescence by the filter is to place some kind of collimator between the filter and the detector, so that only photons traveling near a line of sight back to the sample can reach the detector. An example of such a collimating system is *Soller slits* (Stern and Heald 1979), such as those shown in Figure 8.12. These slits are each oriented so as to allow x-rays emanating from a small region (the location

Often, beamlines will have filters available for use with your experiment. If you are not sure, ask!

Figure 8.12 Soller slits, oriented correctly for a beam coming from the left. It is evident how the slits are focused on a small region of the figurine; this region would be visible through the slits from any angle. An object in any other position would generally be blocked by the slits.

The history of the term 'Soller slits' in this context is worth mentioning. Walter Soller (1924) invented a system of *parallel* slits for collimating x-rays. Stern and Heald (1979), in their work, referred to the focused slit arrangement as 'a Soller-type slit assembly.' For some reason, rather than later practitioners referring to this geometry as 'Stern–Heald slits,' the term 'Soller slits' stuck. From there, some have begun to refer to them as 'solar slits,' presumably via a folk etymology, which visualizes rays diverging from the sample like light from the Sun.

Soller slits have to be made of something! Make sure your Soller slits don't themselves add fluorescent background. Frequently Soller slits are made from, or coated with, silver, since the *K* lines of silver are higher than the most common range for EXAFS measurements, and the *L* lines are lower. But if you were to try to measure a silver edge using silver Soller slits, you'd have a serious problem: not only would the slits contribute substantially to the fluorescent background, but they would absorb the scatter peaks in a way that was strongly energy-dependent, contaminating your signal! Measurements at the cadmium *K* edge would also be contaminated; at the cadmium *K* edge of 26.7 keV, the Compton peak lags 1.3 keV behind the elastic scatter peak and thus would scan across the silver *K* edge (25.5 keV). Even measuring indium with silver Soller slits is probably not a good idea. In these cases, if the Soller slits are silver, either leave them off or replace them with Soller slits made from another material.

of the sample) to pass, while blocking x-rays traveling in most other directions. Soller slits need to be positioned with care, as they are effectively focused on a point and not just a direction; it is necessary to have the sample at the correct distance from the slits (Bewer 2012). The region on which the slits are focused is easy to determine simply by looking through them, as in Figure 8.12.

While filters are a reasonably effective way of suppressing scatter peaks, it can also help to move the detector further away from the sample. As described in Section 6.4.4 of Chapter 6, scattering is at a minimum near 90° from the original beam. Moving the detector back reduces the solid angle it intercepts and thus limits it to angles closer to 90°. Notice that if a filter is used in addition, the filter should be left near the sample, not moved back with the detector, as this will help reduce the fraction of the filter's fluorescence that is directed at the detector.

Moving the detector back is not usually a net improvement for a total-yield detector, since the loss of signal will outweigh the advantage of reducing scatter. But if the number of counts for an energy-discriminating detector needs to be reduced anyway because of dead time, then backing up the detector is preferable to, for example, reducing the amount of beam incident on the sample.

Notice that for a total-yield detector, you don't have a fluorescence spectrum as in Figure 8.10 to work with. Still, if you can guess more or less what that spectrum looks like for your sample, based on composition and concentrations, you can make good choices about filters and slits. In fact, filters and slits are somewhat less important for energy-discriminating detectors, since you get to choose one slice of the fluorescent spectrum to measure anyway. But since energy-discriminating detectors generally have limitations on the number of counts per second that they can record that are low compared to the maximum number of counts that can come off of a sample, it's often useful to be able to reduce unwanted sources of counts.

8.7.2 Suppressing Low-Energy Peaks

Scatter peaks aren't the only source of unwanted counts in Figure 8.10. There is also a prominent calcium peak that we are not interested in measuring and an iron peak as well. Since they are at lower energy than the peak we are interested in, we can use a low-*Z* material to preferentially absorb the lower energy photons. Aluminum foil is quite convenient for this, as it is thin, readily available, and easy to work with. Choose the amount of aluminum foil so as to not have much of an effect on the line you do want to measure. For example, in Figure 8.10, we want to measure the gold *L*α line. At that energy of 9.7 keV, one absorption length of aluminum is 128 μm. Heavy-duty aluminum foil is typically 24 μm (0.9 mil) thick. Using two sheets of that thickness will give 0.38 absorption lengths at the gold line, which will reduce the intensity of the line by 32%. But that same amount of aluminum foil at the calcium line would be 5.9 absorption lengths, resulting in the suppression of 99.7% of the calcium line. Even at the iron *K*α line, that much aluminum foil provides 1.2 absorption lengths, suppressing 70% of the iron signal. The result of adding two layers of aluminum foil is shown in Figure 8.13.

Unlike filters, there is very little fluorescence from aluminum foil used in this way, as most of the energy goes into Auger-Meitner electrons, and any fluorescent photons that do get emitted are absorbed within a short distance (a few centimeters) in the air.

Figure 8.14 shows the effect of the combination of the gallium filter and the aluminum foil.

As with scatter peaks, the suppression of low-*Z* peaks can also be aided by moving the detector back from the sample, this time because the air will act as a low-*Z* absorber, much as the aluminum foil does.

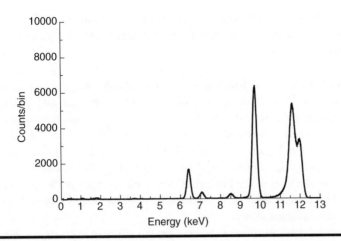

Figure 8.13 Fluorescence spectrum for the sample from Figure 8.10, with two sheets of 24-μm aluminum foil placed between the sample and the detector.

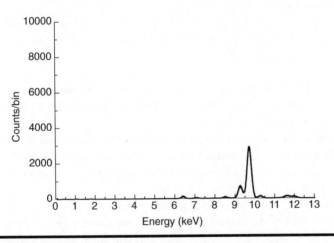

Figure 8.14 Fluorescence spectrum for the sample from Figure 8.10, with a 3 absorption-length gallium filter and two sheets of 24-μm aluminum foil placed between the sample and the detector.

8.7.3 Making the Choice

It is very important to remember that the goal is to increase signal-to-noise ratio, rather than to increase signal to background. The difference is subtle, but important: Poisson statistics tell us that even if there were no fluorescent background at all, there would still be noise given by the square root of the number of signal counts.

With that in mind, the number of counts measured in the fluorescence spectra shown in Figures 8.10 through 8.14 are given in Table 8.1, along with the number

Table 8.1   Quantitative Analysis of Figures 8.10 through 8.14

Filters	Total Counts (million)	Signal Counts (million)	Total-Yield S/N	Partial-Yield S/N
None	2.74	0.59	1010	770
Ga-3	0.50	0.30	1200	770
Al	1.43	0.42	990	900
Ga-3+Al	0.30	0.19	980	620

One particularly annoying problem with dilute fluorescence samples it that, if care is not taken, your reference foil may fluoresce into the fluorescence detector! Some people choose to remove the reference foil in such cases, but it's usually pretty easy to solve the problem by putting some shielding in the line of sight from the reference foil to the fluorescence detector.

A rule of thumb: with a total-yield fluorescence detector, always place the detector as close to your sample as you conveniently can, given the geometries of stages, filters, slits, and so on. With those kinds of detectors, the top priority is to maximize the number of counts for the signal. With energy-discriminating detectors, though, move them away from the rest of the components until the total counts are in a range where the detector is known to behave linearly. If you need to reduce counts anyway, there are added benefits to doing it by increasing the distance of the detector from the sample, filters, slits, and shielding!

of counts in the 9.5–10 keV range ('signal counts'). As an example of how signal-to-noise ratio could differ between these configurations, we make the following assumptions:

- Often, particularly on beamlines with insertion devices, the flux incident on the sample must be reduced to avoid nonlinearity in an unfiltered energy-discriminating detector. For purposes of this example, we assume that, in the unfiltered case, the incident flux had been reduced by a factor of two. Therefore, in the cases shown in Figures 8.11 through 8.14, we calculate the signal-to-noise ratio for the energy-discriminating detector by scaling up the counts by a factor of two or to the level of the unfiltered case, whichever is less. We also assume that, were a total-yield detector to be used instead, we would also scale the incident flux up by a factor of two.
- Total-yield detectors often subtend a greater effective angle than energy-discriminating detectors (Section 2.4 of Chapter 2). For this example, we assume the total-yield detector has four times the effective collecting area as the energy-discriminating detector.
- For the energy-discriminating window, we use 9.5–10 keV. We neglect the contribution of the gallium $K\alpha$ peak to this region, as in these cases it is small compared to the gold $L\alpha$ peak.

Using these assumptions and Poisson statistics, the signal-to-noise ratio can be computed for the case of a total-yield detector ('Total-Yield S/N') and an energy-discriminating detector ('Partial-Yield S/N').

In the particular case described, the Ga-3 filter would be best for the total-yield case and the aluminum foil would be best for the case of the energy-discriminating detector. In this case, the signal-to-noise ratio is actually a bit better for the total-yield detector than the energy-discriminating one.

These results are strongly dependent, however, on the relative numbers of counts in the various peaks of the fluorescent spectrum, on the energies of those peaks, and on the amount by which the incident flux was cut down in the unfiltered energy-discriminating case. It also depends on the thickness of the filter or aluminum foil used. There are no simple 'rules of thumb' on the circumstances when filters are helpful, which kinds to use, and to which thicknesses. In practice, many experimenters simply use trial and error at the beamline to see which works best in a given case. As this section shows, however, it is also possible to compute what combination should be best, either from a fluorescence spectrum or, at least roughly, from estimating the composition of your sample (see Bunker [2010] for further discussion of these kinds of computation).

8.8 Aligning the Sample

This section addresses alignment of a sample for which spatially resolved information is not desired. Microprobe beamlines usually have their own imaging systems, which allow you to choose interesting areas of the sample for measurement.

Of course, it is very important that your beam hit the part of your sample you want it to hit, rather than being partially on a frame or an air gap! On a microprobe beamline, there is an imaging system to help guide you, but on other beamlines, the alignment system might involve any of a number of combinations of video images, motorized stages, phosphorescent cards, registration marks, and 'burn paper,' which turns color when exposed to x-rays.

It is often helpful to have your sample masked or framed, so that the well-prepared part of your sample is surrounded by a material that absorbs more x-rays than the sample does. This makes it relatively easy to tell whether your beam is fully on the sample, just by positioning it so as to maximize I_t. If a motorized stage is

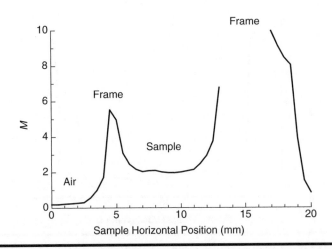

Figure 8.15 **Horizontal scan across an yttrium barium copper oxide transmission sample, measured above the copper *K* edge.**

available, as is often the case, you can scan across the sample, producing a plot like Figure 8.15.

A remarkable amount of information can be deduced from a plot of this type. First, you need to orient yourself to what you are seeing. Positions where there is no sample, frame, or sample holder will have the lowest absorption. Frames usually have higher absorption than the sample, but not always, particularly if the sample is rich in high-*Z* elements. In Figure 8.15, you may wonder why the graph appears missing between 13 and 17 mm. That's because the frame is absorbing so much as to leave no counts in I_t. When we calculate *M*, we thus divide by zero (or a negative number, if the offsets are a bit off) and the result cannot be plotted. This happens quite frequently on scans of this type.

In Figure 8.15, we can easily read off the number of absorption lengths that the sample absorbs at this energy: air is reading about 0.2 and the sample about 2.0, so the difference of 1.8 must be the absorption of the sample. Note that the *values* of *M* are not meaningful, as they depend on things like the gains of the amplifier and the gases in the detectors; it is the *differences* in *M* that matter.

Next, notice that the transitions from air to frame, frame to sample, and sample to frame are not sudden. Each seems to take about 2 mm, and in fact the transition from air to frame is not quite complete before the transition from frame to sample begins. This implies that the beam is about 2 mm wide, and thus, the transitions on the graph correspond to circumstances where the beam is partially on one region and partially on another. We can also conclude that the beam is entirely on the sample from about 7 to 11 mm, which means the sample is 6 mm wide: the 4 mm of travel when the beam is entirely on the sample, plus the 2-mm width of the beam. We also know that the section of frame around the 5-mm mark is not quite 2 mm wide, as the beam cannot quite fit on to it. These figures can all be checked with the dimensions of the actual sample and frame, providing confirmation that the sample is aligned as intended.

Finally, we can see that the sample is quite even, with deviations of no more than 0.2 absorption lengths. Consider, however, Figure 8.16, showing a vertical scan across a sample of calcium nitrate.

We see the typical profile of a sample mounted in a frame. There's a little hitch in the rise up to the frame at around 9 mm; this might indicate a mounting hole or something like that, but it's outside the range we'd think about using for data. There are also thinner spots near both ends of the sample (2 and 8 mm). This sample was

Shouldn't it be 8 mm: the 4 mm of usable area plus the 2 mm for the beam on each side?

No, Simplicio. Picture the leading edge of the beam as it moves on to the sample. It traces out its own width of 2 mm, then it traverses 4 mm more of sample. But then it's on to the frame on the other side; it's now the trailing edge of the beam that is causing the transmission. So the sample must be only 6 mm wide.

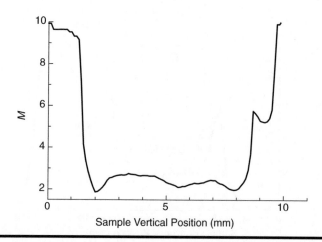

Figure 8.16   Vertical scan across a calcium nitrate transmission sample, measured above the calcium *K* edge.

prepared by the method of thinly brushing powder on to tape, a method that often leaves the edges of the tape with thin spots or pinholes. Again, we would not usually be inclined to collect data from the very edges of the sample anyway. But there is also a distinct thin spot around 5.5 mm. The drop is more than 0.5 compared to the thicker, even region around 2.5–4.5 mm. Since unevenness is the worst enemy of a transmission sample (see Section 6.3 of Chapter 6), it would be much better to place the sample at a position around 3 or 4 mm on the plot than to risk whatever is causing the nonuniformity around 5.5 mm.

Of course, if the sample is irregular on a scale much smaller than the beam, we would not detect that by this method. If irregularities such as thin spots are suspected and you want to test for them, you should narrow the beam to be smaller than the expected scale of irregularities before scanning the position.

So far, we have discussed aligning a sample measured via transmission. Even if you plan your primary data to come from fluorescence or electron-yield measurements, it is often a good idea to align using transmission, assuming you can get any signal through the sample at all. If your sample is too thick to measure any transmission, you can align using fluorescent measurements, but that's harder to do. Frequently frames and sample holders will fluoresce strongly, so it can be difficult (but crucial!) to tell what is sample and what is something else. It often helps to take a motorized scan without the sample present, to try to understand how the other components look. Or you could fall back on a visual method of alignment, using phosphorescent cards, burn paper, or registration marks to assure that the beam is falling on the sample and only the sample. It is also probably not a good idea to hunt around for a high-fluorescence spot if the sample is supposed to be uniform. If you do find such a spot, it means you have found a region that is not characteristic for your sample and, thus, is not what you want to measure. Perhaps there are even microscale concentrations of your element at that spot that could lead to self-absorption, even if the overall concentration of the element is low. Searching for the highest fluorescence by moving the beam around on the sample also tends to be subject to geometrical considerations; for example, the total fluorescence entering the detector will usually be higher if the beam is on the part of the sample nearest the detector. Moving off-center in that way, however, may have unintended consequences, such as creating an uneven distribution of counts across the elements of a multielement detector or interfering with the function of Soller slits. Most of the time, if you cannot get any transmission data for your sample, it is best to align the sample so that the beam hits it in the center.

True story: our author once spent several hours measuring the iron fluorescence from a screw in his custom-designed flow cell, thinking he was measuring the sample. Oops. For the next run, he switched to nylon screws.

BOX 8.3   Which Order?

	Sections 8.5 through 8.8 sort of look like they're written in the order I'd do them at the beamline, but we didn't actually say that. Am I right?
	As a first approximation, friend Simplicio, yes, the order given here is a reasonable one. But the reality is that when we change one aspect, we will often have to readjust others. For example, if, after aligning the sample, we discover that the sample is particularly thin, we may need to adjust the gains on our amplifiers to avoid saturation, which then means we will also need to measure offsets again. Or if the sample were particularly thick, we might decide to detune a little more. Or if our energy-discriminating detector was recording too many counts, we might decide to detune more, or maybe change the slit size. And it's frequently the case that we have to adjust filters, slits, shielding, and detector position after we find the perfect spot on the sample via alignment.

8.9 Scan Parameters

Although the details are different from beamline to beamline, you will be able to choose aspects of how the monochromator steps through energy. There are two main parameters under your control: how big the steps in energy are between points (*step size*) and how long data are collected at each point (*integration time*). Most beamline software will allow you to specify different step sizes and integration times for different regions of the scan and some allow one or both of those to vary smoothly with energy within the defined regions as well.

8.9.1 Scan Regions

If you're collecting EXAFS data, you'll usually divide your scan into three or four regions (for XANES-only scans, you'll skip No. 3 in the below list):

1. *Preedge*. The preedge region is crucial for normalizing the spectrum. You need a long energy range to establish the trend, but there's no detail in this region, so you don't need a finely spaced grid or to spend much time per point. Starting 200 eV below the edge is typical. A step size of 5 eV is reasonable, and unless the signal is unusually noisy, an integration time of 1 second is usually sufficient. (Times below 1 second per point are rarely efficient. The monochromator needs a little time to move and react to the lock-in system between each point, and during that time, the beamline software will instruct the computer not to collect data. This time to move and settle is likely to be several hundred milliseconds. Reducing integration time below 1 second, therefore, doesn't usually result in much time savings.)
2. *XANES*. If you're planning to analyze XANES, this region is more important than if you're just planning to do EXAFS, but even for EXAFS measurements, it is helpful to get good data in this region to aid with alignment. Signal-to-noise

Quick-XAFS doesn't work the same way as traditional XAFS. In quick-XAFS, the monochromator is continuously slewed through the energy range, and the data are collected continuously and binned in a subsequent step. Section 8.9, therefore, does not apply directly to quick-XAFS experiments.

I have never seen anyone actually use k⁴ weighting of integration times, for a practical reason. If the EXAFS region began at 30 eV above the edge and extended to 1000 eV, that would correspond to a k-range of roughly 3–16 Å⁻¹. If the integration time for the start of the region was 1 second, k⁴ scaling would mean the integration time at the end would need to be 800 seconds! That would make for impractically long scans, and the scalers used on most lines cannot count that high. But choosing a more reasonable time at the top of the range, such as 10 seconds, would give about 12 ms at the start of the range, which is pointlessly short considering the time for the monochromator to move and settle. For these practical reasons, plots of k²χ(k) or k³χ(k) will almost always show increasing noise at higher k!

Of course, sometimes you just can't collect data up as high in energy as you'd like. For one thing, there might be another edge up there! That's particularly true if you're trying to measure an L_3 edge—the L_2 edge is always lurking not too far above it!

ratio will be better in the XANES region than the EXAFS region, but features will also generally be sharper. Although energy resolution may be a few eV (see Section 8.5.2), it's a good idea to oversample a bit (i.e., make the step size a few times smaller than the resolution). Glitches, for instance, are often quite narrow in energy, and oversampling can help reduce their impact. Oversampling also acts in a similar way to collecting multiple scans: measuring four points, spaced a half eV apart, for 1 second each may provide a similar contribution to the data as measuring a single point for 4 seconds in the middle of that range, but observing the variation between the four points can give a much better sense of the noise in the data. Since *fractional* broadening owing to core-hole lifetime is roughly independent of energy, the broadening as measured in eV increases with energy. For that reason, while 0.5 eV is a typical step size for the K edges of the first row of transition metals, 1.0 eV is reasonable for the K edges of, for example, arsenic and bromine, and 0.2 eV is appropriate for lower energy edges such as sulfur. Integration times depend on the signal-to-noise ratio and on how important XANES is to your analysis. For concentrated samples, 1 second is often fine, while dilute samples may work well with 10 seconds or even more. For purposes of defining scan parameters, you should begin the XANES region a bit below the nominal edge (perhaps 20 eV) and continue it well past the white line.

3. *EXAFS.* The EXAFS region is most naturally thought of as a function of k. Because E is proportional to the square of k, features will tend to broaden and reduce in amplitude as you get further above the edge. In addition, the signal falls off with increasing energy, further reducing the amplitude of features high above the edge. Ideally, therefore, step size in energy should increase as the scan moves further above the edge, and integration time should increase as well. Many beamlines allow step size in the EXAFS region to be defined in terms of k; if so, a step size of 0.05 Å⁻¹ provides an appropriate amount of oversampling. If not, it is often worthwhile to divide the EXAFS region up into smaller regions and adjust the step size so as to approximate 0.05 Å⁻¹ in k. Since E ≈ $4k^2$ when E is in eV and k is in Å⁻¹ (see Dysnomia's comment in Section 7.4 of Chapter 7), a bit of calculus tells us that $\Delta E \approx 8k\Delta k \approx 4\Delta k \sqrt{E}$. For a spacing of 0.05 Å⁻¹ in k, that means a step size of about $0.2\sqrt{E}$. At 100 eV, this corresponds to a step size of about 2 eV, while at 1000 eV, it corresponds to about 6 eV. Integration times can be considered in terms of noise and k-weighting: to keep constant noise in a plot of $k\chi(k)$, Poisson statistics require the integration time to be weighted by k^2; to keep constant noise in a plot of $k^2\chi(k)$ would require integration times weighted by k^4! As with step size, some beamline software allows weighting integration times by a function of k; for those that do not, the EXAFS region can be divided up into smaller regions and the integration times chosen so as to roughly mimic the desired weighting.

4. *Postedge.* The region above the XANES region is important not just for EXAFS, but also to help determine normalization. Even if you are only interested in XANES, this means you need to collect data over a broad energy range above the edge—perhaps 400 eV. But if you are not interested in EXAFS, this can be treated as the preedge was, perhaps with a step size of 5 eV and integration times of 1 second. And if EXAFS is collected, it is often good to collect a bit beyond the end of your analyzable EXAFS data, to help establish the background function at the top of the range. For this reason, some people add a region about 2 Å⁻¹ in width following the EXAFS region, using a step size of perhaps 5 eV and an integration time of 1 second. This adds very little to the duration of a full EXAFS scan, but can pay off in better determination of the background.

BOX 8.4   XANES Resolution

	I really must object to the implication that there is no point to making the step size in XANES much finer than the energy resolution! Resolution is not the same as accuracy. Suppose that one is fingerprinting a preedge peak. If the resolution of a scan is 1.0 eV, then two peaks 0.1 eV apart could not be resolved, but a peak shift of 0.1 eV might very well still be visible.
	That peak shift would be implicit in data taken at a wider spacing, Robert. You don't really need to collect the energy on such a fine grid.
	Perhaps, but as a practical matter, it is much easier to fit a peak that is explicit in the data than to try to reconstruct it from small differences on a broader grid. Doing so puts one at the mercy of interpolation algorithms, which is a dangerous thing.
	Yeah, and if you like to plot the second derivative, you've got similar algorithmic issues with a grid that's too coarse.
	I concede the point. But those are only an issue for fingerprinting and visual techniques. Linear combination analysis, PCA, and modeling shouldn't need the finer grid.
	'Visual techniques?' If you include any graphs in your papers, Carvaka, the people reading them will be using 'visual techniques' when they look at them! It's easier to understand what's going on in XANES if the features are oversampled. But yeah, it's partly a judgment call and a matter of personal preference. Personally, I'd rather spend a few more minutes per scan to have small shifts in features leap out at me during analysis, but hey, with your data it's your call.

8.9.2 Number of Scans

To get a good sense of what is noise and what is feature, it is important to conduct more than one scan of each sample under each set of conditions. Even with a system undergoing time evolution, it is desirable to have individual scans be short enough that not much has changed about the system from one scan to the next.

To a first approximation, doubling the number of scans has the same statistical effect as doubling the integration time, but there are some subtle differences. In terms

'Desirable?' I guess in a perfect world, sure. But many systems that undergo time evolution show very rapid changes initially, and then slow down. If I'm scanning fast enough so that when the time evolution of the system is at its slowest there's not much difference between scans, I can use that part of the sequence to get an estimate of noise. And sometimes I can't even do that—maybe I have a system that's evolving just a bit faster than the monochromator can keep up with. In that case, I'll just have to be satisfied with estimating noise by running the whole experiment more than once.

of total duration, increasing integration time is more efficient, as it does not increase the number of times the monochromator has to move—including the long move from the highest energy to the lowest energy between each scan. On the other hand, there are disadvantages to long integration times:

Poisson statistics tells us that while you can use integration time and multiple scans to clean up a section of $\chi(k)$ where a noisy signal is evident, it usually isn't worth it to try to make a signal appear where you can't discern one at all.

- Typical beamline scalers can only count at full voltage for perhaps 10 or 20 seconds before they reach their limit.
- XAFS measurements can sometimes have a subtle time dependence—perhaps the gases in the detectors are slowly changing composition, temperature, or pressure, or perhaps the decay of the current in the synchrotron itself has nonlinear effects on the measurement, such as by changing the heat load on the monochromator. These changes are generally slow enough that they show up as subtle differences in the background of different scans or as a shift to higher or lower absorption from scan to scan. Both those effects tend to be removed in the process of data reduction. But if integration times are excessive, these changes can begin to appear on the time scale it takes to measure a single EXAFS oscillation, meaning that they would not be removed by the process of background subtraction.
- The sample itself may change over time, either because of damage caused by the beam or because the sample is air sensitive. Shorter integration times allow changes of this kind to be discovered. If beam damage is occurring, the sample can be moved slightly between each scan, so that the beam is always hitting a fresh spot.
- If, for some reason, you lose beam during a scan (light sources are not 100% reliable!), the scan you are in the process of collecting will need to be restarted when the beam returns. If you're running long scans, that's more of a loss than if the scans are short.

Friend Dysnomia, we should demonstrate your assertion by computation. Suppose that the noise is comparable to the signal in some region of $\chi(k)$, providing a noisy looking, but visible signal. Increasing the product of integration time and number of scans by a factor of 9 will improve the signal-to-noise ratio by a factor of 3, which is quite helpful. But if there is no sign of a signal, it is likely that the signal-to-noise ratio is on the order of 1:10, at best (remember that we are oversampling, so a single oscillation has many points!). It would take an improvement in the product of integration time with number of scans by a factor of 100 just to bring the signal-to-noise ratio to even. If we are willing to spend a very long time, we can therefore sometimes bring a usable signal out of what appears to be only noise, but even this is unlikely to add more than about 2 Å^{-1} to the range of data we can use.

In general, for step scans aim for a minimum of 3 scans and a maximum of 15 seconds of integration time. Within those parameters, whether you prefer a greater number of scans with lower integration times or vice versa is largely a matter of personal preference.

8.9.3 *Time-Resolved Studies*

There are many situations that call for time-resolved studies but do not necessitate quick-XAFS. Cyclic voltammetry, for instance, or *operando* catalysis studies, often have time scales measured in tens of minutes or even hours. If desired, scan parameters can be adjusted to reduce total scan time and allow for finer time resolution, without sacrificing much energy resolution or signal-to-noise ratio. In such cases, here are some recommendations:

- DO trade-off integration time for number of scans
- CONSIDER using a moving average of scans to improve signal-to-noise ratio while still maintaining some time resolution (For example, average each scan with the one before and after it for analysis purposes)
- CONSIDER increasing the step size in preedge and postedge regions
- CONSIDER increasing the step size over the XANES region if you are primarily interested in EXAFS
- CONSIDER reducing how high above the edge you collect EXAFS data
- CONSIDER increasing the step size over the EXAFS region to approximately 0.10 Å^{-1}
- DO NOT reduce the size of the preedge or postedge regions. Poorly normalized data are not of much use, even if the time resolution is good!

With these techniques, you can reduce the time for an EXAFS scan on a conventional line to 10 minutes or less and you can manage a XANES scan in as little as 2 minutes.

8.9.4 Making the Most of Your Beamtime

One of the most important keys to optimizing your use of beamtime is to be willing to change your scan parameters after the first scan or two on a sample. Until you measure your sample, you don't know what the signal-to-noise profile looks like. It may be that your first scan extends to 16 Å^{-1}, but shows no discernible signal above 10 Å^{-1}—an enormous amount of time could then be saved by truncating future EXAFS scans at 10 Å^{-1}, plus an additional 2 Å^{-1} to establish a postedge.

You also might discover that you are spending more than 1 second of integration time on a part of the scan that shows no evident noise—on subsequent scans, that integration time can be reduced in favor of increasing the integration time on noisier portions of the scan.

I'd like to change my number of scans and integration time, but I've already measured my standards, and I've heard that I need to measure my samples exactly the same way as my standards. If I collected five scans for each of my standards, I should have five scans for each of my samples, right?

8.10 'What's That?'

Many features show up in scans that aren't part of the spectrum we're trying to measure. This section will detail some of them.

8.10.1 Noise

8.10.1.1 Cause

Noise may be due to counting statistics in the detectors, electronic noise, or other nonreproducible sources.

8.10.1.2 Identification

Noise does not appear consistent from scan to scan (Figure 8.17).

If scan 2 had been the only scan, it would be difficult to judge if the excursions around 5.5 Å^{-1} were due to noise or some other effect, such as one of the ones described in the remainder of this section. But with four scans, it's clear that they don't recur in the same way from scan to scan—they're due to noise.

No! The only thing that taking more scans or increasing the integration time does is improve signal-to-noise ratio, and that's not something that needs to be matched for standard and sample! When people say that you should measure your samples the same way as your standards, they mean aspects like energy resolution, including the related concept of step size in the XANES region. But feel free to spend, for example, half an hour measuring a standard and 2 hours measuring a sample, if the sample is dilute!

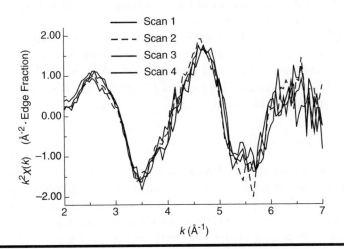

Figure 8.17 $k^2\chi(k)$ for four scans of vanadium K edge of VCl$_3$(THF)$_3$.

The small upward spike around 4.2 Å^{-1} is interesting in that, by chance, it happens to show up in two scans. Once again, though, that is coincidence, and it is due to noise.

8.10.1.3 Effect on analysis

Noise increases the uncertainties resulting from analysis.

8.10.1.4 Prevention

Noise can't be prevented, but it can be reduced, as has been discussed in the previous sections of this chapter.

8.10.1.5 Mitigation

Additional scans and increased integration time will reduce the effect of noise.

8.10.1.6 Silver lining

Noise provides one way of estimating ε, the measurement uncertainty.

8.10.2 Monochromator Glitches

8.10.2.1 Cause

At certain orientations, the diffraction peak being utilized by the monochromator can interfere with multiple reflections associated with another set of crystal planes (Van Der Laan and Thole 1988), resulting in a glitch in I_o. If the harmonic content of the glitch differs from that at the orientations around the glitch, it may show up in M. More commonly, even if the harmonic content is the same, the glitch, which will travel vertically across the slit as the energy changes, can interact with vertical inhomogeneities in the sample to produce variations in M (Bridges et al. 1991; Li et al. 1994).

8.10.2.2 Identification

Monochromator glitches always show up in I_o. They may also show up in M, sometimes inverted.

Figure 8.18 shows an example of monochromator glitches and their effect on a reasonably well-prepared sample.

The glitch structure in I_o between 7630 and 7670 eV is fairly complicated, encompassing as many as seven points. But the sample, an iron oxide measured in transmission, is quite uniform, and only the point at 7637 eV shows much effect from the glitches in M. (Careful examination suggests the next point in M, at 7642 eV, may also be modestly affected by the glitch.)

8.10.2.3 Effect on analysis

Sharp, isolated, narrow glitches have little effect on EXAFS analysis, although for large glitches it may be necessary to remove them to avoid distorting normalization and background. Glitches in the XANES region can be more inconvenient, but oversampling can often allow for their benign removal as well. Glitches sometimes occur in rapid succession; such cases can be more detrimental to analysis.

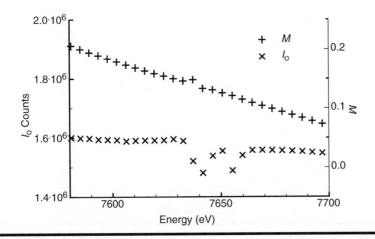

Figure 8.18 Detail of I_o and M for a transmission scan showing monochromator glitches. Individual data points are shown.

8.10.2.4 Prevention

Glitches should divide out of M when samples are uniform and detectors are linear; a well-prepared sample and a well-conditioned beam are your best defense.

8.10.2.5 Mitigation

In some cases, changing the amount of detuning may help reduce the effect of individual glitches. Since sample nonuniformity is a complicating factor, moving the beam to a different part of the sample (or repreparing the sample, if feasible) may help. Because glitches travel vertically across the slit, narrowing the vertical size of the pre-I_o slit can also help reduce the impact of glitches (Li et al. 1994).

8.10.2.6 Silver lining

Since glitches are sensitive to sample nonuniformity, if glitches in I_o are not appearing in M, you can be fairly confident that the sample is uniform on the scale of the beam height. (The sample may still be nonuniform on a smaller scale, however). Glitches can also be used to check energy calibration, particularly in cases where the simultaneous use of a reference spectrum is inconvenient.

8.10.3 Other Edges

8.10.3.1 Cause

An impurity within your sample, or something else in the beam path, including windows, tape, and so on.

8.10.3.2 Identification

Edges are best identified by noting that the energy they occur at corresponds to a known edge and by observing the characteristic shape (Figure 8.19). Even if the edge jump is very small, they tend to show a sharp rise relative to the trend of the background followed by a gradual drop. For small edges, they are often easier to see in $\chi(k)$ than in $M(E)$.

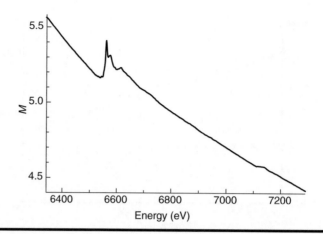

Figure 8.19 **Transmission spectrum of the manganese *K* edge of a lithium manganese oxide cathode. The small hesitation in the background visible just above 7100 eV is an iron edge due to an impurity in the sample.**

8.10.3.3 Effect on analysis

Unless you know the chemical structure of the substance causing the edge, you will probably be unable to rely on any data beyond it, even (especially?) for purposes of establishing background and normalization.

8.10.3.4 Prevention

For dilute samples, it's a good idea to measure a 'blank'—a spectrum with everything in place, including tape (if the sample is to be mounted on tape). This can help you to understand where other edges are coming from and in some cases arrange to reduce their effect.

8.10.3.5 Mitigation

If an unwanted edge cannot be removed just by changing beamline geometry (e.g., it is due to a substance in the sample), then the only mitigation is to shorten the scans so that they stop before the edge. If you can't collect data out further than that, at least you can spend more time collecting data at lower energies.

8.10.3.6 Silver lining

If the edge appears to be from an element in your sample (i.e., it doesn't show up in the spectrum of a corresponding blank), then you have learned additional information about your sample composition.

8.10.4 Other EXAFS

8.10.4.1 Cause

A large edge *below* the energy of the scan can be associated with EXAFS oscillations that extend into the scan region.

8.10.4.2 Identification

Usually when this is the case you know to expect it as a possibility, because you are aware of the large edge responsible. The degree to which the EXAFS oscillations are

What if we're using an energy-discriminating detector? Then we wouldn't be measuring the fluorescence of the new edge.

But the element responsible for the new edge would still be absorbing, thus reducing the number of photons available to be absorbed by the element we are trying to measure. In this way, the new edge would affect the intensity of the line we were interested in.

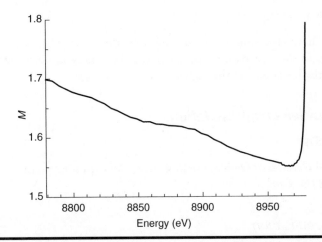

Figure 8.20 Transmission spectrum of the region before the copper *K* edge of a sample of nickel–copper nanoparticles. The spectrum shown is an average of three scans, confirming that the broad peak around 8870 eV is consistently present.

present can be determined by looking at the preedge region—see Figure 8.20 for an example. These oscillations will sometimes also show up as a low-*R* peak in the Fourier transform.

The preedge background in Figure 8.20 appears to wiggle on a scale consistent with EXAFS oscillations 600 eV or so above an edge. Since the sample was known to contain large amounts of nickel (E_o = 8333 eV), these oscillations were suspected to be EXAFS from the nickel edge. Modeling later confirmed this (Carroll et al. 2010).

8.10.4.3 Effect on analysis

In many cases, only the near-neighbor peak from the lower edge has significant amplitude in the scan region. Because E is proportional to k^2, these oscillations will be very broad in energy, and when Fourier transformed relative to the higher, measured edge, they will usually be shifted down below the region of the data and thus removed with the background. There are exceptions, however. For example, if the lower edge is an elemental metal or an alloy, the nearest-neighbor peak may have a half-path length of as much as 3 Å. Even if that was shifted down, it might still appear in the range associated with near-neighbor oxygens for the measured edge.

8.10.4.4 Prevention

This problem is usually better mitigated than prevented. If prevention were truly desired, the sample could be measured at an elevated temperature. This would increase the MSRD for the oscillations from the lower edge, thus suppressing them more quickly. Of course, it would also affect the measured edge, but if only XANES and/or the lower-energy portion of EXAFS was desired, this might be acceptable.

8.10.4.5 Mitigation

In many cases, mitigation is not necessary. But if oscillations from the lower edge are appearing in the range of the Fourier transform of the higher edge that will be used for fitting, it is possible to include the structure of the lower edge in the model (Carroll et al. 2010; Menard et al. 2009).

8.10.4.6 Silver lining

At least you have the comfort of knowing that the lower edge is providing good EXAFS data! In some cases, it is possible to simultaneously fit both edges, including the effect of the lower one on the higher one.

8.10.5 Multielectron Excitations

8.10.5.1 Cause

An additional electron of definite binding energy on the order of tens or hundreds of electron volts is excited.

8.10.5.2 Identification

In some ways, this is similar to another edge, as it will be seen as a sharp jump in the absorption at some energy. Multielectron excitations are generally quite small and are thus best seen in $\chi(k)$, as Figure 8.21 demonstrates. In addition, they can sometimes be detected as a sudden change in the slope of the background.

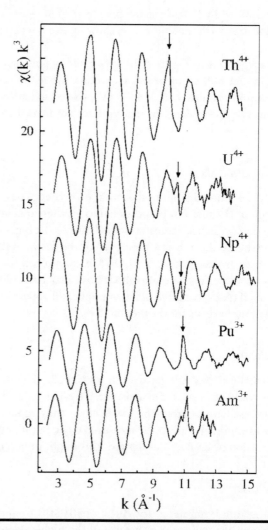

Figure 8.21 $k^3\chi(k)$ for the L_3 edges of a series of actinide species, with multielectron excitations indicated by the arrows. (Reprinted from C. Henning, *Phys. Rev. B* 75, 053120, 2007. With permission.)

Multielectron excitations can be distinguished from noise (Section 8.10.1), in that they are reproducible from scan to scan; from monochromator glitches (Section 8.10.2), in that there is no corresponding peak in I_o; from Bragg peaks (Section 8.10.6), in that rotating the sample has no effect on them; and from edges (Section 8.10.3), in that there is no plausible edge at that energy.

8.10.5.3 Effect on analysis

Once energies are sufficient to excite the second electron to the continuum, there is no longer a single solution for allocating the energy between the two electrons. In effect, amplitude is removed from the single-electron channel from that point in k forward. Thus, including $\chi(k)$ values above that point in a fit could be particularly disruptive to the determination of amplitude parameters such as coordination numbers and MSRDs (Hennig 2007). Background and normalization could also be affected.

8.10.5.4 Prevention

It is not possible to prevent multielectron excitations.

8.10.5.5 Mitigation

A variety of heuristic approaches to compensate for multielectron excitations have been used. The simplest would be to treat the initial rise as a glitch and allow a background spline to handle the slope change. A slowly varying spline, however, is poorly suited to the sharp slope change associated with multielectron excitations. Thus, a variety of phenomenological functions have been used to model (and subsequently subtract) the shape of multielectron excitations, including those based on arctangent (D'Angelo et al. 1993) and a combination of Lorentzian and exponentially asymptotic functions (Kodre et al. 2002). The software package ATHENA (Ravel and Newville 2005) implements a somewhat different approach, allowing the multielectron excitation to be modeled as a shifted and broadened version of the primary channel.

8.10.6 Bragg Peaks

8.10.6.1 Cause

Crystalline samples can themselves cause diffraction. This diffraction will occur at specific energies.

8.10.6.2 Identification

A narrow spike will appear in M, but not in I_o (see Figure 8.22). The spike is reproducible from scan to scan, but changes energy if the sample is placed at a different angle.

Bragg peaks can be seen in the preedge region around 6390 and 6490 eV. Note that they do not appear in I_o. The identification was confirmed by rotating the sample slightly, which caused the peaks to shift in energy.

8.10.6.3 Effect on analysis

As with monochromator glitches, the effect of a single Bragg peak on analysis is modest, and it can be removed by deglitching. Where there's one Bragg peak, however, there are often many, causing the data to be obscured.

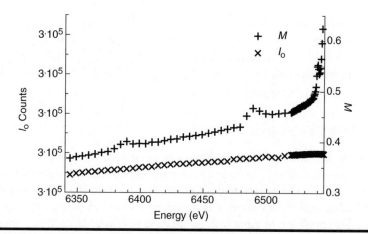

Figure 8.22   Detail of I_0 and M for a fluorescence scan of the region before the manganese K edge of a thin film manganese zinc oxide sample. Individual data points are shown.

8.10.6.4 Prevention

If crystalline samples are spun with a frequency fast compared to the integration time during measurement, most Bragg peaks will be suppressed. One simple way to make a sample spinner is to remove the blades from a miniature electric fan and mount the sample on its hub.

8.10.6.5 Mitigation

Changing the orientation of the sample relative to the beam slightly will cause Bragg peaks to move. If there are only a few, you may be able to find a position in which they are at energies that are less troubling.

8.10.7 Monotonic Time-Dependent Effects

8.10.7.1 Cause

Sometimes, the measured absorption M at a given energy will change gradually with time. This may occur, for example, when the gas in an ion chamber has been changed, and it has not yet completely flushed through the detector. Gradual temperature changes can have a similar effect.

8.10.7.2 Identification

The trend is often visible as a kink in the background when the step size or integration time changes (i.e., when the scan moves from one region to another). Because there is now more (or less) time spent per electron volt, a constant trend now appears as a differing slope. Figure 8.23 shows an example. The cause can be confirmed by stopping the scan and simply observing the detector readouts as a function of time.

In this case, insufficient time was allowed for helium to flow into and through the electron-yield detector before starting the scan. As helium displaced air, the electrons were collected with more efficiency and the signal increased. The kink in M seen at 6519 eV is the boundary between two regions in the scan parameters, where the step size changed from 5 to 0.5 eV and the integration time increased from 2 to 3 seconds. Since much more time was now spent to advance a given amount in energy, the slope of the increase is much greater. The dashed line shows the actual energy of the edge for manganese metal; since this sample was oxidized, the edge would be expected to be shifted a few eV higher.

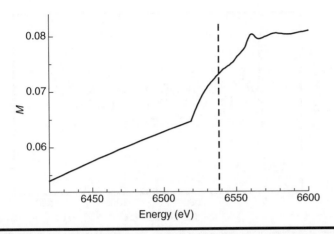

Figure 8.23 **Electron-yield spectrum of the preedge and XANES regions for the manganese *K* edge of a lithium manganese oxide cathode. The dashed line shows E_o for manganese metal.**

8.10.7.3 Effect on analysis

If sharp, kinks of this kind can make normalization and background subtraction difficult.

8.10.7.4 Prevention

Gas changes in detectors should be done well in advance of important measurements. For example, it's usually best to make changes in detector or sample chamber gases before aligning the sample, so that the time it takes to align can allow the gas to finish purging.

8.10.7.5 Mitigation

Often, the effect is temporary. Simply monitor M at a fixed energy, waiting until it is no longer changing to resume measurements.

8.10.7.6 Silver lining

While you're waiting, you can get something to eat. That may sound like a joke, but it's easy to forget to eat often enough at the beamline and that can reduce your efficiency as a data collector!

8.10.8 Oscillatory Time-Dependent Effects

8.10.8.1 Cause

A variety of problems can create changes in M that oscillate with time. Gas flow rates through a detector that are too high, for example, can cause the pressure to vary in this way. Monochromator feedback can be underdamped. Temperatures, either for the monochromator, the sample, or the detector could also suffer from underdamped control.

8.10.8.2 Identification

Figure 8.24 shows an example of a large time-dependent oscillation, in this case due to variations in pressure of a detector gas (Meitzner 1998). If small, however, these kinds of effects can be difficult to spot—the preedge is the easiest place to see them.

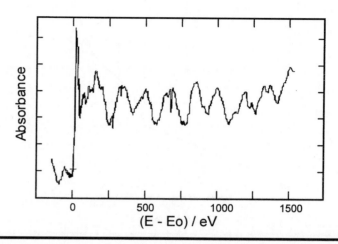

Figure 8.24 **Example of an oscillatory time-dependent effect. (Source: Reprinted from G. Meitzner, *Catalysis Today* 39, 290, 1998, Elsevier. With permission.)**

If suspected, conduct a time scan; that is, measure M as a function of time at constant energy and sample position.

8.10.8.3 Effect on analysis

If the oscillation frequency is such that it is close to the rate at which your scan records EXAFS oscillations, even a small oscillatory component can compromise your data. If it is slower than that, however, it can be removed with the background. Faster oscillations can appear as *jitter*, which looks and acts much like noise but shows correlation from one data point to the next. Jitter can interfere with some methods of estimating the measurement uncertainty of the data (see Box 7.1 of Chapter 7).

8.10.8.4 Mitigation

Time-dependent oscillations are often among the most challenging effects to track down and correct. Try turning down the gas flow through detectors. If that doesn't work, ask your beamline scientist; they'll usually be very interested in this kind of problem. If you still can't make the oscillations go away, choose your scan parameters so that the oscillation is either slower or faster than the rate at which your scan records EXAFS oscillations.

8.10.8.5 Silver lining

You're likely to learn quite a bit about the beamline trying to track this kind of thing down!

A similar kind of phenomenon is an oscillation caused by monochromator moves. Likely attributable to either feedback or temperature control of the monochromator, that kind of oscillation can be especially frustrating. Source effects from an undulator can also have this kind of impact. If something like that appears to be happening (e.g., the oscillation does not show up on a time scan, does not depend on integration time, but does show up with a blank), talk to your beamline scientist!

8.10.9 Electronics Out of Range

8.10.9.1 Cause

An amplifier becomes saturated or goes negative; maximum count of a scaler is exceeded; and so on.

8.10.9.2 Identification

Sometimes, the beamline software will alert you to problems such as this, perhaps only in the sense that it refuses to collect the data. But sometimes you won't get an

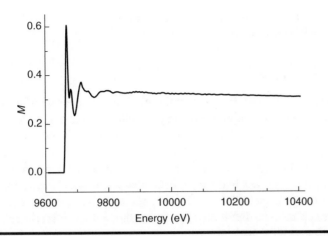

Figure 8.25 **Spectrum of thin film of zinc oxide on quartz, measured in fluorescence at the zinc *K* edge. Note the sharp corner at the onset of the edge.**

alert, and the problem can be seen in the form of a sharp corner in the data, such as in Figure 8.25. Once you suspect this as a problem, it's easy to observe detector readouts and track it down.

While the spectrum in Figure 8.25 looks good above the edge, the preedge looks too perfect; considering the expected energy resolution, the edge should not rise abruptly out of the preedge as it does here. After investigation, it turned out that the problem was that the amplifier attached to the Lytle detector had been inadvertently assigned a negative dark current; that is, its output when there was no signal was a negative voltage. The electronics chain interpreted negative voltages as zero, resulting in *M* being calculated as zero until the signal rose high enough to overcome the dark current.

8.10.9.3 Effect on analysis

Any part of the spectrum for which this occurred is compromised.

8.10.9.4 Prevention

Keep an eye on detector readouts and adjust gains accordingly; measure offsets as needed.

8.10.9.5 Mitigation

Once discovered, these problems are easily corrected by changing gains, changing dark currents, measuring offsets, or reducing integration times, as appropriate.

8.10.10 Sample Motion

8.10.10.1 Cause Sometimes, a sample physically shifts during measurement. A simple example of this is when the sample is not securely attached to the sample holder—it may even fall off altogether! Powder samples sometimes settle during measurement, creating temporary pinholes and then filling them. If the sample is measured at cryogenic temperature and is not completely dry, ice crystals may form and change the density of the sample or trigger settling. Samples measured *operando*, whether batteries or catalysts, may undergo volume changes. Liquid samples, particularly those that flow, may be subject to the formation and motion of gas or air bubbles. Gas samples may change volume or density.

8.10.10.2 Identification

In most cases, *M* will show sharp discontinuities that are not reproducible in energy. In other cases, when the motion is ongoing, the effect will be more like a non-Gaussian form of noise. Figure 8.26 shows an example in which the culprit is bubbles in a liquid sample.

The spectrum in Figure 8.26 is from a time-resolved study measuring a solution as it reacted. Previous scans had shown little noise. This scan started smooth, as can be seen from the preedge and white line, but suddenly became very noisy. The noise also does not appear to be distributed in the way shot noise would; the spikes and dips between 7900 and 7950 had not been present on previous scans and did not correspond to monochromator glitches.

The explanation was that a bubble had formed in the flow cell, and the motion of the flowing solution was jiggling it in and out of the beam.

8.10.10.3 Effect on analysis

Scans with these kinds of discontinuities are not generally usable, unless the scan can be truncated before the discontinuity.

8.10.10.4 Prevention

Think about possible changes in the sample before mounting it, and if possible, try to allow for them in a way that won't change the part of the sample exposed to the beam. If bubbles have a place to go, for example, they are less likely to overlap with the beam. For powder samples in a well (Section 6.6.1 of Chapter 6), it is often helpful to tap them on a table a few times to encourage the powder to settle *before* measurement.

8.10.10.5 Mitigation

Visual inspection of the sample will usually reveal the problem and suggest a method of fixing it: perhaps the sample position can be changed or it can be gently shaken to restore uniformity.

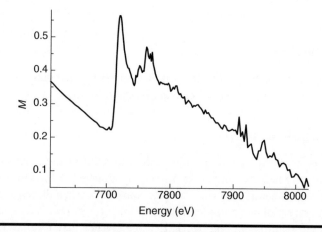

Figure 8.26 **Transmission spectrum of cobalt acetate solution undergoing reduction, measured at the cobalt *K* edge. Previous scans in the time-resolved series had exhibited little noise on this scale.**

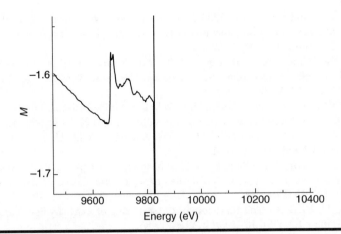

Figure 8.27 Transmission spectrum at the zinc *K* edge of a sample of nickel zinc ferrite nanoparticles. The scan was supposed to extend to approximately 10,400 eV.

8.10.11 Loss of Beam

8.10.11.1 Cause

The synchrotron may unexpectedly lose beam, the feedback system on the monochromator may lose lock, a safety interlock may be tripped, and so on.

8.10.11.2 Identification

Beamline software may note the absence of a reading on I_o or the spectrum may abruptly flatline. Figure 8.27 shows a typical example, with the beam dumping at about 9830 eV.

8.10.11.3 Effect on analysis

The part of the spectrum after loss of beam is, of course, not usable.

WHAT I'VE LEARNED IN CHAPTER 8, BY SIMPLICIO

- Signal-to-noise ratio isn't everything. It has to be balanced against distortion and speed.
- Signal-to-noise ratio generally improves as you go from transmission to total-yield fluorescence to energy-discriminating fluorescence, but the possibilities for systematic error also increase.
- Harmonics can be suppressed by *harmonic rejection mirrors* or by *detuning* the monochromator but can never be completely eliminated.
- Fluorescent background can be reduced by a combination of filters, shielding, and slits.
- Careful choice of *scan parameters* can improve signal-to-noise ratio without increasing the total scanning time.

References

Bewer, B. 2012. Soller slit design and characteristics. *J. Synchrotron Radiat.* 19:185–190. DOI:10.1107/S0909049511052319.

Bilderback, D. H. and S. Hubbard. 1982. X-ray mirror reflectivities from 3.8 to 50 keV (3.3 to 0.25 Å) Part II—Pt, Si and other materials. *Nucl. Instrum. Methods Phys. Res.* 195:91–95. DOI:10.1016/0029-554X(82)90763-7.

Bridges, F., X. Wang, and J. B. Boyce. 1991. Minimizing "glitches" in XAFS data: A model for glitch formation. *Nucl. Instrum. Methods Phys. Res. A.* 307:316–324. DOI:10.1016/0168-9002(91)90199-Z.

Bunker, G. 2010. *Introduction to XAFS.* New York: Cambridge University Press.

Carlon, H. R. 1988. Electrical properties of atmospheric moist air: A systematic, experimental study. Report No. CRDEC-TR-88059. Aberdeen, MD: U. S. Army Armament Munitions Chemical Command.

Carroll, K. J., S. Calvin, T. F. Ekiert, K. M. Unruh, and E. E. Carpenter. 2010. Selective nucleation and growth of Cu and Ni core/shell nanoparticles. *Chem. Mater.* 22:2175–2177. DOI:10.1021/cm1004032.

Compton, A. H. 1923. A quantum theory of the scattering of x-rays by light elements. *Phys. Rev.* 21:483–502. DOI:10.1103/PhysRev.21.483.

D'Angelo, P., A. Di Cicco, A. Filipponi, and N. V. Pavel. 1993. Double-electron excitation channels at the Br *K* edge of HBr and Br_2. *Phys. Rev. A.* 47:2055–2063. DOI:10.1103/PhysRevA.47.2055.

Darwin, C. G. 1914. The theory of x-ray reflexion. Part II. *Philos. Mag.* 27:675–690. DOI:10.1080/14786440408635139.

DuMond, J. W. 1937. Theory of the use of more than two successive x-ray crystal reflections to obtain increased resolving power. *Phys. Rev.* 52:872–883. DOI:10.1103/PhysRev.52.872.

Hart, M. and A. R. D. Rodrigues. 1978. Harmonic-free single-crystal monochromators for neutrons and x-rays. *J. Appl. Crystallogr.* 11:248–253. DOI:10.1107/S0021889878013254.

Hart, M. and L. Berman. 1998. X-ray optics for synchrotron radiation: Perfect crystals, mirrors, and multilayers. *Acta Crystallogr. A.* 64:850–858. DOI:10.1107/S0108767398011283.

Hennig, C. 2007. Evidence for double-electron excitations in the L_3-edge x-ray absorption spectra of actinides. *Phys. Rev. B.* 75:035120 (7). DOI:10.1103/PhysRevB.75.035120.

Kirkpatrick, P. and A. V. Baez. 1948. Formation of optical images by x-rays. *J. Opt. Soc. Am.* 38:767–774. DOI:10.1364/JOSA.38.000766.

Kodre, A., I. Arflon, J. Padežnik Gomilšsek, R. Prešeren, and R. Frahm. 2002. Multielectron excitations in x-ray absorption spectra of Rb and Kr. *J. Phys. B.* 35:3497–3513. DOI:10.1088/0953-4075/35/16/310.

Krause, M. O. 1979. Atomic radiative and radiationless yields for *K* and *L* shells. *J. Phys. Chem. Reference Data.* 8:307–327.

Li, G. G., F. Bridges, and X. Wang. 1994. Monochromator induced glitches in EXAFS data II. Test of the model for a pinhole sample. *Nucl. Instrum. Methods Phys. Res. A.* 340:420–426. DOI:10.1016/0168-9002(94)90121-X.

Lugovskaya, O. M. and S. A. Stepanov. 1991. Calculation of the polarizabilities of crystals for diffraction of x-rays of the continuous spectrum at wavelengths of 0.1–10 Å. *Sov. Phys. Crystallogr.* 36:478–481.

Meitzner, G. 1998. Experimental aspects of x-ray absorption spectroscopy. *Catalysis. Today.* 39:281–291. DOI:10.1016/S0920-5861(97)00114-4.

Menard, L. D., Q. Wang, J. H. Kang, A. J. Sealey, G. S. Girolami, X. Teng, A. I. Frenkel, and R. G. Nuzzo. 2009. Structural characterization of bimetallic nanomaterials with over-lapping x-ray absorption edges. *Phys. Rev. B.* 80:064111(11). DOI:10.1103/PhysRev B.80.064111.

Parratt, L. G., C. F. Hempstead, and E. L. Jossem. 1957. "Thickness effect" is absorption spectra near absorption edges. *Phys. Rev.* 105:1228–1232. DOI:10.1103/PhysRev.105.1228.

Ravel, B. and M. Newville. 2005. ATHENA, ARTEMIS, HEPHAESTUS: Data analysis for x-ray absorption spectroscopy using IFEFFIT. *J. Synchrotron Radiat.* 12:537–541. DOI:10.1107/S0909049505012719.

Soller, W. 1924. A new precision x-ray spectrometer. *Phys. Rev.* 24:158–167. DOI:10.1103/PhysRev.24.158.

Stern, E. A. and S. M. Heald. 1979. X-ray filter assembly for fluorescence measurements of x-ray absorption fine structure. *Rev. Sci. Instrum.* 50:1579–1582. DOI:10.1063/1.1135763.

Thompson, A. C. 2009. X-ray detectors. In *X-Ray Data Booklet*, ed. A. C. Thompson, 4–32 to 4–39. Berkeley, CA: Lawrence Berkeley National Laboratory.

Tran, C. Q., Z. Barnea, M. D. de Jonge, B. B. Dahl, D. Paterson, D. J. Cookson, and C. T. Chantler. 2003. Quantitative determination of major systematics in synchrotron x-ray experiments: Seeing through harmonic components. *X-Ray Spectrom.* 32:69–74. DOI:10.1002/xrs.630.

Tran, C. Q., M. D. de Jonge, Z. Barnea, B. B. Dahl, and C. T. Chantler. 2004. Quantitative determination of the effect of the harmonic component in monochromatized synchrotron x-ray beam experiments. In *Developments in Quantum Physics*, eds. F. Columbus and V. Krasnoholovets, 255–260. Hauppaugue, New York: Nova.

Van Der Laan, G. and B. T. Thole. 1988. Determination of glitches in soft x-ray monochromator crystals. *Nucl. Instrum. Methods Phys. Res. A.* 263:515–521. DOI:10.1016/0168-9002(88)90995-3.

XAFS ANALYSIS

Chapter 9

Fingerprinting

Identification without Physics

No physics? But how can I understand XAFS without understanding physics?	
You can't. But you don't have to understand something to use it. Do you understand what causes one person to have a whorl on their finger and another an arch? But you can still use the pattern to identify whether a particular person was at a crime scene.	
Ugh. This is not going to be my favorite chapter. I don't like 'pattern matching' without knowing where the patterns come from.	
There will be plenty of physics later, friend Carvaka. Fingerprinting of XAFS has been used for nearly a century and is still used today—let us give it its due before moving on to more recently developed techniques.	

9.1 Matching Empirical Standards

The simplest kind of fingerprinting is comparing the spectrum of a sample to that of a known substance—an *empirical standard*.

Figure 9.1 shows the normalized absorption, k-weighted $\chi(k)$, and magnitude of the Fourier transform for a nickel foil and a sample of nickel–zinc ferrite nanoparticles that had been annealed at high temperature (this system is discussed in Swaminathan et al. [2006]).

While the spectra of the sample are clearly not *identical* to those of the nickel foil, the difference is largely in amplitude. This difference might, in part, be due to issues

DOI: 10.1201/9780429329555-12

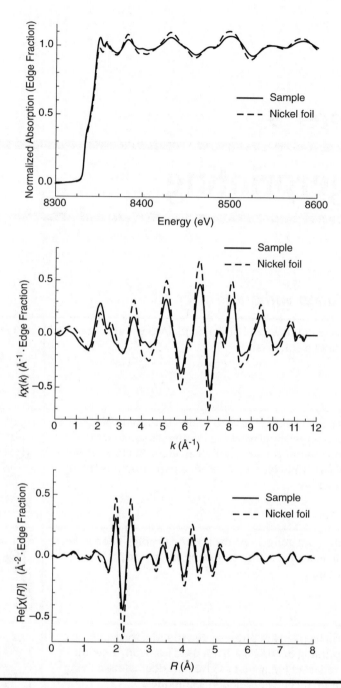

Figure 9.1   Comparison of annealed nickel–zinc ferrite nanoparticles to nickel foil measured in transmission. (*Top*) Normalized absorption. (*Middle*) *k*-weighted χ(*k*). (*Bottom*) Real part of the Fourier transform from 3 to 10 Å⁻¹.

in sample preparation, as it is difficult to prepare a powder to be as uniform in thickness as a metal foil (see Section 6.3.2 of Chapter 6). In addition, it may suggest a lower average coordination number for the sample, perhaps due to vacancies or a nanoscale structure. Since the sample was prepared by annealing nanoparticles, such differences from bulk nickel are not surprising.

Nevertheless, the resemblance to the spectra of the nickel foil is so striking that we can state with confidence that the nickel in the sample is in a form 'similar to' ordinary nickel metal. If much of it were oxidized, for example, we would see significant changes in the shape and position of the edge. Significantly different local structures, such as body-centered cubic (ordinary nickel metal is face-centered cubic), can also be ruled out. (Face-centered cubic structures have 12 near neighbors, while body-centered cubic have 8 near neighbors and 6 a bit further out.) But just glancing at these spectra might not be enough to eliminate other nickel spectra with 12 near neighbors, such as hexagonal close-packed or icosahedral nickel, or close-packed alloys of nickel with iron or zinc. Other evidence would have to be used to distinguish between those choices, either deduced from the spectra using the methods of the remainder of this book or based on other knowledge about the system.

9.2 Fingerprinting Spectral Features

Sometimes, a particular feature within the normalized XANES, $\chi(k)$, or Fourier transform can be correlated with particular structural information.

As an example, we will consider the *spinel* crystal structure. This structure, commonly adopted by a class of materials called *ferrites* with molecular formula AB_2O_4, is interesting in that the cations A and B may sit in sites that are either tetrahedrally or octahedrally coordinated with oxygen atoms. The spinel structure includes twice as many octahedral sites as tetrahedral ones, so one might be tempted to guess that the A cations sit in tetrahedral sites, while the B cations sit in octahedral sites, but that is often not the case. Depending on the identity of A and B and how the sample was synthesized and processed, some or all of the A atoms may sit in octahedral sites and some of the B atoms may sit in tetrahedral sites. The distribution of cations between tetrahedral and octahedral sites has important ramifications for the magnetic properties of these materials (Smit and Wijn 1959).

Figure 9.2 shows *theoretical standards* (i.e., computed spectra) for the magnitude of the Fourier transforms of the Mn K edge of manganese ferrite ($MnFe_2O_4$), assuming several distributions of manganese cations between the sites. '40% Tetrahedral,' for example, means that 40% of the manganese ions are in tetrahedral sites and 60% are in octahedral sites.

It is probably not surprising that the first peak in the Fourier transform, around 1.5 Å, becomes somewhat larger when a greater percentage of the manganese is in

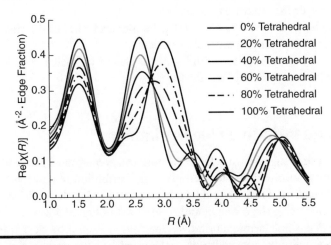

Figure 9.2 (See color insert.) Magnitude of the Fourier transform for the manganese K edge of theoretical spectra of manganese ferrite, computed assuming a variety of site occupancies for the manganese atoms. The transforms were taken on data with a k-weight of 1 from 3 to 10 Å^{-1}, using Hanning windows with sills of 1.0 Å^{-1} on each side.

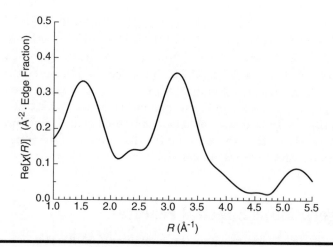

Figure 9.3 **Magnitude of the Fourier transform for the zinc *K* edge of a sample of manganese–zinc ferrite nanoparticles. The transform was taken on data with a *k*-weight of 1 from 3 to 10 Å⁻¹, using Hanning windows with sills of 1.0 Å⁻¹ on each side.**

In fact, in this case, the first peak is made up of nearest-neighbor scattering for cations in tetrahedral sites and in octahedral sites. The two bond lengths are somewhat different, and so the two contributions will interfere. The way that interference translates into peak height and shape depends on the *k* range used for the Fourier transform. One can, therefore, understand why it is unwise to presume that the height of the first peak is proportional to the average coordination number.

octahedral sites. That first peak is largely due to scattering from the oxygen near neighbors, and there are more of them in octahedral sites than in tetrahedral. But we wouldn't be able to judge the occupancy of a spinel cation just by looking at the first peak in the Fourier transform of its spectrum—there are too many other factors, such as disorder or Fourier truncation effects, that could affect the size of that first peak. We could try to use modeling (see Chapter 12) to decide based on the first peak, but just a glance wouldn't tell us much.

The region between 2.0 and 3.5 Å, on the other hand, shows a difference between tetrahedral and octahedral occupancy that is much starker: for octahedral occupancy, there is a large peak around 2.6 Å, while for tetrahedral, the peak is around 3.1 Å. Calculations of theoretical standards show that this is true for all first-row transition metal spinels, regardless of the particular cations involved.

We thus have a feature that fingerprints site occupancy in a spinel. No modeling is necessary; we can just glance at a Fourier transform (given the right *k*-weighting and *k* range) and know whether the cation is present mainly in tetrahedral sites, mainly in octahedral sites, or in a mixture of both.

For example, see Figure 9.3, which is the Fourier transform of the zinc *K* edge of a sample of manganese–zinc ferrite nanoparticles.

It is evident from comparison with Figure 9.2 that, since there is a peak around 3.1 Å, the zinc is almost entirely in tetrahedral sites.

9.3 Semiquantitative Fingerprinting

It is crucial, when fingerprinting using features in the Fourier transform, that the same *k* range and weighting are used! This is particularly a risk if you are comparing your data to a figure in a publication!

The spinel example suggests that fingerprinting can sometimes be extended to yield semiquantitative information—for example, the distribution of a cation between tetrahedral and octahedral states could be estimated by finding the position of the 2.0–3.5 Å peak when using the *k*-weight and region used in Figure 9.2.

This practice is only semiquantitative, however; as should be clear from Box 9.1, the position, amplitude, and width of peaks often have a complex dependence on structure. When analyzing EXAFS for quantitative information, it is better to use techniques such as modeling (Chapter 12 and Section III).

BOX 9.1   Nothing Out of Something

	So I guess the peak in the spinel Fourier transforms around 1.5 Å is due to the first-shell oxygen atoms. That's shorter than the bond length, but I learned in Section 1.4.2 of Chapter 1 that a Fourier transform is not a radial distribution function—the distances are shifted down some. But the difference in where the second peak is for tetrahedral and octahedral must correspond to a difference in cation–cation distance between the two cases, right?
	You haven't thought that through, Simplicio. Some of the cation–cation scattering for absorbing atoms in tetrahedral sites is from scattering atoms in octahedral sites and vice versa. The distance from a tetrahedral site to an octahedral site must be identical to that from an octahedral site to a tetrahedral site. What peak represents that distance?
	Why not find out, my friends? Let us use the theoretical standard to show us the contribution of each single-scattering path. Figure 9.4 shows this for the case where all the absorbing atoms are in octahedral sites.
	Wait—the tetrahedral scatterers on their own yield a good size peak around 3.1 Å, but the total doesn't show a peak there! How is that possible?
	Recall, friend Simplicio, that Fourier transforms are complex valued, and both the real and imaginary parts can be positive or negative. In this case, the contribution from the tetrahedral scatterers must be out of phase with some of the contributions from the other single scatterers, and possibly the multiple-scattering paths that are not explicitly shown in Figure 9.4.
	So it turns out that the Fourier transform of the octahedral absorber doesn't show a peak at 3.1 Å because there are a bunch of paths contributing peaks that end up canceling out? Weird!

Particularly in the 1980s and 1990s, many works were published detailing the XANES signatures of common compounds of particular elements, with the aim of facilitating these kinds of analyses.

9.3.1 Example: Vanadium XANES

As an example, we will consider a work by Wong et al. (1984) on the *K* edge of vanadium compounds. In that study, the effect of oxidation state, coordination geometry, and bonded element are all examined. In some cases, reliable trends are found, such as those shown in Figure 9.5.

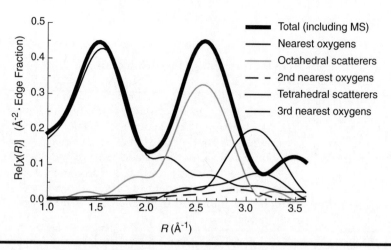

Figure 9.4 Magnitude of the Fourier transform for the manganese *K* edge of theoretical spectra of manganese ferrite with manganese entirely in octahedral sites, along with the contributions from single-scattering paths.

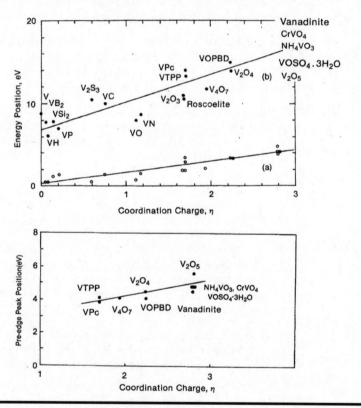

Figure 9.5 Position of features in the vanadium *K*-edge XANES of various vanadium compounds as a function of the coordination charge of the vanadium. In the top graph, the top line is for the steepest inflection point in the XANES region, while the bottom line is for the first inflection point in that region. The bottom graph is for the peak of the sharp preedge feature seen in many vanadium *K*-edge spectra (Source: Reprinted from J. Wong, F. W. Lytle, R. P. Messmer, and D. H. Maylotte, *Phys. Rev. B*, 30, 5596–5610, 1984. Copyright 1984 by the American Physical Society).

Coordination charge is a concept defined by Bastanov (1967) to quantify the charge associated with an atom, including the net charge it receives from bonding. Other correlations found in Wong et al. and other XAFS studies depend on conventional oxidation state, bond length, or other structural parameters.

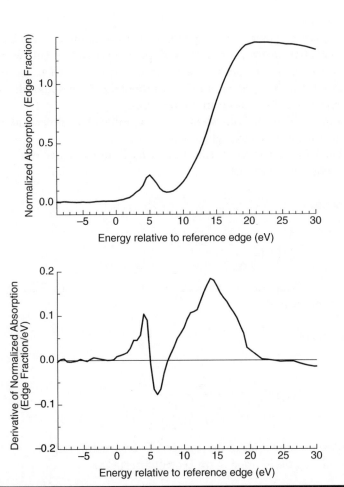

Figure 9.6 Vanadium *K*-edge spectrum of an uncharacterized sample. (*Top*) Normalized absorption. (*Bottom*) Derivative of normalized absorption. The scale on the *x*-axes are relative to the first inflection point of a vanadium reference foil.

To see how this works, consider Figure 9.6 that shows the vanadium *K*-edge XANES and derivative spectra for a vanadium compound not among those studied by Wong et al.

The first inflection point in Figure 9.6 is given by the first peak of the derivative spectrum, which is about 4 eV above the reference edge. According to Figure 9.5, that corresponds to a coordination charge of about 2.8. The steepest inflection point is at about 14.0 eV above the reference edge, which corresponds to a coordination charge of about 2.1. And the peak of the preedge feature in the normalized spectrum is at about 5.0 eV, which corresponds to a coordination charge of about 2.9. Using three features provides us an estimate of uncertainty; it's reasonable to say that the coordination charge of the compound is 2.5 ± 0.5. By comparing to the computed coordination charge of candidate compounds, this kind of semiquantitative fingerprinting can tell us something about the measured sample.

9.3.2 Fitting Features

The example in Section 9.3.1 raises an issue: the data are a little bit noisy, particularly when we look at the derivative spectrum. XANES fingerprinting is often used in cases when the data are noisy, since in those cases EXAFS, which is often easier to interpret, might not be available. But if the data are noisy, how can we be sure we identified a particular inflection point, or the top of a peak? In addition, consider the case where we want to analyze changes over a series of spectra, perhaps because we are studying time-resolved data or a temperature series or a series of samples that differ in a synthetic parameter such as doping fraction. In those cases, we would like to be able to detect small changes from spectrum to spectrum.

One way of addressing both issues is to fit a peak with a simple mathematical line shape rather than just find its highest point.

Individual, isolated peaks, such as the preedge feature in vanadium, are often fit by one of the following functions:

■ **Gaussian:**

$$M(E) = \frac{A}{\sigma\sqrt{2\pi}} e^{-\frac{(E-E_c)^2}{2\sigma^2}}$$

(9.1)

A is related to the amplitude of the peak and σ to its width, with E_c being the location of the center of the peak.

■ **Lorentzian:**

$$M(E) = \frac{A}{\pi} \frac{\frac{\Gamma}{2}}{(E-E_c)^2 + \left(\frac{\Gamma}{2}\right)^2}$$

(9.2)

A is related to the amplitude of the peak and Γ to its width, with
E_c again being the location of the center of the peak.

■ **Voigt:** The convolution of a Gaussian with a Lorentzian.
■ **Pseudo-Voigt:** The sum of a Gaussian and a Lorentzian. E_c is the same for both functions, but the amplitudes can be different.

Figure 9.7, for instance, shows the preedge peak from Figure 9.6 fit by a combination of a Voigt function and a quadratic background.

This fit yields a peak position of 4.95 ± 0.03 Å, a precision which could be useful for working with a series of spectra. But notice that it does not help us determine the

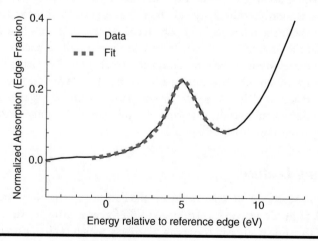

Figure 9.7 **Fit to preedge feature from Figure 9.6. Fit was performed by pro Fit 6.2.8 software from −1 to 8 eV, using a quadratic background and a Voigt peak.**

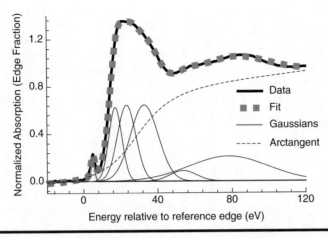

Figure 9.8 **Fit to XANES from Figure 9.6, along with the Gaussian peaks and arctangent edge used to construct the fit. The arctangent 'edge' incorporates a quadratic background as well.**

coordination charge using Figure 9.5. If we had left out the background, for instance, the peak position would have been 5.09 ± 0.06 Å. Using the background but fitting from –1 to 10 eV yields a peak position of 4.83 ± 0.06 Å, while 3–7 eV yields 5.00 ± 0.06 Å. These fits, in other words, do no better with a single spectrum than what we can judge by eye; their merit lies in allowing small changes between spectra to be quantified. If we wish to use the precision achievable by peak fitting to allow us to obtain precise estimates of correlated values from trends like that shown in Figure 9.5, we must measure and analyze our own set of standards on the same beamline as and with similar conditions to those used for our sample, in effect redoing the work of the authors who originally alerted us to the trend. In addition, the correlations exploited by these trends aren't that strong anyway; the bottom graph in Figure 9.5 shows a half an eV or so of scatter.

This idea can be taken a step further, at least aesthetically. Multiple simple functions can be fitted to the XANES at once: a polynomial background, various peak functions, and an arctangent or error function to represent the rounded step function associated with the edge. A fit of that kind for the spectrum from Figure 9.6 is shown in Figure 9.8.

The take-home message here is that using functions to fit peaks makes the process less subjective, allowing you to compare series of spectra, even when the signal-to-noise ratio is not great. But fingerprinting is only a semi-quantitative technique anyway, so *precision* in determining a peak position doesn't correspond to *accuracy* in determining structural parameters.

BOX 9.2	**The Physical Basis**
	I'm not a big fan of 'fits' like Figure 9.8. There's no physical meaning to those individual peaks—they're just chosen to reproduce the shape.
	In papers, I've often seen peaks like that identified with particular transitions, splittings, and the like. For example, the preedge in vanadium is the *forbidden transition* 1s → 3d. The fact that we get a forbidden transition tells us that there is no center of inversion (Wong et al. 1984).

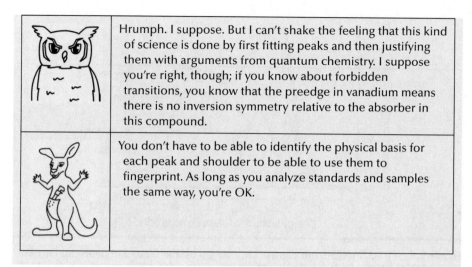

This section describes fingerprinting using theoretical XANES standards. Section 9.2 gave an example of fingerprinting using theoretical EXAFS standards.

Consult Chapter 12, friends, for more about the methods and software used to compute theoretical standards.

9.4 Theoretical XANES Standards

In recent years, the ability to calculate XANES spectra from structural information has improved significantly. Fingerprinting XANES, therefore, can now be based on theoretical standards as well as empirical ones.

For example, suppose we think we have a compound that has a vanadium atom coordinated to six oxygen atoms, but that the vanadium may be off-center. We can use software (in this case FDMNES) to create theoretical standards corresponding to placing the vanadium atoms at various points within the oxygen octahedron. The result is shown in Figure 9.9.

The forbidden transition 1s → 3d, discussed in Box 9.2, can be clearly seen as a preedge peak around –1 eV in the off-center standards. It is evident that the amplitude of the peak correlates to the distance the vanadium atom is displaced from the center and that the effect is nonlinear, with 0.432 Å displacement producing a preedge peak more than twice as large as a 0.216 Å displacement.

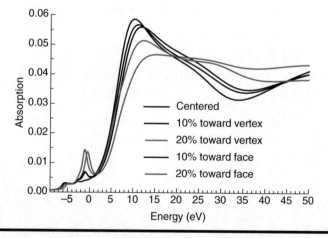

Figure 9.9 **(See color insert.) Theoretical standards calculated by FDMNES for vanadium atoms at each of five positions within an octahedron of oxygen atoms 2.16 Å from center to vertex. In addition to placing the vanadium atom in the center of the octahedron, spectra are shown in which the vanadium atom is placed 10% or 20% of the way (i.e., 0.216 or 0.432 Å) toward one vertex or the same distances toward the center of a face. The values on axes are as reported by FDMNES.**

While the direction of displacement doesn't have a significant effect on the size of the preedge peak, moving toward a vertex causes a greater suppression of the white line than moving the same distance toward a face and also shifts most features, including the preedge peak, to higher energies.

These kinds of calculations can be very helpful in understanding the structural factors contributing to differences between spectra. In recent years, great strides have been made in using theoretical standards in the XANES regions for fully quantitative analyses. Doing so constitutes *modeling*, not just fingerprinting, and will be discussed further in Chapter 12.

By using theoretical standards, we can investigate structural changes even when we don't have empirical standards for them. In Figure 9.9, we move a single atom. We could also easily try distorting the unit cell, changing bond lengths, switching out ligands, etc. Finding empirical standards for all those things would be a chore, and in some cases might be impossible!

WHAT I'VE LEARNED IN CHAPTER 9, BY SIMPLICIO

- Fingerprinting works by comparing spectra to standards. That means we don't necessarily have to understand where the features in the spectra come from.
- Entire spectra can be compared, or just individual features.
- By identifying correlations between feature position, area, or width and structural parameters, semiquantitative characterizations can be made.
- Theoretical XANES standards can help us understand differences between spectra

REFERENCES

Bastanov, S. S. 1967. Effective charges of atoms and oxidative reactions. *Russ. Chem. Bull.* 16:1175–1180. DOI:10.1007/BF00908268.

Smit, J. and H. P. J. Wijn. 1959. *Ferrites.* New York: Wiley.

Swaminathan, R., J. Woods, S. Calvin, J. Huth, and M. E. McHenry. 2006. Microstructural evolution model of the sintering behavior and magnetic properties of NiZn ferrite nanoparticles. *Adv. Sci. Tech.* 45:2337–2344. DOI:10.4028/www.scientific.net/AST.45.2337.

Wong, J., F. W. Lytle, R. P. Messmer, and D. H. Maylotte. 1984. *K*-edge absorption spectra of selected vanadium compounds. *Phys. Rev. B.* 30:5596–5610. DOI:10.1103/PhysRevB.30.5596.

Chapter 10

Linear Combination Analysis

Making It All Add Up

Oh, good. I already understand a little bit about this kind of analysis, just from what I read in Section 1.6.2 of Chapter 1. This should be an easy chapter!	
Not so fast, Simplicio. It's easy to do linear combination analysis (LCA) sloppily, and the results can be crazy wrong if you're not careful.	
I always try to be careful, dear Dysnomia.	
Well, Robert, you usually work with stuff like 400-year-old priceless paintings, right? You can afford to take your time. Me, I'll cut corners if it lets me get more done, as long as it doesn't louse up my results too much. But linear combination analysis is especially sensitive to corner-cutting, so I'll try to learn from the master of carefulness: lead on, Robert!	

10.1 When LCA Works

10.1.1 A Simple Example

Imagine a thin film sample that is made up of two layers: the front layer is gypsum (hydrated calcium sulfate) and the back layer cinnabar (mercury sulfide). Let's further

DOI: 10.1201/9780429329555-13

suppose that the gypsum layer is half the thickness of the cinnabar layer. For measurements made at the sulfur K edge in the transmission geometry, what would the resulting spectrum look like?

According to Bouguer's Law (Equation 6.3 of Chapter 6), the intensity I_{1g} of the x-ray beam after traveling through the gypsum portion of the sample would be given by

$$I_{1g} = I_o e^{-\mu_{mg}\rho_g x_g} \tag{10.1}$$

where the g in the subscripts refers to the properties of the gypsum layer.

After passing through the cinnabar layer, the transmitted intensity would be given by

$$I_1 = I_{1g} e^{-\mu_{mc}\rho_c x_c} = \left(I_o e^{-\mu_{mg}\rho_g x_g} \right) e^{-\mu_{mc}\rho_c x_c} = I_o e^{-\left(\mu_{mg}\rho_g x_g + \mu_{mc}\rho_c x_c \right)} \tag{10.2}$$

(The subscript c refers to the properties of the cinnabar layer.)

Thus, the absorption of the sample is just the sum of the absorption due to each layer separately.

BOX 10.1   Measures of Percent Composition

	Oh! So the result would be a linear combination of the sulfur spectrum of gypsum and the sulfur spectrum of cinnabar. If I took a normalized sulfur spectrum of gypsum, scaled by one-third, and added it to a normalized sulfur spectrum of cinnabar, scaled by two-thirds, I should get what was measured, right?
	No! You're right that the result will be a linear combination of the gypsum and cinnabar spectra, but the percentages aren't right. The beam is traversing a material that is one-third gypsum and two-thirds cinnabar *by volume*. But the normalization procedure explained in Section 7.3.4 of Chapter 7 gives a spectrum proportional to the absorption per atom of sulfur, not per cubic centimeter of material!
	Here is how we can do the calculation correctly, friends. Gypsum is $CaSO_4 \cdot 2H_2O$, so it is 18.6% sulfur by weight. The density of gypsum is about 2.32 g/cm³, so there is about 2.32 times 0.186 = 0.432 g of sulfur per cubic centimeter of gypsum. Dividing by the atomic mass of sulfur, we have 0.0134 moles of sulfur per cubic centimeter of gypsum. Cinnabar (HgS) is 13.8% sulfur by weight and has a density of 8.18 g/cm³, giving 0.0532 g of sulfur per cubic centimeter of cinnabar. In our example, the ratio of cinnabar to gypsum by volume is 2:1, so the ratio of moles of sulfur in the cinnabar environment to total moles of sulfur is $$\frac{2 \cdot 0.0532}{2 \cdot 0.0532 + 1 \cdot 0.0134} = 89\%.$$ That leaves us 11% for gypsum.
	So, Simplicio, if you took a normalized sulfur spectrum of gypsum, scaled it by 0.11, and added it to a normalized sulfur spectrum of cinnabar, scaled by 0.89, then you should get what was measured.
	There are many ways to measure percent composition, including percent by weight, percent by volume, parts per million of a mineral, and so on. *X-ray absorption from a given phase is proportional to the percentage of the absorbing atom present in that phase*, which does not correspond to any of the measures commonly used in other fields. It is, therefore, usually necessary when comparing XAFS to the results of other techniques either to convert the XAFS percentages to a more familiar measure or to convert the familiar measure to the one that XAFS uses.

10.1.2 Intimate Mixtures in Transmission

Next, consider an intimate mixture of two phases—perhaps gypsum and cinnabar again, but now mixed on a scale much smaller than the beam or the thickness of the sample. For instance, nanoscale gypsum inclusions might be present in a cinnabar matrix. In that case, the sample is still uniform in the sense discussed in Section 6.3.2 of Chapter 6, and each section of beam will traverse alternating layers of cinnabar and gypsum. Once again, the total absorption will be the sum of the absorption due to the gypsum and the absorption due to the cinnabar, weighted by the percentage of sulfur atoms present in each phase.

10.1.3 Intimate Mixtures in Fluorescence

A review of the derivations in Section 6.4 of Chapter 6 reveals that the result from Section 10.1.2 also holds in either the thin or dilute limits for fluorescence, as long as the mixing scale is also much smaller than the penetration depth of the x-ray beam into the sample.

Just because a sample has a uniform thickness (i.e., has the same thickness in micrometers across its width) does not mean it is uniform in the sense discussed in Section 6.3.2 of Chapter 6! If a 30-μm-thick film is pure copper on the left side of the beam, and pure Cu_2O on the right side of the beam, the absorption of the left side will differ significantly from the right side, leading to distortions!

BOX 10.2 Fluorescence for 'Surface' Analysis

I'm not sure that I'm comfortable with saying that the mixing scale has to be much smaller than the penetration depth. Some people choose fluorescence for its surface sensitivity. Saying they want surface-sensitive data means their sample is inhomogeneous on a scale larger than the penetration depth, doesn't it?

That's a good point, Kitsune. It would be better to say that we don't want the mixing scale to be similar to the penetration depth. If the penetration depth is 4 μm, and you're interested in the composition of a 100 μm thick 'surface' layer, there will be no problem. Or, if the penetration depth is 4 μm, and the material consists of 10 nm (= 0.01 μm) nanocrystals embedded at random in a matrix, that's fine too. The problem appears if the penetration depth is 4 μm and there is a surface layer that is itself a few micrometers thick. It is very difficult to interpret the data quantitatively in that kind of situation.

But, if there's a surface layer a few micrometers thick, it wouldn't be thin or dilute anyway, right? I learned from Chapter 6 that if I have, for example, 20 μm iron oxide grains in a silica matrix, that an iron-edge spectrum will show self-absorption no matter what the iron to silicon ratio is.

I'm glad you learned that lesson, Simplicio, but inclusions like that aren't the only kind of inhomogeneity. Suppose you had clay that included a small amount of Cr^{3+} substituted for Al^{3+}. If there were a surface layer that was only 10 μm or so thick, you'd still have no self-absorption at the chromium edge, because the sample really is dilute, even on the nanoscale. But, the composition would be changing on roughly the scale of the x-ray penetration, so it would be difficult to evaluate a LCA quantitatively.

It's difficult quantitatively, sure. But, you can still get a qualitative sense of what's going on from LCA. The percentages would be sketchy, but there'd be nothing wrong with the evaluation of what minerals are present.

10.2 When LCA Doesn't Work

For LCA to be feasible, the total absorption of the sample must be the sum of the contributions of each of its constituents, weighted by the fraction of the absorbing atoms found in each constituent. This was the case in the circumstances listed in Section 10.1, but that doesn't mean it is *always* true.

10.2.1 Nonuniform Samples in Transmission

In Section 6.3.2 of Chapter 6, we discussed the problems with measuring nonuniform samples in transmission. The distortions introduced by nonuniform samples will affect LCAs.

This kind of distortion only applies to nonuniformity perpendicular to the beam. As Section 10.1.1 shows, it is OK in transmission for the sample to be nonuniform front to back.

10.2.2 Surface Gradients in Thick Fluorescence Samples

Fluorescence spectra taken from thick samples are weighted more heavily toward the front of the sample. Therefore, LCA ceases to be an accurate quantitative measure if there is a surface gradient. It may still be qualitatively useful, particularly when used in conjunction with transmission analyses. If, for instance, the fluorescence measurement shows an enhancement of a highly oxidized species relative to what was found in transmission, it may be inferred that the highly oxidized species tends to lie nearer to the surface. Similarly, *K*-edge fluorescence data sample a deeper region than that in the *L* edge of the same element, and thus a comparison of the two can yield some qualitative information regarding surface gradients.

Electron yield experiments probe an even shallower region than fluorescence, generally reaching less than 100 nm into the material. Indeed, this surface sensitivity is one of the primary reasons that electron yield measurements are made. The ramifications of this shallow penetration depth should be considered carefully when interpreting LCAs of electron yield measurements.

10.3 An Example of LCA

Consider the cadmium *K*-edge spectrum shown in Figure 10.1. Is the sample a mixture of monteponite (CdO) and greenockite (CdS)? And, if so, how much of each is present?

The spectrum of the sample does look something like that of greenockite, and the deviations look as if they could be caused by the addition of a small amount of monteponite. We'll use a computer program to find the percentage of each standard that will match the data as well as possible.

Surface gradients aren't always a concern. Measuring standards in electron yield, for instance, may be a perfectly sensible method of avoiding self-absorption effects.

Most of the 'unknown sample' data in this chapter are simulated. We took measured spectra of standards, combined them, and added some noise and other distortions. For teaching LCA, we don't lose a lot by doing that, and it means we can bring you a wider range of examples. For examples with honest-to-goodness real data, see the case studies in Chapter 19.

Estimating how much of each constituent is present by eye helps develop the sense of XAFS spectra that most experts possess. Before reading further, let us guess how much of each mineral is present in the sample. 5% Monteponite? 30%? What do *you* think?

Figure 10.2 shows the best fit as determined by a computer program.

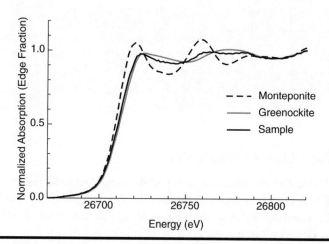

Figure 10.1   Normalized *K*-edge cadmium spectrum of sample, along with monteponite and greenockite standards.

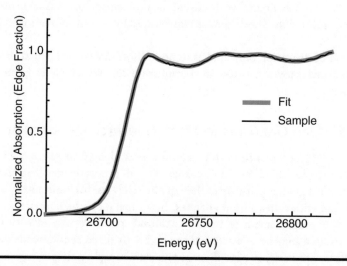

Figure 10.2   Normalized sample spectrum together with a linear combination of 24% monteponite and 76% greenockite. (As in Box 10.1, percentages refer to the percentage of cadmium atoms present as each phase.)

The data are almost indistinguishable from the fit, which ended up combining 24% monteponite with 76% greenockite. We can be fairly confident that the sample actually is a mixture of monteponite and greenockite.

As explained in Box 10.1, it's straightforward to calculate that if 24% of the cadmium atoms are present in monteponite and 76% as greenockite, then by mass the sample is 22% monteponite and 78% greenockite. You might want to try that calculation yourself, to make sure you understand how to go back and forth between different measures of percent composition.

BOX 10.3   Confidence

'Fairly confident?' That fit looks perfect! I'd bet anything that we've got the right mixture

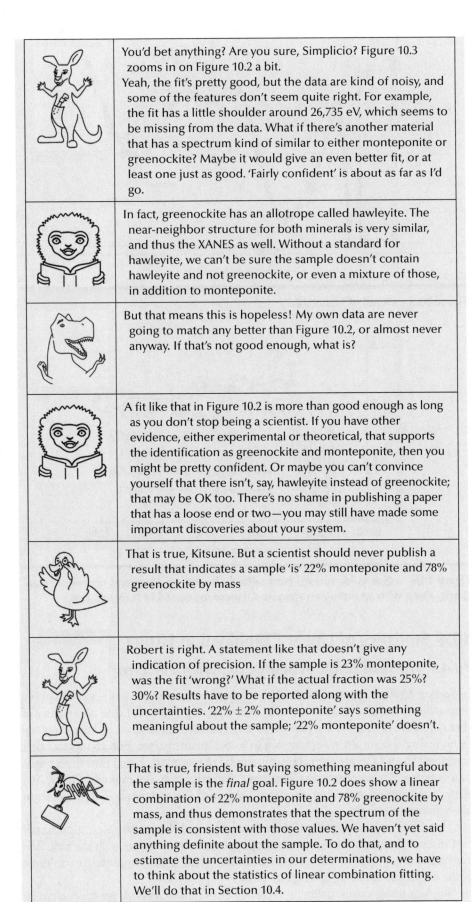

You'd bet anything? Are you sure, Simplicio? Figure 10.3 zooms in on Figure 10.2 a bit.
Yeah, the fit's pretty good, but the data are kind of noisy, and some of the features don't seem quite right. For example, the fit has a little shoulder around 26,735 eV, which seems to be missing from the data. What if there's another material that has a spectrum kind of similar to either monteponite or greenockite? Maybe it would give an even better fit, or at least one just as good. 'Fairly confident' is about as far as I'd go.

In fact, greenockite has an allotrope called hawleyite. The near-neighbor structure for both minerals is very similar, and thus the XANES as well. Without a standard for hawleyite, we can't be sure the sample doesn't contain hawleyite and not greenockite, or even a mixture of those, in addition to monteponite.

But that means this is hopeless! My own data are never going to match any better than Figure 10.2, or almost never anyway. If that's not good enough, what is?

A fit like that in Figure 10.2 is more than good enough as long as you don't stop being a scientist. If you have other evidence, either experimental or theoretical, that supports the identification as greenockite and monteponite, then you might be pretty confident. Or maybe you can't convince yourself that there isn't, say, hawleyite instead of greenockite; that may be OK too. There's no shame in publishing a paper that has a loose end or two—you may still have made some important discoveries about your system.

That is true, Kitsune. But a scientist should never publish a result that indicates a sample 'is' 22% monteponite and 78% greenockite by mass

Robert is right. A statement like that doesn't give any indication of precision. If the sample is 23% monteponite, was the fit 'wrong?' What if the actual fraction was 25%? 30%? Results have to be reported along with the uncertainties. '22% ± 2% monteponite' says something meaningful about the sample; '22% monteponite' doesn't.

That is true, friends. But saying something meaningful about the sample is the *final* goal. Figure 10.2 does show a linear combination of 22% monteponite and 78% greenockite by mass, and thus demonstrates that the spectrum of the sample is consistent with those values. We haven't yet said anything definite about the sample. To do that, and to estimate the uncertainties in our determinations, we have to think about the statistics of linear combination fitting. We'll do that in Section 10.4.

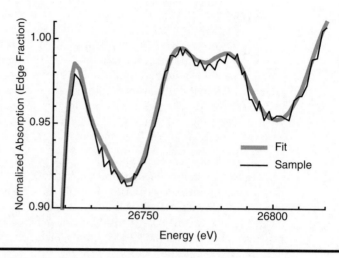

Figure 10.3   Detail of Figure 10.2.

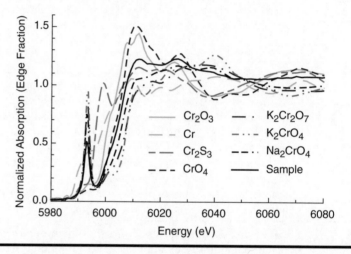

Figure 10.4   (See color insert.) Normalized absorption of several chromium standards, along with an unknown sample. All were measured in transmission.

10.4 Statistics of Linear Combination Fitting

10.4.1 Normalization: A Source of Systematic Error

Be warned, my friends: Section 10.3 showed a simple example of LCA. Some real examples are equally simple, but some are not. Beginning in Section 10.4, we will examine a more complicated case. This example will have moderate levels of noise, a small amount of systematic error similar to what is often present in real measurements, and several possible solutions. It will require judgment calls and caution befitting a careful scientist. We will follow this example for the remainder of this chapter.

In Section 7.3.4 of Chapter 7, we looked at a spectrum where we were uncertain of the proper normalization to within ±12%. In Chapter 9, we learned that uncertainty of that type does not have a big impact on fingerprinting techniques, as long as we try to normalize spectra we are comparing in a consistent manner. But, the spectra that contribute to an LCA are not necessarily similar to one another, and thus there might be a question of what constitutes 'consistent' normalization. For example, Figure 10.4 shows a number of chromium standards along with the chromium K-edge spectrum of an unknown sample.

There's no way of knowing if these standards, and the sample, were all normalized in a consistent manner, even if a single person took great care with the task. The normalization of these standards relative to each other, or the sample relative to some of the standards, could easily be off by 10%.

The best linear combination fit to the data using the standards shown is given in Figure 10.5.

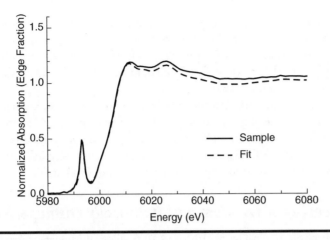

Figure 10.5 Linear combination fit of data in Figure 10.4, comprising 0.149 Cr_2O_3, 0.231 CrO_4, and 0.620 Na_2CrO_4. (Total = 1.00, so weights may be considered fractional abundances, but see Box 10.1.)

BOX 10.4   How to Avoid (Some) Normalization Problems in the First Place

	In actuality, anyone who 'took great care with the task' would never collect data over such a limited energy range. Even if one is only interested in the XANES, it only adds a little scan time to gather enough data to establish the postedge trend of the background. Points spaced every 5 eV out to 400 eV above the edge should be sufficient.
	Sure, Robert, if you can, you'd collect points well above the edge. But there are a lot of reasons you might not be able to. Maybe there's another edge not too far above the XANES— that happens a lot with *L* edges. Maybe there are some really, really bad glitches up there. Or the monochromator might not be able to go much higher in energy. And sometimes the time it takes to measure those extra points does matter. In quick-XAFS measurements, the monochromator slews through the full energy range, so a much bigger energy range means much more time, and you don't want that if good time resolution is your goal. Sure, you could still collect careful spectra on the standards, but the spectra of the sample might still not extend very far above the edge.
	You raise good points, my dear Dysnomia. But the circumstances you describe are less likely to apply to standards than to an unknown sample. I, for one, would prefer to use standards that had been properly measured out well above the edge to ones that are truncated, even if it means collecting the data on a different beamline at a different time.

The fit in Figure 10.5 looks pretty good, except that it tracks below the data above the edge. As the discussion between Robert and Dysnomia in Box 10.4 indicates, there are many realistic situations where the standards are normalized in a reasonably

What if, for some reason, you lack confidence in the normalization of your standards *and* data? In that case, you won't be able to trust your linear combination results to better than the uncertainty in your normalization. That's OK, as long as you acknowledge that uncertainty when reporting your results!

consistent manner, but the proper normalization of the sample is less clear. If we suspect that to be the case, we can improve the fit by allowing the weightings to add up to something other than 1; that's equivalent to fitting the edge jump of the sample spectrum along with the fraction of each standard present. A fit adding that degree of freedom is shown in Figure 10.6.

The fit in Figure 10.6 matches the sample spectrum very well, but adding up the weights of each constituent gives 1.034. This implies that the sample spectrum should be divided by 1.034 to be normalized consistently with the standards, giving a fit of 14.6% Cr_2O_3, 20.8% CrO_4, 3.2% $K_2Cr_2O_7$, and 61.4% Na_2CrO_4. Inspection of Figure 10.4 shows that the resulting normalization is not implausible.

10.4.2 Degrees of Freedom and Statistically Distinguishable Fits

A comparison of the fits in Figures 10.5 and 10.6 raises an important issue. The fit in Figure 10.5 is completely specified by two quantitative parameters, and a qualitative one. The qualitative parameter is which of the 120 possible combinations of the 7 standards shown in Figure 10.4 is used; in this case, it's the combination of Cr_2O_3, CrO_4, and Na_2CrO_4. The two quantitative parameters could be the percentage of Cr_2O_3 and the percentage of CrO_4. Once those two percentages are known, the percentage of Na_2CrO_4 could be calculated, because for the fit in Figure 10.5 we required the total to come to 100%.

The fit in Figure 10.6, in contrast, required the one qualitative parameter and *four* quantitative ones. That's a lot more freedom—is the fit closer because it's closer to the truth, or is it just closer because we're giving it more possibilities to choose from?

A rigorous statistical analysis of XANES fits turns out to be hard to do. For one thing, suppose we made a combination using the exact physical percentage of each constituent. Would the fit then be perfect? No, in part because of noise, but also because of systematic errors such as normalization and other sources we'll discuss in Section 10.6. In many cases, systematic errors are the more important source of mismatch. Most of the familiar machinery of statistics is not well-adapted to cases where systematic errors dominate. For instance, it is difficult to estimate what the 'measurement uncertainty' of a XANES spectrum is.

Another difficulty with traditional statistics is that it is not clear how many 'independent points' a XANES measurement includes. Compare, for instance, a spectrum collected with an energy spacing of 0.5 eV to one collected with a spacing of 1.0 eV.

Don't forget that these percentages give the percentage of chromium atoms present in each phase, as explained in Box 10.1! They are neither mole fractions nor percent by mass!

Mathematically inclined readers might think that there are 128 possible combinations of the 7 standards—after all, each one might be present or absent, giving 2^7 possibilities. But the case where no standard is used doesn't produce anything, and if only one standard were needed, simple fingerprinting would have provided the answer. That gives us $128 - 1 - 7 = 120$ feasible combinations.

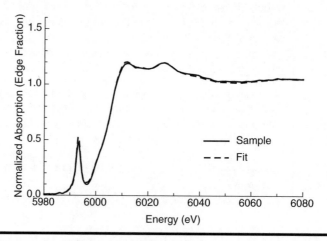

Figure 10.6 Linear combination fit of data in Figure 10.4, also allowing normalization of sample spectrum to vary. Result is 0.151 Cr_2O_3, 0.215 CrO_4, 0.033 $K_2Cr_2O_7$, and 0.635 Na_2CrO_4. (Total = 1.034.)

The former may give us a little more information about sharp features, but it certainly does not provide us *twice* the information of the latter, despite including twice as much data. In other words, the absorption at any particular energy is correlated to the absorption at neighboring energies, and the points are, therefore, not independent.

The combination of these difficulties makes it very hard to use statistics to answer the question 'Is this linear combination fit good enough that it might be the right one?' by statistical methods alone.

It turns out we *can*, however, use statistics to address a more limited, but still very useful, question: 'Is fit B significantly more likely to be right than fit A?' This question does not depend on our estimates of measurement uncertainty, and only modestly on our estimate of the number of independent points. Walter Hamilton addressed this problem in 1965 for the context of crystallographic modeling (Hamilton 1965). It is straightforward to adapt the *Hamilton test* of statistical significance to other forms of modeling, including XAFS (Downward et al. 2007).

10.4.3 Quantifying Fit Mismatch

Mismatch between data and fit can be measured in a number of ways. One common method is the *XAFS R factor*, which we define as the sum over all N measured points of the squared difference between each data point and the fit, normalized by the sum over all measured points of the data:

$$R = \frac{\sum_{i=1}^{N}\left(\text{data}_i - \text{fit}_i\right)^2}{\sum_{i=1}^{N}\left(\text{data}_i\right)^2} \tag{10.3}$$

Unfortunately, there are multiple definitions of the R factor in use. One common alternative is to use the square root of the expression in Equation 10.3:

$$R_{alt1} = \sqrt{\frac{\sum_{i=1}^{N}\left(\text{data}_i - \text{fit}_i\right)^2}{\sum_{i=1}^{N}\left(\text{data}_i\right)^2}} \tag{10.4}$$

XAFS R factor is our term, but it is not in wide use. For historical reasons, it is usually referred to as the EXAFS R factor.

Here's another, somewhat less common alternative:

$$R_{alt2} = \frac{\sum_{i=1}^{N}\left|\text{data}_i - \text{fit}_i\right|}{\sum_{i=1}^{N}\left|\text{data}_i\right|} \tag{10.5}$$

We should also note the recommendation of the International XAFS Society Standards and Criteria Committee (IXS Standards and Criteria Committee 2000):

$$R_{IXS}^2 = 100 \times \frac{\sum_{i=1}^{N}\left(\text{data}_i - \text{fit}_i\right)^2}{\sum_{i=1}^{N}\left(\text{data}_i\right)^2}\% \tag{10.6}$$

While the superscripted 2 may appear to make the IXS definition similar to Equation 10.4, context makes it clear that the 2 is purely notational, like the 2 in the symbol for the statistical measure χ^2. When the IXS document refers to the R factor, it means the measure we have adopted (Equation 10.3), expressed as a percentage.

Speaking of χ^2, another common method of quantifying fit mismatch is to use the definition of that statistical quantity:

$$\chi^2 = \sum_{i=1}^{N_{mad}} \frac{\left(data_i - fit_i\right)^2}{\varepsilon_i^2} \tag{10.7}$$

where ε_i is the measurement uncertainty associated with point i and the points i are taken so as to make $data_i$ represent N_{ind} independent measurements. As we discussed in Section 10.4.2, it is difficult to know the measurement uncertainty ε_i for XANES data, and even more difficult to know how to pick points so as to make the $data_i$ independent. One option that has been adopted is to simply use Equation 10.7 on every data point, and to assign some convenient value to ε_i, such as 1. While the symbol χ^2 may still be used for the resulting quantity, a calculation using every data point and an ε_i unrelated to measurement error has more in common with the XAFS R factor than with a true χ^2.

10.4.4 Degrees of Freedom

Statistically, the degrees of freedom for a fit is given by the difference in the number of independent data points and the number of free parameters. Because we do not know how many independent points are in a XANES spectrum, it may appear that we are stuck. Fortunately, we can put lower and upper limits on the number of independent points.

Consider Figure 10.5 in terms of the kind of peak analysis we discussed in Chapter 9. How many different features are visible? There's a preedge at around 5993 eV, a shoulder around 6001 eV, and a pair of peaks around 6012 and 6026 eV. There are probably at least a couple of more subtle features, perhaps at 6017 and 6037 eV.

Each peak or shoulder can be characterized by an amplitude, a centroid, and a width. If we think there are around 6 features visible, that means there are 18 quantitative parameters extractable from the data. Add in 2 to specify the edge itself (energy and width), and we can think of this spectrum as having at least 20 independent points.

The upper bound is easy, since it's the number of distinct energies at which the data are measured; in this case, that turns out to be about 90.

'Somewhere between 20 and 90' may not sound very precise, but it turns out to be good enough to get information from the Hamilton test.

The method suggested here for placing a lower bound on the number of independent points in a XANES spectrum is, unfortunately, quite subjective. The procedure for EXAFS is, at least, more objective, as we can see in Section 14.1.1 of Chapter 14.

10.4.5 The Hamilton Test

The Hamilton test is relatively insensitive to which measure of closeness of fit is used (Hamilton 1965), except that the types that are not 'squared' (Equations 10.4 and 10.5) have to be squared first. (In the case of Equation 10.4, that immediately transforms it to Equation 10.3.)

The procedure for the Hamilton test for linear combination fits on XANES data is then as follows:

1. Compute the ratio of the R factor for the closer fit to that for the other fit. Call the result r
2. Take half of the lower bound of the number of degrees of freedom of the closer fit, and call that a
3. Take half of the number of free parameters that have to be added to switch from one fit to the other. Call the result b

4. Calculate the regularized incomplete beta function $I_r(a,b) = \dfrac{\int_0^r t^{a-1}(1-t)^{b-1}\,dt}{\int_0^1 t^{a-1}(1-t)^{b-1}\,dt}$.

The result is a lower bound on the probability one fit would show the observed improvement over the other due to chance alone.

Let's apply this procedure to the case discussed in Section 10.4.1 and see what happens.

According to the software used to produce the fits, which uses definition 10.3 to compute the R factor, the fit shown in Figure 10.5 gives an R factor of 0.00116, while that in Figure 10.6 gives 0.00017. The ratio is, therefore, $r = 0.00017/0.00116 = 0.147$.

The lower bound on the number of independent points for our data is 20. The closer fit has four quantitative free parameters, so it has at least 16 degrees of freedom. That means the lower bound for a is 16/2 = 8.

The closer fit adds two free parameters: the amounts of $K_2Cr_2O_7$ and Na_2CrO_4. So, in this case, $b = 2/2 = 1$.

Calculating $I_{0.147}(8,1)$ gives 0.0000002, meaning that there is less than one chance in a million that the difference between these fits would arise by random variation. The fit in Figure 10.6 is, therefore, significantly better (in a statistical sense) than the fit in Figure 10.5.

10.4.6 Uncertainties

Numbers corresponding to physical systems should be reported with uncertainties (or at least the implicit uncertainty associated with the number of significant figures shown). That is a basic rule in the practice of science. How do we determine the uncertainties associated with the fraction of each constituent as determined by a linear combination fit?

The uncertainty in a fitted parameter (we'll call it δ) can be defined as the amount by which the parameter can be changed, while allowing the other fitted parameters to vary, without causing the fit to be significantly worse. This means that determining δ, like the determination of whether a fit is 'good' statistically, requires consideration of both the measurement uncertainty ε and the number of independent points in the data.

Software that performs the least-squares minimization used for linear combination fitting is commonly available: you may, for instance, use a stand-alone application, a feature that's part of a graphical analysis suite, a routine in a mathematical analysis language, or a software package designed specifically for XAFS analysis. Any of these should provide an estimate of the δs for fitted parameters or a way of computing them—if what you're using doesn't, then start using something else! But in almost all cases, the uncertainty estimate provided by software is not an appropriate uncertainty for XAFS linear combination fitting, and needs to be modified.

To understand how to modify the reported uncertainty, we first need to understand how we would identify a statistically good fit if we did know the measurement uncertainties ε_i and the number of independent points. We would start by computing the *reduced* χ^2, represented by the symbol χ_v^2:

$$\chi_v^2 = \frac{N_{ind}}{vN} \sum_{i=1}^{N} \frac{\left(\text{data}_i - fit_i\right)^2}{\varepsilon_i^2} \tag{10.8}$$

where v is the number of degrees of freedom of the fit and N_{ind} is the number of independent points.

As Dysnomia explains, for a fit to be statistically good, the reduced χ^2 should be around 1 (for a more careful discussion of this criterion, see Bevington and Robinson [2003]).

To help us adjust the δs reported by a fitting routine, we'll first note that almost all software that would be used for XANES linear combination fits will assume $N_{ind} = N$, and thus will also have too large a value for v. We discussed in Section 10.4.2

b is calculated from half the *total* number of free parameters added, not half of the *net* number of free parameters added! Suppose fit alpha uses the amounts of constituents A, B, and C as free parameters, while fit beta uses the amounts of constituents C, D, and E. Even though there are the same number of free parameters (three each), to get from fit alpha to fit beta we need to add two free parameters (the amounts of D and E). Thus *b* would be 2/2 = 1.

If the mismatch between data and fit were due just to the uncertainty ε, we'd expect the average term in the sum in Equation 10.8 to be about 1, so summing the terms and dividing by N should give us 1. Of course, each constraint increases the mismatch a little bit. The factor N_{ind}/v accounts for that, so that the reduced χ^2 is around 1 for a good fit.

Since we are using a statistical model where ε_i takes on the same value for all data points, we will drop the subscript *i* for the remainder of this discussion.

how to place a lower bound on N_{ind}. For ε_i, let's start by using an arbitrary value for all *i*—typically 1.

For the fit shown in Figure 10.6, the lower bound on N_{ind} is 20, and $\nu = 16$. Substituting those values, choosing $\varepsilon = 1$, and taking the rest from the report of the fitting software, we get an initial estimate for $\chi^2_{\nu(\varepsilon=1)}$ of 0.00015.

But we believe our best fit to be a 'good' fit—we chose a reasonable set of standards to conduct our fitting process, and believe that the result we've arrived at is probably also reasonable. This means we expect χ^2_ν to be around 1; the value of 0.00015 we got is presumably so much smaller because we made up an arbitrary value for ε. Examining the proportionalities in Equation 10.8, we can see that to get a χ^2_ν of 1, we need an ε of $\sqrt{0.00015} = 0.012$. Since the δs should be proportional to ε, that means we should multiply all of the δs reported by our fitting software for $\varepsilon = 1$ by 0.012.

Following that procedure for the fit from Figure 10.6 gives Cr_2O_3 of 0.151 ± 0.013, CrO_4 of 0.215 ± 0.015, $K_2Cr_2O_7$ of 0.033 ± 0.038, and Na_2CrO_4 of 0.635 ± 0.015. It is interesting to note that we, therefore, should not be confident that there is any $K_2Cr_2O_7$ at all, as the uncertainty in its weighting is greater than the value.

Here is a summary of the method we just used to estimate uncertainties:

1. Compute the average squared difference $\frac{1}{N}\sum_i (data_i - fit_i)^2$. Some software calls this χ^2, although it is not the statistical quantity of the same name.
2. Estimate the number of independent points N_{ind}.
3. Use the number of fitted parameters f to calculate the degrees of freedom $\nu = N_{\text{ind}} - f$
4. Multiple the quantity from step 1 by N_{ind}/ν
5. Multiply the uncertainties reported by your fitting software by the square root of the result from step 4. (If your software already performs this scaling, then multiply the uncertainties it reports by the square root of N_{ind}/ν instead of by the quantity from step 4. Read your software's documentation carefully so that you know what it is doing!)

10.5 Combinatoric Fitting

You need to make sure you understand what your software is doing before reporting uncertainties! For instance, some XAFS-oriented software performs the trick for you of assuming χ^2_ν should be 1 to compute uncertainties, but still assumes $N_{\text{ind}} = N$.

We have established that the fit shown in Figure 10.6 is a statistical improvement over the fit shown in Figure 10.5, but that doesn't mean we can't do better, and it doesn't mean it's ready for publication. For example, should we just leave off the 3% $K_2Cr_2O_7$?

We could try a fit without $K_2Cr_2O_7$, and see what the statistics tell us. But why stop there? Given that we have a computer to do the number-crunching for us, let's try fitting all 120 possibilities. That's called combinatoric fitting. Since we've already established that the normalization of the sample was likely a bit different than that for the standards, we won't force the percentages to total to 100%.

The R factors of the resulting combinations range from the 0.00017 we already found up to 0.06025 for a combination of Cr and Cr_2S_3. The fits corresponding to the lowest values are shown in Table 10.1.

Now we can use the Hamilton test to understand what the combinatorics results tell us. A good principle is to begin by comparing fits with relatively few parameters because they are the easiest to distinguish. In this case, let's compare fit number 5 (lowest R factor with two free parameters) to fit number 2 (lowest R factor with three free parameters). The ratio of R factors is $0.00017/0.00052 = 0.33$, fit number 2 has at least $20 - 3 = 17$ degrees of freedom, and fit number 2 adds one free parameter to fit number 5. So we calculate $I_{0.33}(8.5, 0.5) = 0.00002$. That means that fit number 2

Table 10.1 Best Fits to Data from Section 10.4, in Ascending Order of *R* Factor

Fit No.	R Factor	Cr_2O_3	Cr	Cr_2S_3	CrO_4	$K_2Cr_2O_7$	K_2CrO_4	Na_2CrO_4
1	0.00017	0.151	—	—	0.215	0.033	—	0.635
2	0.00017	0.146	—	—	0.212	—	—	0.675
3	0.00032	—	0.056	—	0.350	—	—	0.629
4	0.00040	—	—	0.035	0.345	—	—	0.654
5	0.00052	—	—	—	0.373	—	—	0.661

represents a statistically significant improvement over fit number 5 at the 99.998% confidence level, and we thus reject fit number 5.

What about fit number 3? Comparing it to fit number 2, the ratio of *R* factors is 0.00017/0.0032 = 0.53. Fit number 2 still has at least 17 degrees of freedom, and fit number 2 again adds one free parameter to fit number 3. (Again, note that it is the total number of free parameters added that matters; in this case, we need to fit the amount of Cr_2O_3. The fact that we no longer need to fit the amount of Cr does not enter into the computation.) $I_{0.53}(8.5, 0.5) = 0.001$, indicating a 99.9% confidence level that fit number 2 is better not just by chance, so we reject fit number 3.

Since fit number 4 has a worse *R* factor and the same relative number of parameters, we can reject it as well.

Fit numbers 1 and 2 cannot be distinguished, as they have the same *R* factor. Both must be considered possible.

Of course, we have assumed there is little systematic error in our analysis, beyond the allowance we have already made for the normalization of the sample's spectrum. Systematic error could conceivably change our conclusions. We'll discuss ways in which it could originate next.

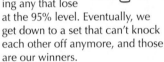

There's no magic order to doing these Hamilton tests. We're just playing the fits off in pairs and rejecting any that lose at the 95% level. Eventually, we get down to a set that can't knock each other off anymore, and those are our winners.

10.6 Sources of Systematic Error

We've already discussed inconsistent normalization as a source of systematic error. This section will discuss additional sources.

10.6.1 Energy Alignment

Suppose that, because of inconsistent energy calibration, the sample spectrum we've been fitting in the last two sections had been shifted to the right by 1 eV relative to the standards. Rerunning the analysis, it turns out that the *R* factor for the best fit rises all the way to 0.0050— that's nearly 30 times worse than before! Even worse, the closest fit turns out to be 0.47 CrO_4, 0.45 $K_2Cr_2O_7$, and 0.10 K_2CrO_4, which is radically at odds with the conclusions in Section 10.5. The danger is clear: even a slight error in relative energy calibration can wreak havoc with LCA of XANES spectra.

Fortunately, it is usually easy to prevent this from happening. If reference spectra are collected simultaneously with the sample data, those can be used for alignment as described in Section 7.2.1 of Chapter 7. Even when simultaneous collection of a reference isn't feasible, reference data can be collected periodically during the experiment, and carefully scrutinized for drift.

Still, it is possible that at some point you will be working with a data set for which the energy calibration is not reliable. It is not uncommon, for example, to find that a

Inspecting encoder values can correct for some kinds of energy drift!

constituent is something you did not anticipate in advance, and, therefore, to have to use a standard collected by someone else, at least until you get more beam time and a chance to measure it yourself. If the person who measured that standard didn't take good care with energy calibration, you may be stuck.

In cases like that, it's not difficult to allow the fitting routine to fit an energy shift for any spectrum where calibration is suspect. Continuing with our example, when we allow our computer software to treat the energy calibration of the sample as a free parameter, we immediately retrieve something very close to the closest fit found in Section 10.4.

So what's the big problem? Can't we just rely on our software to save us?

The problem is, of course, that each distinct energy shift allowed in a fit represents one more free parameter, and one fewer degrees of freedom. That means the ability to statistically reject fits will drop, and we'll become less confident in our conclusions.

To take it to an extreme, suppose that we had no confidence in any of our energy calibrations, and allowed a separate shift for each of the standards. When the fits from Section 10.4 are rerun, but with every energy allowed to vary independently, we do get a closest fit something a bit like the best fits from before: 0.24 Cr_2O_3, 0.13 CrO_4, 0.02 $K_2Cr_2O_7$, 0.02 K_2CrO_4, and 0.62 Na_2CrO_4. The R factor for this fit is 0.00010, which is better than the old fit number 2. But is it a statistical improvement? We can use the Hamilton test to find out.

In this case, we are adding seven free parameters: the old fit number 2 didn't have any $K_2Cr_2O_7$ or K_2CrO_4, and we have five new parameters for energy calibration. Our new fit has a total of 10 free parameters, and thus 20 − 10 = 10 degrees of freedom. That brings a down to 5, and b up to 3.5. With a ratio of R factors of 0.00010/0.00017 = 0.59, we compute $I_{0.59}(5, 3.5) = 0.49$, meaning the improvement in the R factor is quite likely to have occurred by chance alone.

It is also worth mentioning that, in this case, the fit turns out to give energy shifts for the standards that vary by nearly 5 eV. If we knew that we (or someone else) had collected the standards fairly carefully, and had used a reference foil for calibration, then a shift of 5 eV should not have occurred. So, the fit in which all the energy shifts allowed to vary is, in this case, *less* realistic than our old one.

The moral: don't allow parameters to vary just because your software happens to allow it. Careful planning during measurement and data reduction can cut down on the number of free parameters, which in turn should increase the confidence you place in your results.

Why, if we have no confidence in the energy calibrations of five standards and a sample, does that represent five free parameters rather than six? It is because it is only the relative shift that matters. Varying the shift of each standard relative to the sample gives five parameters.

10.6.2 Background

For XANES and energy-space EXAFS analyses, if the spectra being used for standards and sample have different slopes in the preedge or postedge region, it can introduce systematic error. This can be addressed either during data reduction (by some sort of 'flattening' routine) or during fitting (by introducing additional free parameters).

If this is a problem in your data set, it can also be advisable to limit the range over which the data are fit (see Section 10.7).

10.6.3 Attenuation: Self-Absorption, Inhomogeneous Transmission Samples, Harmonics, Dead Time, and So On

As discussed in Chapters 6 and 8, there are several sample preparation and data collection issues that can cause spectra to be distorted. While they differ in detail, these distortions generally result in attenuation of high-absorption features, most notably white lines. They can also result in exaggerated amplitude for preedge features.

Distortions of this kind have a substantial effect on LCAs, particularly of XANES. Even in cases where the shapes of the standards are very different, such as the greenockite/monteponite example in Section 10.3, distortions can cause substantial changes in the percentages of each constituent found by a linear combination fit.

With the partial exception of self-absorption corrections (Section 7.3.5 of Chapter 7), not much can be done about distorted spectra after the fact. When planning on LCAs of XANES, use the information from Chapters 6 and 8 to minimize the extent of distortion before analysis.

10.6.4 Energy Resolution

As discussed in Section 4.1 of Chapter 4, the shape of an XAFS spectrum depends on the energy resolution of the measurement; this is particularly true for sharp features such as the white line and some preedge features. While it is possible to mathematically modify a spectrum after collection to deliberately simulate a coarser energy resolution, it is better to try to have the same energy resolution for all measurements.

10.6.5 Glitches

Because the minimization routines used in LCA minimize the *square* of the difference between the data and the fit, even a single-point glitch, if large enough, can introduce systematic error into a XANES or energy-space EXAFS linear combination fit. If possible, deglitch all the standards and the sample before analysis (see Section 7.3.2 of Chapter 7). If large glitches happen to lie near the boundaries of the region you're fitting (e.g., in the linear preedge region), simply choose the region so that they're not included.

10.6.6 Noise

Noise might seem as if it should contribute only to random error, not systematic error. That would be right if individual scans were each fit separately. But since we are usually working with merged data (see Section 7.2.2 of Chapter 7), the noise remaining after averaging gets 'frozen in.' If shot noise causes one of your standards to have a little blip up or down at, say, the top of the white line, then that will affect all analyses using that standard. In the sense that it biases all analyses in the same way, the error is systematic, even though it is random in origin.

BOX 10.5   Measuring Standards for Linear Combination Analysis

	I guess the message of Section 10.6 is: 'Measure your standards under the same conditions as the sample.' If I do that, then problems I have with self-absorption or energy resolution or whatever will apply to the standards and the sample, and I'll be fine.
	Careful! If, by trying to measure 'under the same conditions,' you mean that you'd take a standard that is pure gypsum and measure it in the same fluorescence geometry as a sample that is 80 parts per million sulfur, then your standard would suffer severely from self-absorption, but your sample would not!

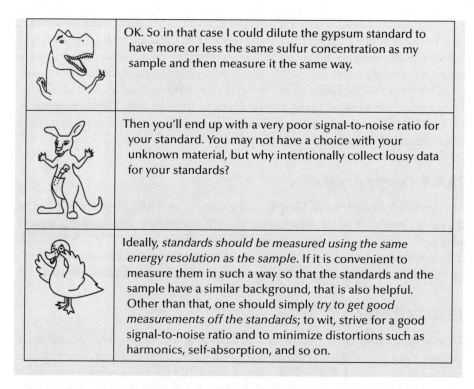

OK. So in that case I could dilute the gypsum standard to have more or less the same sulfur concentration as my sample and then measure it the same way.

Then you'll end up with a very poor signal-to-noise ratio for your standard. You may not have a choice with your unknown material, but why intentionally collect lousy data for your standards?

Ideally, *standards should be measured using the same energy resolution as the sample.* If it is convenient to measure them in such a way so that the standards and the sample have a similar background, that is also helpful. Other than that, one should simply *try to get good measurements off the standards;* to wit, strive for a good signal-to-noise ratio and to minimize distortions such as harmonics, self-absorption, and so on.

10.7 Choosing Data Range and Space For LCA

So far, this chapter has focused primarily on LCAs of XANES data in energy space. We've done that for three reasons: because it's one of the most common forms of LCA, because it's particularly sensitive to several types of systematic error, and because we thought that sticking with one or two examples for most of the chapter was less distracting than looking at all the varieties of LCA together.

In this section, we will discuss each of the common forms of LCA of XAFS.

10.7.1 XANES in Energy Space

As discussed in Chapter 9, XANES is sensitive to valence, the symmetry of the local environment, and nearest-neighbor bond length. It is largely insensitive to 'disorder' of the kind discussed in Section 1.2.12 of Chapter 1, and is not usually as good a tool as EXAFS for investigating structure beyond the first coordination shell.

The signal-to-noise ratio will be better for the XANES region than for the EXAFS region; for low-concentration samples, this may make XANES the only usable option.

 The insensitivity of XANES to disorder can be a good thing, since the sample can then be more disordered than the standards without affecting the analysis much.

To avoid systematic error introduced by differences in the slope of the preedge, it is best to choose the low end of the fitting region to be about 10 eV below the first preedge feature (or the beginning of the main rise, if there are no preedge features).

It is also possible to perform linear combination fits on particular features within the XANES region; this is closely related to some of the advanced fingerprinting techniques discussed in Chapter 9.

10.7.2 XANES in Derivative Space

Any XANES fit in energy space is dominated by the overall shape of the edge. By performing the analysis on the derivative of the normalized absorption, the emphasis is shifted to features such as peaks and shoulders.

Derivative spectra are often noisy, however. As indicated in Section 10.6.5, noise can act as systematic error in certain cases. Rebinning (Section 7.1.1 of Chapter 7) or smoothing can reduce that noise, but also reduces energy resolution unless the data were oversampled to begin with.

10.7.3 *EXAFS in Energy Space*

The EXAFS region of the spectrum, as discussed in Section 1.2 of Chapter 1, is sensitive to the distances, coordination numbers, chemical identity, and disorder of scattering shells from the nearest neighbors out to 5 Å or more. It thus provides somewhat complementary information to the XANES region. If, for example, the sample is disordered in comparison with the standards, a linear combination of standards may fit the XANES region much better than the EXAFS region. If, however, a local distortion changes the symmetry of the nearest-neighbor environment but has minimal impact on the longer-range crystal structure, then the EXAFS may fit better than the XANES.

EXAFS is also somewhat less severely affected than XANES by measurement problems such as self-absorption, inhomogeneous transmission samples, harmonics, or instability in energy calibration. In the EXAFS range, modest amounts of attenuation due to those problems tend to mimic normalization errors. While they would thus still contribute to systematic error in the quantitative results (how much of each constituent is present), the qualitative results (which constituents are present) are less affected.

The fits in Sections 10.4 and 10.5 utilized both the XANES region and a bit of what many would consider the EXAFS region, extending to 80 eV above the edge. If we had wanted to investigate much farther above the edge than that, however, it's generally best to fit the XANES and EXAFS regions separately. That's because the features in the EXAFS region are usually much smaller than those in the XANES region, and are certainly much smaller than the edge jump itself. If a preedge peak, for instance, were 10 times larger in amplitude than a particular EXAFS oscillation, then a 5% mismatch in the preedge feature would count as much as a 50% mismatch in the EXAFS feature. Thus, a single fit across the entire spectrum could prioritize reducing small fractional errors in the XANES part of the spectrum at the expense of entire wiggles in the EXAFS. If, as is often the case, systematic error at the 5% level is present in the XANES region, a single fit across the entire XANES and EXAFS regions might discount EXAFS information in favor of trying to improve a fit in a XANES region which is somewhat distorted.

In addition, fitting the two regions separately provides a partially independent check of the results. It may even act as a way of distinguishing between fits which are otherwise difficult to distinguish, such as fit numbers 1 and 2 shown in Table 10.1, or of narrowing the uncertainty in quantitative results.

Application of the Hamilton test to an EXAFS fit once again requires estimating the number of independent points present in the data, but the method discussed in Section 10.4.4 seems somewhat less well-suited to the oscillations of EXAFS data. Instead, we'll take inspiration from Robert's argument in Section 7.5.8 of Chapter 7, and assert there can't be more than one independent point per 0.16 Å⁻¹, and that's only if we think we can extract data out to 10 Å in the Fourier transform. That's very rarely the case, except for the most highly ordered crystals—5 Å might be more typical, in which case we're getting one independent point per 0.32 Å⁻¹. In the spirit of conservatism, we'll round that up to one independent point per 0.4 Å⁻¹.

10.7.4 *EXAFS in χ(k)*

If we want to focus on the EXAFS, linear combination fits can also be performed in $\chi(k)$. As described in Chapter 7, this requires background subtraction and a consistent

The second derivative spectrum is like the derivative spectrum, but more so—peaks become strongly emphasized, at the cost of additional noise.

R factors for different spaces (e.g., energy vs. derivative) or for different regions cannot be directly compared! That's part of the price we pay for not being able to calculate true statistical measures of closeness of fit. When we talk about the XANES fit or the EXAFS fit being better, we are relying on judgment and experience, not quantitative measures.

We will consider the statistics of EXAFS more rigorously in Section 14.1 of Chapter 14.

choice of E_o. Background subtraction can be beneficial, in that it removes a potential source of nonstructural variation between spectra. However, if background subtraction is done inconsistently, that can introduce a new source of systematic error!

Of course, 'fitting in $\chi(k)$' generally actually refers to fits on $k\chi(k)$, $k^2\chi(k)$, or $k^3\chi(k)$, in accord with Section 7.4.5 of Chapter 7.

BOX 10.6   Choosing E_0 for LCA

Section 7.3.3 in Chapter 7 describes different ways of choosing E_o. Do I just pick one method and use it to find an E_o for the sample and each of the standards?

No! You have to use the *same* value of E_o for all the spectra—that is, assuming you've aligned them using a reference or some other method.

Wait—I'm confused. I thought that the oxidation state of the absorbing atom could shift E_o by an eV or two. If my standards are in different oxidation states, shouldn't I account for that somehow?

Absolutely not! Suppose you have a sample that is a mixture of metallic zinc and zinc oxide. You have to pick an E_o for the spectrum of the sample, right? But the spectrum is a linear combination of the spectrum of a metal and an oxide. So for the sample, you will necessarily be choosing E_o incorrectly for at least one of its constituents—there's no way around that. To make LCF work properly, the standards must have the same errors. So for an LCF analysis, you need to assign the same E_o to every spectrum involved, even if they have different oxidation states!

Table 10.2 shows the result of combinatorial fits to $\chi(k)$ data drawn from the same spectra used in Sections 10.4 and 10.5.

After looking at the R factors in Table 10.1, the R factor 0.138 sounds large. But we should expect the R factors to be larger for $\chi(k)$ fits than for energy-space XANES, because the background, which is in theory the aspect of the spectrum most

Table 10.2   Best Fits to Data from Section 10.4, Using a Fit to k-Weighted $\chi(k)$ From 5 to 12 Å^{-1}, in Ascending Order of R Factor

Fit No.	R Factor	Cr_2O_3	Cr	Cr_2S_3	CrO_4	$K_2Cr_2O_7$	K_2CrO_4	Na_2CrO_4
1	0.138	0.206	—	0.007	0.125	—	—	0.701
2	0.138	0.205	—	—	0.124	—	—	0.707
3	0.190	0.258	0.008	0.001	—	—	—	0.585
4	0.190	0.258	0.008	—	—	—	—	0.587
5	0.191	0.262	—	—	—	—	—	0.581

likely to be shared by all the spectra, has been stripped away. In addition, we're using k-weighted fits, which will amplify the noise at high k. But should we expect it to be as large as 0.138?

The easiest way to answer that is to look at one of the fits graphically. Look ahead a bit to Figure 10.8, which shows the data along with a couple of fits with R factors comparable to those in Table 10.2. (In fact, the fit labeled 'EXAFS best fit' is one of the fits from Table 10.2. The fit labeled 'XANES best fit' will be discussed later in this chapter.)

The k-weighted $\chi(k)$ data are quite noisy, and that noise appears to be the greatest contributor to the R factor for these fits. In this case, an R factor above 0.10 appears to be consistent with reasonably good fits, given the noise in the data. This would certainly not have been the case for the XANES data, for which the signal-to-noise ratio is much better.

Let's apply the Hamilton test to fit numbers 5 and 2 in Table 10.2 (as before, we start with a pair that has relatively few free parameters, because they are easier to distinguish statistically). The ratio of R factors is $0.138/0.191 = 0.72$. Taking $0.4\ \text{Å}^{-1}$ per independent point, there are about 18 independent points in the data. Fit number 2 has three free parameters, and thus 15 degrees of freedom and $a = 7.5$. Since one parameter needs to be added to go from fit number 5 to 2, $b = 0.5$ and $I_{0.72}(7.5, 0.5) = 0.029$. Using a 95% confidence level and particularly keeping in mind that we used a lower bound on the number of independent points, we tentatively reject fit number 5.

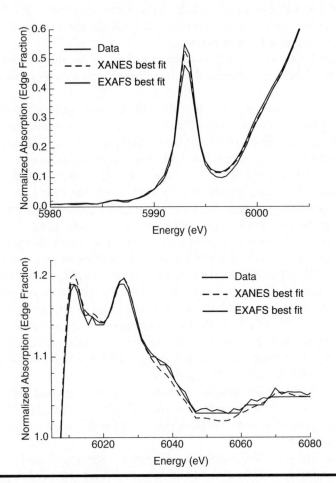

Figure 10.7 **(See color insert.) Comparison of two linear combination fits in energy space. The graph has been split into two regions to facilitate examination of the differences between the two fits and the data.**

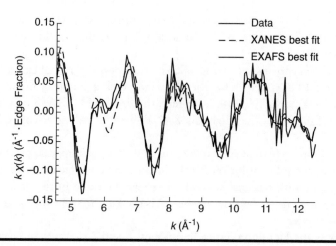

Figure 10.8 **(See color insert.) Comparison of two linear combination fits in *k* space.**

Next, let's compare fit numbers 3 and 4 with fit number 2. The ratio of R factors is $0.138/0.190 = 0.73$. Once again, only one parameter needs to be added to go to fit number 2 (the amount of CrO_4), so we compute $I_{0.73}(7.5, 0.5) = 0.033$. These fits are also tentatively rejected.

Tables 10.1 and 10.2 now each include two fits that we have not rejected on statistical grounds. But only one of those is common to both: the one with just Cr_2O_3, CrO_4, and Na_2CrO_4, numbered 2 in both tables.

Those two fits are not entirely in agreement, however, as the percentage of each constituent they show are different. Are they close enough that we could say they're 'consistent?'

At this point, we can assume each of those fits is 'good,' and use the method discussed in Section 10.4.6 to compute the uncertainties in each amount. For fit number 2 in Table 10.1, we get Cr_2O_3 of 0.146 ± 0.012, CrO_4 of 0.212 ± 0.014, and Na_2CrO_4 of 0.675 ± 0.006. For fit number 2 in Table 10.2, on the other hand, we get Cr_2O_3 of 0.205 ± 0.018, CrO_4 of 0.124 ± 0.018, and Na_2CrO_4 of 0.707 ± 0.035. Since the weights of some of the constituents differ by much more than their uncertainties, the fits must be considered different. But is one statistically superior to the other?

To answer that, we construct an energy-space spectrum using the constituent weights found by the k-space fit, that is, the weights given in fit number 2 of Table 10.2. We'll refer to that as the 'EXAFS best fit,' meaning that it uses the weights found by fitting the EXAFS portion of the spectrum. When the R factor for the resulting energy-space model is found, it comes out to 0.00026. We can compare this to fit number 2 in Table 10.1 ('the XANES best fit'), which had an R factor of 0.00017. The ratio is thus $0.00017/0.00026 = 0.65$. The energy-space 'fit' using the values from the EXAFS best fit has no free parameters—everything is constrained. Since the XANES best fit has three free parameters, we compute $I_{0.65}(8.5, 1.5)$ and get 0.057; the fits are not quite distinguishable at the 95% confidence level.

One option at this point would be to consider ourselves done, and just declare that our sample has somewhere between 12% and 22% Cr_2O_3, between 10% and 22% CrO_4, and between 64% and 72% Na_2CrO_4, with possible trace amounts of other chromium compounds.

But we do have one more weapon in our arsenal—the magnificent ability of the trained human brain to synthesize and analyze complicated information, particularly when presented visually. Let's compare the two fits in both energy space Figure 10.7 and k space Figure 10.8. To aid the examination of fine detail, the energy-space plot has been split into the preedge and postedge regions.

The EXAFS best fit appears to follow the data better in both the *k*-space graph, which covers the EXAFS region, and over the peaks and shoulders from 6010 to 6050 eV, which comprise the white line. The one place where the XANES best fit has an advantage is at the sharp preedge feature around 5993 eV; both fits overestimate the size of that peak, but the overestimate of the XANES best fit is slightly less.

Is that one slight-looking difference enough to account for the better *R* factor of the XANES best fit? Notice the difference in scale between the *y*-axis of the top and bottom graphs in Figure 10.7; that 'slight' difference would look nearly three times bigger if the scales were the same. Also recall that we are computing the *R* factor using the *square* of the difference between data and fit; this creates an emphasis on improvements at the energies where the mismatch between fit and data is greatest. So yes, that one peak is the source of most of the difference in *R* factor between the two fits.

But why does a fit that looks pretty good above the edge show that kind of mismatch on a preedge feature? It suggests a systematic error of some sort. If there were harmonics in the beam, that would emphasize preedge peaks at the expense of features above the edge, the opposite of what we see. But, as Section 4.1 of Chapter 4 shows, poor energy resolution results in sharp peaks being reduced in amplitude, and the preedge peak is the sharpest feature in the spectrum. (Merging individual scans that are slightly misaligned could also cause this kind of effect, but is easy to check for.)

In the scenario Robert outlines in Box 10.7, we would identify the EXAFS best fit as superior and report it in any publications arising from the experiment, perhaps broadening the uncertainty ranges a bit to err on the side of conservatism.

BOX 10.7   Good Data Are Not Perfect

Wait a minute! Are you telling me we went through all this, and the data are messed up? I'm going to sound like Robert here; I'd just try to get good data in the first place!

Alas, dear Simplicio, these data are fairly good. There are always trade-offs involved, especially considering the limited beamtime that is available. Suppose, for instance, that the experimenter collected the standards shown above on a previous trip to the synchrotron; there are too many to realistically expect them to be recollected every time. When attempting to collect the spectrum of the sample, our experimenter notes how poor the signal-to-noise ratio is, and decides more photons are needed. The experimenter tries widening the vertical slits. Mindful of the risk of degrading the energy resolution, the experimenter collects spectra on a chromium foil before and after changing the slit size, and finds the spectra to be indistinguishable. But the spectrum of chromium metal has no features as sharp as the preedge feature in question, as one can see from inspection of Figure 10.4. It would be eminently possible to degrade the preedge peak slightly without realizing it.

In such a case, the experimenter did nothing wrong. Each decision was defensible, given the options available. As long as the experimenter kept good notes, we could, at this point in the analysis, identify the likely source of systematic error and select the superior fit.

10.7.5 The Back-Transform of EXAFS

In some cases, it might be desirable to use Fourier filtering before performing a linear combination fit in the EXAFS region. Filtering out low R has the benefit of reducing the effect of difference in background subtraction between the data and the standards. Filtering out high R can be useful if the standards are likely to deviate most from the constituents of the sample beyond the nearest neighbor; for example, the sample is made of nanoscale phases, and the standards are bulk crystalline materials.

Filtering also allows a more rigorous definition to be used for the number of independent points present (see Section 14.1.1 of Chapter 14).

If filtering is desired, it is best to perform LCA on the back-transform, and not directly on the Fourier transform $\chi(R)$. While $\chi(R)$ of a mixture is in theory a linear combination of the contributions from the individual constituents, the interactions between the systematic errors described in Section 10.6 and the Fourier transform effects detailed in Section 7.5 of Chapter 7 can be difficult to understand intuitively.

WHAT I'VE LEARNED IN CHAPTER 10, BY SIMPLICIO

- X-ray absorption from a given phase is proportional to the percentage of the absorbing atom present in that phase. To work with other common measures of percent composition, we need to convert.
- It's best to collect data well before and well after the edge to aid with normalization, even if only the XANES is going to be used.
- The *Hamilton test* can be used to determine whether one fit gives a statistically significant improvement over another.
- We should always think about how to compute and report uncertainties.
- Sources of systematic error in LCA can include normalization, energy alignment, background subtraction, attenuation, energy resolution, glitches, and noise.
- Standards and samples should be measured using the same energy resolution. Other than that, we should focus on getting good spectra, rather than making sure the standards and sample are measured in exactly the same way.
- LCA can be done on XANES in energy space, XANES in derivative space, EXAFS in energy space, EXAFS in k-space, or on the back-transform of EXAFS.
- When performing LCA on EXAFS in k-space, or on the back-transform of EXAFS, the same E_0 should be chosen for all spectra.

References

Bevington, P. R. and D. K. Robinson. 2003. *Data Reduction and Error Analysis for the Physical Sciences*, 3rd ed. Boston: McGraw-Hill.

Downward, L., C. H. Booth, W. W. Lukens, and F. Bridges. 2007. A variation of the F-test for determining statistical relevance of particular parameters in EXAFS fits. *X-Ray Absorption Fine Structure—XAFS13: 13th International Conference. AIP Conf. Proc.* 882:129–131. DOI:10.1063/1.2644450.

Hamilton, W. C. 1965. Significance tests on the crystallographic R factor. *Acta Crystallogr.* 18:502–510. DOI:10.1107/S0365110X65001081.

IXS Standards and Criteria Committee. 2000. Error reporting recommendations: A report of the standards and criteria committee. http://ixs.iit.edu/subcommittee_reports/sc/err-rep.pdf.

Chapter 11

Principal Component Analysis

Don't Sweat the Details

And let me guess—it's all details, right?	
Nope. In fact, separating what's important from what's not is the main idea of principal component analysis (PCA).	
Even when one is focused on the important things, Dysnomia, one must still get the details right.	
Some of them, yeah. But as we'll see, Robert, PCA lets us get by with less …	

DOI: 10.1201/9780429329555-14

We're going to use the data from Figure 11.1 through the rest of the chapter. We'll call it the *gold series* for short.

I thought the energy resolution of typical x-ray absorption fine structure (XAFS) scans is more than an eV. If an isosbestic point is smeared out over a half eV or so, couldn't that just be an effect of energy resolution?

Energy resolution means what it says: it has to do with the ability to *resolve* features such as double peaks. It doesn't mean that the energy of a peak or an intersection of spectra or whatever differs by that amount from scan to scan or spectrum to spectrum.

There is some truth to what Simplicio says, my dear Dysnomia. If the spectra in the gold series were measured using *different* energy resolutions, by, for example, changing the width of the pre-I_o vertical slit width (Section 8.5.2 of Chapter 8), then that could cause features such as peaks to appear to shift energy.

11.1 Introduction

In this chapter, we will learn about principal component analysis (PCA). We will begin by introducing an example system. Next, we will explore the ideas of PCA informally, without dwelling on formal definitions. After introducing the formalism, we will explore some additional ways to use the technique.

11.1.1 An Example from the Literature

Figure 11.1 shows the gold L_3 edge as a function of time for a gold (III) chloride solution in the presence of cyanobacteria (details available in Lengke et al. [2006]). Clearly, the bacteria are interacting with the gold. But how many gold-containing species are present? Is all of the original gold (III) chloride consumed or only some of it? Is any gold metal produced in the process?

In their paper, Lengke et al. used linear combination analysis to address those questions. That's a challenging problem, however, as they had to measure the right standards, some of which have spectra that appear different in solution than in their solid form.

Is there a way to get a sense of what is going on in that series of spectra without having to identify all the constituents?

11.1.2 Isosbestic Points

Consider two normalized spectra of pure constituents, $f(E)$ and $g(E)$. Because they are normalized, any spectrum made by mixing them can be expressed as a weighted average of the two; that is, by $xf(E) + (1-x)g(E)$. At any point where $f(E)$ and $g(E)$ have the same value, then, any weighted average will have to take on that same value—the average of two identical numbers does not depend on how the average is weighted! These points are called *isosbestic points*. Figure 11.2 shows this idea graphically, using linear combinations of monteponite and greenockite.

Most spectra representing different mixtures of two constituents will reveal several isosbestic points. It is unlikely, however, for different mixtures of three or more constituents to show isosbestic points, as that would require three (or more!) constituents to happen to have the same normalized absorption at some energy. Thus, isosbestic

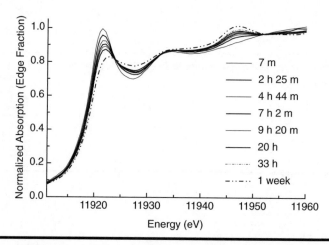

Figure 11.1 (See color insert.) Time series of gold L_3 XANES for a sample of gold (III) chloride exposed to cyanobacteria, as a function of time of exposure. Data were from Lengke et al. (2006).

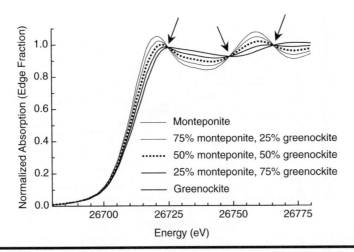

Figure 11.2 **Linear combinations of the normalized spectra of monteponite and greenockite measured at the cadmium *K* edge. Isosbestic points are indicated by arrows.**

Sure, and if I had wings, I could fly. The scientists who collected these data were very familiar with XAFS. I'm sure they didn't do anything loopy like changing slit width in the middle of collecting a related series of spectra.

points can be used as a clue that two (and only two) constituents are being mixed in different amounts to make the series of spectra.

Does the gold series show isosbestic points? At first glance, it appears there may be one around 11,924 eV, another around 11,934 eV, and a third around 11,952 eV. But that's worth a closer look, as we can see in Figure 11.3.

On closer inspection, none of the points appear to be isosbestic; the plots do not intersect at all at 11,934 eV, and in the other two cases, the intersections span more than half an eV. Unless the alignment of the spectra through references or the normalization was sloppy, it appears the fractions of more than two constituents are varying in this set.

In the following several sections, we will introduce the concepts of PCA, without saying anything about *how* these quantities are computed, deferring that to Section 11.5.

11.2 The Idea of PCA

Suppose we want to focus on the differences between the eight spectra in the gold series. Our first step might be to average the spectra, so that we have a baseline with which to compare them.

Next, we might want to construct a function of energy that captures most of the differences that we see. More precisely, we could choose a function that, in linear combination with the average, captures the largest possible fraction of the variance between the original eight spectra. Figure 11.4 shows the average, along with such a function.

One of the most notable changing features in Figure 11.1 is the white line; over time, it shrinks and shifts to the right. The function labeled 'component 2' in Figure 11.4 has a dip at a slightly lower energy than the white line in the average; by adding it to the average, it will decrease the amplitude of the white line and shift it to the right, whereas subtracting it will increase the amplitude of the white line and shift it to the left.

Figure 11.5 demonstrates how well linear combinations of the average and component 2 reproduce the original spectra.

On this scale, the match appears very good. But there are hints of modest differences, as can be seen more clearly in Figure 11.6.

The two functions, on their own, do not perfectly reproduce the high-energy side of the white line for some of the spectra.

Sure, in Section 11.5, we'll give some of the theory behind the computation. But the honest answer is that there is software that will do these computations for you!

We will explain the convention for the scaling of component 2 in our graphs in Section 11.5, once we understand more about how PCA works.

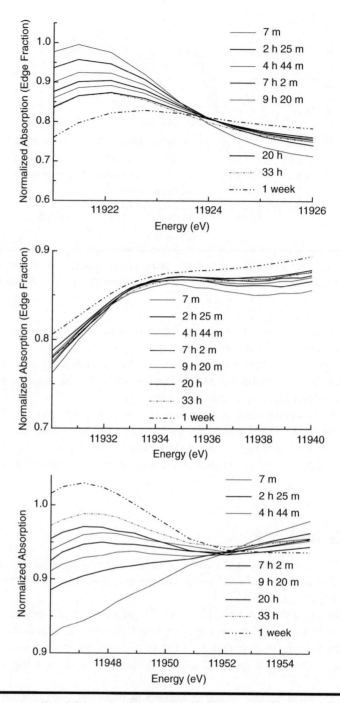

Figure 11.3 Details of three possible isosbestic points from Figure 11.1. (Top) Near 11,924 eV. (Middle) Near 11,934 eV. (Bottom) Near 11,952 eV.

In PCA, it is also a requirement that each component be orthogonal to all of the previous components. So it is not *quite* true that each component accounts for the maximum amount of remaining variance.

We can, therefore, seek a third component (i.e., in addition to the two shown in Figure 11.4) that will account for the majority of the remaining variance between scans. This component is shown in Figure 11.7.

This component is capable of making the drop in the spectra from the white line to 11,927 eV steeper or less steep, depending on whether a positive or a negative multiple of it is added to the spectra from Figure 11.6. The drop in this component from 11,927 to 11,960 eV also suggests it can have an effect on the slope of the background.

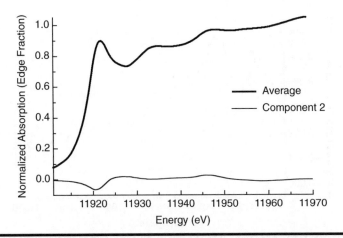

Figure 11.4   The average of the spectra from Figure 11.1, along with a function (labeled 'component 2') that, in linear combination with the average, accounts for the maximum possible fraction of the variance between spectra.

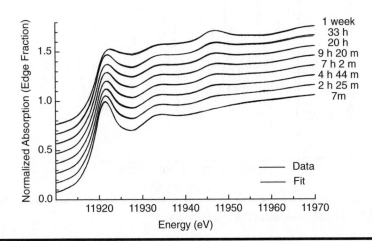

Figure 11.5   Fits to the spectra from Figure 11.1 using the two functions from Figure 11.4. Spectra have been displaced from each other vertically for clarity.

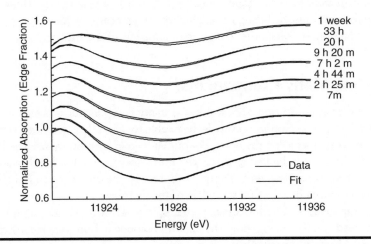

Figure 11.6   Detail from Figure 11.5.

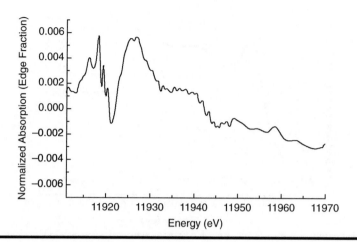

Figure 11.7   Third component for gold series.

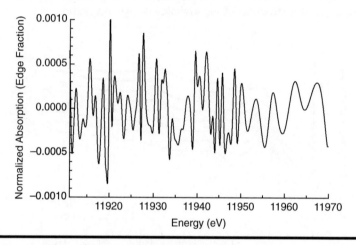

Figure 11.8   Eighth component for gold series.

Unlike the first two components (shown in Figure 11.6), however, there are high-frequency features visible on the scale of a single eV. We know that real spectra don't have structural features that vary that quickly—for one thing, the energy resolution of the measurement just isn't good enough (Section 4.1 of Chapter 4). Those rapidly changing features must be fitting noise, glitches, Bragg peaks, artifacts generated during data reduction, or some other nonstructural aspects of the spectra.

11.3 How Many Components?

As we will discuss in Section 11.5, it takes as many components as there are spectra to account for *all* of the variance; that is, to be able to perfectly reproduce every point of each spectrum, including the noise. For our eight gold spectra, we need eight components. We've already looked at three (the initial average counts for this purpose). Let's jump straight to the eighth (Figure 11.8); that is, the one that handles all of the remaining variance after the first seven have done their best.

This component is *all* high frequency; that is, it is only fitting noise and other nonstructural features. Throwing away the eighth component from our reconstructions would mean throwing away only noise, not signal.

So how do we know which components include signal and which are only noise? There is no single fixed, reliable rule. Several approaches, however, yield valuable clues.

The original scan parameters for these data switch from a step size of 0.5 to more than 1 eV at 11,950 eV. Because the points are measured further apart above 11,950 eV, the frequency of the noise also appears to change.

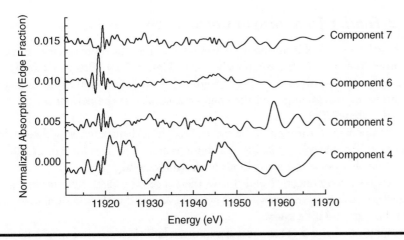

Figure 11.9   Fourth through seventh components for gold series.

11.3.1 *Appearance of Components*

As implied in earlier sections, we can use our understanding of the appearance of XAFS spectra to distinguish features related to noise or other nonstructural features to those related to the structure of the material. Figure 11.9 shows the four components of the gold series that we have not yet examined.

The fourth component appears to have some structure to it on the scale of XANES features. The remaining three components appear to consist primarily of noise-like features. It appears that four components are enough to account for all the structural information in the spectra.

BOX 11.1   Magic?

	Wait—that sounds almost like magic! We can throw our spectra in machinery called PCA, find some components that are due to noise, and then rebuild the spectra without those noisy parts? Doesn't that violate some rule of statistics or something? How can we improve signal to noise just by processing the data?
	Suppose we had five scans of the same material under the same conditions, and we tried performing PCA on those five scans to eliminate noise. We would find that the first component—the average—was the only one that wasn't just due to noise, and we could throw the other four away. And in fact, that's what we do—we average the five scans and use that average for analysis! The best PCA can ever do at eliminating noise is the improvement specified by Poisson statistics. But because we usually operate on spectra that aren't supposed to be identical, it doesn't do quite that well; some of the variance is due to structure rather than noise.
	Kitsune is correct, but that does not mean you have not identified a useful technique, dear Simplicio. Suppose one has a sequence of related, noisy scans, such as is often the case for time-resolved studies. One could use PCA on the series and remove components that appear to correspond solely to noise. The idea is similar to a moving average, but more sophisticated, as the commonalities in all scans, not just subsets, are contributing to the signal.

One's eye might be drawn to the peak around 11,959 eV in component 5. It is true that it appears to be of large amplitude compared to the rest of the peaks in that component. Because of the increased step size for data collection in that part of the spectrum, however, it does not appear to be more than a data point or two in width and is certainly far narrower than any structural feature that would appear that far above the edge. It might be associated with a glitch or a transient phenomenon such as a bubble. It is not, however, structural.

Yeah, and those jiggles around 11,920 eV in components 5 and 6 have something to do with the edge, but seeing as they go up and down so fast, they are probably an artifact of the way the data were measured, aligned, and averaged. They're not structural either.

11.3.2 Fourier Transform of Components

We can make our visual intuition from Section 11.3.1 a little more concrete by taking the Fourier transform of the components, as in Figure 11.10. This is *not* an EXAFS Fourier transform—there's been no background subtraction, conversion to k-space, k-weighting, or windowing, and the conjugate variable is the periodicity in energy space rather than $2R$.

For components 6 through 8, the spectral weight is distributed roughly equally as a function of frequency; that is, the distributions are *white*. This kind of distribution is typical for shot noise and many kinds of electronic noise.

In contrast, components 2 and 3 have their spectral weight concentrated at low frequencies in energy space, suggesting they are associated with structural features, or perhaps background differences.

Components 4 and 5 are intermediate. About half their spectral weight in Figure 11.10 is associated with oscillations larger than 5 eV.

Our conclusion, then, is similar to that from Section 11.3.1: from three to five components are necessary to account for all the structural information in the spectra.

11.3.3 Compare to Measurement Error

Another way to evaluate which components are structural is to compare the fraction of the variance between spectra that is explained by a component with the estimated measurement error of the individual spectra.

This, of course, raises a problem we discussed in Section 10.4.2 of Chapter 10 (and will cover in more detail in Section 14.1.2 of Chapter 14): it is not clear how to determine measurement uncertainty for XAFS spectra! But even the random part of the error (i.e., the noise) is often difficult to determine for PCA series, as we'll see.

Section 7.2.2 of Chapter 7 provides four methods for estimating noise identified by the Standards and Criteria Committee of the International X-ray Absorption Society (IXS Standards and Criteria Committee 2000). The second method in the list requires EXAFS data, which we frequently do not have when performing PCA. The last method requires multiple scans under identical conditions, a luxury that we do not have for a time-resolved series. That leaves, in this case, either a comparison to a smoothed version of the data or estimation from theory. For this example, we'll use a comparison of the preedge region of each sample to a quadratic fit (although not shown in Figure 11.1, data for each spectrum were collected starting 200 eV below the

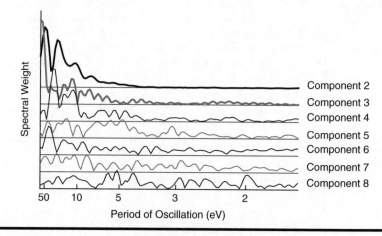

Figure 11.10 Fourier transform of gold series components, normalized to have the same area.

edge). That yields an average root-mean-square error of about 1×10^{-3} for the earlier spectra and about 4×10^{-4} for spectra beginning at 20 hours. The decrease in measurement error for the later scans is expected; time evolution had slowed sufficiently for the longer measurements to allow Lengke et al. to average five scans each for the later measurements, suggesting that the error in the normalized spectrum should decrease by a factor of $\sqrt{5}$.

The average variance per point between the eight gold spectra over the region used for PCA was 8×10^{-5}. This should be compared to the square of the measurement error, or about 1×10^{-6} for the earlier spectra. In round numbers, only about 1% of the variance between samples is due to noise.

Because the idea behind PCA is to account for as much of the remaining variance as possible with each additional component, the fraction of variance accounted for by each component is a natural byproduct of the process. For the gold series, the results are summarized in Table 11.1.

We don't know what fraction of the variance between spectra is due to systematic errors such as energy alignment, normalization, and differences in background, but they surely play some role. If the systematic error is comparable to the amount of noise in this sample set, then components 5 through 8 are all dominated by noise and systematic errors, in agreement with our conclusions from earlier sections.

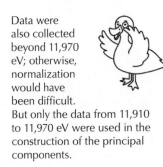

Data were also collected beyond 11,970 eV; otherwise, normalization would have been difficult. But only the data from 11,910 to 11,970 eV were used in the construction of the principal components.

11.3.4 Scree

The last paragraph of the preceding section probably feels a bit unsatisfying. Because we don't know how much systematic error there is, we really don't know where to draw the line between structural information and the rest.

But we do have another piece of information: we're pretty sure that component 8 is dominated by nonstructural differences (noise and systematic error), and it accounts for 1.2% of the variance. (It's comforting that this is on the order of the percentage we expected to be due to noise from our estimates in Section 11.3.3.) The nature of PCA requires each previous component to account for progressively larger chunks of the variance. So component 7 should account for a greater fraction of the variance than component 8, even if they're both due to noise.

Figure 11.11 plots the percentage of variance against component number. Component 7 accounts for a bit more of the variance than component 8, component 6 a bit more than that, and component 5 a bit more in turn. But then, the differences between components become much greater, and the trend line established by components 5 through 8 is broken.

Our language has become a bit of a mess here, I'm afraid. In some fields, PCA is performed by first *centering* the data; that is, by subtracting the mean of each data set from that set in advance of applying the PCA algorithm. It has not been the tradition in XAFS to do so. The result is that the quantities that are reported as 'variance' are not truly variance! In fact, the first component, which is related to the mean, is given the lion's share of this 'variance.' Some practitioners, therefore, do as we have done and compute the fraction of 'variance' accounted for after the first component. As always, one should be clear regarding the procedure one has used when describing one's results.

Table 11.1 Percentage of Variance (After the First Component) Accounted for by Each Component

Component	Variance (%)	Cumulative Variance (%)
2	76.7	76.7
3	10.2	86.9
4	5.5	92.4
5	2.4	94.9
6	2.0	97.0
7	1.7	98.7
8	1.2	100.0

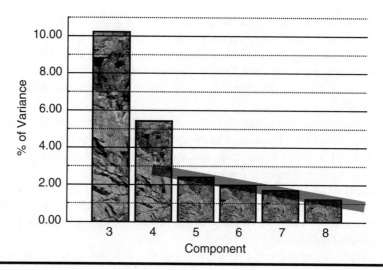

Figure 11.11 The percentage of variance due to components 3–5 in the gold series. Component 2 would be far off scale. Components 5–8 fit a trend that is not shared by components 3 and 4.

While you're thinking about that, Figure 11.12 is a scenic picture of Yamnuska, a mountain in the Canadian Rockies.

Because of the similarities between the shapes of Figures 11.11 and 11.12, a plot like Figure 11.11 is called a *scree plot*. (No, we're not kidding.)

Many times, as is the case here, the scree plot ends with a nearly constant slope. The components that fall on that nearly constant slope are likely to be predominately due to noise and systematic errors, whereas the components before the break are more likely to be structural.

In this case, we can conclude that the first four components incorporate the majority of the structural information, whereas the last four include primarily noise and systematic error.

11.3.5 Objective Criteria

The criteria in Sections 11.3.1–11.3.4 are all somewhat subjective. Several objective, numerical criteria have therefore been proposed. We won't try to detail all of them here, in part because they can become somewhat redundant. The advantage of objective criteria is, of course, their objectivity, but the disadvantage is that they are more likely to include a component that is not structural, or exclude one that is, than a judgment call based on multiple criteria. Making a decision subjectively also has the advantage that it is clear how easy or difficult the decision is in a given case.

One criterion that is sometimes used in other fields is to include components until they cumulatively explain some percentage of the variance, perhaps 90% or 95%. For XAFS analysis, this kind of criterion is not a good idea. Conceptually, what is important is the contribution of a component relative to the level of noise, not relative to the total amount of variation. To see this more clearly, imagine adding an additional spectrum to the original gold series that was quite different from the other spectra, including a constituent that wasn't present in the rest of the series (this sometimes happens when the spectrum of a quickly-consumed starting material is included in a series). That would increase the total variance and would probably mean that the number of structural components needed to recreate the data would increase by one. And yet, using a criterion based on cumulative percent of variance explained would

One will sometimes see it stated that the component at the 'elbow' of the scree plot—in this case, component 5—should be considered structural. That does not seem sensible to me; it clearly belongs to the less significant components.

I agree, Robert, but the more important point is the one made in Section 11.3.5: all hard-and-fast rules about selecting criteria are imperfect. Because of that, I'd rather look at the components in several ways and then make a judgment call.

Figure 11.12 Yamnuska, a mountain in the Canadian Rockies. Note how the steep mountain side gives ways to a gentler slope made up of scree, which is a material made of rock fragments weathered from the mountain. Kevin Lenz. This photo is licensed under the Creative Commons Attribution-Share Alike 2.5 Generic license.

make components that explained the small variations between the rest of the data less likely to make the cut.

Another common criterion is to include components that explain an amount of variance at least 70% of the average amount explained by a component (Everitt and Dunn 2010). Although this criterion is designed for centered data, we can adapt it for the kind of analysis we've done here, by removing the first component from the computation. Under this criterion, with eight total components, we would be looking for components that explain at least 70/(8-1) = 10% of the variance beyond the first component. For the gold series, it would mean three components.

Finally, we should mention Malinowski's *factor indicator function* (IND), described by Malinowski (2002), which is supposed to reach a minimum when the ideal number of structurally meaningful components are included. Figure 11.13 shows IND values for the gold series. To three significant figures, components 3 and 4 yield identical values, and form the minimum of the series, suggesting that 3 or 4 components should be included.

The work by Malinowksi (2002) is a treasure trove of information about principal component analysis. He describes 26 distinct methods for determining the number of components that should be considered structural! By 2008, he'd cut that down to 15, including *determination of rank by median absolute deviation* (DRMAD), which has the benefit of having a sound statistical rationale (Malinowski 2008).

Trying to determine the number of meaningful components is a fun game, but in the end it's a judgment call. The methods we tried in Section 11.3 have told us to keep, in the order we tried them, four components, or three to five, or four, or four, or three, or three to four. Let's say it's four and move on …

11.4 How Many Constituents?

11.4.1 Relationship to Number of Components

Hopefully, it is clear that components (with the possible exception of the first one) are not spectra of specific materials. Instead, the second component is related to the difference between spectra in the original series, the third is related to differences of the differences, and so on.

But the number of meaningful components (i.e., those not associated primarily with noise or systematic errors) does provide us information about the number of degrees of freedom of the system of spectra. If a series of spectra has four meaningful components, then it must have three independent free parameters specifying the differences of the individual spectra from the average; that is, there are three parameters that differ within the set of materials.

The simplest way a series of spectra could have three free parameters between them is if the materials being measured are themselves mixtures of four constituents that are present in varying amounts. And in fact, the linear combination analysis performed by Lengke et al. modeled the spectra as being a linear combination of four constituents.

This is one of the major uses of PCA in XAFS: to provide an estimate of how many different constituents are present in a series of mixtures. This can be very helpful for distinguishing between linear combination fits—just think how much easier the analysis of the chromium sample in Chapter 10 would have been if we knew it consisted of three constituents! Similarly, knowing the number of constituents can help cut down the time-consuming process of constructing models (Chapter 12 and Part III).

11.4.2 Energy Misalignment

Let's consider the effect that energy misalignment has on PCA. Figure 11.14 shows the XANES of the strontium K edge of strontium carbonate.

To see the effect of misalignment, we made four more copies of this spectrum and displaced them by 0.5, 1.0, 1.5, and 2.0 eV from the original spectrum. We then applied PCA to the set of five spectra.

The second component of this set accounted for 93% of the variance, and is shown in Figure 11.15.

The second component in Figure 11.15 looks broadly similar to the second component in Figure 11.4 because they both accomplish the same thing: they allow the white line to be shifted to higher or lower energy. But in Figure 11.4, the shift occurs because

That is, 93% of the variance after the first component, with the caveat I noted in Section 11.3.3.

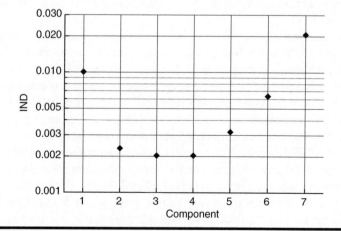

Figure 11.13 Factor indicator function values for the gold series.

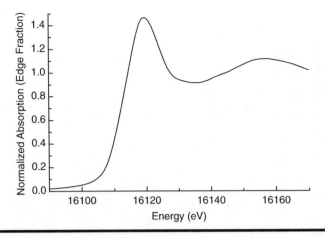

Figure 11.14   Normalized strontium *K* edge spectrum of strontium carbonate.

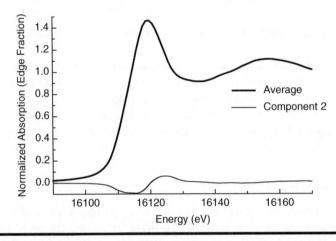

Figure 11.15   Average and second PCA component for series generated from strontium carbonate scans duplicated with energy shifts.

the oxidation state of the gold is changing between spectra, whereas in Figure 11.15, it occurs because the spectra were misaligned in energy. This is a very dangerous kind of systematic error, because it masquerades as a structural effect!

As a second example, we've shifted the spectra in the gold series by values ranging from 0 to 3.5 eV, randomly assigned so as not to correlate with the time evolution. Figure 11.16 shows the fifth component that results; recall that the fifth component in Figure 11.9 shows little evidence of structural information.

This time, the component looks like it may include structural information. Note how energy misalignment could mislead us into thinking there were more constituents present than was actually the case.

11.4.3 Other Structural Free Parameters

In some cases, the question as to the number of constituents is semantic. Suppose, for example, that the distribution of cations between tetrahedral and octahedral sites in a spinel (Section 9.2 of Chapter 9) gradually changes across a series. Is a spinel in which 30% of the manganese atoms are in tetrahedral sites a different constituent than a spinel in which 25% are? No, because, then we would think of a series that went from 25% to 30% to 35% to 40% as having four different constituents. Is a 50% tetrahedral sample a mixture of 0% and 100% materials? No, because the local structure of the

Don't think that the component shown in Figure 11.16 accounts for energy misalignment! Removing component 5 would not somehow realign the data! PCA always works to try to take care of as much variance as possible with each successive component. So the second component already incorporates a lot of the energy misalignment, although some of it will end up in components further down the line too. We've added an independent variable—the amount of misalignment, therefore it does take one more component to fit the data well, but we can't identify any specific component as being due to the misalignment!

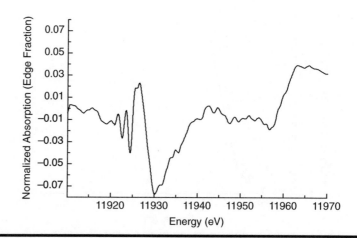

Figure 11.16   Fifth PCA component for series generated from data in Figure 11.1, but with the spectra misaligned.

50% sample is not exactly given by averaging the local structure of the 0% and 100% cases.

But we certainly can say that a series of manganese ferrites in which the distribution of the manganese ions between tetrahedral and octahedral sites varies has a free parameter associated with that variation, and thus it should add to the number of components. If stoichiometry also varied, so that the chemical formula was $Mn_xFe_{3-x}O_4$, that would add another free parameter—*if* the difference in distribution were not solely dependent on the stoichiometry.

11.4.4 Coupled Constituents

Suppose a painter were working with an ochre that included a blend of hematite and goethite, and a Prussian blue (another iron-based pigment). The artist might have blended the ochre with the Prussian blue in different proportions at different points on a painting. If that were the case, PCA performed on spectra collected from different areas of the painting would reveal two components, and thus one free parameter, corresponding to the ratio of one pigment to another. And yet, the samples would comprise three mineral phases: hematite, goethite, and Prussian blue. This kind of coupling can complicate the interpretation of PCA.

11.5 PCA Formalism

Up until this point, we've tried to motivate the ideas and uses of PCA. In this section, we'll talk a little bit about the formalism; consult the references if you'd like to know more.

PCA is closely related to a technique known as *singular value decomposition* (SVD)—the results of SVD can easily be transformed to yield the results of PCA (Malinowksi 2002; Shlens 2003; Wall et al. 2003). SVD is conveniently implemented on computers, so typically that is the method actually used, with the corresponding PCA results then being reported.

To briefly compare the two, suppose we arrange the normalized absorption values from our spectra in columns. Notice that we throw away the information of the energies that those normalized absorptions correspond to. It doesn't matter, for instance, if the step size changes across the scan—each measurement gets one value in the column. It *does* require that the different spectra be measured at the same energies or be interpolated on to the same energy grid.

Each spectrum now has a column associated with it. Putting the columns together, we can form a matrix **X**.

In PCA, we form the covariance matrix $\mathbf{XX}^T$ (Everitt and Dunn 2010; Malinowksi 2002; Shlens 2003; Smith 2002). The eigenvectors of this matrix are the principal components. If the data are *centered*, that is, if the mean of each column of **X** is zero, then the eigenvalues of each are proportional to the amount of variance the component explains. As Robert has noted in earlier sections, we do not usually center data in PCA analysis of XANES, and thus, the eigenvalues are not proportional to the true variance. Nevertheless, the term variance is almost universally used in this context, even though the data are not centered. In this chapter, we have gone one step further and only used eigenvalues after the first to compute percentage of variance. This latter practice is not universal; much of the literature assigns a fraction of the 'variance' to the first component, even though the first component is closely related to the mean of the data!

In SVD, we write (Malinowksi 2002; Ressler et al. 2000),

$$\mathrm{X} = \mathrm{P}\mathrm{S}w^T \tag{8.1}$$

where **P** is the matrix consisting of the principal components expressed as columns and **S** is a diagonal matrix. **w** consists of principal components of a different kind, taken from the rows of **X** rather than the columns (e.g., the nth element of the first component is proportional to the average of spectrum n across all points; the nth element of the second component is proportional to how much spectrum n differs from the mean of all spectra, and so on).

Equation 11.1 provides some flexibility as to how to distribute scaling factors between the matrices **P**, **S**, and **w**. Commonly, the columns in **P** and **w** are normalized, so that the sum of the squares of each is 1. If this is done, it turns out that the eigenvalues of $\mathbf{XX}^T$ are proportional to the square of the values on the diagonal of **S**.

Inspection of Equation 11.1 also shows us that $\mathbf{S}\mathbf{w}^T$ tells us the weighting for reconstructing the original spectra from the principal components, sometimes known as *principal component scores*.

When the principal components are plotted, however, they are often scaled by their eigenvalues; in effect, it is **PS** that is plotted, rather than **P**. This scaling makes the first component a multiple of the mean, but not equal to it; frequently it will even come out inverted. In this chapter, we have therefore applied an additional scaling factor to plots of components, so that the first component is the mean of the spectra.

11.6 Cluster Analysis

SVD tells us how each spectrum depends on the components. This can provide suggestive information. For example, Figure 11.17 plots the weighting of each of the first four components in each spectrum.

Component 2 correlates very strongly with time and presumably has to do with the reduction of the gold by the cyanobacteria. Component 3, on the other hand, is telling us there is something different with the 33-hour spectrum. Component 4 is particularly interesting, as it peaks for times of a few hours and then drops back off—perhaps there is an intermediate formed and then was consumed?

In this case, we were able to plot the weightings as a function of time. In other systems analyzed by PCA, we might be able to use an analogous way of ordering our spectra—perhaps depth in a soil sample or doping fraction in a materials science problem. But sometimes there is no obvious dimension on which to rank the spectra (see,

We are glossing over some technicalities having to do with making the number of rows and columns in Equation 11.1 work out—consult the references for more.

There is the potential for confusion here! $\mathbf{S}\mathbf{w}^T$ provides the weighting for the normalized components, but just $\mathbf{w}^T$ is the weighting if the components have been scaled by their eigenvalues.

Our additional scaling should be used for graphing purposes only. The weighting of the first component is not quite the same for each spectrum. If one were to apply the PCA method to unnormalized spectra, the weightings of the first component would reflect differences in normalization. Because PCA does not 'understand' XAFS normalization, differences in, for example, the white line can cause modest differences in the weights associated with the first component.

Figures like 11.17 are lot of fun, but don't get carried away identifying components with constituents. The fourth component suggests that there may be a constituent appearing after a few hours and then getting consumed, but that doesn't mean the fourth component should be identified with that constituent—the contributions from that constituent have probably worked their way into other components as well.

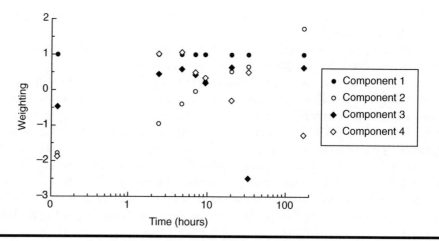

Figure 11.17   Weightings for components in the gold series.

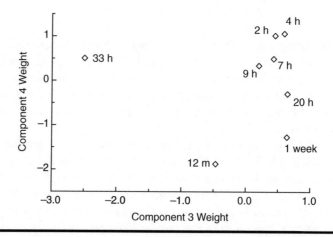

Figure 11.18   Cluster analysis of gold series, showing weights for fourth component plotted against the weights for the third component. Time labels for points in the upper right quadrant have been rounded to the nearest hour for clarity.

for instance, the case study in Section 19.6 of Chapter 19)—perhaps the spectra are a set of flakes from the paintings of an artist, a set of soil samples from different sites, or a set of organometallics synthesized using different protocols. Is there another way to look for patterns among the weightings?

In those cases, we can plot one of the components as the *x*-axis, a technique some-times known as cluster analysis. In the case of the gold series, using the second compo-nent as the *x*-axis would look much like the time series, because the second component is so strongly correlated with time. Instead, let's use the third component as the *x*-axis and see what we get (Figure 11.18).

There is a cluster of spectra in the upper right, suggesting that the spectra from 2 to 20 hours form a group. The remaining spectra are scattered across the plot, suggesting they should not be thought of as being grouped together.

Cluster analysis often helps suggest priorities for the time-consuming process of modeling (Chapter 12 and Part III). One model (with free parameters) is likely to suffice for the cluster of spectra in Figure 11.18, but a different model would likely be needed to accurately fit the 33-hour sample. Whether this would mean that the 33-hour sample is particularly important to model or can be discounted as an outlier depends on the particular scientific questions the study is trying to address.

Does the 20-hour spec-trum belong with the cluster, or not? What about the 1 week—maybe that cluster should be thought of as a big blob on the right side of the plot. You don't have to make definitive decisions about that kind of stuff; cluster analysis is meant to take advantage of our brain's ability to see patterns, not to generate rigid criteria.

BOX 11.2 How Many Experimental Spectra?

	I'm starting to really like this PCA. And it seems to me it would work best if I measure a lot of spectra. That way I'll get good statistics, and I'll be more easily able to tell which components are meaningful.
	Please exercise caution, my dear Simplicio. If you collect many spectra of similar samples, and have only a few examples of differences, a large proportion of the variance between spectra will be attributable to noise, interfering with the ability to identify meaningful components.
	Yeah, for PCA it's a lot better to have a diverse set of spectra than a lot of them. For example, if the time evolution of your system is pretty much done after 40 minutes, don't collect spectra every 10 minutes for 5 hours. Most of your spectra will correspond to the end state, and your PCA might be dominated by noise.

11.7 Target Transforms

Often, we are interested in knowing whether specific substances are present in our series. For instance, is one of the constituents of the solutions processed by the cyanobacteria metallic gold? Although in some cases fingerprinting might be sufficient for addressing questions of that kind, it can be difficult to apply if the concentrations of other constituents are also varying from spectrum to spectrum.

With PCA, however, this kind of question is remarkably easy to address. Suppose we call a standard spectrum of the constituent we wish to test for $\mathbf{T}$ (for 'target'). We can then compute $\mathbf{PP}^{\mathrm{T}}\mathbf{T}$, using only the principal components that we believe contain structural information. If the target can be expressed in terms of those components, $\mathbf{T}$ will be left essentially unchanged by this *target transform*, except for alterations consistent with the systematic errors and noise present in the set of spectra (Ressler et al. 2000).

Figure 11.19 shows the result of a target transform using the first four components of the gold series on a foil of a gold spectrum.

The result of the target transform is suggestively close to the gold foil spectrum but shows some small differences. Is that just due to noise and systematic error or does it mean that metallic gold is not present in the mixture? To answer that, we'll use another quantitative function developed by Malinowski, called SPOIL (Malinowski 1978; Malinowski 2002). Although we'll leave the method of computation of SPOIL to the references, the idea is that if the target belongs as part of the set, it should be reproduced by the components as well as the original spectra. If not, though, it should add error to the mix, 'spoiling' it. Malinowski classifies SPOILs using the thresholds shown in Table 11.2.

The SPOIL for the target transform in Figure 11.19 is 1.74, confirming the presence of metallic gold in the series, perhaps in a form modestly different from the bulk foil (e.g., it might be nanoscale).

SPOIL is not an acronym; Malinowski just likes using all caps when he defines functions!

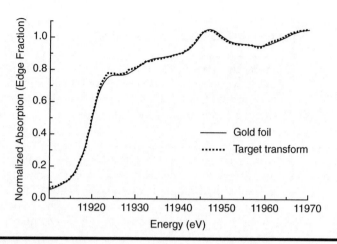

Figure 11.19 Target transform of a gold foil spectrum using the first four components of the gold series.

Table 11.2 Malinowski's Guidelines for Assessing SPOIL

SPOIL	Interpretation
<1.5	Excellent
1.5–3.0	Good
3.0–4.5	Fair
4.5–6.0	Poor
>6.0	Not acceptable

Source: Malinowski, E. R. *Anal. Chim. Acta.*, 103, 339–354, 1978.

We can explore a bit further. Are the cyanobacteria cranking out metallic gold from the first hours of the experiment, or are they first producing an intermediate and only later converting it to gold? To answer this, we can repeat the PCA using only the samples that aged less than 10 hours, and try the target transform again.

Running PCA on only the samples aged less than 10 hours yields only two components that definitely have structural information, with the third being borderline. And in fact, careful examination of Figure 11.3 suggests that the points near 11,924 and 11,952 eV might be isosbestic, if only the less-aged samples are included.

Figure 11.20 shows the target transform of the gold foil to the less-aged set, using three components.

The transform certainly appears noisy this time but still captures the general contours of the foil. And indeed, the SPOIL is 1.43; we *expect* a bit more noise when we cut down on the number of spectra, and so Malinowski's measure still confirms the presence of metallic gold.

This, perhaps, is a good time to show what happens when we attempt a transform to a material that is *not* present. For this purpose, we'll use gold cyanide (Figure 11.21).

The SPOIL is 7.40, but we don't need the quantitative measurement to tell us what our eyes can clearly see from Figure 11.20. There is no question; gold (I) cyanide is not one of the constituents responsible for the differences between these samples.

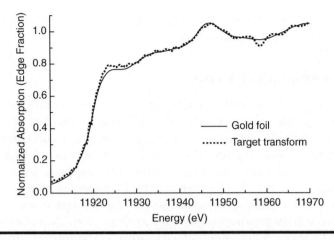

Figure 11.20 **Target transform of a gold foil spectrum using the first three components generated from samples younger than 10 hours in the gold series.**

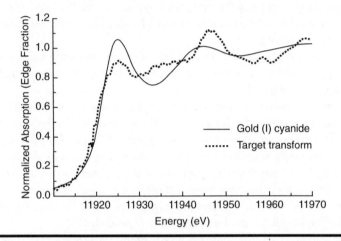

Figure 11.21 **Target transform of a gold (I) cyanide spectrum using the first three components generated from samples younger than 10 hours in the gold series.**

11.8 Blind Source Separation

Target transforms can be used to identify whether a given constituent is contributing to a set of spectra, but is there a way to determine the spectra of constituents without guessing them one by one (*blind source separation*)?

The answer is—sort of. Multiple methods exist for turning PCA data into a set of spectra that *could* correspond to constituents. But that doesn't mean they *do* correspond to the constituents that are actually present. The problem is that there just isn't enough information content in an XAFS spectrum to choose from all the possible constituents if none of them are known. The set of possible constituent spectra include not just every substance containing the element corresponding to the edge being measured but also every conceivable structural modification to those substances, including ones that aren't stable enough to exist. More than one set of superficially feasible spectra could be used to construct the series of measured spectra, and thus, there is no way for PCA to reliably provide the correct set.

This shouldn't surprise us: every technique covered in this book requires some sort of guess as to structures that are, or might be, present in the sample; whether those are the comparison structures used in fingerprinting (Chapter 9), the set of standards used in linear combination analysis (Chapter 10), or the theoretical standards used in modeling (Chapter 16). PCA is a very useful technique, but it is still working with the same limited information as those others.

11.8.1 Transformation Matrix

 It should also be recalled that the number of components do not always correspond to the number of constituents. For example, a system may have a degree of freedom associated with the growth of nanoparticles during a synthesis. There is no constituent corresponding to 'nanoparticle growth.'

The simplest approach for blind source separation is to use a transformation matrix (TM). In essence, this consists of trial-and-error combinations of PCA components until the results look like a set of spectra. But since the PCA components were generated by singular value decomposition, choosing one mixture of components for one constituent imposes constraints on how the other constituents can be chosen. Mathematically, this is considered a rotation of the PCA components through a multidimensional space until they appear to correspond to a feasible set of constituent spectra. In practice, this has been implemented as a set of sliders which can be manipulated to produce different combinations of candidate spectra (Martini et al. 2020).

Often, additional constraints can be provided to reduce the number of sliders. For example, we can require that each candidate spectrum be normalized. In addition, one (or perhaps two) of the measured spectra used to generate the PCA components might correspond to pure, known substances. This might occur in a time series, for example, if the initial state is known.

Martini et al. (2020) provide the number of sliders necessary, depending on the constraints applied and the number N of meaningful components generated by PCA. Their results are summarized in Table 11.3.

For example, suppose there are two meaningful components in a time series, and we know the identity of the starting material. There would be $N^2 - 2N + 1 = 2^2 - 2(2) + 1 = 1$ slider. This makes sense: since we have a spectrum of one of the two constituents in the series, and we have spectra representing the combination of the two constituents, we can find the 'shape' of the spectrum of the unknown constituent. We don't know, however, whether the constituent contributes a large amount to a given spectrum because the constituent is present in large amounts, or because the constituent has particularly strong features. Thus, one slider is needed.

It should be noted that this approach is only useful if the number of sliders is fairly small, as otherwise the amount of trial and error becomes very large. A mixture with three meaningful components and one known constituent already requires four sliders; if there are four meaningful components and one known constituent the number rises to nine, which is likely impractical.

11.8.2 Iterative Target Transform Factor Analysis (ITTFA)

The TM approach described in the last section relies on a human being to go through a trial-and-error process looking for an appropriate 'rotation' of the PCA components to produce a set of feasible spectra. Computers are very fast at trial and error, but are not as good at evaluating whether a candidate constituent spectrum is 'feasible.'

Table 11.3 Number of Sliders Needed for N Meaningful Components

Component	Number of Sliders
No constraints	N^2
Spectra are normalized	$N^2 - N$
Spectra are normalized and one pure spectrum is known	$N^2 - 2N + 1$
Spectra are normalized and two pure spectra are known	$N^2 - 3N + 2$

Iterative target transform factor analysis (ITTFA) allows a computer to find plausible rotations using the following guidelines (Martini and Borfecchia 2020):

- Candidate constituent spectra must be normalized.
- Spectra in the original series used to create the PCA components can never include a negative amount of a candidate constituent.
- The sum of the fractional contributions from the constituents must be one for each spectrum in the original series (*mass balance*).
- Each of the original spectra is expressed, as much as possible, as being largely due to as few of the constituents as possible. This is accomplished by a mathematical technique known as a varimax rotation (Kaiser 1958).

The last guideline is, of course, the one that will select a unique solution, although not necessarily the correct one. Still, in many circumstances, particularly those involving time evolution, it is plausible. If a system evolves from an initial state through a series of intermediates to a final one, each individual spectrum in the series is likely to be dominated by one or two constituents. That might also be the case for some spatially resolved samples, such as those taken at different depths relative to a weathered surface. It is not as clear that the criterion would be appropriate for samples which are not easily ordered, such as samples taken from leaves of different specimens of the same species of plant.

11.8.3 Evolving Factor Analysis (EFA)

While ITTFA is more reliable with a series of spectra that have a definite order (e.g. a time series), the use of *evolving factor analysis* (EFA) requires it. This approach works by identifying the number of meaningful components if PCA is done on just the first two spectra in the series, then on just the first three spectra, the first four, and so on. An increase in the number of meaningful components is treated as evidence of the introduction of a new constituent (Maeder 1987). The process is then repeated in reverse order, starting with just the last two spectra in the series, and then the last three, and so on. In this case, an increase in the number of meaningful components suggests that a component disappeared in the subsequent spectrum.

By examining the rise and fall of a component associated with a particular constituent, the concentration of the constituent in each of the spectra in the original series can be estimated. This estimation is necessarily very rough at this stage, since it is not known how prominent the features are in the spectrum of the constituent! In addition, by analyzing the difference in the last spectrum prior to the appearance of the new constituent and the first spectrum in which it appears, an estimate may be made of a spectrum that would correspond to the new constituent, particularly if changes in concentration of the previous constituents are also taken into account. This procedure can also be done by analyzing the difference in the last spectrum for which a constituent is present and the first one in which it is absent (disappearance), or even the difference between two spectra in which the concentration of a constituent changes.

If an initial estimate for each constituent is in place, then the process of examining differences, including differences between spectra where the constituent is present in each selection but its concentration rises and falls, can be used to further refine the estimated spectrum of each pure constituent. The process, aided by the constraints of mass balance, normalization, and nonnegative contributions, continues iteratively until a stable set of constituent spectra are found. Taken together, this process is sometimes known as *multivariate curve resolution—alternating least squares* (MCR-ALS).

Manne (1995) has derived an important pair of theorems applicable to MCR-ALS. For the purposes of these theorems, 'window' can either refer to a set of spectra starting

Instead of gradually increasing the number of spectra starting with the first two, and then going in the reverse direction starting with the last two, a fixed window of some number of spectra can be scanned across the series: *fixed sized window evolving factor analysis* (FSWEFA). For an example of applying FSWEFA to XANES, see Conti et al. 2010.

with the first analyzed for EFA, a set of spectra ending with the last analyzed for EFA, or a set of spectra included in a window for FSWEFA (see Kitsune's comment).

- If constituent A occurs in some set of windows, and every other constituent appears at least once outside of those windows, then the amount constituent A contributes to each spectrum in the original series can be determined.
- If constituent A has at least one window where constituent B is absent, at least one window where constituent C is absent, and so on for all of the other constituents, then the spectrum of constituent A can be found.

Note that if, for example, there are two constituents that always appear together, then the spectrum of neither might not be able to be unambiguously determined by MCR-ALS alone, even if they do not always appear in the same ratio. Note also that if there is a constituent that appears in *all* spectra, then it may not be possible to unambiguously determine the spectrum of any other constituent by MCR-ALS alone.

In the same article in which he introduces these theorems, Manne points out that in real investigations, experimenters have additional knowledge about the system being studied. In fact, if the process is treated as a purely mathematical manipulation, without even the constraints of normalization, mass balance, and nonnegative contributions, then the two theorems not only provide sufficient conditions for the determinations described but also necessary ones. But in reality, at least those three constraints are likely to be appropriate, and in some cases, additional constraints (e.g. the identity of the starting material) may be known.

11.8.4 Simple-to-Use Interactive Self-Modeling Mixture Analysis (SIMPLISMA)

In contrast to EFA, *simple-to-use interactive self-modeling mixture analysis* (SIMPLISMA) does not require the original spectra to be part of an ordered series, such as a time series. For SIMPLISMA to work, however, one or more of the original spectra must include only a single constituent, or at least be heavily dominated by only one.

When that is the case, a mathematical procedure can be used to find which spectrum is most likely to be due to a single constituent, by finding the spectrum that is most extreme compared to the others. For example, in Figure 11.22, we've drawn a

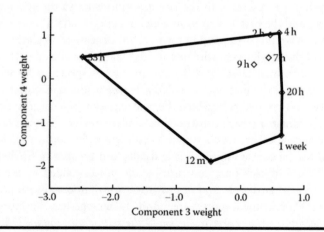

Figure 11.22 We have drawn a convex polygon on the data from Figure 11.18 so that it passes through or encloses each of the data points.

convex polygon through five of the spectra from Figure 11.18, enclosing the other spectra. Since the 2-hour, 7-hour, and 9-hour spectra are enclosed by the polygon, they are likely to represent mixtures of constituents, rather than pure samples. The 20-hour sample, while technically a vertex of the polygon, is very close to being on the line joining the 4-hour and 1-week samples, and thus also likely to be a mixture. If we had to guess based on this graph alone, the 33-hour spectrum is the most likely to represent a pure sample.

Of course, Figures 11.18 and 11.22 only show the third and fourth component. The second component is even more important to the series and, as we saw in Figure 11.17, it rises monotonically with time.

Visualizing the contributions of the second, third, and fourth component to each spectrum would require a three-dimensional plot. Doing so would reveal the 12-minute sample as one of the most 'extreme,' and thus one of the most likely to be a pure sample. That is not surprising, as in the experiment we expect that the 12-minute sample is mostly composed of the starting material, gold (III) chloride.

Of course, in the actual SIMPLISMA algorithm, determining the spectrum most likely to represent a pure spectrum is done mathematically, but the principle is similar to what we have just described (Windig and Guilment 1991).

Once the spectrum of a pure sample is described, the next pure sample would be assumed to explain the maximum amount of remaining difference between samples, and so on.

For XANES analysis, SIMPLISMA is usually followed by further refinement using MCR-ALS (Martini and Borfecchia 2020), as was the case with EFA in the previous section. This iterative process can often correct for cases where the 'pure' spectrum is not entirely pure but merely dominated by one constituent. The use of SIMPLISMA followed by MCR-ALS is thus governed by Manne's theorems, which were discussed in Section 11.8.3.

In short, the distinction between the EFA and SIMPLISMA approaches lie in how the initial 'guesses' for constituent spectra are developed. In EFA, they are based on changes between spectra in a time series (or other ordered series), while in SIMPLISMA, they are based on the spectra which are the most 'extreme' relative to the others.

Looking back to Figure 11.17, the first component does not vary much between spectra. That does *not* mean that there is a constituent that is present in large amounts in all samples! The first component merely provides the general shape of the edge and background which is then modified by the other components.

11.9 PCA of EXAFS

So far, we've discussed PCA of normalized x-ray absorption near edge structure (XANES). But it can be used with EXAFS $\chi(k)$ as well. For example, Beauchemin et al. (2002) is an excellent article discussing and using the technique, and Wasserman (1999) is another prominent, if brief, example.

We do *not*, however, recommend the use of PCA with derivative spectra. Because derivative spectra sharpen features, small misalignments in energy will have a greater impact on the analysis.

We also recommend against applying PCA to back transforms. PCA works best when it can extract difference between spectra; Fourier filtering tends to reduce those differences.

11.10 How PCA Is Used

PCA is most commonly used as an exploratory technique, preliminary to either linear combination analysis or modeling. In fact, this use doesn't always make it into the literature—if PCA leads you to try modeling your spectra with three constituents, and the modeling is convincing, why mention the PCA?

'Other approaches' could mean other XAFS analysis techniques, but it could also mean other structural probes, or even functional measurements such as in *operando* experiments.

Because it is often exploratory, be aggressive! Try separate PCA analyses on subsets of your spectra. Try cluster plots. Try target transforms on the whole set and on subsets. See if blind source separation can help suggest constituents.

But don't overinterpret! Components aren't constituents and it's difficult to draw defensible conclusions from cluster plots. Even blind source separation is subject to ambiguity. Use PCA to guide you but try to confirm what you think it's telling you with other approaches.

BOX 11.3   PCA Sounds Scary

	I thought I liked PCA, but now I'm not so sure. It sounds like it might give me the wrong answer!
	Fingerprinting might give you the wrong answer, Simplicio. Same with linear combination analysis and modeling. You're training to be a scientist, and that means learning to work with uncertainty and imperfect techniques.
	Perhaps we should remind ourselves of the lesson from Section 1.5 of Chapter 1: XAFS is not a black box. We wrote that 'If you thought that you could put XAFS data through some procedure—a "black box"—which would yield the structure of your material, you were mistaken.' Although PCA, and particularly blind source separation, can at times seem like magic, it is no more of a black box than our other techniques.
	I agree, Mandelbrant! I suggest adding PCA to your tool set, but treat the results just as you would the results gleaned from any other technique in isolation: with skepticism.

WHAT I'VE LEARNED IN CHAPTER 11, BY SIMPLICIO

- *Isosbestic points* suggest that a series of spectra were generated from different mixtures of only two constituents.
- PCA components are not constituents.
- There are many criteria that can be used to decide how many PCA components include structural information; there are advantages to making this decision subjectively.
- The number of PCA components with structural information is one more than the number of free parameters in the series of spectra being analyzed. Often, this means the number of significant PCA components is equal to the number of constituents represented in the series, but not always.
- *Target transforms* can be used to discover whether particular standards are present in the series.

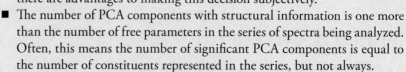

- *Blind source separation* can be attempted by a number of different techniques, none of which is foolproof.
- The *transformation matrix* approach works best when there are very few components, *iterative target transform factor analysis (ITTFA)* works best when there aren't very many constituents contributing to each of the original spectra, *evolving factor analysis (EFA)* works well for a time series with distinct intermediates that appear and disappear one at a time, and *simple-to-use interactive self-modeling mixture analysis* (SIMPLISMA) works best when at least one of the original spectra is dominated by one constituent.
- PCA is not magic, and it's not a black box, but it's a useful technique.

References

Beauchemin, S., D. Hesterberg, and M. Beauchemin. 2002. Principal component analysis for modeling sulfur K-XANES spectra of humic acids. *Soil Sci. Soc. Am. J.* 66:83–91. DOI:10.2136/sssaj2002.8300.

Conti, P., S. Zamponi, M. Giorgetti, M. Berrettoni, and W. H. Smyrl. 2010. Multivariate curve resolution analysis for interpretation of dynamic Cu K-edge x-ray absorption spectroscopy spectra for a Cu doped V_2O_5 lithium battery. *Anal. Chem.* 82:3629–3635. DOI:10.1021/ac902865h.

Everitt, B. S. and G. Dunn. 2010. *Applied Multivariate Data Analysis*, 2nd ed. Chichester: Wiley.

IXS Standards and Criteria Committee. 2000. Error reporting recommendations: A report of the standards and criteria committee. http://www.ixasportal.net/ixas/images/ixas_mat/Error_Reports_2000.pdf.

Kaiser, H. F. 1958. The varimax criterion for analytic rotation in factor analysis. *Psychometrika* 23:187–200. DOI:10.1007/BF02289233.

Lengke, M. F., B. Ravel, M. E. Fleet, G. Wanger, R. A. Gordon, and G. Southam. 2006. Mechanisms of gold bioaccumulation by filamentous cyanobacteria from gold (III)-chloride complex. *Environ. Sci. Technol.* 40:6304–6309. DOI:10.1021/es061040r.

Maeder, M. 1987. Evolving factor analysis for the resolution of overlapping chromotographic peaks. *Anal. Chem.* 59:527–530. DOI:10.1021/ac00130a035.

Malinowski, E. R. 1978. Theory of error for target factor analysis with applications to mass spectrometry and nuclear magnetic resonance spectrometry. *Anal. Chim. Acta.* 103:339–354. DOI:10.1016/S0003-2670(01)83099-3.

Malinowski, E. R. 2002. *Factor Analysis in Chemistry*, 3rd ed. New York: Wiley.

Malinowski, E. R. 2008. *J. Chemometr.* 23:1–6. DOI:10.1002/cem.1182.

Manne, R. 1995. On the resolution problem in hyphenated chromatography. *Chemometr. Intell. Lab.* 27:89–94. DOI:10.1016/0169-7439(95)80009-X.

Martini, A. and E. Borfecchia. 2020. Spectral decomposition of x-ray absorptionspectroscopy datasets: Methods and applications. *Crystals* 10:664. DOI:10.3390/cryst10080664.

Martini A., S. A. Guda, A. A. Guda, G. Smolentsev, A. Algasov, O. Usoltsev, M. A. Soldatov, A. Bugaev, Y. Rusalev, C. Lamberti, and A. V. Soldatov. 2020. PyFitit: The software for quantitative analysis of XANES spectra using machine-learning algorithms. *Comput. Phys. Commun.* 250:107064. DOI:10.1016/j.cpc.2019.107064.

Ressler, T., J. Wong, J. Roos, and I. L. Smith. 2000. Quantitative speciation of Mn-bearing particulates emitted from autos burning (methylcyclopentadienyl)manganese tricarbonyl-added gasolines using XANES spectroscopy. *Enrviron. Sci. Technol.* 34:950–958. DOI:10.1021/es990787x.

Shlens, J. 2003. A tutorial on principal component analysis: Derivation, discussion, and singular value decomposition, Version 1. http://www.cs.princeton.edu/picasso/mats/PCA-Tutorial-Intuition_jp.pdf.

Smith, L. I. 2002. A tutorial on principal components analysis. http://www.cs.otago.ac.nz/cosc453/student_tutorials/principal_components.pdf.

Wall, M. E., A. Rechsteiner, and L. M. Rocha. 2003. Singular value decomposition and principal component analysis. In *A Practical Approach to Microarray Data Analysis*, ed. D. P. Berrar, W. Dubitzky, and M. Granzow, 91–109. Norwell, MA: Kluwer.

Wasserman, S. R., P. G. Allen, D. K. Shuh, J. J. Bucher, and N. M. Edelstein. 1999. *J. Synchrotron Radiat.* 6:284–286. DOI:10.1107/S0909049599000965.

Windig, W. and J. Guilment. 1991. Interactive self-modeling mixture analysis. *Anal. Chem.* 63:1425–1432. DOI:10.1021/ac00014a016.

Chapter 12

Curve Fitting to Theoretical Standards

I Always Wanted to Model

'Curve fitting to theoretical standards' is a lot of words. Why not just call it 'fitting'? I've heard other people do that.	
True, my dear Simplicio. But did we not 'fit' spectra to a linear combination of standards in Chapter 10? And 'fit' line shapes in Chapter 9? 'Fitting,' as a description, is only a little more precise than words like 'analyzing' or 'computing.'	
I've heard people call it 'modeling.' That's a little more precise than 'fitting' because it suggests we've got an underlying guess as to the structure, but not as big a mouthful as 'curve fitting to a theoretical standard.'	
In papers and that kind of thing, you need to be specific anyway, so you'll probably end up saying something like 'the data were fit to a constrained model based on ab initio calculations by the FEFF code'—filling in details always makes for a mouthful.	
Whatever we decide to call it, dear friends, let us begin. It is a large topic and will take us some time to cover.	

DOI: 10.1201/9780429329555-15

You may be wondering how this chapter fits in with Section 1.6.4 of Chapter 1, which spends a page on curve fitting to theoretical standards, and with Part III, which spends five chapters on it. The answer is that in this chapter we'll examine a bit more closely the theoretical underpinnings of theoretical standards and how they are used. Although the primary focus is on EXAFS, the distinction between EXAFS and XANES is not sharp, so we'll touch on some aspects of XANES as well. In the final section of the chapter, we will also provide an introduction to strategies for modeling, which in turn provides a framework for the material in Part III.

12.1 Fitting

In Chapter 9, we looked at the process of fitting simple mathematical functions, such as Gaussians, to data. In the process, we used a computer program to vary several free parameters (perhaps the amplitude, centroid, and width) so as to achieve the closest least-squares fit between the function and the data. The best-fit values for these parameters were used to infer information about the sample, such as oxidation state.

In Chapter 10, we used empirical standards, that is, spectra of possible constituents. By allowing a computer program to vary the contribution from each spectrum, we once again achieved a closest least-squares fit to the data. In addition to the amount of each constituent, free parameters sometimes included energy shifts.

These chapters teach us that free parameters serve two purposes: they allow us to compensate for unknown information (e.g., inconsistent energy calibration) and they reveal useful physical and chemical information about the sample (e.g., the relative amount of each constituent).

It would be nice if we could design free parameters that corresponded in an easily understood way to the physical arrangement of atoms in our material: to bond lengths, for instance, or coordination numbers. Then, we could use XAFS to help discover the structure of novel materials and not just mixtures of known substances.

For EXAFS, the key to doing so was discussed in Section 1.2 of Chapter 1 and is printed on the inside cover of the book: the EXAFS equation. This equation expresses $\chi(k)$ as a sum of scattering paths, some of which are *single* (the photoelectron scatters elastically off of a neighboring atom and then returns to the absorber) and some of which are *multiple* (any other sequence of elastic scattering, for example, the photoelectron scatters off of one neighboring atom, then another, and then returns to the absorber). According to our version of the EXAFS equation, each path is completely specified by three parameters and three functions of k. The three parameters N_i, D_i, and σ^2_i, correspond directly to physically useful concepts such as coordination number, bond length, and disorder. $f(k)$, $\delta(k)$, and $\lambda(k)$, however, have to do with the detailed physics of the EXAFS process: how likely is a photoelectron to scatter off of a given atom? How much of a phase shift would it accumulate in doing so? How likely are inelastic processes to occur?

In the last three decades of the 20th century, dramatic progress was made in computing those functions of k, given a structure. And, as the EXAFS equation implies, those functions of k are fairly insensitive to small changes in the parameters N_i, D_i, and σ^2_i.

The outline for extracting physical information about a novel structure should now be clear:

1. Make an educated guess as to the structure of the sample being measured. In many cases, the sample will be suspected to be slightly different from the guess (e.g., the sample is doped while the guess is the structure of the undoped material). Some aspects of this step will be discussed in Chapter 16. The case studies in Chapter 19 also provide good examples of how this is done in practice.

2. Use theory to calculate the functions of k that appear in the EXAFS equation for each important scattering path. This is sometimes described as generating a theoretical standard. This will be discussed further later in this chapter, as well as in Chapter 13.

3. Fit the theoretical standard to the data by allowing for small changes in the parameters N_i, D_i, and σ^2_i. A number of chapters have additional information on this part of the process, including Chapters 13, 15, and 17.

4. If the resulting fit is poor, then the initial guess is probably wrong and a different guess should be made. If the fit is good, then that gives support to the initial guess and provides physically meaningful information such as estimates of the bond lengths and coordination numbers. The question of how we decide whether a fit is 'good' or 'poor' is covered in Chapter 14.

This process can be described formally as *curve fitting to a theoretical standard*. Less formal terms are *modeling* or *fitting*.

BOX 12.1   EXAFS Modeling Software

This might be a good time to briefly mention some of the common software packages used for modeling EXAFS. The list is not exhaustive, but it does include the most commonly used software. Citations are to papers describing the methods used by each package and don't always refer to the most recent version. More comprehensive lists, which also include software for data analysis and related tasks, can be found at xafs.xrayabsorption.org/software.html.

FEFF (Rehr and Albers 2000): The most commonly used code for generating theoretical standards. A 'lite' version is freely available; complete versions require a license. Many pieces of software have been written to use those empirical standards in modeling:

ARTEMIS (Ravel and Newville 2005): Part of the DEMETER package of analysis tools (previously, part of the older HORAE package). This is the primary software used by the author of this book. While there may, therefore, be some tendency for examples to be framed in a way familiar to ARTEMIS users, the principles discussed are applicable to any FEFF-based analysis. In addition, the case studies in Chapter 19 highlight analyses conducted using a variety of software packages.

EDA (Kuzmin 1995).

EXAFSPAK: Developed by Graham George at the Stanford Synchrotron Radiation Laboratory.

LASE (Emmanuel and Simone 2000).

RSXAP (Booth and Bridges 2021).

SIXPACK (Webb 2005).

VIPER (Klementev 2001).

WINXAS (Ressler 1998).

XDAP (Vaarkamp et al. 1995): Licensed software available for MS-Windows only.

DL-EXCURV (Tomic et al. 2005): This licensed package combines the calculation of theoretical standards with modeling and fitting. Analysis using DL-EXCURV and FEFF is done in a similar manner, so the material in this chapter and Part III are largely applicable to both packages.

GNXAS (Filipponi et al. 1995; Filipponi and Di Cicco 1995): This package uses an expansion of the EXAFS signal that is somewhat different from the usual path expansion. While we won't specifically cover the GNXAS approach, many of the principles we will discuss still apply.

LARCH (Newville 2013): The successor to IFEFFIT; it can act as the 'engine' for other software packages such as ARTEMIS.

12.2 Theoretical EXAFS Standards

The calculation of theoretical standards for EXAFS has become quite sophisticated. For the most part, the computation of a theoretical standard can be treated as a black box: if you enter the positions and species of atoms in a structure, software will produce

Figure 12.1   An ice tray with symmetry properties analogous to muffin-tin potentials.

estimates of $f(k)$, $\delta(k)$, and perhaps $\lambda(k)$ for a set of important scattering paths. If you would like to understand more about how they do that, good places to start are the review article by Rehr and Albers (2000) and the book by Bunker (2010).

Nevertheless, it's helpful to know about some of the approximations that underlie the theory.

12.2.1 Muffin-Tin Potentials

When theoretical standards for EXAFS are generated by FEFF, DL-EXCURV, or GNXAS, atoms are usually treated as creating spatially localized, spherically symmetric potentials. These potentials may be treated as not quite touching each other, barely touching, or overlapping, but in any case that will leave some space not assigned to any atom. The potential in this space is assumed to be constant.

The result is referred to as a muffin-tin potential. Figure 12.1 shows an ice tray rather than a muffin tin, but the idea is the same. There are circular regions, some of which overlap slightly, in which the potential drops, with a constant height surface in between. This ice tray, along with most muffin and cupcake tins, thus provides a two-dimensional analog of a muffin-tin potential.

BOX 12.2   Full Potential Methods

	I thought you said the calculation of theoretical standards is sophisticated. Sophisticated? The muffin-tin approximation seems pretty ridiculous to me. Shouldn't the atoms bond to each other? That means the potential shouldn't be spherically symmetric. And that flat potential in the interatomic regions is crazy! Surely, it's got to vary in some smooth manner.
	Right on both counts, Simplicio. But how big an azimuthal dependence of the potential do you expect? And how much do you expect the potential to vary in the interatomic region?

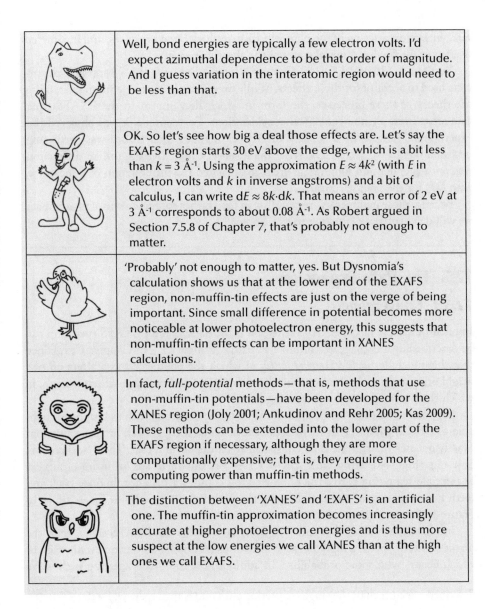

	Well, bond energies are typically a few electron volts. I'd expect azimuthal dependence to be that order of magnitude. And I guess variation in the interatomic region would need to be less than that.
	OK. So let's see how big a deal those effects are. Let's say the EXAFS region starts 30 eV above the edge, which is a bit less than $k = 3$ Å^{-1}. Using the approximation $E \approx 4k^2$ (with E in electron volts and k in inverse angstroms) and a bit of calculus, I can write $dE \approx 8k \cdot dk$. That means an error of 2 eV at 3 Å^{-1} corresponds to about 0.08 Å^{-1}. As Robert argued in Section 7.5.8 of Chapter 7, that's probably not enough to matter.
	'Probably' not enough to matter, yes. But Dysnomia's calculation shows us that at the lower end of the EXAFS region, non-muffin-tin effects are just on the verge of being important. Since small difference in potential becomes more noticeable at lower photoelectron energy, this suggests that non-muffin-tin effects can be important in XANES calculations.
	In fact, *full-potential* methods—that is, methods that use non-muffin-tin potentials—have been developed for the XANES region (Joly 2001; Ankudinov and Rehr 2005; Kas 2009). These methods can be extended into the lower part of the EXAFS region if necessary, although they are more computationally expensive; that is, they require more computing power than muffin-tin methods.
	The distinction between 'XANES' and 'EXAFS' is an artificial one. The muffin-tin approximation becomes increasingly accurate at higher photoelectron energies and is thus more suspect at the low energies we call XANES than at the high ones we call EXAFS.

12.2.2 Final State Rule

Theoretical standards are usually calculated using the *final state rule*; that is, the absorbing atom is treated in the calculations as if the photoelectron had been removed from it and then the resulting ion allowed to reach a relaxed state, but with the hole corresponding to the missing electron still present.

In reality, however, the incoming x-ray doesn't interact with the atom for just an instant. (If it did, then Heisenberg's uncertainty principle tells us that it wouldn't have a definite energy.) Instead, the electric field from the x-ray polarizes the atom, changing the potential around it and placing the resulting ion in a state different from the relaxed state assumed by the final state rule. This is called a *local field effect* (Zangwill and Soven 1980). As with the muffin-tin approximation, local field effects are generally negligible for EXAFS but may have an effect on XANES (Ankudinov and Rehr 2005).

12.2.3 Losses

There are a variety of phenomena that cause EXAFS signals to be smaller than typical theoretical calculations predict. Among those are the limited lifetime of the core hole, inelastic scattering of the photoelectron, multielectron processes, and incomplete overlap of the final and initial states of the absorbing atom.

Sections 1.2.10 and 1.2.11 of Chapter 1 introduced us to the S_o^2 and $\lambda(k)$ parameters used to account for these effects. While progress is being made in understanding the theory of these processes, the form in which they appear in the EXAFS equation arose from phenomenological considerations. For example, it was observed that experimental data showed smaller EXAFS oscillations than theory was predicting, necessitating an S_o^2 factor. Likewise, the discovery that EXAFS, unlike diffraction, was not dependent on long-range order necessitated the introduction of a mean-free path $\lambda(k)$ for the photoelectron.

Because these parameters are standing in for a rather complex array of phenomena, we will defer further discussion until Chapter 13.

12.3 The Path Expansion

12.3.1 Convergence

Implicit in the use of the EXAFS equation is the hope that an EXAFS spectrum can be described by a manageable number of paths. Section 1.2.7 of Chapter 1 explained that these include *single-scattering paths*, in which the photoelectron scatters off of a neighboring atom and returns to the absorbing atom, and *multiple-scattering paths*, in which there is more than one scattering event.

Suppose, then, that we had an isolated diatomic molecule, such as bromine monochloride (BrCl) gas and that we measure the bromine K edge. Each bromine atom has exactly one neighbor; there is no consistent structure beyond that. There is, therefore, only one single-scattering path: Br → Cl → Br. But there are an infinite number of multiple-scattering paths: the photoelectron could, for instance, bounce back and forth between the two atoms, following the path Br → Cl → Br → Cl → Br, as in Figure 12.2.

Intuitively, a path of that sort will contribute less to the EXAFS signal than the direct-scattering path, simply because it involves additional scattering events each of which occurs with some probability. In addition, the multiple-scattering path we've described is longer than the single-scattering path and is thus reduced by the $\dfrac{1}{D^2}e^{-\frac{2D}{\lambda(k)}}$ from the EXAFS equation. Higher-order multiple-scattering paths, involving more bouncing back and forth, will have even larger values of D. The exponential dependence on the mean free path, therefore, assures that the EXAFS spectrum will be described well by a small number of paths in this simple case.

In contrast, consider the more common example of a solid. A typical length for a mean free path in the EXAFS region is 10 Å (Section 13.2.5 in Chapter 13), so we

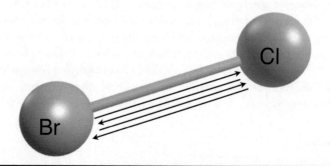

Figure 12.2   A triple-scattering path within a diatomic molecule.

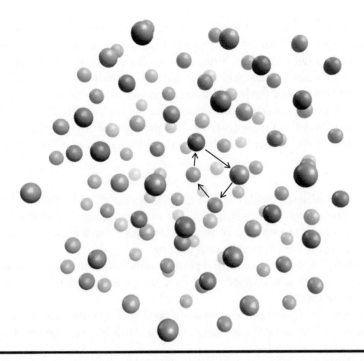

Figure 12.3 **A cluster of about 100 atoms, showing one triple-scattering path originating at the central atom. There are hundreds of thousands of triple-scattering
paths that can be drawn within this cluster, and millions of quadruple-scattering
paths.**

would have to consider scattering off of hundreds of different atoms. For simplicity,
suppose there were a hundred atoms within range. There would then be a hundred possible single-scattering paths, but how many multiple-scattering paths would there be?
Each of the hundred scatterers could be paired with any of the other 99 atoms for a second scattering event, giving us nearly 10,000 possibilities for double scattering. Some of
those paths would be quite long—for instance, traveling to one side of the cluster and
then all the way to the opposite side of the cluster, and then back to the absorbing atom.
Since we've chosen the radius of the cluster to be about the mean free path of the photoelectron, those long multiple-scattering paths would contribute very little. But *most*
of the double-scattering paths would be shorter than the longest single-scattering path
in our cluster and would thus need to be considered. If we add in triple scattering, we're
thinking about hundreds of thousands of paths. (Figure 12.3 shows one example of a
triple-scattering path in a cluster of about a hundred atoms.) And quadruple scattering
puts us into the millions. Continuing along with all the multiple-scattering paths that
are shorter than some cutoff based on the mean-free path, we find that we are dealing
with an enormous number of paths—on the order of a billion (Zabinsky et al. 1995).
While modern computing power puts this kind of computation within reach, it would
be impractical for a human being to consider those paths one by one, and thus, part of
the conceptual appeal of the path formulation would be lost.

Fortunately, it turns out that the vast majority of those paths contribute very little
to the EXAFS spectrum (Zabinsky et al. 1995). This is for several reasons:

The sheer number of low-amplitude multiple-scattering paths means that all
phases are likely to be well represented at each value of k. The statistics of this kind of
combination is similar to counting statistics (Section 6.3.3 of Chapter 6). N paths of
amplitude A will tend to sum to a signal of amplitude roughly $\sqrt{N}A$. Thus, 10,000
small paths end up contributing an amount roughly equal to 100 times what a single
one of them would.

For those who
might be curi-
ous, Figure 12.3
is based on the
copper–indium
selenide struc-
ture, with the
central atom being copper. Atoms
up to 8.5 Å from the central
atom are shown for a total of 103
atoms.

■ In most cases, each additional scattering event lowers the amplitude of the event because only part of the photoelectron wave is scattered at each atom (see Box 12.3 for an exception). Thus, each double-scattering path tends to be less important than an otherwise similar single-scattering path, triple scattering less important than double scattering, and so on. On the basis of this aspect and path length considerations alone, Zabinsky et al. (1995) estimated the relative contributions of more than a billion paths to the XAFS of metallic copper and concluded that just 56 paths accounted for more than 90% of the spectral weight.

The $e^{-2k^2\sigma^2}$ disorder factor tends to affect multiple-scattering paths more strongly than otherwise similar single-scattering paths. For example, thermal disorder will cause each scattering atom to vary a bit in position; the more scattering atoms we add to a path, the greater the variance in the path length.

Thus, while it is difficult to know a priori exactly how many paths will be necessary to accurately reproduce a spectrum of a given species, the number is unlikely to exceed a few hundred.

In addition, fitting is often done on just a portion of a Fourier transform (Section 7.5 of Chapter 7), allowing the exclusion of paths with lengths outside of a chosen range.

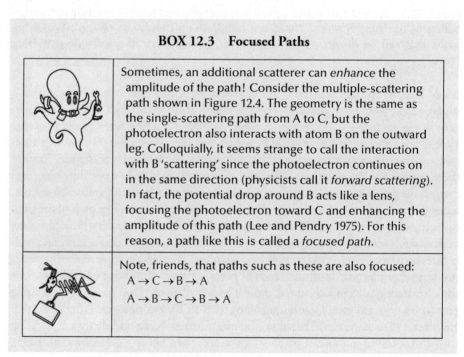

BOX 12.3   Focused Paths

Sometimes, an additional scatterer can *enhance* the amplitude of the path! Consider the multiple-scattering path shown in Figure 12.4. The geometry is the same as the single-scattering path from A to C, but the photoelectron also interacts with atom B on the outward leg. Colloquially, it seems strange to call the interaction with B 'scattering' since the photoelectron continues on in the same direction (physicists call it *forward scattering*). In fact, the potential drop around B acts like a lens, focusing the photoelectron toward C and enhancing the amplitude of this path (Lee and Pendry 1975). For this reason, a path like this is called a *focused path*.

Note, friends, that paths such as these are also focused:
A → C → B → A
A → B → C → B → A

12.3.2 *Full Multiple Scattering*

While EXAFS is generally described well by a manageable number of paths, the same is not necessarily true of XANES. Since the photoelectron has less kinetic energy

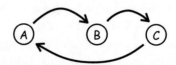

Figure 12.4   A focused path.

in the XANES region, the probability of scattering off of each atom is greater, and thus, the additional scattering events in multiple-scattering paths do less to suppress the contributions from those paths. In addition, the effect of disorder is less since the σ^2 factor is weighted by k^2. Finally, Fourier filtering is not generally applied to the XANES region, in part because of the difficulty of background subtraction. Taken together, these effects mean that in the XANES region the path expansion may not converge well (Zabinsky et al. 1995) or, in some cases, may not converge at all (Ravel 1997).

Fortunately, the lower energy of the photoelectron in the XANES region allows for *full multiple-scattering* calculations to be practical, that is, methods that calculate total spectra directly rather than relying on the enumeration of individual scattering paths (Rehr and Albers 2000).

Even if the path expansion fails to converge, that doesn't necessarily mean it's useless. It could be, for instance, that a manageable number of paths could account for 90% of the variations in a XANES spectrum, but that an infinite number of paths would be necessary to account for 98%. And depending on what you're doing, that 90% might be good enough.

BOX 12.4 Full Multiple Scattering

Computers keep getting more powerful, and theorists are really smart. I bet it won't be long until it's no big deal to perform full multiple scattering all the way through the EXAFS region.

I'd take the other side of that bet, Simplicio. The computational power required goes up very rapidly with the energy (Ravel 1997), so even big improvements in computers will only buy us another inverse angstrom or two.

And even if we could use full multiple scattering for the whole EXAFS spectrum, we wouldn't want to. The path expansion and the EXAFS equations are nice because they give us a window into what is physically going on in our material: how far apart atoms are, how much disorder there is, how many atoms there are. It is easy, with the path expansion, to see qualitatively and quantitatively the effect of, for example, changing out one ligand or stretching one bond length a little bit. Each individual contribution to the complete spectrum can be identified, graphed, and modified. Full multiple scattering takes away that ability; you put a proposed structure in, and a spectrum comes out, but you don't really know what features in the spectrum correspond to what aspects of your structure. In the regimes where the path expansion works well, this makes it preferable to a full multiple-scattering calculation.

In recent years, a number of software packages have been written that allow for XANES modeling to be done with full multiple-scattering computations, including FDMNES (Joly 2001), MXAN (Benfatto and Della Longa 2001), and FITIT (Smolentsev and Soldatov 2006). They do this by computing the spectrum for one structure (in the case of FITIT, this calculation is performed by other software such as FEFF), modifying the structure a little bit, computing the new spectrum, and seeing if the result is closer to the data. If it is, the structure is modified a bit more in the same direction; if not, the opposite direction is tried. While computationally expensive, Simplicio is right that modern computing power and clever interpolation schemes makes this kind of thing possible.

 It's still not the same thing as the path expansion—it's more like a high-powered version of the kind of fingerprinting we did in Chapter 9. It doesn't allow us to directly point at a feature in the spectrum, for instance, and say 'this feature is due to the second-shell oxygen.' Instead we can change the oxygen to another element, or move it, or do something else like that, and observe where the spectrum changes. Furthermore, the path expansion handles modest disorder easily through the σ^2 parameter, something full multiple scattering doesn't do.

 That's kind of the nature of XANES, though. This isn't a theory problem waiting for a breakthrough. When we Fourier transform EXAFS data, it kinda-sorta looks like a radial distribution function and it's that correlation that the path expansion is exploiting. That doesn't happen in XANES. I know you keep saying that there isn't a sharp distinction between XANES and EXAFS, Carvaka, but if it takes a million paths to reproduce some XANES feature well, then I'm OK with saying 'the path expansion doesn't work in that case because it's XANES.' Insisting the underlying physics is the same is like saying that there's no real difference in transportation across town or across the country. Sure, it might be theoretically possible to fly a plane across town or to walk across the country, but no one would do either of those things except as a stunt. So full multiple scattering is useful for XANES because the path expansion might be impractical, but the path expansion is better for EXAFS because it gives you more insight into what is actually going on.

 While we're on the subject, let me emphasize that, despite the similar names and the similar regimes in which they are useful, *full potential* and *full multiple scattering* are two completely different things. It is quite possible to make full multiple-scattering muffin-tin calculations or use a full-potential path expansion!

12.4 EXAFS Fitting Strategies

While 50 or 100 paths are more manageable than billions, it still means there are a lot of parameters that could potentially be fit. Each path needs to be assigned a value of N, D, and σ^2. If each of those were treated as a free parameter to be found by least-squares fitting for each significant path, there could easily be hundreds of free parameters. While we'll leave a careful discussion of the statistics of EXAFS fitting until Section 14.1 of Chapter 14, our discussion of linear combination fitting in Chapter 10 should warn us that we are unlikely to be able to fit that many variables successfully. We need to do something to constrain our fits! Broadly speaking, there are two common strategies that are used.

12.4.1 Bottom-Up Strategy

In the *bottom-up* strategy, you start with a very small number of paths, often only one or two. This can be done by

- Fitting only the part of the Fourier transform that is dominated by nearest-neighbor paths (i.e., single-scattering paths from the absorbing atom to one of the nearest-scattering atoms)
- Assuming all nearest-neighbor atoms are at the same distance
- If consistent with what is known about the sample, beginning with structures where all the nearest neighbors are the same species

By doing this, you will likely be left fitting a single path: the single-scattering path off of the nearest neighbors. N, D, and σ^2 for that path can then be determined by fitting. This is known as a *single-shell fit*.

Single-shell fits can be a good place to start, but they often leave a lot of data on the table. Therefore, the goal in the bottom-up strategy is to start with the simple nearest-neighbor fit and then gradually add more information.

How to do that depends on the result of the single-shell fit and on your preexisting knowledge about the system. Perhaps, for instance, the fit reports a larger value of σ^2 than is usual for the kind of bond being fit. Perhaps, there is also discussion in the literature about the particular system that you are studying, debating whether or not there are two different bond lengths for the near-neighbors, such as an equatorial and an axial length. If there were two different lengths, then your initial fit (which only allowed for one length) would likely report a large σ^2, since that is a measure of the variance of the absorber–scatterer distance. In the case just described, you could then proceed by trying to fit *two* paths, allowing for the scatterers to be at slightly different distances.

Likewise, a poor fit might suggest you have the identity of the scatterers wrong; perhaps you tried all oxygen atoms, but in reality, some of the oxygen atoms were substituted by sulfur atoms.

If, on the other hand, the single-shell fit was good, but the Fourier transform shows data at higher R than the range you fit, you could try to expand the fitting range and incorporate paths corresponding to more distant atoms.

Hopefully, information soon starts to emerge that dovetails with your preexisting knowledge in some way. For example, you might find that your nearest neighbors are nitrogen atoms, and you might also have a prior reason for suspecting that there are ethylenediamine ($NH_2CH_2CH_2NH_2$) ligands in your system. If that were the case, you might hypothesize that each pair of nearest-neighbor nitrogen atoms was attached to a pair of carbons further out; those carbons might show up in the EXAFS. Thus, the degeneracy N of the second-nearest-neighbor path to carbon could be required to be the same as the coordination number for the nearest-neighbor path to nitrogen. A requirement of this type that is imposed during fitting is called a *constraint*. The list of paths and constraints you are using during a fit comprises the model you are using.

Continuing in this way, we could gradually add more paths and more constraints, gradually building up a detailed model of our material.

12.4.2 *Top-Down Strategy*

You may have noticed that the bottom-up approach starts with very few assumptions—you may not even know with confidence what the nearest-neighbor atoms are. In contrast, the *top-down* strategy works well when you are looking for modifications to a well-known structure or to choose between a handful of structural possibilities.

For this strategy, begin with a complete structure, including multiple shells of scatterers. For instance, if you were working with a thin-film-doped zinc oxide, you could begin with the structure of bulk zinc oxide. This would mean including many paths in your model—perhaps as many as a hundred. To keep the number of free parameters down, the initial model would be highly constrained. For example, you might assume that all the atoms in your sample occupy exactly the same positions as they do in the bulk, and with the same coordination numbers. You might further make simplifying assumptions regarding thermal disorder, requiring many paths to adopt the same value of σ^2. Your first fit, therefore, while including many paths, would only have a few free parameters.

If you're trying to decide between a few alternatives, such as two completely different allotropes of a crystal, then a highly constrained fit may be enough to provide you your answer. But frequently, you expect your material to differ from the structure in your initial model in some way—perhaps local distortions, or vacancies, or subtle changes in symmetry. If that's the case, then you'll need to relax your model, bit by

Single-shell fits can be a good place to start, and sometimes they can be helpful when used on a series of spectra; for example, one sample measured under different conditions or a set of related samples. But it is not much of an exaggeration to say that a single single-shell fit on a single sample under a single set of conditions is worthless on its own. Curve fitting to theoretical standards has too many potential sources of systematic error to blindly trust an isolated result. But once you start to extend the fit, or to make it part of a series, your confidence can grow.

Note, friends, that bottom-up and top-down approaches share another similarity: they both start with a small number of free parameters, gradually adding more to improve or extend the model.

bit, to allow for those possibilities. Randomly distributed oxygen vacancies could be modeled, for instance, by replacing the constraint that all oxygen coordination numbers must be *the same* as in your starting structure with the constraint that all oxygen coordination numbers must be *reduced by the same unknown percentage* from those in the starting structure. In this way, a free parameter is added to the fit.

It is important to note that the bottom-up and top-down strategies end up in the same place: a multiple-path fit with enough free parameters to reveal the desired structural information about the sample, but enough constraints to keep the problem from being statistically untenable. The bottom-up approach accomplishes this by beginning with few constraints and few paths, and gradually adding both. The top-down approach, in contrast, starts with many paths and strong constraints, which are then gradually relaxed.

12.5 Theoretical XANES Standards

In recent years, several approaches to developing theoretical XANES standards have been developed, allowing quantitative modeling of XANES data in much the same way as has been done for EXAFS.

12.5.1 Real-Space Multiple-Scattering

The real-space multiple-scattering (RSMS) approach, also known as the real-space Green's function (RSGS) approach, is an extension of the EXAFS methods described in Section 12.2 to XANES. The inclusion of full-potential methods (Box 12.2), vibronic coupling, and local field effects improves the calculation of XANES spectra, yielding standards suitable for use in semi-quantitative applications (Ankudinov and Rehr 2005). The accuracy of XANES spectra calculated directly by this method is mixed, generally reproducing qualitative features of experimental spectra but often with significant quantitative mismatches, both in regard to the energy at which XANES features are found and the amplitudes of those features (Bosman and Thieme 2009; Nakanishi and Ohta 2009).

Of course, the same can be said of theoretical standards for EXAFS. In order for the fits to be useful quantitatively, phenomenological parameters such as S_o^2 (see Section 13.1.4 of Chapter 13) and shifts in E_o (see Section 13.1.5 of Chapter 13) are often fit along with structural parameters. In recent years, efforts have been made to allow similar flexibility in fitting data to theoretical XANES standards (Hayakawa et al. 2007). This method shows promise for allowing quantitative modeling of XANES data.

12.5.2 Finite Difference Method

The finite difference method (FDM) imposes a spatial grid on a cluster of atoms, connecting the derivatives of the wave function of the photoelectron at each grid point to its neighbors using the Schrödinger equation (Bourke et al. 2016). While this works well in interstitial regions, near atomic cores the grid would need to be very closely spaced, making the procedure computationally difficult. Fortunately, near atomic cores the wave functions are expected to be nearly spherically symmetric, allowing the wave function to be written in terms of spherical harmonics (Joly et al. 2022). Similarly, the grid spacing used may be different for different photoelectron energies (Bourke et al. 2016), or even be nonuniform for a single energy calculation.

Theoretical standards found by FDM are accurate enough to be used for quantitative modeling in the XANES region.

BOX 12.5 Full Potential Methods

 In recent years, the use of FDM has been extended into the EXAFS region. (Bourke et al. 2016)

 Wait—in Box 12.4 I bet Kitsune that full multiple scattering methods could some day be extended into the EXAFS region. Did I just win the bet?

 I am afraid not, dear Simplicio. While FDM is not a partial multiple-scattering approximation, neither it is a full one. It is not based on a path expansion at all.

 Oh. So just like Carvaka said in Box 12.4, I can't extract the effect of individual scattering paths on the spectrum using FDM. Oh, well.

BOX 12.6 XANES Modeling Software

 As we did in Box 12.1, we'll provide a partial list of software used for modeling XANES.

FEFF (Rehr and Albers 2000): The licensed version can be used for XANES.

    LASE (Emmanuel and Simone 2000).

    MAX (Michalowicz et al. 2009): An updated version of the EXAFS POUR LE MAC package, now available for multiple platforms.

    SIXPACK (Webb 2005).

    WINXAS (Ressler 1998): Licensed software available for MS-Windows only.

    XDAP (Vaarkamp et al. 1995): Licensed software available for MS-Windows only.

    VIPER (Klementev 2001): Available for MS-Windows only.

FDMNES (Joly et al. 2022): This free package can use the finite difference method to generate theoretical standards for XANES, RIXS, and more.

FITIT: A free package for interpolating fits using theoretical standards generated by FDMNES, FEFF, or other sources.

MXAN: A free package to perform quantitative modeling on XANES data using theoretical standards it generates.

WHAT I'VE LEARNED IN CHAPTER 12, BY SIMPLICIO

- EXAFS theoretical standards are usually calculated using *muffin-tin* potentials, but *full-potential* methods are sometimes needed for accurate XANES calculations.
- Accurate EXAFS spectra can be constructed from at most a few hundred scattering paths. For some XANES spectra, path expansion methods may be impractical and *full multiple-scattering* approaches may be needed.
- Modeling can be done using a *bottom-up* strategy. To do that, we would start with very few paths and very few constraints, gradually adding more of each as we learned about our sample.
- Modeling can also be done using a *top-down* strategy. To do that, we would start with a highly constrained model with lots of paths. If the structure seems to be on the right track, we would then gradually relax constraints to understand the details of our material better.
- Most of all, I've learned that I want to read Part III to learn more!

References

Ankudinov, A. L. and J. J. Rehr. 2005. Nonspherical potential, vibronic and local field effects in x-ray absorption. *Physica Scripta T.* 115:24–27. DOI:10.1238//Physica.Topical.115a00024.

Benfatto, M. and S. Della Longa. 2001. Geometrical fitting of experimental XANES spectra by a full multiple-scattering procedure. *J. Synchrotron Radiat.* 8:1087–1094. DOI:10.1107/S0909049501006422.

Booth, C. H. and F. Bridges. 2021. Real-space x-ray absorption package (RSXAP). *International Tables for Crystallography*, Vol. 1. DOI:10.1107/S1574870720003444.

Bosman, E. and K. Thieme. 2009. Modeling of XANES-spectra with the FEFF program. *J. Phys. Conf. Ser.* 186:012004. DOI:10.1088/1742-6596/186/1/012004.

Bourke, J. D., C. T. Chantler, and Y. Joly. 2016. FDMX: Extended X-ray absorption fine structure calculations using the finite difference method. *J. Synch. Rad.* 23:551–559. DOI:10.1107/S1600577516001193.

Bunker, G. 2010. *Introduction to XAFS*. New York: Cambridge University Press.

Emmanuel, C. and B. Simone. 2000. *Développement d'Outils Informatiques et Statistiques pour l'Analyse des Spectres EXAFS—Application aux Systèmes Bioinorganiques [Development of Informatic and Statistic Tools for EXAFS Spectra Analysis—Application to Bioinorganic Systems]*. Orsay, France: Université de Paris, 11.

Filipponi, A. and A. Di Cicco. 1995. X-ray-absorption spectroscopy and *n*-body distribution functions in condensed matter. II. Data analysis and applications. *Phys. Rev. B.* 52:15135–15149. DOI:10.1103/PhysRevB.52.15135.

Filipponi, A., A. Di Cicco, and C. R. Natoli. 1995. X-ray-absorption spectroscopy and *n*-body distribution functions in condensed matter. I. theory. *Phys. Rev. B.* 52:15122–15134. DOI:10.1103/PhysRevB.52.15122.

Hayakawa, K., K. Hatada, S. D. Longa, P. D'Angelo, and M. Benfatto. 2007. Progresses in the MXAN fitting procedure. *AIP Conf. Proc.* 882:111–113. DOI:10.1063/1.2644444.

Joly, Y. 2001. X-ray absorption near-edge structure calculations beyond the muffin-tin approximation. *Phys. Rev. B.* 63:125120. DOI:10.1103/PhysRevB.63.125120.

Joly, Y., A. Y. Ramos, and O. Bunău. 2022. Finite-difference method for the calculation of X-ray spectroscopies. *International Tables for Crystallography, Vol. I, X-ray Absorption Spectroscopy and Related Techniques*. DOI:10.1107/S1574870722001598.

Kas, J. 2009. *Toward Quantitative Calculation and Analysis of X-Ray Absorption Near Edge Spectra*. Seattle, WA: University of Washington.

Klementev, K. V. 2001. Extraction of the fine structure from x-ray absorption spectra. *J. Phys. D Appl. Phys.* 34:209. DOI:10.1088/0022-3727/34/2/309.

Kuzmin, A. 1995. EDA: EXAFS data analysis software package. *Physica B Condens. Matter.* 208/209:175–176. DOI:10.1016/0921–4526(94)00663–G.

Lee, P. A. and J. B. Pendry. 1975. Theory of the extended x-ray absorption fine structure. *Phys. Rev. B.* 11:2795–2811. DOI:10.1103/PhysRevB.11.2795.

Michalowicz, A., J. Moscovici, D. Muller-Bouvet, and K. Provost. MAX: Multiplatform applications for XAFS. *J. Phys. Conf. Ser.* 190:012034. DOI:10.1088/1742–6596/190/1/012034.

Nakanishi, K. and T. Ohta. 2009. *J. Phys. Condens. Mat.* 21:104214. DOI:10.1088/0953-8984/21/10/104214.

Newville, M. 2013. Larch: An analysis package For XAFS and related spectroscopies. *J. Phys. Conf. Ser.* 430:012007. 10.1088/1742-6596/430/1/012007.

Ravel, B. 1997. *Ferroelectric Phase Transitions in Oxide Perovskites Studied by XAFS.* Seattle, WA: University of Washington.

Ravel, B. and M. Newville. 2005. ATHENA, ARTEMIS, HEPHAESTUS: Data analysis for X-ray absorption spectroscopy using IFEFFIT. *J. Synchrotron Radiat.* 12:537–541. DOI:10.1107/S0909049505012719.

Rehr, J. J. and R. C. Albers. 2000. Theoretical approaches to x-ray absorption fine structure. *Rev. Mod. Phys.* 72:621–654. DOI:10.1103/RevModPhys.72.621.

Ressler, T. 1998. WinXAS: A program for x-ray absorption spectroscopy data analysis under MS-Windows. *J. Synchrotron Radiat.* 5:118–122. DOI:10.1107/S0909049597019298.

Smolentsev, G. and A. V. Soldatov. 2006. FitIt: New software to extract structural information on the basis of XANES fitting. *Comp. Mater. Sci.* 39:569–574. DOI:10.1016j.commatsci.2006.08.007.

Tomic, S., B. G. Searle, A. Wander, N. M. Harrison, A. J. Dent, J. F. W. Mosselmans, and J. E. Inglesfield. 2005. New tools for the analysis of EXAFS: The DL EXCURV package. Council for the Central Laboratory of the Research Councils.

Vaarkamp, M., J. C. Linders, and D. C. Koningsberger. 1995. A new method for parameterization of phase shift and backscattering amplitude. *Physica B Condens. Matter.* 208/209:159–160. DOI:10.1016/0921–4562(94)00658–I.

Webb, S. M. 2005. SIXpack: A graphical user interface for XAS analysis using IFEFFIT. *Physica Scripta T.* 115:1011–1014. DOI:10.1238/Physica.Topical.115a01011.

Zabinsky, S. I., J. J. Rehr, A. Ankudinov, R. C. Albers, and M. J. Eller. 1995. Multiple-scattering calculations of x-ray-absorption spectra. *Phys. Rev. B.* 52:2995–3009. DOI:10.1103/PhysRevB.52.2995.

Zangwill, A. and P. Soven. 1980. Density-functional approach to local-field effects in finite systems: Photoabsorption in the rare gases. *Phys. Rev. A.* 21:1561–1572. DOI:10.1103/PhysRevA.21.1561.

Plates

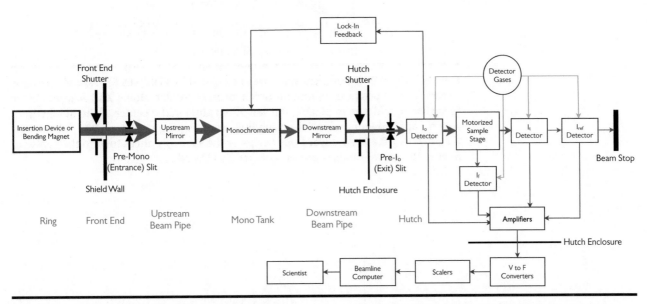

Plate 1. Fig. 8.2. Schematic of a typical beamline. X-rays are shown in magenta, signals in blue, and gases in green. Brown labels refer to regions used to identify the location of components; e.g., a beamline scientist might refer to some particular control as being 'at the front end,' or 'in the hutch.'

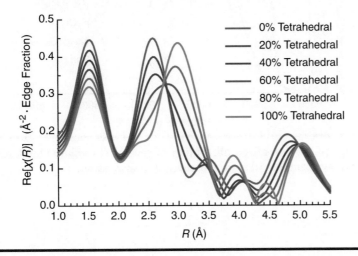

Plate 2. Fig. 9.2. Magnitude of the Fourier transform for the manganese K edge of theoretical spectra of manganese ferrite, computed assuming a variety of site occupancies for the manganese atoms. The transforms were taken on data with a k-weight of 1 from 3 - 10 Å^{-1}, using Hanning windows with sills of 1.0 Å^{-1} on each side.

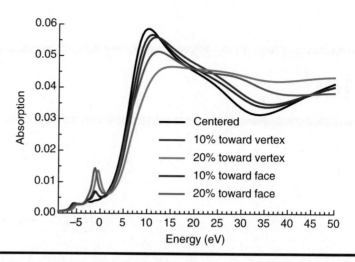

Plate 3. Fig. 9.9. Theoretical standards calculated by FDMNES for vanadium atoms at each of five positions within an octahedron of oxygen atoms 2.16 Å from center to vertex. In addition to placing the vanadium atom in the center of the octahedron, spectra are shown in which the vanadium atom is placed 10% or 20% of the way (i.e., 0.216 or 0.432 Å) toward one vertex or the same distances toward the center of a face. The values on axes are as reported by FDMNES.

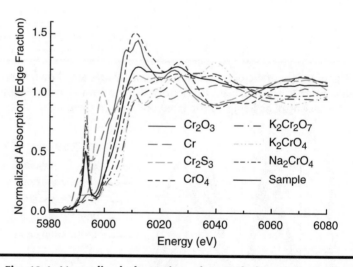

Plate 4. Fig. 10.4. Normalized absorption of several chromium standards, along with an unknown sample. All were measured in transmission.

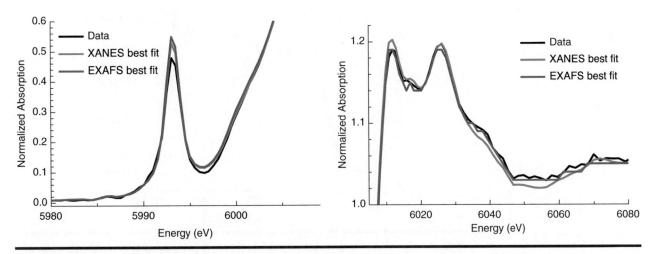

Plate 5. Fig. 10.7. Comparison of two linear combination fits in energy space. The graph has been split into two regions to facilitate examination of the differences between the two fits and the data.

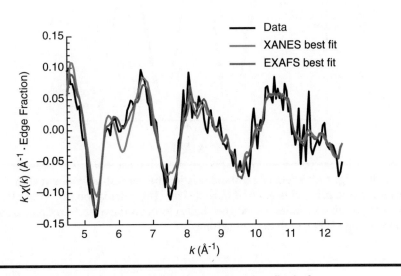

Plate 6. Fig. 10.8. Comparison of two linear combination fits in *k* space.

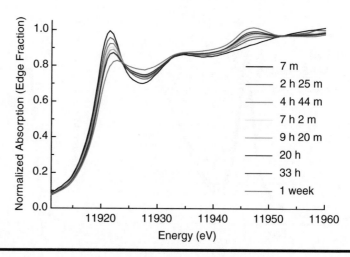

Plate 7. **Fig. 11.1. Time series of gold L_3 XANE S for a sample of gold (III) chloride exposed to cyanobacteria, as a function of time of exposure. Data were from Lengke et al. (2006).**

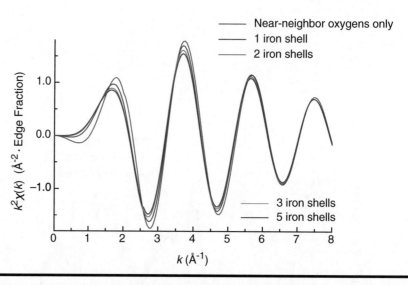

Plate 8. **Fig. 16.2 Theoretical standards for the nearest-neighbor scattering path of iron(II) oxide using clusters of different sizes. The calculation, for example, for 'two iron shells' also includes all oxygen shells closer to the absorber. Calculations were made using FEFF6L with default settings.**

MODELING

Chapter 13

A Dictionary of Parameters

Something for Everyone

A dictionary of EXAFS parameters? That sounds like a good idea! But what should it include?	
It must certainly include symbols and terminology for each parameter.	
Symbols are meaningless unless we understand the concepts they represent. We need to discuss what each parameter represents physically.	
Be real! The point is to learn to fit. To do that, we need to know how each parameter affects the squiggly lines on the graphs.	
We should also let people know what values are typical for each parameter, so that they know when something has gone wrong!	

DOI: 10.1201/9780429329555-17

And, friends, we should discuss some common constraint schemes, so that we shall have some idea where to start.	
And you, Kitsune? This chapter is called 'Something for Everyone.' I like it all so far, but what do *you* want to see?	
To tell the truth, I'm not all that interested in this chapter. Unlike Dysnomia, the point for me is not 'learning to fit.' The point is to answer questions I have about the materials I study.	
Come on, Kitsune. There must be something you'd like to know!	
Perhaps I'd like to know how the parameters correlate. That way I might understand better why I need to worry about parameters that don't directly answer my scientific questions.	
Yay! Something for everyone!	

13.1 Common Fitting Parameters

There are five parameters related to the EXAFS equation (on the inside front cover!) that are frequently allowed to vary when fits are performed: the *half path length D*, the *degeneracy N*, the *mean square relative displacement* (MSRD) σ^2, the *amplitude reduction factor S_o^2*, and E_o. We'll deal with those first.

13.1.1 Half Path Length

13.1.1.1 Symbol

The symbol used for half path length in this book is D. In the literature, R is often used.

13.1.1.2 Nomenclature

For single-scattering paths, the half path length may also be called the *absorber–scatterer distance*. In cases where it makes chemical sense to do so, it is often simply referred to as the *bond length*.

13.1.1.3 Physical interpretation

For single-scattering paths, the half path length is simply the average distance between the absorbing and scattering atoms. It's half the path length because a single-scattering path must travel from absorber to scatterer *and* back.

One should not use the term *bond length* too freely. For example, an oxygen atom 4 Å from an iron atom is not bonded to it!

There are two reasons we specify that it is an average. First, individual atoms move around due to thermal motion, and the distance between a particular absorbing atom and a particular scattering atom varies a bit. This is called *thermal disorder*. But it is also true that not all absorber–scatterer pairs represented by the path are necessarily identical. There may, for instance, be defects such as vacancies or dopants in the material. There may also be phase boundaries or surfaces. In addition, you might be modeling a distorted shape as if it were more symmetrical, perhaps choosing not to make a distinction between equatorial and axial atoms. These are examples of *static disorder*.

For multiple-scattering paths, the half path length is just what it sounds like: half the average length of the scattering path. It is defined in this way to make it consistent with the definition for single-scattering paths. Note that a multiple-scattering path may have a value of D that is greater than the distance from the absorber to any of the individual scattering atoms.

You may have noticed that we treat the EXAFS process as if it is instantaneous relative to any atomic motions that occur. This is equivalent to saying that EXAFS photoelectrons travel much faster than atoms undergoing thermal motion in a material. Room temperature thermal motion corresponds to a kinetic energy of about 1/40 of an electron volt. Thus, even 10 eV above the edge, a photoelectron has roughly 400 times the kinetic energy of the atom it scatters off of. Add to that the fact that even a hydrogen atom is about 1800 times heavier than an electron, and you can see that the scattering atom is essentially stationary relative to the electron that scatters off of it.

BOX 13.1 Average Distance Is Not the Same as Distance between Average Positions

EXAFS can allow us to find the average distance between atoms and XRD can give us their average positions. But the average distance between atoms is not quite the same thing as the distance between their average positions. To see why, consider three kinds of relative motion a pair of atoms could make:

1. Perfectly correlated motion, in which they both move the same distance in the same direction (Figure 13.1). This does not change the distance between them.
2. Anticorrelated motion along the line joining them, in which the atoms alternately move toward and away from each other (Figure 13.2). The average distance between them for this kind of motion is the same as the distance between their average positions.
3. Anticorrelated motion perpendicular to the line joining them (Figure 13.3). The distance between the atoms in this case is always at least as large as the distance between their average positions, and usually it's larger.

Any relative motion can be made up of combinations of those three types. In two of them, the EXAFS and XRD distances are the same, but in the third, the EXAFS distance is larger. Therefore in a real pair of atoms, the EXAFS distance is larger than the XRD distance.

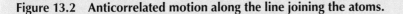

Figure 13.1 Perfectly correlated motion.

Figure 13.2 Anticorrelated motion along the line joining the atoms.

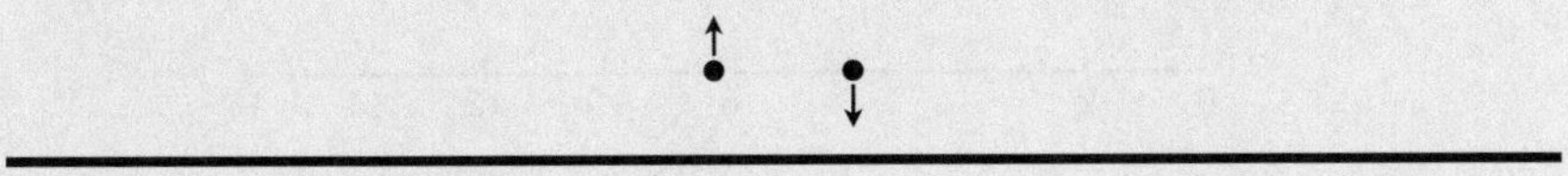

Figure 13.3 Anticorrelated motion perpendicular to the line joining the atoms.

Unless otherwise specified, the graphs presented in this chapter are FEFF calculations of the sulfur nearest-neighbor path of pyrite (FeS₂) measured at the iron *K* edge, with the following parameters: $D = 2.26$ Å, $N = 6$, $\sigma^2 = 0.003$ Å², $S_o^2 = 0.90$, $E_o = 7112$ eV, $C_3 = C_4 = 0$. Fourier transforms are taken from 3 to 15 Å⁻¹ using a *k*-weight of 2 and Hanning windows with sills of 1.0 Å⁻¹ on each side.

This effect is fairly small and often not explicitly discussed in EXAFS analyses. But it can be important to think about when comparing spectra for which both interatomic distance and thermal disorder differ significantly, such as a temperature series designed to measure thermal expansion.

13.1.1.4 Typical values

As a scientist, studying whatever it is that you study, you know much more about what the possible distances between atoms in your samples are than we do! It is worth noting, though, that if a fit yields a value of *D* that is more than about 0.1 Å different from the value in the standard you are using, then the standard is not entirely appropriate. If you think the fit is on the right track, then you should get a new standard. For theoretical standards, that simply means redoing the ab initio calculation with atomic positions more similar to those found by your fit.

13.1.1.5 Effect on χ(k)

The larger *D* is, the smaller the spacing of the peaks for that path in χ(*k*). Figure 13.4 shows that the difference in phase due to a modest shift in *D* is more pronounced at high *k*.

Scattering atoms heavier than the 3d transition metals may cause the Fourier transform to show a split peak, even if there is only one absorber–scatterer distance (see Section 13.3).

13.1.1.6 Effect on Fourier transform

The magnitude of the Fourier transform will show a peak centered a few tenths of an angstrom below the value of *D*. For a well-isolated peak, changing *D* will shift the peak by an amount about equal to the change (Figure 13.5).

The real part of the Fourier transform won't generally show a single peak, but it is affected by shifts in a similar way (Figure 13.6).

13.1.1.7 Common constraints

If your material is cubic, one highly constrained option is to assume that the entire structure is expanded or contracted uniformly relative to the standard, that is, each half path length differs by the same percent from the standard.

Watch out! Because the Fourier transforms used in EXAFS are complex valued, the effect of a small shift in one path on the magnitude of the Fourier transform is often nonintuitive. If the shifted path overlaps significantly with another path, the shift may cause a substantial change in amplitude of the overall magnitude! It may also cause the position of a peak to shift by considerably more or less than the shift of the individual path!

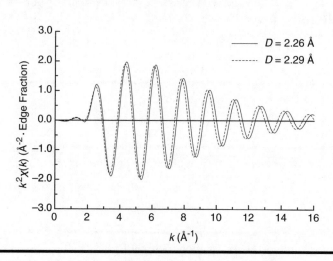

Figure 13.4 *k²-weighted χ(k) dependence on half path length.*

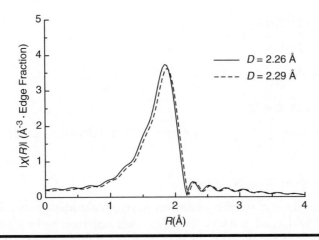

Figure 13.5   **Fourier transform magnitude dependence on half path length.**

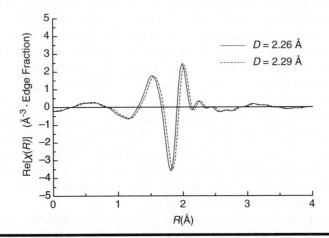

Figure 13.6   **Dependence on half path length of real part of the Fourier transform.**

Other constraints for half path lengths depend on your knowledge about the material and how it behaves. For instance, you might have a ligand that is relatively stiff, but not know how it is oriented. It would then be possible to use geometry to parameterize the distances from the absorbing atom in terms of a small number of free variables that describe the orientation. Another example is a material for which you know the long-range structure from x-ray diffraction. In that case, you might suspect local distortions that need to be fit but constrain them in such a way that the average corresponds to the XRD data.

13.1.1.8 Correlations

Half path length can correlate to a number of other parameters, depending on the particular model you are using. In simple fits, it often shows a substantial correlation to E_o. But the half path length of particular paths may also correlate to just about any fitted parameters, including the MSRD, degeneracies, and the half path lengths of other paths. This is because, as Mr. Handy pointed out earlier in this section, the amplitude of a peak including contributions from more than one path may be affected by a shift in just one of them. Thus, even though D is primarily a phase parameter, in complicated models, it can also correlate strongly with amplitude parameters.

13.1.2 Degeneracy

13.1.2.1 Symbol

Note that if three atoms are involved in a multiple-scattering path, then going $1 \rightarrow 2 \rightarrow 3 \rightarrow 1$ counts separately from going $1 \rightarrow 3 \rightarrow 2 \rightarrow 1$! For that reason, double-scattering paths have even degeneracies.

The symbol for degeneracy is N.

13.1.2.2 Nomenclature

For single-scattering paths, the degeneracy is often referred to as the *coordination number*.

13.1.2.3 Physical interpretation

Theorists refer to that as *time-reversal symmetry*, Mr. Handy.

N is simply the identical number of distinct ways, per absorbing atom, that the scattering defined by the path can take place. For single-scattering paths, it's therefore the number of atoms of a given type at a given distance: if there are six carbon atoms at a distance of 4.7 Å from a uranium absorber, then the degeneracy of that path is 6.

To see how degeneracy works for multiple-scattering paths, imagine a manganese atom tetrahedrally coordinated to four identical oxygen atoms. The shortest possible multiple-scattering path is one that goes from the absorbing atom, to one of the oxygens, on to another oxygen, and then back to the absorber. There are four possible oxygen atoms for the first leg. In each case, there are then three possible oxygen atoms to pick from for the second leg. Thus, there are 4×3 possibilities contributing to the path, giving a degeneracy of 12.

Consider what happens if there is more than one phase present and the scattering path is only associated with one of the phases. For instance, imagine a mix in which 30% of the iron atoms are present as iron phosphide and 70% are present as iron oxide. Each iron atom in iron phosphide has six phosphorous near neighbors, but only 30% of the iron atoms are in the phosphide. Therefore, the degeneracy for the iron–phosphorous near-neighbor path is 30% of 6, or 1.8.

Similarly, nanoscale materials often exhibit reduced degeneracies, because atoms near the edge of a nanoparticle or nanocrystal are 'missing' some of the nearby atoms, lowering the average degeneracy per absorbing atom.

A little thought about your system may yield additional causes for degeneracies of individual paths to be lower than in the equivalent pure bulk material.

13.1.2.4 Typical values

Degeneracies must never be negative.

For single-scattering nearest neighbors, chemical sense limits the number of coordinated atoms. There aren't many oxides with more than eight oxygens coordinated to the metal atoms, for instance, nor do metals in alloys usually have more than twelve nearest neighbors. Degeneracies for multiple-scattering paths, however, can often be quite high. 96, for instance, is the degeneracy of some multiple-scattering paths found in face-centered cubic (fcc) crystals.

Degeneracies can also be quite small, because they are calculated per absorbing atom. A zinc oxide doped with 5% aluminum, for instance, might very well have a degeneracy of less than 0.20 for a particular zinc–aluminum single-scattering path.

13.1.2.5 Effect on χ (k)

The amplitude of the path scales with the degeneracy; increasing the degeneracy of a path from 4 to 6 increases the amplitude of that path by a factor of 6/4 (Figure 13.7).

13.1.2.6 Effect on Fourier transform

Again, the amplitude of the path scales directly with degeneracy (Figures 13.8 and 13.9).

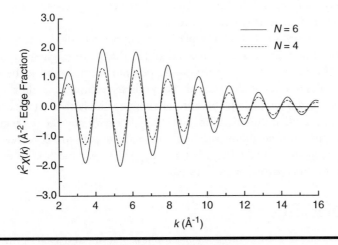

Figure 13.7 **k^2-weighted $\chi(k)$ dependence on degeneracy.**

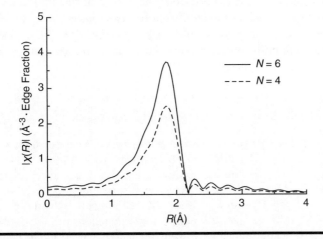

Figure 13.8 **Fourier transform magnitude dependence on degeneracy.**

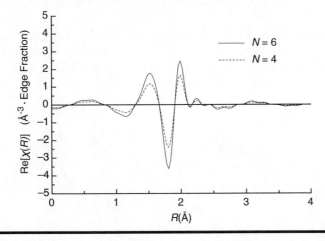

Figure 13.9 **Dependence on degeneracy of real part of the Fourier transform.**

13.1.2.7 Common constraints

Degeneracies should be constrained in a manner that is consistent with your understanding of the material.

If, for instance, you are working with a bulk crystal, the degeneracies are given by the crystal structure and may not need to be fit.

As another example, consider a platinum–copper fcc alloy. The total number of nearest neighbors in an fcc compound is 12. But there might be a question as to whether the metals show clustering—do platinum atoms tend to be near other platinums, perhaps as a consequence of the synthesis technique, or do they tend not to be nearest neighbors, perhaps because of strain in the lattice? In such a case, the degeneracy of the platinum–platinum near-neighbor path and the degeneracy of the platinum–copper near-neighbor path could be made to sum to 12. Although neither degeneracy is known individually, only one independent parameter is needed to fit the degeneracies of these paths, rather than two.

13.1.2.8 Correlations

Degeneracies can correlate strongly to other 'phase' parameters, such as the amplitude reduction factor and the MSRD for the path. The degree to which this is the case depends in part on how highly constrained the fit is. A fit that uses knowledge about the chemical system to reduce the number of free parameters related to degeneracy to a handful, for instance, may show only weak correlations. At the opposite extreme, it is impossible to fit both coordination number and S_0^2 if only a single shell is being fit—they are perfectly correlated!

BOX 13.2   An Example of a Degeneracy Constraint

Friends! Let us consider an example of a copper atom with two nitrate ligands (right structure in Figure 13.10). Let us further suppose the copper atom is the absorber. The degeneracy of the path scattering off of the nearest-neighbor oxygens is two, and the degeneracy of the path scattering off of the more distant nitrogen atoms is also two. The degeneracy of the double-scattering path involving the nearby oxygen and its neighboring nitrogen is four, because for each nitrate, the electron could scatter off of the oxygen first or off of the nitrogen first.

But what if we didn't know the number of nitrate ligands but did know that the ligands were some mix of sulfur and nitrate? The copper might have only one nitrate ligand, for instance, as in the left structure in Figure 13.10. Now we

Figure 13.10   Copper atom with sulfur and nitrate ligands. (Left) One nitrate ligand. (Right) Two nitrate ligands.

can see that the nearest-neighbor oxygen path has a degeneracy of one, the more distant nitrogen path has a degeneracy of one, and the double-scattering path involving both has a degeneracy of two.

No matter how many nitrate ligands there were, the degeneracy for the nitrogen path would always be equal to the degeneracy of the near-neighbor oxygen path, and the double-scattering path would always have a degeneracy twice that.

So rather than three independent free parameters (the degeneracy of the near oxygen, the degeneracy of the more distant nitrogen, and the degeneracy of the double-scattering path), we only need to have one because if we know one of those coordination numbers, then we know the other two.

13.1.3 Mean Square Relative Displacement

13.1.3.1 Symbol

Most commonly, the symbol for mean square relative displacement is σ^2. The symbol C_2 is also occasionally used, to emphasize that the mean square radial displacement is also the second cumulant in the cumulant expansion.

13.1.3.2 Nomenclature

Also known as the EXAFS Debye–Waller factor, the pseudo-Debye–Waller factor, and the second cumulant.

13.1.3.3 Physical interpretation

The MSRD is the variance in the half path length, that is, the square of the standard deviation of the half path length:

$$\sigma^2 = \overline{(r - \bar{r})^2} \qquad (13.1)$$

As described under the entry for half path length, an individual path will have contributions from scattering pairs that are farther apart or closer than the average because of static and/or thermal disorder. The MSRD is therefore a measure of both kinds of disorder.

As a practical matter, σ^2 also serves another purpose: it is used to correct for the difference between the experimentally determined and theoretical $\chi(k)$ functions that were discussed in Section 7.4.2 of Chapter 7. It turns out that the necessary correction is equivalent to an additive term to the MSRD (Rehr et al. 1991). This additive term, called the *McMaster correction*, must be subtracted off if the true MSRD is desired. Either the uncorrected or corrected σ^2 are acceptable to report for publication, as it is straightforward for a reader to convert from one to the other. It is therefore important to make clear when presenting results whether this correction was applied or not.

13.1.3.4 Typical values

Because the MSRD is the square of a standard deviation, it must be positive. Typical values range from about 0.002 to 0.03 Å². If an MSRD adopts a value much larger than that range, the path is so disordered that it is contributing little to the fit—often a sign that the model is wrong in some way.

13.1.3.5 Effect on χ(k)

Increasing the MSRD reduces the amplitude of $\chi(k)$. Because the MSRD enters into the EXAFS equation through the factor $e^{-2k^2\sigma^2}$, the effect is much more pronounced at high k (Figure 13.11). This should make it readily distinguishable from changes in amplitude due to either S_0^2 or the degeneracy.

The term *Debye–Waller factor* is borrowed from x-ray diffraction. But in XRD, it refers to the variance in the position of an atom relative to its mean position in the lattice, whereas in EXAFS, it is the variance in the distance between two atoms. To see how this can make a difference, imagine a pair of near-neighbor atoms. Long wavelength vibrations will tend to move both in the same direction, so that the distance between them doesn't change very much. The EXAFS Debye–Waller factor may therefore be much smaller than the XRD factor. On the other hand, if two atoms are well separated, they may move completely independently. In that case, the EXAFS factor will be roughly twice that of XRD because each atom's motion contributes to the variation in the distance between them in an uncorrelated fashion.

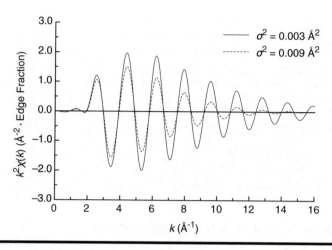

Figure 13.11   *k²*-weighted χ(*k*) dependence on MSRD.

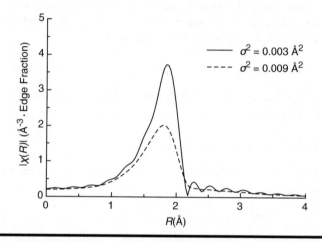

Figure 13.12   **Fourier transform magnitude dependence on MSRD.**

If σ² is much smaller than $(1/k_{max})^2$, where k_{max} is the top limit of the range you are trying to fit, there's a good chance the distribution is roughly Gaussian to within the accuracy of your fit. If it's bigger than that, consider trying a third cumulant or splitting the path into paths at different distances!

Some may feel surprise that the larger MSRD in Figure 13.12 does not result in a much broader peak. The reason is that most of the width of a peak in the Fourier transform of EXAFS data comes from having only a finite range of *k* data to work with; that is, there is limited resolution in *R*-space. If you look at Figure 13.12, the 0.009 Å² curve *is* a bit wider relative to its height than the 0.003 Å² curve, but the effect is modest.

13.1.3.6 Effect on Fourier transform

The primary effect a change in MSRD has on the Fourier transform is to change the amplitude of that path (Figure 13.12).

There's also a slight shift to lower *R* with increasing MSRD. This is because both the $1/D^2$ and the mean free path factors in the EXAFS equation cause high-*R* contributions to be weighted somewhat less heavily than low-*R* contributions. Thus, as the distribution is broadened by increasing the MSRD, the new low-*R* contributions are weighted more heavily than the new high-*R* contributions, and the peak shifts slightly to the left (Figure 13.13).

It is also worth noting that the area under the curve is much smaller, even though the number of scatterers is the same. This isn't surprising, as the functions at the low-*D* end of the distribution are not completely in phase with the functions at the high-*D* end of the distribution. This is another reminder that the Fourier transform is not a radial distribution function, as for a radial distribution function, the area under the curve would be proportional to the number of scatterers!

13.1.3.7 Common constraints

A wide variety of constraint schemes have been employed with MSRD's, some of which will be treated further in Chapter 17. A simple heuristic scheme is to constrain 'similar' paths to have the same MSRD, with similarity defined by length and/or

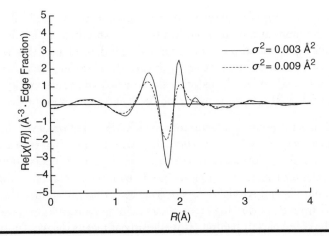

Figure 13.13 **Dependence on MSRD of real part of the Fourier transform.**

species of scattering atom. Other constraint schemes use simple physical models such as those of Einstein or Debye (Sevillano and Rehr 1979). Finally, some use the results of calculations such as those employed by density functional theory.

13.1.3.8 Correlations

Because MSRD primarily affects the amplitude of a path, it can correlate strongly with other amplitude parameters such as coordination number. If the path is particularly prominent in the fit, its MSRD may also correlate significantly with S_o^2. For fits with more than one free parameter affecting half path lengths, there may also be strong correlations with those parameters.

Because MSRD is the only parameter appearing in the EXAFS equation weighted by k^2, it is often possible to reduce correlations with other parameters by using a wider range of k data or by fitting multiple k-weights simultaneously.

13.1.4 Amplitude Reduction Factor

13.1.4.1 Symbol

S_o^2 is the symbol for amplitude reduction factor.

13.1.4.2 Nomenclature

Also known as the many-electron overlap reduction factor or the passive electron reduction factor.

13.1.4.3 Physical interpretation

To understand the physical interpretation of S_o^2, we need to understand a little about its history.

As the first successful theories of EXAFS were developed in the early 1970s, it was recognized that experimental EXAFS spectra showed oscillations with roughly half the amplitude the theories predicted (Kincaid and Eisenberger 1975). This was not a surprise, as the theories were known to include several simplifying assumptions.

Although theorists were willing to shoulder much of the responsibility for the disagreement between theory and experiment, in some cases, experimental effects contributed to the reduced amplitude as well. As we saw in Chapter 6, inhomogeneity and harmonics both act to suppress the amplitude of EXAFS oscillations for transmission measurements, while self-absorption reduces the oscillations for fluorescence mode.

From the beginning of modern EXAFS theories, it was recognized that, all else being equal, the more distant an atom was from the absorber, the less it contributed to the EXAFS signal. In addition to the usual $1/D^2$ dependence, this was correctly attributed both to phenomena that could disrupt the electron's wave function (such as inelastic scattering) and to the fact that the core hole in the absorbing atom would eventually get filled. These effects were modeled by assigning a mean free path to the electron, and thus introducing a factor into the EXAFS equation of the form $e^{-2D/\lambda(k)}$. These are sometimes referred to as *extrinsic losses*, with the idea that they are caused by events subsequent to the formation of the core hole (Rehr and Albers 2000).

That left *intrinsic* losses: those that had to do with the formation of the core hole, and thus did not depend strongly on path length. For example, once the photoelectron is ejected from the absorbing atom, the remaining *passive* electrons adjust to the presence of the core hole left behind. This adjustment means the overlap between the initial and final quantum states is imperfect, and this suppresses the EXAFS oscillations somewhat. Other intrinsic losses are due to *shake-up* processes, in which outer electrons are excited when the core hole is formed, and *shake-off* processes, in which outer electrons are ejected from the atom. Most x-ray absorption events do not include shake-up or shake-off processes, so in most cases the energy of the photoelectron is uniquely correlated to the energy of the incoming photon, a fact we take advantage of during data reduction by converting from the energy of the x-ray to the wave number of the photoelectron. But if a shake-up or shake-off event does occur, then that robs the photoelectron of energy, and the k we calculate is wrong. Because there are many different shake-up and shake-off processes that can occur, the net effect is that the amplitude of the EXAFS oscillation is smaller than what we would measure in the absence of these processes (Rehr et al. 1978).

Theoretical investigations of intrinsic processes are ongoing. Although it is beyond the scope of this text to go through the arguments, it turns out that intrinsic processes result in a reduction of amplitude that does not vary much over the ranges of k and R used in EXAFS analysis. This reduction in amplitude is generally on the order of 10% (Campbell et al. 2002; Kas et al. 2007; Li et al. 1995; Rehr et al. 1991; Rehr and Ankudinov 2005).

Examining the EXAFS equation, you can see that S_o^2 represents a reduction in amplitude that is independent of k and R. A good physical interpretation, therefore, is that it is an approximation of the effect of intrinsic losses on the EXAFS spectrum.

The distinction between intrinsic losses due to imperfect overlap between initial and final states in the absence of excitation, and losses due to shake-up and shake-off processes, is somewhat artificial. The fact is that when the photoelectron is removed, the new orbitals of the atom do not correspond exactly to the old orbitals. According to the rules of quantum mechanics, the passive electrons will occupy the new orbitals with a probability related to the degree of overlap with the original orbitals. If all of the passive electrons end up in the new orbitals that have the most overlap with the originals, we tend to think of the atom as having 'relaxed.' If any end up in orbitals with less overlap, we talk about the atom having been 'excited.' Aside from the fact that in the first case, the energy of the passive electrons is modestly reduced, whereas in the second case, at least one of the passive electrons has its energy increased, the physics of the processes is the same.

BOX 13.3 IS S_o^2 Independent of k and R?

Some leading theorists argue that intrinsic effects are weakly dependent on both k and R (Campbell et al. 2002; Li et al. 1995; Rehr et al. 1978; Rehr and Ankudinov 2005). To see why, imagine a shake-up process that robs the photoelectron of 10 eV. At low k and high R, 10 eV is a large fraction of an EXAFS oscillation and can result in a contribution that is significantly out of phase with the spectrum. But at high k or small R, the effect is barely noticeable.

Sure, Carvaka, but for most EXAFS measurements, any k or R dependence is likely to be less than a 5% difference between paths, or between low k and high k. So almost all EXAFS analyses based on modeling ignore this issue.

Alternatively, one can treat S_o^2 as a phenomenological parameter that accounts for any amplitude suppression independent of k and R, regardless of physical cause (Krappe and Rossner 2004). Under this view, S_o^2 does not have any particular physical meaning, and the k or R dependence of intrinsic losses can be assigned to other parameters.

In practice, determinations of S_o^2 are very sensitive to measurement and data reduction issues such as harmonics, self-absorption, sample inhomogeneity, detector nonlinearity, and incorrect normalization. Thus, S_o^2 can act as a 'canary in a coal mine,' with low values signaling that the spectrum may also be distorted in other ways.

13.1.4.4 Typical values

Both theoretical and experimental studies suggest S_o^2 should be 'a bit less than 1'; 0.70 to 1.05 is a reasonable range (Li et al. 1995). Values outside of that range suggest the data are distorted, the model is imperfect, or the fit is questionable. If the values are slightly outside the range, 'phase' parameters such as bond lengths may still be quite accurate, but 'amplitude' parameters such as MSRD and coordination number should be treated with considerable caution.

13.1.4.5 Effect on $\chi(k)$

Under the usual assumption that S_o^2 is independent of k and R, its effect on $\chi(k)$ is simply to change the amplitude uniformly (Figure 13.14).

13.1.4.6 Effect on Fourier transform

Once again, if we assume that S_o^2 is independent of k and R, its effect on the Fourier transform is to change the amplitude uniformly (Figures 13.15 and 13.16).

13.1.4.7 Common constraints

Because S_o^2 is due to intrinsic effects, it should be the same for every scattering path associated with a given absorbing atom, at least for a given value of k and R.

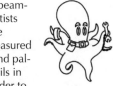

In 2009, beamline scientists across the world measured copper and palladium foils in part in order to determine how much determinations of S_o^2 were dependent on 'beamline specific parameters' (Kelly et al. 2009). The values found for copper foil on 11 different beamlines ranged from 0.91 ± 0.02 to 1.04 ± 0.04. For palladium foil, the range on 6 different beamlines was 0.83 ± 0.04 to 0.85 ± 0.04. The variation from beamline to beamline of roughly 10% for the copper foil gives a sense of the limits for how well S_o^2 can be determined experimentally, even by experts measuring near-ideal samples with simple, known structures.

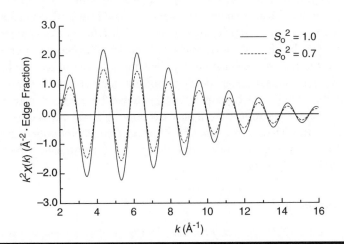

Figure 13.14 k^2-weighted $\chi(k)$ dependence on S_o^2.

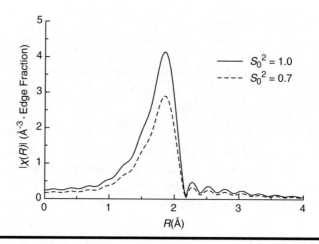

Figure 13.15 **Fourier transform magnitude dependence on S_o^2.**

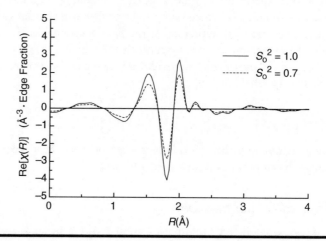

Figure 13.16 **Dependence on S_o^2 of real part of the Fourier transform.**

But what if the same element is present in different environments? For instance, you might be comparing silver metal and silver oxide. Should both be given the same value of S_o^2? This is the question of *chemical transferability* and is a matter of ongoing investigation. Given that most materials are expected to exhibit values of S_o^2 in the range 0.8 to 1.0, the variation between environments of a given element in a given measurement should be *at most* 0.2. Measurement, data reduction, and fitting effects make it difficult to determine to an accuracy much better than that. Therefore, we recommend initially considering S_o^2 to be chemically transferable between different environments of the same element. But it's also reasonable to try allowing each environment to have its own S_o^2, and see if that results in a statistically significant improvement to the fit.

On the other hand, much greater caution should be exercised before assuming that different *edges* of a sample should have the same S_o^2, even when those edges refer to the same atoms, such as in the case of a K and an L edge. Not only is the electronic structure of the resulting core hole very different, but measurement effects may also vary considerably with energy.

13.1.4.8 Correlations

For a single-shell fit, S_o^2 is perfectly correlated with the degeneracy N; it is impossible to determine one without knowing the other. It is also highly correlated with the MSRDs of the most prominent paths.

13.1.5 E₀

13.1.5.1 Symbol

There are no common alternatives to the symbol E_o.

13.1.5.2 Nomenclature

Also known as reference energy, edge energy, and inner potential.

13.1.5.3 Physical interpretation

Formally, E_o is the energy used to calculate the momentum of the photoelectron:

$$k_{electron} = \sqrt{\frac{2m_{electron}}{\hbar^2}\left(E_{photon} - E_o\right)} \qquad (10.2)$$

But what does it correspond to physically?

It is generally best for us to start out fits by treating S_o^2 as independent of k, R, and chemical environment but to allow it to be different for different edges. In most cases, that will be good enough even for our final fits, but it's good to be aware that we are making some simplifying assumptions when we do that.

BOX 13.4 What Does E_o Correspond to Physically?

That's easy. Below E_o, the core electron doesn't absorb any photons. E_o is just the edge energy.

But, my dear Simplicio, that won't work. Edges do not start abruptly. In Section 7.3.3 of Chapter 7, we discussed some arbitrary choices that can be made for E_o at the data reduction stage, promising that we would leave it as a fitting parameter later.

In fact, there may not even be a unique E_o. If channels including shake-up and shake-off events do contribute significantly to the spectrum, then some photoelectrons have different energy origins than others. In addition, theoretical standards are usually based on a muffin-tin potential (Section 12.2.1 of Chapter 12), which is certainly a simplification of the actual interatomic potential. If we've got the potential the electron is traveling through a little bit wrong, it means we also have E_o a little bit wrong.

E_o also doesn't have to be precisely the same as any of the usual energy levels tabulated for a material; it's not quite the same thing as the Fermi energy, or the ionization energy, or the energy of the conduction band.

But the bottom line is that it WORKS! In the EXAFS region, the spectrum behaves as if there are electrons with a well-defined momentum, meaning E_o is well defined too.

It is not easy to distinguish the effect of a shift in E_o from a shift in D when one looks at a *single* Fourier transform plot. But if the shift gets smaller when the k-weight is increased or when the Fourier transform is started at a greater value of k, then the shift is probably due to E_o.

There are few mistakes more common in published EXAFS work, and more unambiguously wrong, than publishing fits where every path has a unique E_o with values varying by tens of electron volts. There is no physical justification for large E_o differences between paths! Take it from Mandelbrant, my friends: we should start all our fits with E_o for all paths constrained to be the same. Once we have a fit that seems reasonable, we can try introducing a second E_o, particularly if there are strongly ionized species present, and see whether the fit improves.

Hey, the spectrum is only sensitive to small differences in E_o at low k, right? So although it might make sense that you'd need an extra E_o to successfully extend a fit an extra inverse angstrom or two lower in k-space, is it worth it? That depends on the scientific question you're trying to answer, your data quality, and your personal style.

Although it's difficult to ascribe a rigorous physical meaning to E_o, *differences* in E_o, even when smaller than 1 eV, can still be meaningful. Oxidized materials, for instance, typically have values of E_o an electron volt or two higher than the corresponding element because the electrons are held more tightly.

13.1.5.4 Typical values

E_o should always be on or near the rising portion of the edge. A little past the white line might be acceptable, but E_o should never be in the smooth preedge region or the EXAFS region of the spectrum.

13.1.5.5 Effect on χ(k)

Shifting E_o results in a shift of $\chi(k)$ in the same direction that is most evident at low k (Figure 13.17). Notice how this makes it easy to distinguish this shift from one caused by a change in D.

13.1.5.6 Effect on Fourier transform

The effect on the Fourier transform looks superficially similar to the effect of a shift in D (Figures 13.18 and 13.19).

13.1.5.7 Common constraints

One common constraint is to have only one E_o for each edge. Another would be to have one E_o for each oxidation state of the absorbing element that is present.

A more contentious question is whether using different values of E_o for different scattering paths from the same absorbing atom is ever justified. There is some evidence that theoretical standards may not always account properly for charge transfer, and that allowing for a slightly different E_o for each type of scattering atom can compensate for that (Haskel et al. 1995), as well as for some other types of errors present in theoretical standards.

13.1.5.8 Correlations

E_o correlates strongly with the D of the most prominent paths. This correlation can be broken somewhat by using a wide k range, or a combination of different k-weights.

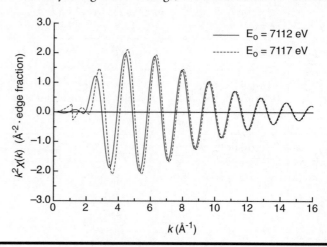

Figure 13.17 k^2-weighted χ(k) dependence on E_o.

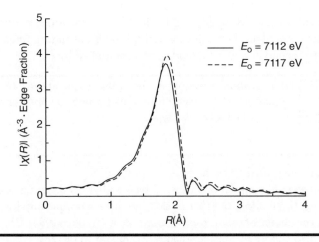

Figure 13.18 Fourier transform magnitude dependence on E_0.

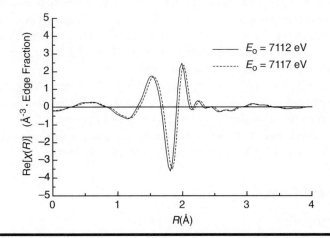

Figure 13.19 Dependence on E_0 of real part of the Fourier transform.

13.2 Less Common Fitting Parameters

These parameters are less commonly fit; in fact, not all fitting software allows all of them. But for some systems, they can still be important.

13.2.1 Cumulants

We've already alluded a few times to the fact that an EXAFS spectrum samples a large number of absorbing atoms (typically millions), and the distribution of scattering atoms around each absorber is not identical, both because of static and thermal effects. The *cumulant expansion* is a method for approximately characterizing half path length distributions that are reasonably narrow, such as when the absorber–scatterer distance in a solid varies primarily because of thermal vibration, or when a single path is used to model a modestly distorted near-neighbor environment. The effect of those kind of distributions on $\chi(k)$ can be described by introducing factors of the form $e^{\frac{(2ik)^n}{n!}C_n}$ into the EXAFS equation, where C_n is the nth *cumulant* (Bunker 1983). Inspection of the exponential suggests that odd cumulants should affect the phase of $\chi(k)$ and even cumulants its amplitude. In addition, each cumulant appears weighted by k^n. It turns out that for reasonably narrow distributions, the series converges pretty rapidly, so that $\chi(k)$ is well approximated by an expression including just the first few cumulants.

Recall that earlier we noted that the decrease in EXAFS amplitude with increasing D causes σ^2 to have a modest effect on peak position in the Fourier transform. For the same reason, odd cumulants do have some effect on amplitude, and even cumulants have some effect on phase (Kas et al. 2007). So perhaps it is better to say that odd cumulants *primarily* affect phase and even cumulants *primarily* affect amplitude.

Are these cumulants already lurking in the EXAFS equation we've presented? If they are, the first cumulant should appear multiplied by k as the argument of an exponential, and the second cumulant should appear multiplied by k^2.

 Friends! Before reading further, please look at the EXAFS equation on the inside front cover now and consider if we can see any factors of this type.

You've hopefully identified σ^2 as the second cumulant, C_2.

It turns out that all cumulants with $n \geq 2$ have a physical meaning associated with the shape of the distribution, but that C_1 does not have a unique physical meaning as it depends on the choice of origin used to compute it (Bunker 1983). Therefore, some people in the field choose to consider the first cumulant to be zero.

Others, however, note that there is already a factor in the EXAFS equation that affects the phase of $\chi(k)$ and appears weighted by k, just as the first cumulant would: namely, the D that appears in the factor $\sin(2kD + \delta(k))$. They therefore choose C_1 to be equal to the average value of D (Fornasini et al. 2001). Finally, some choose the first cumulant to represent the average difference from some reference length, such as the classical length of the bond at 0 K (Wenzel et al. 1990).

For a Gaussian distribution, all cumulants with $n > 2$ are zero, yielding the EXAFS equation we printed on the inside front cover. But if the distribution deviates from a pure Gaussian distribution, higher cumulants may be important.

Because we've already met the second cumulant, and the definition of the first is arbitrary, let's take a closer look at a few of the higher cumulants.

13.2.2 Third Cumulant

13.2.2.1 Symbol

Symbols for the third cumulant include C_3 and $\sigma^{(3)}$.

13.2.2.2 Physical interpretation

The third cumulant turns out to be the mean cubed deviation of the distribution from its mean value.

$$C_3 = \overline{\left(r - \bar{r}\right)^3} \tag{13.3}$$

Note that for a symmetric distribution it is zero. The third cumulant is therefore a measure of the asymmetry of the distribution: positive values mean it has a long tail toward high D, whereas negative values mean it has a long tail toward low D.

13.2.2.3 Typical values

The third cumulant may be negative, positive, or zero, although for near neighbors, it's frequently positive. The magnitude is usually less than 0.001 Å³. Unusually large magnitudes provide a warning that the distribution is broad enough that the cumulant expansion may not converge well; consider other methods of modeling the distribution, such as using more than one path.

13.2.2.4 Effect on $\chi(k)$

A change in the third cumulant results in a shift in the same direction in $\chi(k)$. Because the third cumulant appears weighted by k^3 in the EXAFS equation, the effect is much more evident at high k (Figure 13.20).

 The many different definitions in use for the first cumulant can make reading the literature on the topic confusing! In any paper discussing the first cumulant, be sure to figure out how they're choosing to define it.

 An ordinary chemical bond is usually a bit easier to stretch than to compress. To see why, notice that it only takes a few electron volts to break a bond; that is, to stretch it infinitely far. But compressing a bond very far becomes much more difficult, because you are trying to place two atoms on top of each other. Thus, bonded atoms will often exhibit a modestly positive third cumulant. Although the bond may behave roughly symmetrically near equilibrium (perhaps allowing you to approximate the third cumulant as zero), once the excursions get further from equilibrium (e.g., at higher temperatures), the asymmetry will become more pronounced and the third cumulant will rise.

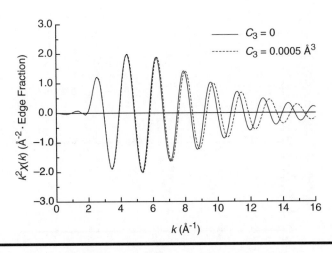

Figure 13.20 k^2-weighted $\chi(k)$ dependence on the third cumulant.

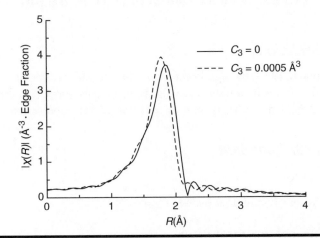

Figure 13.21 Fourier transform magnitude dependence on the third cumulant.

13.2.2.5 Effect on Fourier transform

Changing the third cumulant will shift the peak in the Fourier transform in the opposite direction (e.g., making the third cumulant more positive will shift the Fourier transform peak lower), as Figure 13.21 demonstrates.

The change in the real part of the Fourier transform is less easily described than is the case for changes in D or E_o, as the shift is most visible for the sharpest features (Figure 13.22).

13.2.2.6 Common constraints

The most common constraint on the third cumulant is simply to assume it is zero. This constraint is most questionable when comparing spectra exhibiting different amounts of disorder. For example, if the spectra being compared were taken at different temperatures or if the samples had different doping fractions or different dimensions on the nanoscale, then caution is warranted.

In most materials, the nearest-neighbor third cumulant is larger than the third cumulant for more distant single-scattering paths. Thus, it may be reasonable to constrain the third cumulant for more distant paths to zero even when it is allowed to vary for nearest neighbors.

Assuming that the third cumulant is zero when measuring a temperature series sometimes leads to results that incorrectly show the nearest-neighbor bond distance *decreasing* with increasing temperature! Without the third cumulant as a free parameter, the fit is trying to account for the Fourier peak shifting lower in the best way allowed to it: by decreasing D. But because of the difference in the k-dependence of D and C_3, the fit will likely be able to correctly assign the shift to the change in the third cumulant if you let it.

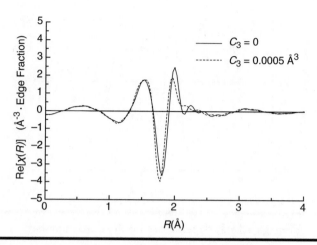

Figure 13.22   Dependence on the third cumulant of the real part of the Fourier transform.

13.2.2.7 Correlations

The third cumulant can show significant correlations to D for paths on which it is fit, particularly at low k-weight or if the data range being fit in k is short. A multiple k-weight fit (Section 15.2.2 of Chapter 15) is often successful at reducing troublesome correlations.

13.2.3 Fourth Cumulant

13.2.3.1 Symbol

Symbols for the fourth cumulant include C_4 and $\sigma^{(4)}$.

13.2.3.2 Physical interpretation

The fourth cumulant is given by

$$C_4 = \left(r - \bar{r}\right)^4 - 3C_2^2 \tag{10.4}$$

 This odd-looking definition is used so that it yields $C_4 = 0$ for a Gaussian distribution. Thus, it is expected to yield only a small correction for distributions that are nearly Gaussian.

It is primarily related to the size of the tails of the distribution; a positive value for the fourth cumulant implies that the distribution falls off less quickly at both low and high D than would a Gaussian distribution with the same value of σ^2.

13.2.3.3 Typical values

The fourth cumulant may be negative, zero, or positive. The magnitude is typically less than 0.0001 Å^4.

BOX 13.5   Convergence of the Cumulant Series

 Wait—the fourth cumulant will enter the EXAFS equation through a factor that looks something like $e^{\frac{2^4 k^4}{4!} C_4}$, right? But if C_4 is positive, then that factor will blow up at high k, and $\chi(k)$ will go to infinity. That doesn't seem right!

But it is right, my dear Simplicio. It simply reminds us that one needs even more terms in the cumulant expansion to go out to an arbitrarily high k.

Almost nobody includes terms beyond the fourth cumulant anyway. So Simplicio's observation provides a rough way of testing for convergence. If the fourth cumulant is overpowering the second cumulant (σ^2) *inside the fitting range*, then the cumulant expansion isn't converging very well and it would be better to consider some other approach, like splitting the path up. In other words, we want

$$\frac{2^2 k_{max}^2}{2!}\sigma^2 > \frac{2^4 k_{max}^4}{4!}C_4 \tag{10.5}$$

Rearranging gives us:

$$C_4 < \frac{3\sigma^2}{k_{max}^2} \tag{10.6}$$

13.2.3.4 Effect on $\chi(k)$

A positive fourth cumulant increases the amplitude at high values of k (Figure 13.23); a negative fourth cumulant decreases it.

13.2.3.5 Effect on Fourier transform

A positive fourth cumulant acts to increase the magnitude of the Fourier transform peak for the path (Figure 13.24), whereas a negative fourth cumulant would reduce its amplitude.

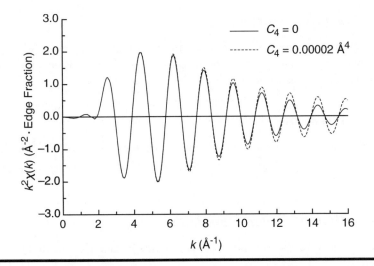

Figure 13.23 k^2-weighted $\chi(k)$ dependence on the fourth cumulant.

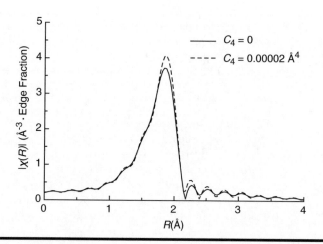

Figure 13.24   Fourier transform magnitude dependence on the fourth cumulant.

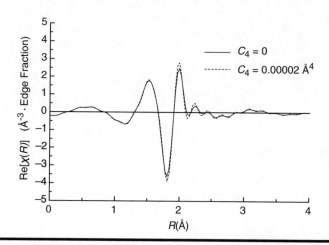

Figure 13.25   Dependence on the fourth cumulant of the real part of the Fourier transform.

As with the third cumulant, examination of the real part of the Fourier transform (Figure 13.25) shows that sharp features are affected more strongly.

13.2.3.6 Common constraints

It is very common to constrain the fourth cumulant to zero. You are only likely to use this parameter if you want to carefully compare the MSRD as a function of temperature or some other measure of disorder, such as nanoparticle size.

13.2.3.7 Correlations

The fourth cumulant can show significant correlations to the MSRD for the same path. Fitting multiple k-weights or extending the k range being fit can help to reduce the correlation.

13.2.4 Fifth and Higher Cumulants

Some fitting software allows you to vary cumulants beyond the fourth, but it is very rarely done.

13.2.5 Mean Free Path

13.2.5.1 Symbol

The symbol for mean free path is $\lambda(k)$.

13.2.5.2 Physical interpretation

As discussed earlier in Section 13.1.4, extrinsic losses in the EXAFS process are phenomenologically modeled by introducing a k-dependent *mean free path* for the photoelectron.

13.2.5.3 Typical values

The solid line in Figure 13.26 shows the mean free path as a function of k for the example path we have been using in this chapter and is fairly typical. (We'll get to the dashed line in a moment.)

13.2.5.4 Effect on χ (k)

Different fitting software will allow the mean free path to be altered in different ways, so we won't be too specific about parameters here. Instead, we'll use the mean free path shown by the dashed line in the previous graph as a concrete example. Note that this represents a considerably shorter mean free path than the default. Although the effect is particularly pronounced for $k < 2$ Å^{-1}, that region is not generally used for EXAFS analysis.

The effect of shortening the mean free path is to reduce the amplitude of $\chi(k)$. Even after shortening the mean free path significantly, at high k the mean free path is much longer than the path length, so almost no reduction in amplitude is visible in that part of the spectrum (Figure 13.27).

13.2.5.5 Effect on Fourier transform

Shortening the mean free path reduces the amplitude of the equivalent peak in the Fourier transform (Figures 13.28 and 13.29). If the mean free path is reduced for all scattering paths, then peaks at higher R will be reduced more than paths at lower R.

13.2.5.6 Common constraints

Most scientists do not attempt to fit adjustments to the mean free path, preferring to use empirical (Seah and Dench 1979) or theoretical (Penn 1987) models. In practice, these models are built into fitting packages and are largely invisible to the user.

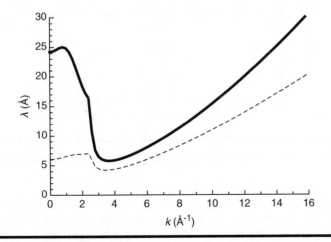

Figure 13.26 **Two possibilities for mean free path as a function of *k*.**

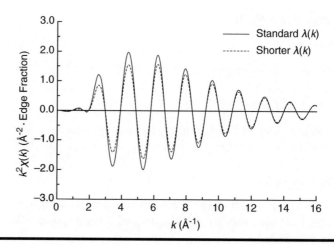

Figure 13.27 k^2-weighted $\chi(k)$ dependence on mean free path.

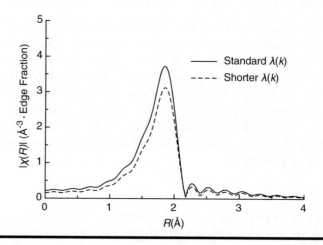

Figure 13.28 Fourier transform magnitude dependence on mean free path.

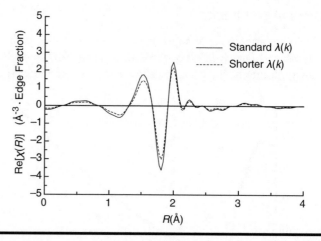

Figure 13.29 Dependence on the mean free path of the real part of the Fourier transform.

The extent to which mean free path models are transferable is, however, questionable, perhaps particularly in the case of nanoparticles (Zhao and Montano 1989). Experimenting with varying the mean free path can at least provide an estimate of how much systematic error the assumption of transferability is introducing into your analysis.

13.2.5.7 Correlations

As an amplitude variable, mean free path changes correlate strongly with the amplitude reduction factor and coordination numbers. The k dependence is different, however, as the mean free path tends to have the most pronounced effect at low to moderate k. Correlations can therefore be reduced by extending the fit to higher k and/or by fitting multiple k-weights simultaneously (Section 15.2.2 of Chapter 15).

13.3 Scattering Parameters

An examination of the EXAFS equation reveals the following two parameters we have not yet discussed in this chapter: $f(k)$ and $\delta(k)$. You will see the EXAFS equation written in a variety of ways in the literature and in tutorials, depending on the features the presenter wants to emphasize. For example, higher cumulants are usually left off. What we have written as $f(k)$ and $\delta(k)$ show a particularly large amount of variation as to notation because they represent parameters that are not varied during a fit but are instead drawn from either theory or experiment. In that sense, they are of less interest to someone interested in practical applications of EXAFS. But it is still wise to know something about them.

To understand what $f(k)$ and $\delta(k)$ represent, consider the journey of a photoelectron leaving the central atom, scattering off of a neighboring atom, and then returning (Rehr et al. 1991). We'll think about this semi-classically, so that we get an idea of roughly what happens. As the electron departs the absorbing atom (often called the *central atom* in this context), it begins its journey near the bottom of the potential valley created by the atom (Figure 13.30). Conservation of energy tells us that it's therefore traveling faster than it will be once it makes it all the way out and so has a higher wave number k than it will once it's out. Because we don't account for that when we write the interference term $2kD$, this introduces an additional phase shift—the electron's wave function oscillates a bit more than it would if it spent the whole time

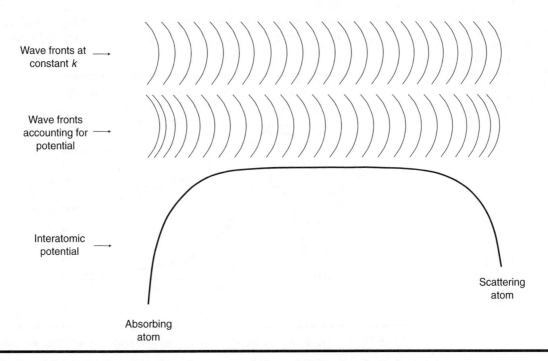

Figure 13.30   Schematic representation of a source of phase shift for the photoelectron.

The notation used in this book, in which $f(k)$ represents the amplitude factor from scattering and $\delta(k)$ represents the phase factor, is one convention in use in the literature. Another convention is to have $f(k)$ represent the effect of the scattering atom, and $\delta(k)$ represent the effect of the absorbing atom. Each must then be complex, so that it can affect both phase and amplitude. Under the latter convention, the real part of $f(k)$ affects amplitude and the imaginary part phase, whereas for $\delta(k)$, the real part affects phase and the imaginary part amplitude—a state of affairs that admittedly could cause confusion. There is no good reason for the strange convention; it is just one of those historical accidents that happens from time to time.

Don't worry too much about all the different ways the EXAFS equation gets written. It is enough to know that there is a phase shift and an amplitude effect due to both scatterer and absorber—who cares what symbol a particular person chooses to use, as long as they tell you what they're doing?

traveling at the speed corresponding to the k we calculate from E_o. This phase shift is part of what we've written as $\delta(k)$.

Sooner or later, the electron encounters the scattering atom. As it descends into that potential, it speeds up again, adding an additional shift to $\delta(k)$. In addition, there can be a phase shift of π (or $-\pi$, if you prefer … the physical result is the same) associated with reversing direction. Once again, we lump that in as part of $\delta(k)$.

Of course, the electron does not necessarily scatter at all—it may go sailing on by. (As a wave, we can think of it as 'partially' scattering, which is perhaps a better description, given the interference that subsequently occurs.) So we need to include a probability for scattering. That probability is part of what we've written as $f(k)$.

After scattering, the electron once again descends into the central atom's potential, causing an additional phase shift identical to what it experienced on the way out. Sometimes you'll see these various phase shifts written separately in the EXAFS equation, instead of lumping them all together the way we did.

Finally, it should be noted that the central atom's potential 'pulls in' the scattered electron a bit as it returns, somewhat as the Sun pulls in a comet. This increases the probability of a successful round trip. On the other hand, it's possible the returning electron will scatter off of the central atom, spoiling its chances to complete the round trip. These two factors affect the probability of the particular round trip path being complete, and we thus incorporate them into $f(k)$.

In olden days, scientists used to look at the spectra of known materials similar to a sample of interest to extract $f(k)$ and $\delta(k)$, and some still do. But nowadays most people rely on software such as FEFF to calculate $f(k)$ and $\delta(k)$.

Below we'll provide a few examples calculated by FEFF, so that you can get a sense of what the functions look like.

(The phase variables can, of course, have integer multiples of 2π added to them without changing the physical meaning. For the purpose of these graphs, this ambiguity was resolved by requiring $0 < phase < 2\pi$ at $k = 16$ Å^{-1}.)

To start with, let's look at a calculation for a typical path with an iron absorbing atom and an oxygen scatterer (Figure 13.31), and an iron absorber with an iron scatterer (Figure 13.32).

Next, an iron absorber and a bismuth scatterer (Figure 13.33), and silver as both the absorber and scatterer (note the change of axis scale on the left in Figure 13.34).

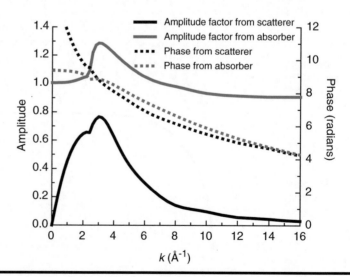

Figure 13.31 **Scattering parameters for an oxygen scatterer with an iron absorber.**

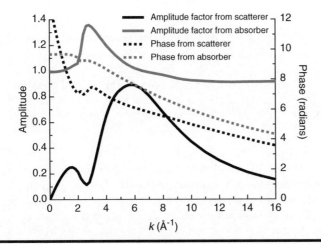

Figure 13.32   Scattering parameters for an iron scatterer with an iron absorber.

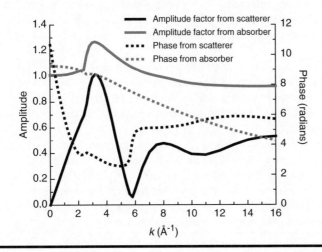

Figure 13.33   Scattering parameters for a bismuth scatterer with an iron absorber.

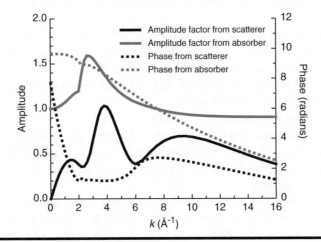

Figure 13.34   Scattering parameters for a silver scatterer with a silver absorber.

These kinds of graphs allow for some nice fingerprinting tricks. Suppose you've got an iron oxide, and you're not sure if a peak in the Fourier transform is due mainly to scattering off of oxygen atoms, iron atoms, or a combination. If you're plotting the Fourier transform from, say, 3 to 10 Å⁻¹, Figures 13.31 and 13.32 tell you that if you switch from k-weight 1 to k-weight 3, the iron scattering peaks in the Fourier transform should get bigger relative to the oxygen scattering peaks because iron scatters much more strongly at high k than oxygen does. It's only qualitative, but it can help you get a preliminary understanding of your system.

BOX 13.6   Why the Fourier Transform of Paths Peak Below Their Half Path Length

Friends! We can now see why the Fourier transform of a path peaks at a lower value than the D of that path, although it takes a few steps to reach our destination:

In the semiclassical explanation for the phase shift highlighted in Figure 13.30, we attributed much of the phase shift to the change in kinetic energy of the electron as it climbs into and out of potential wells caused by atoms. At high k, this has less of an effect, because a given change in E corresponds to a smaller and smaller fraction of a complete oscillation in $\chi(k)$. (This is an effect we've now discussed in several contexts, stemming ultimately from the fact that kinetic energy is proportional to k squared.) Thus, the phase shift should become less positive as k increases. We can confirm by looking at these graphs that the phase shift does usually fall off with k, although for high-Z scatterers, resonances occasionally disrupt that trend.

Because $\delta(k)$ generally falls off with increasing k, each peak in $\chi(k)$ will be less and less advanced in phase; in other words, the peaks will come later and later relative to what would be suggested by the $2kD$ part alone. This mimics the effect of a smaller D, and the peak in the Fourier transform thus shifts to the left.

It's sometimes possible, when a narrow k range near a resonance of a high-Z scatterer is used, to have $\delta(k)$ mostly rising with increasing k. In that case, we can actually have the Fourier transform of a path peak a little higher than the D for the path. If a reasonably broad k range is used, however, the downward trend of the part of $\delta(k)$ attributable to the absorbing atom will prevail, shifting even high-Z scatterers down in R.

The most important things for us to know about $f(k)$ and $\delta(k)$ are

- $f(k)$ and $\delta(k)$ are fairly gentle functions of atomic number. This is the origin of the conventional wisdom that EXAFS cannot easily distinguish scattering species that are near each other on the periodic table.
- For low-Z scatterers, $f(k)$ drops off with k. Oxygen scatterers, for instance, contribute to $\chi(k)$ primarily below 6 Å⁻¹.
- At high enough k, $f(k)$ always drops off.
- As Z increases, peaks in $f(k)$ shift toward higher k. As scatterers, 3d transition elements give peak values of $f(k)$ between 5 and 8 Å⁻¹. $\chi(k)$ for an iron oxide, therefore, will be dominated by oxygen backscatterers at low k, but by iron backscatterers at higher k.
- For scatterers heavier than the 3d metals, $f(k)$ shows multiple peaks in the EXAFS region. This, in turn, can yield split peaks in the magnitude of the Fourier transform even when only a single absorber–scatterer distance is present.

For a further discussion of the trends shown by the scattering parameters, see Teo and Lee (1979). Although theory has advanced considerably since that paper was written, rendering the detailed calculations largely obsolete, the general trends are shown particularly clearly in that paper.

BOX 13.7 We Never Said This Was Easy

I think I'm OK with the phase parameters: D, E_o, and the third cumulant. The effect each has on $\chi(k)$ is different enough that I can sort them out. But the amplitude parameters! There are so many! And except for the degeneracy, the physical interpretation of each is either abstract or muddled. And some of them seem to affect the spectrum in similar ways, or even the exact same way. Help!

Sometimes you might not care all that much about the phase variables. If you can find the D for each path, you can get a lot of information about speciation, which might be what you're after. As you pointed out, except for coordination number, most of the amplitude parameters are pretty abstract.

Except for coordination number? That's a pretty big 'except,' Kitsune! What if that's the thing I want to know?

Start simple, my friend! Heavily constrain all the amplitude parameters except for the coordination numbers you'd like to know, and then gradually change or relax your constraints to see what effects they have on your conclusions. Incorporate what you find into the uncertainty you report in your final paper.

And be sure to measure a standard: a known material with a similar structure. Measure it under conditions as similar to those you measure your sample under as you can conveniently manage. Then fit the standard. By understanding how amplitude variables interact for your standard, you'll learn how to handle them for your sample too.

And although some amplitude parameters, like S_o^2 and N, have the same k dependence, others do not. Fitting multiple k-weights and comparing the effect of different k ranges and weights are both effective tactics for sorting out these parameters. One shouldn't just look at the parameters that come out of a fit; much can often be learned by carefully looking at $\chi(k)$ for different fits and different samples in a series.

Well, I'll try, but I'm not filled with confidence.

All any of us can do is try, Simplicio. Everyone who fits EXAFS data is faced with the same sorts of problems when it comes to these parameters. And yet we've managed to successfully discover many, many nifty things about the materials we've studied. Chapters 14–17 will hopefully give you some idea as to how this is done. But maybe the case studies in Chapter 19 will be the most helpful: there's nothing like seeing how experts have dealt with these issues in their publications.

WHAT I'VE LEARNED IN CHAPTER 13, BY SIMPLICIO

- Parameters can be divided into those that primarily affect the phase of $\chi(k)$ and those that primarily affect its amplitude.
- *Half path length* shifts $\chi(k)$ more at high k than low, whereas E_o shifts $\chi(k)$ more at low k than high.
- *MSRD* suppresses amplitude, particularly at high k.
- The *amplitude reduction factor* has complicated theoretical underpinnings, but can be thought of as a phenomenological parameter accounting for k- and R-independent amplitude suppression regardless of cause.
- Similarly, E_o is the effective energy origin for photoelectrons in the EXAFS region, but it is difficult to assign a more precise physical meaning to that energy.
- The third cumulant can be used when moderate disorder is present to address asymmetry in the radial distribution function for a path.
- For low-Z scatterers, $f(k)$ drops off with k. Oxygen scatterers, for instance, contribute to $\chi(k)$ primarily below 6 Å^{-1}.
- As Z increases, peaks in $f(k)$ shift toward higher k, eventually resulting in multiple peaks in the EXAFS region for scattering elements beyond the 3d metals.

References

Bunker, G. 1983. Application of the ratio method of EXAFS analysis to disordered systems. *Nucl. Instrum. Methods Phys. Res.* 207:437–444. DOI:10.1016/0167-5087(83)90655-5.

Campbell, L., L. Hedin, J. J. Rehr, and W. Bardyszewsi. 2002. Interference between extrinsic and intrinsic losses in x-ray absorption fine structure. *Phys. Rev. B.* 65:064107(13). DOI:10.1103/PhysRevB.65.064107.

Fornasini, P., F. Monti, and A. Sanson. 2001. On the cumulant analysis of EXAFS in crystalline solids. *J. Synchrotron Radiat.* 8:1214–1220. DOI:10.1107/S0909049501014923.

Haskel, D., B. Ravel, M. Newville, and E. A. Stern. 1995. Single and multiple scattering XAFS in BaZrO$_3$: A comparison between theory and experiment. *Physica B Condens. Matter.* 208 & 209:151–153. DOI:10.1016/0921-4526(94)00654-E.

Kas, J. J., A. P. Sorin, M. P. Prange, L. W. Cambell, J. A. Soininen, and J. J. Rehr. 2007. Manypole model of inelastic losses in x-ray absorption spectra. *Phys. Rev. B.* 76:195116(10). DOI:10.1103/PhysRevB.76.195116.

Kelly, S. D., S. R. Bare, N. Greenlay, G. Azevedo, M. Balasubramanian, D. Barton, S. Chattopadhyay, S. Fakra, B. Johannessen, M. Newville, J. Pena, G. S Pokrovski, O. Proux, K. Priolkar, B. Ravel, and S. M. Webb. 2009. Comparison of EXAFS foil spectra from around the world. *J. Phys. Conf. Ser.* 190:012032. DOI:10.1088/1742-6596/190/1/012032.

Kincaid, B. M. and P. Eisenberger. 1975. Synchrotron radiation studies of the *K*-edge photoabsorption spectra of Kr, Br$_2$, and GeCl$_4$: A comparison of theory and experiment. *Phys. Rev. Lett.* 22:1361–1364. DOI:10.1103/PhysRevLett.34.1361.

Krappe, H. J. and H. H. Rossner. 2004. Bayesian approach to background subtraction for data from the extended x-ray absorption fine structure. *Phys. Rev. B.* 70:104102(7). DOI:10.1103/PhysRevB.70.104102.

Li, G. G., F. Bridges, and C. H. Booth. 1995. X-ray absorption fine-structure standards: A comparison of experiment and theory. *Phys. Rev. B.* 52:6332–6348. DOI:10.1103/PhysRevB.52.6332.

Penn, D. R. 1987. Electron mean-free-path calculations using a model dielectric function. *Phys. Rev. B.* 35:482–486. DOI:10.1103/PhysRevB.35.482.

Rehr, J. J., E. A. Stern, R. L. Martin, and E. R. Davidson. 1978. Extended x-ray-absorption fine-structure amplitudes—Wave-function relaxation and chemical effects. *Phys. Rev. B.* 17:560–565. DOI:10.1103/PhysRevB.17.560.

Rehr, J. J., J. Mustre de Leon, S. I. Zabinsky, and R. C. Albers. 1991. Theoretical x-ray absorption fine structure standards. *J. Am. Chem. Soc.* 113:5135–5140. DOI:10.1021/ja00014a001.

Rehr, J. J. and R. C. Albers. 2000. Theoretical approaches to x-ray absorption fine structure. *Rev. Mod. Phys.* 72:621–654. DOI:10.1103/RevModPhys.72.621.

Rehr, J. J. and A. L. Ankudinov. 2005. Progress in the theory and interpretation of XANES. *Coord. Chem. Rev.* 249:131–140. DOI:10.1016/j.ccr.2004.02.014.

Seah, M. P. and W. A. Dench. 1979. Quantitative electron spectroscopy of surfaces: A standard data base for electron inelastic mean free paths in solids. *Surf. Interface Anal.* 1:2–11. DOI:10.1002/sia.740010103.

Sevillano, E., H. Meuth, and J. J. Rehr. 1979. Extended x-ray absorption fine-structure Debye-Waller factors. I. Monatomic crystals. *Phys. Rev. B.* 20:4908–4911. DOI:10.1103/PhysRevB.20.4908.

Teo, B.-K. and P. A. Lee. 1979. Ab initio calculations of amplitude and phase functions for extended x-ray absorption fine structure spectroscopy. *J. Am. Chem. Soc.* 101:2815–1832. DOI:10.1021/ja00505a003.

Wenzel, L., D. Arvantis, H. Rabus, T. Lederer, and K. Baberschke. 1990. Enhanced anharmonicity in the interaction of low-Z adsorbates with metal surfaces. *Phys. Rev. Lett.* 64:1765–1768. DOI:10.1103/PhysRevLett.64.1765.

Zhao, J. and P. A. Montano. 1989. Effect of the electron mean free path in small particles on the extended x-ray-absorption fine-structure determination of coordination numbers. *Phys. Rev. B.* 40:3401–3404. DOI:10.1103/PhysRevB.40.3401.

Chapter 14

Identifying a Good Fit

Are We There Yet?

I've tried some fitting on my data, and I think I'm getting somewhere. But I'm not sure how to tell when I'm done.	
When we are fitting, friend Simplicio, we are never done. It is always possible to work a little longer and to make the fit a little better.	
Oh, c'mon, Mandelbrant! Sure, you can always make a fit a little better, but once you hand it to your boss or send it to a publisher, it's done.	
But how do we know when to send it to a publisher, Dysnomia? And please don't tell me it's when your boss tells you to. Some of us do research on our own schedule.	
A fit always has a point of diminishing returns. Maybe the first 10 hours of work makes your fit a lot better, and the next 10 quite a bit better than that, but the next 10 only gets a little bit better. At some point it just doesn't seem worth it to keep going.	

DOI: 10.1201/9780429329555-18

OK. But it must also be important to know why we are using EXAFS in the first place. There is some question we are trying to answer. Once it is answered to our satisfaction, that is also an indication we are done.	
Allow me to point out, if I may, that this conversation assumes that one can properly identify which of two fits is 'better,' and whether a fit has provided a 'satisfactory' answer to a question. Those issues form the gist of Simplicio's question, I believe.	
And those, friends, are the issues this chapter shall explore.	

In this chapter, we shall discuss eight criteria for deciding whether one fit is better than another. We have grouped them so that related ideas are presented sequentially; they are *not* presented in order of importance. At the end of the chapter, we shall briefly discuss how to use them in combination.

14.1 Criterion 1: Statistical Quality

The most obvious method for comparing two fits is probably to use a statistical measure. But as discussed in Section 10.4.2 of Chapter 10, it can be difficult to apply traditional statistical tests to XAFS analyses. It turns out, however, that EXAFS has some advantages over XANES in this area.

14.1.1 Number of Independent Points in EXAFS

In EXAFS modeling, we are usually fitting either the Fourier transform or the back-transform. According to the Rayleigh criterion (Section 7.5.2 of Chapter 7), we can just barely resolve peaks in the Fourier transform with a spacing of

$$\Delta R = \frac{\pi}{2\left(k_{max} - k_{min}\right)} \tag{14.1}$$

If we fit over a range from R_{min} to R_{max}, that means we could consider there to be a number of independent points given by

$$N_{ind} = \frac{\left(R_{max} - R_{min}\right)}{\Delta R} = \frac{2\left(k_{max} - k_{min}\right)\left(R_{max} - R_{min}\right)}{\pi} \tag{14.2}$$

This is often known as the *Nyquist criterion*, due to its relationship to an analysis of the information content of telegraph signals by Harry Nyquist (Nyquist 1928).

In the 1980s and 1990s, there was some discussion as to whether Equation 11.2 should actually have a '+1' or '+2' added to it—Stern (1993), for instance, argued for the +2. But all of these formulas assume the data to be *ideally packed*, and that's unlikely.

To explain what is meant by ideally packed information, let's consider an attempt to model a chlorine- and methyl-substituted ferrocene, such as the one shown in Figure 14.1.

In the first three sections of this chapter, we will frequently cite the Error Reporting Recommendations of the International XAFS Society (IXS Standards and Criteria Committee 2000). This is in part because it is a frank summary of statistical issues related to EXAFS, and in part because the IXS has made suggestions as to how to standardize the treatment of statistical issues in EXAFS.

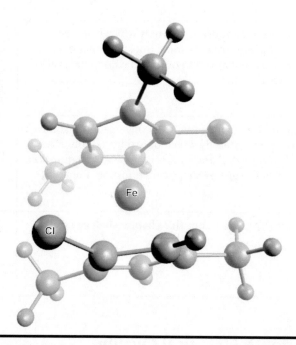

Figure 14.1   One possibility for the substituted ferrocene. The iron and chlorine atoms are labeled; large unlabeled atoms are carbon, and small unlabeled atoms are hydrogen.

Suppose we plan to use iron K-edge data from 3.0 to 10.0 Å^{-1} and to fit from 1.0 to 3.0 Å in the Fourier transform. According to the Nyquist criterion, that gives us eight independent points (rounding down), or ten if you use the '+2' from the Stern formula. Let us suppose we are trying to fit the following parameters:

- S_o^2 and E_o
- The distance from the iron atom to the carbons in the rings and an MSRD for that distance
- The number of chlorines substituted on to the rings and the number of methyl groups
- The distance from the iron atom to the chlorine atoms and an MSRD for that distance
- The distance from the iron atom to the carbon atoms in the methyl groups and an MSRD for that distance

That's ten free parameters in all, so we have just enough independent points by the Stern formula. But the first two parameters affect the fit overall, the next two the near neighbors only, and the remaining six the scatterers further out. To see how the scatterers distribute themselves through the Fourier transform, we use FEFF to generate theoretical standards for each of the scatterers (Figure 14.2).

Information about the chlorine atoms and the methyl carbons is almost entirely limited to the range from 2.0 to 3.0 Å. Because that's half the range being fitted, we only have four or five independent points that apply to that part of the Fourier transform—not enough to fit the six parameters that have to do solely with the chlorines and the methyl carbons. That is what is meant by nonideal packing.

In general, it is unlikely that the free parameters distribute evenly across the Fourier transform, and thus the Stern formula, or even the more conservative Nyquist formula, can mislead us. It's better to have a bit of a cushion to account for this nonideal packing; dropping the +2 is one way of doing that.

This argument has nothing to do with the relative amplitude of the peaks! There isn't 'more information' between 1.0 and 2.0 Å just because the peak is bigger!

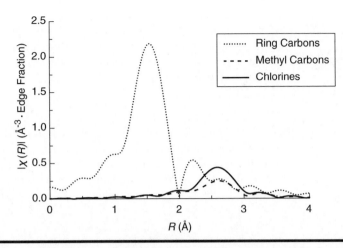

Figure 14.2 FEFF calculation of the Fourier transform of the single-scattering paths in the substituted ferrocene. The calculation shown here is scaled for ten ring carbons, two chlorines, and four methyl groups, for consistency with Figure 14.1. This plot uses k^2 weighting.

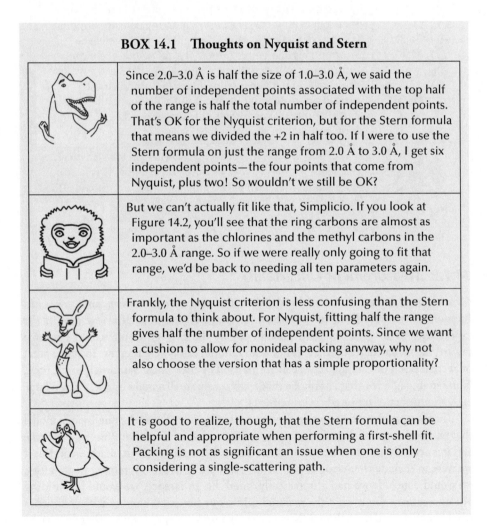

BOX 14.1 Thoughts on Nyquist and Stern

Since 2.0–3.0 Å is half the size of 1.0–3.0 Å, we said the number of independent points associated with the top half of the range is half the total number of independent points. That's OK for the Nyquist criterion, but for the Stern formula that means we divided the +2 in half too. If I were to use the Stern formula on just the range from 2.0 Å to 3.0 Å, I get six independent points—the four points that come from Nyquist, plus two! So wouldn't we still be OK?

But we can't actually fit like that, Simplicio. If you look at Figure 14.2, you'll see that the ring carbons are almost as important as the chlorines and the methyl carbons in the 2.0–3.0 Å range. So if we were really only going to fit that range, we'd be back to needing all ten parameters again.

Frankly, the Nyquist criterion is less confusing than the Stern formula to think about. For Nyquist, fitting half the range gives half the number of independent points. Since we want a cushion to allow for nonideal packing anyway, why not also choose the version that has a simple proportionality?

It is good to realize, though, that the Stern formula can be helpful and appropriate when performing a first-shell fit. Packing is not as significant an issue when one is only considering a single-scattering path.

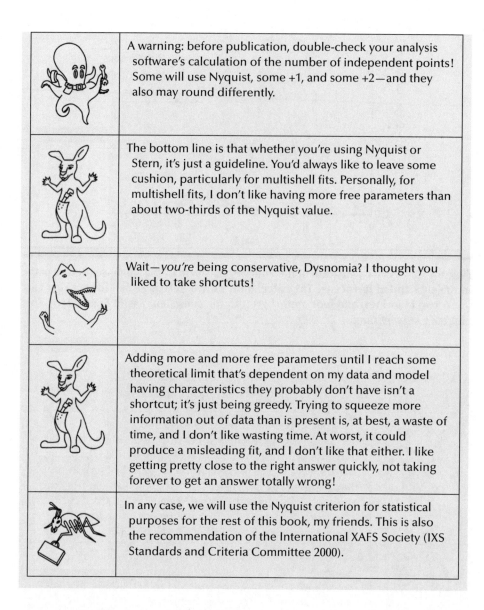

	A warning: before publication, double-check your analysis software's calculation of the number of independent points! Some will use Nyquist, some +1, and some +2—and they also may round differently.
	The bottom line is that whether you're using Nyquist or Stern, it's just a guideline. You'd always like to leave some cushion, particularly for multishell fits. Personally, for multishell fits, I don't like having more free parameters than about two-thirds of the Nyquist value.
	Wait—*you're* being conservative, Dysnomia? I thought you liked to take shortcuts!
	Adding more and more free parameters until I reach some theoretical limit that's dependent on my data and model having characteristics they probably don't have isn't a shortcut; it's just being greedy. Trying to squeeze more information out of data than is present is, at best, a waste of time, and I don't like wasting time. At worst, it could produce a misleading fit, and I don't like that either. I like getting pretty close to the right answer quickly, not taking forever to get an answer totally wrong!
	In any case, we will use the Nyquist criterion for statistical purposes for the rest of this book, my friends. This is also the recommendation of the International XAFS Society (IXS Standards and Criteria Committee 2000).

14.1.2 Measurement Uncertainty

Measurement uncertainty can be divided into two types: *random* and *systematic*. Random errors vary from measurement to measurement in such a way that, if the experiment were to be repeated enough times, the average result would converge toward the result that would be found if no random error were present. In the context of Section 6.3.3 of Chapter 6, we called this kind of error *noise*. Systematic error, then, is any source of error that cannot be made arbitrarily small simply by averaging a large enough number of repeated measurements.

There's a problem, however, in that we are trying to evaluate the quality of a model that has been fitted to data. Suppose, for instance, that we model our material as rutile and it is actually anatase. We are, in some sense, then making a systematic error. But if we were to include that error as part of the estimate of the measurement uncertainty, we would conclude we had a statistically 'good' fit. In essence, we would be building our uncertainty about what the material is into our statistics. Used that way, statistics would be a method for telling us how uncertain our process was, rather than how good the fit was.

To avoid this, we exclude errors in the model from being counted as 'measurement uncertainty.' Thus, if the fitted model deviates from the data by an amount much greater than what we've classified as measurement uncertainty, it must be because the model is poor.

But that distinction is not clear-cut. While shot noise is certainly part of measurement uncertainty, and the decision of whether to model anatase or rutile is not, what about if we used an inaccurate mean free path for the electron (Section 13.2.5 of Chapter 13)? What about errors caused by the muffin-tin approximation (Section 12.2.1 of Chapter 12)?

Some items, such as thickness effects and energy-dependent absorption, are in both lists!

The International XAFS Society has considered this question and concluded that the distinction should be between errors that can be identified and fixed as opposed to those that are 'poorly understood' (IXS Standards and Criteria Committee 2000). Thus, the IXS recommends that the following not be considered as contributing to measurement uncertainty:

■ Thickness effects, such as those described in Section 6.3.2 of Chapter 6
■ Self-absorption effects (Section 6.4 of Chapter 6)
■ Energy-dependent normalization (Section 13.1.3.3 of Chapter 13)
■ Inadequate structural models

The IXS report also includes a long list of systematic errors that should be included in the measurement uncertainty, including the following:

The idea is that if the effect can be calculated for and then corrected, it should not be considered part of the measurement uncertainty. That is why the IXS places self-absorption in the first category, for example (although, in practice, it is usually not possible to completely correct for self-absorption, and it should, therefore, have some contribution to the measurement uncertainty as well). Some kinds of thickness effects are fairly easy to correct for, while others are not.

■ Sample inhomogeneity (Section 6.3.2 of Chapter 6)
■ Beam damage (Box 2.1 of Chapter 2)
■ Thickness and particle size effects (Section 6.3.2 of Chapter 6)
■ Harmonics (Sections 6.2.1 and 8.5.3 of Chapters 6 and 8, respectively)
■ Nonlinear detectors (Chapter 8)
■ Glitches (Section 8.10.2 of Chapter 8)
■ Sample alignment issues (Section 8.5.1 of Chapter 8)
■ Errors in normalization (Section 7.3.4 of Chapter 7)
■ Imperfect standards, including errors in ab initio calculations such as FEFF (Chapter 12)
■ Difficulties associated with S_o^2 and energy-dependent normalization of ab initio standards (Section 13.1.4 of Chapter 13)
■ Preprocessing errors (Sections 7.1 and 7.2 of Chapter 7)

Determining the contribution of random error to the measurement uncertainty is relatively straightforward. Section 7.2.2 of Chapter 7 outlines four different methods suggested by the IXS; for more details, consult IXS Standards and Criteria Committee (2000) directly.

The truth is, there really isn't a sharp separation between the two categories. Just be as clear as you can as to how you handled this issue in any papers you write.

Estimating the contribution of systematic error is much more difficult. While it might seem that the measurement and analysis of a known standard could provide a good estimate, several potential sources of systematic error discussed earlier in the book might affect a sample differently from a standard. A uniform standard, for instance, is less affected by a nonuniform beam profile than a sample which incorporates some inhomogeneities. Ab initio calculations are more likely to be based on unjustified assumptions for a novel material than for a known one. And radiation damage, which is rarely a problem with standards, may cause significant changes in a sample.

In their report, IXS identified 'round-robin' comparisons of beamline measurements as one way to improve this situation (IXS Standards and Criteria Committee 2000):

At the present time the magnitude of the systematic error in a 'typical' XAFS experiment is not known. Planned future activities of the IXS Standards and Criteria committee include round-robin type measurements at various XAFS beamlines around the world and modeling of various analytical procedures. The goal of these activities will be to determine the magnitude and distribution of the major systematic errors.

Kelly et al. (2009) performed one such set of measurements; Chantler et al. (2018) have proposed a series of follow-up experiments.

Until then, the IXS recommends that measurement uncertainty be initially calculated as if it were due solely to random error. In Section 14.3, we'll discuss how this recommendation is used to compute uncertainties, and in Chapter 18 we'll indicate how statistical quantities should be reported in a paper.

14.1.3 Reduced χ^2

We are now able to compute a preliminary reduced χ^2 for an EXAFS fit with N_{ind} independent points and ν degrees of freedom, given by Equation 14.3:

$$\chi_\nu^2 = \frac{N_{ind}}{\nu N} \sum_{i=1}^{N} \frac{\left(data_i - fit_i\right)^2}{\varepsilon_{i\,\text{random}}^2} \tag{14.3}$$

 When fitting the Fourier transform (the most usual case for EXAFS), it is noteworthy that Equation 14.3 is applied to the real and imaginary parts of the Fourier transform separately, and the results added together to give χ_ν^2. This is a different result than what one would get by using the magnitudes of the data and the fit!

This is identical to Equation 10.8 in Chapter 10, with the exception that we have explicitly indicated that ε_i represents the measurement uncertainty due to random error only. That is also why we are referring to this quantity as 'preliminary.'

Since χ_ν^2 is a statistically meaningful quantity, we can use it to compare fits on the same set of data. If, for example, adding a free parameter to a fit causes χ_ν^2 to increase, then the addition of the parameter has not improved the fit.

Using the random error in place of the total measurement error has important ramifications, however.

 And fitting the magnitude of the Fourier transform rather than the real and imaginary parts would be a terrible idea! You would be throwing much of the information in the spectrum away!

First of all, χ_ν^2 should not be compared for two fits for which the measurement uncertainty is expected to be significantly different. This is true even if it is only the random contribution that differs. Thus, χ_ν^2 should not be used to compare two otherwise identically constructed fits that are made on different ranges of $\chi(k)$ taken from the same data set.

To understand this better, imagine performing a fit on a range of k^2-weighted data from 3.0 to 10.0 Å⁻¹, and that the measurement error at 3.0 Å⁻¹ is 5% from random sources and 95% from systematic errors. How do those contributions vary with k? That depends on many factors, including data collection decisions such as integration time (Section 8.9 of Chapter 8). But, for the sake of argument, let's assume that the random error in the unweighted $\chi(k)$ is independent of k, while the systematic error in the weighted $k^2\chi(k)$ is independent of k. This is not unreasonable if the k^2 weighting was chosen so as to keep the amplitude roughly constant across the fitting interval. In that case, the contribution to the measurement error at 10.0 Å⁻¹ from random sources would be 11 times as large as at 3.0 Å⁻¹. But the total measurement error would only increase by 50% because most of the error still comes from the systematic contribution, which is decreasing in relative importance as k increases.

While it is possible to use one of the methods suggested by the IXS to estimate the random error as a function of k, this example reveals why, in cases for which systematic error is expected to dominate over random error, it is better to estimate an average random error and apply it throughout the range; that is, to set $\varepsilon_{i\,\text{random}}^2 = \varepsilon_{\text{avg random}}^2$ in Equation 14.3. Otherwise, you may overestimate the importance of the low end of the fitting range relative to the high end. If your data are very noisy, on the other hand, then random error is more likely to dominate, and it might be appropriate to use a method that estimates $\varepsilon^2_{i\,\text{random}}$ at each point.

In cases where systematic error is important, it should now also be clear why χ^2_v cannot be used to compare two fits to different k ranges from the same data set. Continuing our above example, if we extended the k range so that it included from 3.0 to 14.0 Å^{-1}, the average random noise over the entire interval would increase by about 78%, even though the true measurement uncertainty would only have increased by about 17%. Using the random error alone to estimate the measurement uncertainty would result in a dramatic overestimate of the ratio between the measurement uncertainties for the two ranges, and a corresponding bias toward the second fit.

While χ^2_v cannot be used to compare fits on different data, that doesn't mean it's useless. It is a very good tool for comparing two fits to the same data. Those fits may differ in the number or identity of free parameters, for instance, or in the range of Fourier transform used.

We've included this example for those of you who want some sense of why you shouldn't compare the reduced χ^2 of fits on different data. To show how that might cause problems, we've made a lot of iffy assumptions, such as the idea that the systematic error is independent of k in the k^2-weighted data. The point here is not whether that assumption is good or not—it's that you have no way of knowing, in a given case, how systematic error varies from one part of $\chi(k)$ to another or from one sample to another. And if you don't know that, then any statistical claims you make in comparing the two are clearly unjustified!

14.2 Criterion 2: Closeness of Fit

14.2.1 R Factor

If we can't use a statistical measure to compare fits on different data, what can we do?

We can use the XAFS R factor, defined in Section 10.4.3 of Chapter 10:

$$R = \frac{\sum_{i=1}^{N}\left(data_i - fit_i\right)^2}{\sum_{i=1}^{N}\left(data_i\right)^2} \tag{14.4}$$

If the R factor is small, then the fitted model is very similar to the data. In such a case, does it really matter whether the remaining mismatch is a random error, a systematic error, or a failure in the model? It may well be that in some particular case that a 1% mismatch between fitted model and data is primarily due to an error in the model; but it is likely to be a small error, such as constraining two bond distances to be the same when actually there is a very slight splitting. In such a case, the major features of the model are still accurate. In another case, the 1% mismatch might be due primarily to noise—but once again, we would be likely to conclude that the major features of the model are accurate.

But how close is 'close enough'? 1%? 10%? The IXS (IXS Standards and Criteria Committee 2000) says 'as long as the signal-to-noise ratio (S/N) of the data is good, the R factor of adequate fits can be expected to be not more than a few percent.'

We'll supplement that with our own informal guidelines, provided in Table 14.1.

When using Equation 14.4 on a fit of the Fourier transform, the real and imaginary parts of the numerator are computed separately and then summed, and the magnitude of the data is used in the denominator.

Table 14.1   Informal guidelines for Assessing *R* Factor in EXAFS Fits

R *Factor*	*Interpretation*
<0.02	Good enough
0.02–0.05	Model has some details wrong or data are of low quality. Nevertheless, consistent with a broadly correct model.
0.05–0.10	Serious flaws in model or very low data quality
>0.10	Model may be fundamentally incorrect

The *R* factor really isn't telling you anything that you can't quickly judge by looking at a plot of the data and the fit. For example, Figures 14.3 through 14.6 show four fits to the same data, one for each of the four ranges identified in Table 14.1.

14.2.2 Hamilton Test

In case you're curious, these fits are to the nickel edge of a sample of nickel–zinc ferrite nanoparticles. The fits are conducted on the Fourier transform of the data from 3.0 to 9.0 Å⁻¹ using k^2 weighting.

There is a difference between preferring one fit as 'better' than another and drawing conclusions about the physical structure of the material based on the improvement seen.

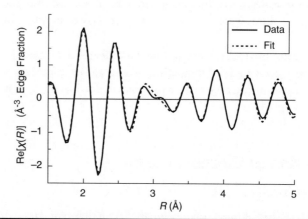

Figure 14.3 **Fit showing an *R* factor of 0.01. Only the fitted range is shown.**

Our eye is a better guide than the *R* factor alone, my friends! Does the fit reproduce the features of the model, although off a bit in amplitude or phase, as is the case in Figure 14.4? Or does it fail to reproduce features altogether, as is the case above 3.5 Å in Figure 14.6? The former suggests an overconstrained model, while the latter suggests that the fit is fundamentally failing to find a structure that matches the data.

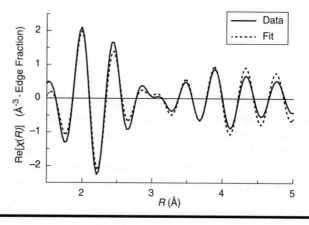

Figure 14.4 **Fit showing an *R* factor of 0.04.**

Don't chase *R* factor! When we say 0.02 is good enough, we mean what we say! One of the most common mistakes beginners make is to try everything they can to reduce the *R* factor, to the point of preferring 0.001 to 0.002. If the fit is that close, it's close enough for nearly all purposes, and you should be focusing on the other seven criteria!

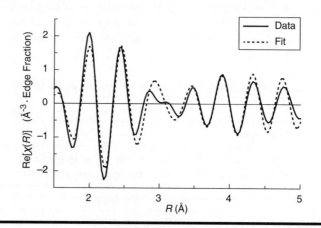

Figure 14.5 **Fit showing an *R* factor of 0.09.**

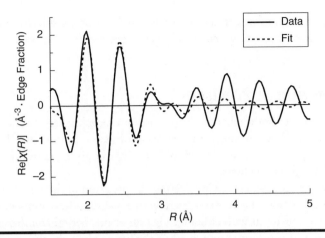

Figure 14.6   Fit showing an *R* factor of 0.17.

For example, suppose that, for some set of data, the introduction of a splitting into the nearest-neighbor distances causes the preliminary χ_v^2 (i.e., χ_v^2 calculated on the assumption that all error is random) to drop from 77 to 75. Certainly, we might then 'prefer' the fit with the splitting and choose to use it as the primary fit in any resulting literature. But does that mean that we should report that the data tell us that the first shell *is* split? Or, if we are uncomfortable with being that definite, should we use some more cautious wording, such as writing that the data 'suggest' that the first shell is split? Or can we only write something even more tentative, such as that the data 'are consistent' with the first shell being split?

These are exactly the kinds of questions the Hamilton test, which we described in detail in Section 10.4.5 of Chapter 10, can address. If the Hamilton test does not show a statistically significant difference between the fits, then we should not assert anything stronger than consistency. If it does show a statistically significant difference, then we can temper our language on the basis of the degree of significance.

While the procedure for applying the Hamilton test is given in Section 10.4.5 of Chapter 10, there are a few notes concerning its application to EXAFS modeling that are worth making:

■ The Hamilton test is only appropriate for comparing two fits on the same *k* and *R* ranges; that is, two fits that use the same data but fit it with different models. To compare fits on two different *R* ranges, use reduced χ^2. Fits on different *k* ranges cannot be compared statistically, since the systematic error is unknown; other criteria from this chapter must be used to select between them.

■ Since the Hamilton test depends on the ratios of closeness-of-fit parameters, there is no difference between applying it to χ^2, χ_v^2, or the *R* factor, as long as each point is weighted the same (e.g., a constant measurement uncertainty is applied across the fitted range). Conventionally, EXAFS fits have applied it to the *R* factor (Downward et al. 2006).

■ Changes in the theoretical standards used by the model can often be accounted for as if they were changes in parameters. For example, if fit A assumed that the nearest neighbors were oxygen and included coordination number, bond length, and MSRD for those oxygen atoms as free parameters, while fit B assumed the nearest neighbors were sulfur and used the same three parameters, then a Hamilton test could be applied with the assumption that those three parameters were 'changed'—instead of applying to oxygen, they now apply to sulfur. If E_o

In fact, if the fit is that close, trying to make it closer may mean that you are trying to fit systematic errors better, or even noise. Fitting details of the noise is certainly a waste of time.

We haven't actually given the criterion for determining if a change in reduced χ^2 is statistically significant. That's because it's usually not an important question—if you're tinkering with the *k* range, you're almost always looking for the *best* fit, and not trying to show that it allows you to reject some other fit. And if you're not tinkering with the *k* range, then the Hamilton test is the way to go.

If you would like to know more about statistical significance and reduced χ^2, the work done by Bevington and Robinson (2003) is a good place to start.

and S_o^2 were also free parameters, they would not be considered to be changed, as they are primarily a property of the absorber, not the scatterer.

14.3 Criterion 3: Precision

Suppose you'd like to know whether the iron in your sample is tetrahedrally or octahedrally coordinated. You, therefore, make the coordination number for the iron a free parameter and get a fit with an R factor of 0.01. That fit gives a coordination number of 4.6 ± 2.1. Are you done fitting?

The answer, of course, is no. Considering the precision of the result, you have not answered your question; the fit is consistent with a coordination number of either 4 or of 6, that is, it's consistent with either tetrahedral or octahedral coordination. In this case, you'd want to continue trying to refine the fit to reduce the uncertainty.

14.3.1 Finding Uncertainties in Fitted Parameters

Of course, this raises the question of how to find the uncertainties in fitted parameters. The IXS recommends that the uncertainty be defined as the amount by which the fitted parameter would need to be changed (while still allowing all other fitted parameters to vary) so as to cause the χ^2 to increase by 1.

Note that it is χ^2 that must increase by 1, not χ_ν^2. Recall that $\chi^2 = \nu\chi_\nu^2$, where ν is the number of degrees of freedom of the fit.

But didn't Section 14.1 teach us that we don't have a good way of computing χ^2 because we don't have a good way of estimating systematic error?

Not quite. Recall that there are three categories of phenomena that contribute to a mismatch between fitted model and data: random error, systematic error, and poor fit due to, for example, a poor choice of model. Therefore, if we believe a fit is 'good,' (i.e., the model chosen is appropriate and we've corrected for other quantifiable effects), then all of the mismatch between fitted model and data is due to measurement error, and since we can estimate the random contribution to the measurement error, we know the rest must be systematic.

And we necessarily believe our final fit is pretty good or we wouldn't consider it our final fit! In fact, this chapter provides several criteria for judging the quality of a fit, most of which don't depend on statistical methods.

For a good fit, χ_ν^2 *should* be around 1 (see Section 10.4.6 of Chapter 10). So, if our *preliminary* value for χ_ν^2, calculated by neglecting the contribution of systematic errors to the measurement uncertainty ε, were, say, 100, we would know that our preliminary value for χ_ν^2 was 100 times too large, meaning that ε^2 was 100 times too small, and thus ε was 10 times too small. With the new, more accurate estimate of ε, we could then apply the rule given in the first paragraph of this section to find the uncertainty for each fitted parameter.

Remember that this procedure is only valid if the fit is good! If the fit is not good by the other criteria presented in this chapter, then this rescaling procedure does not produce meaningful uncertainties!

14.3.2 Calculations of Uncertainties by Analysis Software

Almost all software designed for fitting will report uncertainties in the fitted parameters. It can be shown that the definition of uncertainty provided in the first paragraph of Section 14.3.1 is roughly equivalent to taking the square root of the diagonal elements of the covariance matrices used in fitting routines (IXS Standards and Criteria Committee 2000). A few software packages, including those based on IFEFFIT and LARCH, automatically scale the uncertainties by $\sqrt{\chi_\nu^2}$, which is equivalent to the scaling procedure for accounting for systematic error in good fits described in the last paragraph of Section 14.3.1. Other software packages allow for the user to provide a scaling factor, which could then be chosen to be $\sqrt{\chi_v^2}$. Some simply report the square root of the diagonal elements of the covariance matrix, leaving any scaling to be applied manually by the user. And at least one package introduces an additional factor of 2 to produce a more conservative estimate of uncertainty. An out-of-date but still informative comparison of the methods used by several of the more packages can be found in (IXS Standards and Criteria Committee 1998). Be sure to check the documentation of the software you are using so that you know what additional scaling, if any, you should apply.

14.3.3 How Precise?

While it is necessary to determine fitted parameters with sufficient precision to answer the scientific questions you are posing, this doesn't mean that every fitted parameter needs to be found to the same level of precision. For example, if you are interested in determining a *nearest*-neighbor bond length to within 0.02 Å, it doesn't matter if the same fit only determines the distance to the *next*-nearest neighbor to ±0.04 Å. Likewise, the uncertainty in the fitted value of S_o^2 is largely irrelevant unless you plan to use the value you have found to constrain another fit.

For the parameters that do help answer the questions you are studying, it is important to realize that, if all errors are normally distributed, the true answer will lie outside the specified uncertainty range nearly one-third of the time, and more than twice the specified uncertainty away from the best-fit value nearly 5% of the time. Thus, a coordination number of 4.1 ± 1.1 is only good enough to tentatively choose tetrahedral over octahedral coordination, since an octahedral coordination of 6.0 is less than twice the stated uncertainty away from the best-fit value. A fit that satisfies the other seven criteria in this chapter well but reduces the uncertainty from, say, ± 1.1 to ± 0.8, is therefore preferable.

14.3.4 Correlations

Many software packages report the correlations between fitted parameters. If the procedure in Section 14.3.1 is used to compute uncertainties, then the effects of these correlations have already been included in those uncertainties.

So, why report correlations at all? A high correlation between a pair of parameters can explain why the uncertainties for those parameters are large, and thus may provide clues as to how to reduce those uncertainties. For example, constraining one of the parameters by using information from another source (e.g., another characterization technique, a fit to an empirical standard, or a theoretical computation) will likely reduce the uncertainty in the other.

14.4 Criterion 4: Size of Data Ranges

Even without knowing anything about someone's EXAFS investigation, we can likely say a few things about the structure being studied. There is probably a big peak between about 1 and 2.5 Å in the Fourier transform due to the near neighbors. Above that first peak, multiple scattering by near neighbors becomes a factor. For many materials, there will also be a signal from more distant atoms (in some cases, such as metal carbides and high-Z oxides, the peak associated with the second-shell scatterers may even exceed the amplitude of the peak associated with near neighbors). As we go higher and higher in R, the Fourier transforms of EXAFS spectra of different substances show

The square root of a diagonal element of the covariance matrix is exactly equivalent to the definition provided in Section 14.3.1 (to wit, the amount by which the corresponding fitted parameter must be changed to increase χ^2 by 1 when still allowing other fitted parameters to vary) if the response of χ^2 to changes in that parameter is parabolic. Near a minimum in χ^2 that is generally a reasonably good assumption.

Of course, it is not OK to simply arbitrarily constrain one correlated parameter to achieve an apparent reduction in the uncertainty of another! For example, fixing the value of a parameter to the result of a prior fit to the same spectrum is a definite no-no!

a greater range of behavior; some may be nearly featureless except for a bit of noise and sidebands from truncation effects on lower peaks, while others may show strong structure from ordering of distant scatterers, perhaps amplified by focused multiple scattering.

This means that a fit on the near-neighbor peak alone is reasonably likely to be 'good' in the sense of criteria 1 through 3, even if the model is incorrect. While some of the later criteria in this chapter (notably Criterion 7) will help to weed out incorrect models, there is still generally the potential for a fit using a poor model to seem OK when only the first peak is being fit. Extending the fit to higher R, as long as there is structural signal, will increase the probability that a good fit indicates a correct model. This improvement is not just the 'statistical' improvement that comes from increasing the number of independent points—it is because beyond the first shell, the candidate models tend to produce Fourier transforms that differ from each other more dramatically than they do at low R.

While so far we've been arguing that, all else being equal, a fit that includes more of the Fourier transform is better, this argument does not apply in the same way to the amount of k-space data that is selected. $\chi(k)$, unlike the Fourier transform, includes data from the whole structure throughout its range. It is true that low-Z scatterers are emphasized more at low k than higher-Z scatterers (Section 13.3 of Chapter 13), but each still contributes at all k. On the other hand, we cannot use reduced χ^2 to guide us as to whether we are gaining a statistical improvement by extending the k range because the relative importance of random and systematic error is likely to change (Section 14.1.3). This means, all else being equal, we should prefer a fit with a larger k range to a smaller one, even if the R factor rises slightly.

14.5 Criterion 5: Agreement Outside the Fitted Range

Consider the fit shown in Figure 14.7.

The fit looks reasonably close within the fitting range of 1.2–5.0 Å. But, above the 5.0 Å fitting range, the model doesn't match at all, either in phase or in amplitude. Now compare to Figure 14.8, which shows another fit to the same data.

This time, the fit follows reasonably close to the data through about 6.0 Å, well above the end of the fitting range. It even catches the fact that the peak around 5.4 Å is a little larger than those around it.

Recall from Section 9.2 of Chapter 9, for instance, how tetrahedral or octahedral occupancy in a spinel is most easily fingerprinted by data from 3 to 4 Å, even though the nearest neighbors are in the 1–2 Å range.

If you extend the fit beyond the point in the Fourier transform for which there are data, so that you are fitting only noise, then you are intentionally making a signal that is very far from being ideally packed (see Box 14.1)! The extra independent points reported by the Nyquist criterion aren't legitimate because they don't correspond to data. So while it is good to extend a fit to include more data, you should never extend a fit *beyond* the data!

The reason for preferring to fit a larger range in the Fourier transform is a bit different than the reason for preferring to fit a larger amount of k-space, but the bottom line is the same: fitting more data is better.

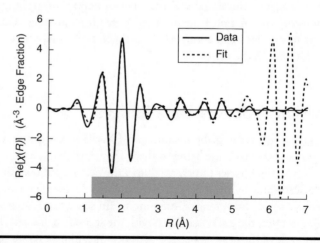

Figure 14.7 **Fourier transform of data and fitted model for the iron K edge of a novel iron nanomaterial containing phosphorous and carbon. The 1.2–5.0 Å range used by the fitting model is highlighted in gray above the *x*-axis.**

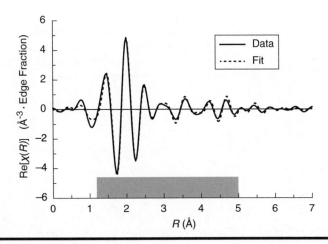

Figure 14.8 **Fourier transform of the data shown in Figure 11.7, with a different model. The 1.2–5.0 Å range used by the fitting model is highlighted in gray above the x-axis.**

How does the fit 'know' what to do with data it was not instructed to fit? Presumably because the structural model is essentially correct. Behavior similar to that shown in Figure 14.7, on the other hand, raises questions as to the accuracy of the structural model being used.

A similar check can be made in *k*-space; a good fit should follow the data for a bit above the end of the *k* range being fit.

It is less important that the fit agrees with the data below the fitted range in the Fourier transform or below the data range in *k*-space, however. In both cases, the region below the fitted range is heavily influenced by details of background subtraction, and thus a mismatch is not necessarily indicative of a faulty model.

BOX 14.2   Balancing Criteria 4 and 5

 Don't criteria 4 and 5 conflict? In Figure 14.8, I could surely extend the fitting range a little higher in *R* and still get a good fit. But at some point, I'm going to get past the range of paths I've included, and then the fit won't follow above the fitted range anymore.

 I think it's more a semantic conflict than a substantive one. A fit similar to the one in Figure 14.8 shows readers that the Fourier transform could be extended to higher *R*, but that the model can follow that part of the spectrum without it being included in the fit. It's like saying you want a car that can go fast and also looks good sitting in your driveway. Obviously, while it's going fast it's not sitting in your driveway, but once you've looked at it parked and test driven it on the highway, you know that it satisfies both of your criteria.

 When I fit, I keep increasing the *R* range until the fit stops working. At that point, I lower the top end of the *R* range by 1.0 Å so as to show that the fit continues above the top of the *R* range.

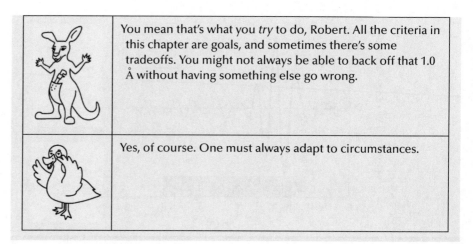

You mean that's what you *try* to do, Robert. All the criteria in this chapter are goals, and sometimes there's some tradeoffs. You might not always be able to back off that 1.0 Å without having something else go wrong.

Yes, of course. One must always adapt to circumstances.

14.6 Criterion 6: Stability

For someone learning to fit, it can at times feel as if every decision is crucial; exactly what range of data should you fit? Which paths to include? What background subtraction? Which constraints? k-weight? Window?

But when you think about it, most of those things can't be crucial. If they were, but were also choices that a reasonably well-trained researcher could have made differently, then the technique is useless! It must be that well-trained researchers can arrive at the same conclusion about the scientific questions being addressed, even if they make some different choices along the way. That means if the fit is good, the results must be *stable* when some of those choices are made differently.

A good fit, therefore, should give the same results for the parameters of interest (i.e., those which answer the scientific questions the experiment is intended to address) when any of the following are changed slightly:

- Slightly (perhaps 0.5 or 1.0 Å⁻¹) different low end of the k range. This probably cannot be changed too drastically, as both background and theoretical approximations become important at low k.
- Different high end of the k range. This can probably be changed somewhat more than the low end of the k range.
- Different low end of the R range. As with the k range, this probably can't be changed by much—perhaps only 0.2 Å—because it should remain between the background region and the peak of the first-shell data.
- The high end of the R range. The amount by which this can be changed depends in part on whether you are employing a bottom-up (Section 12.4.1 of Chapter 12) or top-down (Section 12.4.2 of Chapter 12) strategy. For the top-down strategy, you may be able to change the high end by an angstrom or more.
- Different k-weight
- Modestly different background subtraction
- Modestly different constraint scheme. For example, grouping together or splitting up the MSRD of similar paths

Of course, asking for a fit to be stable under these stresses does not mean that nothing changes at all. Some fits will have higher R factors, for instance, and as you move away from your best fit, the R factor may degrade significantly. Precision may also vary substantially, so that some fits, while giving values that are consistent with the best fit, may be insufficiently precise to be considered good themselves. The best-fit values of the parameters of interest will, of course, drift around some but hopefully in such a way that the range defined by the new values with their uncertainties overlaps the range defined by the best-fit values with their uncertainties. Finally, it is also possible that for relatively severe stresses, the fit may flip into a completely different 'false minimum,' giving significantly different values for the parameters of interest. As long as this false minimum can be identified as significantly inferior by other criteria described in this chapter, the better fit can still be considered to have passed its stability check.

How does a fit fail its stability check? A fit fails if the parameters of interest change in such a way that the range as defined by the reported uncertainties of one of the fits from the stability test does not overlap the range from the best fit, and the new fit is not significantly inferior given the other criteria in this chapter. If that's the case, more work has to be done so that the two fits can be distinguished, allowing one to be rejected as a false minimum.

14.7 Criterion 7: Are the Results Physically Possible?

If the parameters found by a fit are not consistent with physically possible values, then the fit is not good, regardless of what the rest of the criteria in this chapter say. Chapter 13 identifies 'typical values' for each of the parameters in the EXAFS equation; parameters far outside that range are generally indications of a bad fit. For example, if a fit gives values that are unambiguously negative for an MSRD, the fit should be rejected.

You may also be fitting parameters that don't directly correspond to values in the EXAFS equation but are instead used to calculate those values. Since those parameters likely are chosen to correspond to physical quantities or concepts, they will have a known acceptable range of values. For example, a bond angle must be between 0° and 180°, and a site occupancy must be between 0 and 1.

BOX 14.3   'Surprising' Is Different from 'Impossible'

	That doesn't sound scientific to me. I'm going to throw out a fit because I don't like what it tells me? Doesn't that mean I'll always just keep prodding at a fit until it tells me what I want to hear?
	There is a difference between something you don't expect and something that's physically impossible. You may be expecting six sulfurs coordinated to your absorbing atom, and find only four. That may be surprising, but it's within the realm of accepted chemical behavior, and you could not reject the fit out of hand. But if a fit told you there were 20 sulfurs directly coordinated to the absorber, you'd know that was wrong—there just isn't room for them. Likewise, you might find a bond between iron and sulfur that's a bit longer or shorter than what you expected—but if it were 1.0 Å, you would know it was wrong—only bonds to hydrogen are anywhere near that short.

14.8 Criterion 8: How Defensible Is the Model?

Finding a good fit usually involves a lot of trial and error. If done thoughtlessly, that trial and error can amount to additional free parameters.

Suppose, for instance, that the Nyquist criterion indicates that a fit of a nanoscale lead sulfide has ten independent points, and that the fit includes seven explicit free parameters: E_o, S_o^2, a lattice parameter, the MSRD for the near-neighbor sulfur scattering, a third cumulant for the near-neighbor sulfur scattering, an MSRD for the fifth nearest-neighbor sulfur scattering, and an MSRD for all of the other scattering in the material. Looking at the list of free parameters, a question naturally arises: why assign the *fifth* nearest-neighbor sulfurs a different MSRD, when that's not done for any other scattering except for the near neighbors? The near neighbors are special because they will tend to vibrate in sync with the absorbing atom, but what's special about the fifth nearest-neighbor sulfurs? If the answer is 'there's nothing special about them; I just kept trying different combinations of free parameters until I got a good fit,' then there are, in effect, unreported free parameters in the fitting process. The process of searching for a fit by arbitrarily releasing constraints on parameters can itself be a kind of minimization process, with the person helping the computer change one parameter at a time until a 'best fit' is found. In the example discussed, this could very well mean that more than ten parameters were effectively allowed to vary during the process. When more parameters are allowed to vary than there are independent points, we are exceeding the information content of the data, with the result that a 'good' fit may actually be a meaningless coincidence. On the other hand, suppose some theories suggested that our nearest-neighbor distances were split, while others suggested that all of our nearest-neighbor distances are the same. We test both models and find that allowing the nearest-neighbor distances to be split does not give a significantly different result in any of our parameters or in the quality of the fit. Because we tested the possibility of two different nearest-neighbor distances in one fit, does that mean we must forever after count the amount of splitting as a free parameter? The answer is no because we are testing two conceptually defensible models against each other, either of which would have been considered a reasonable possibility a priori.

This is, admittedly, a subtle distinction to make, and one that allows for gray areas. While we would have been very unlikely to speculate about a special MSRD for the fifth nearest-neighbor sulfurs a priori, what about assigning different MSRD's to lead and sulfur scatterers? That sounds less arbitrary, and more like a model that might be considered a priori. Thus, there is no sharp distinction between a 'defensible' model and an 'arbitrary' one—it's a matter of degree.

Defensible models usually have at least one of these characteristics:

1. *Simplicity.* Simple models have very few free parameters and incorporate constraints that rely on simplifying assumptions. For instance, there might be a single parameter representing an overall lattice expansion. For many materials, such a model is likely to be a simplification of the actual behavior of the material as it changes temperature, but it might work as a first approximation. A top-down strategy usually starts with a simple model.
2. *Flexibility.* Flexible models can fit a wide variety of candidate structures. Such a model will have few constraints, but many free parameters. A bottom-up strategy often starts with a flexible model.
3. *Physical accuracy.* A model gains physical accuracy when it is constrained by a physical understanding of the material. This understanding may be derived from other characterization techniques or from theoretical tools such as density functional theory.

14.9 Evaluating a Fit

Evaluating your fits based on these criteria will necessarily involve tradeoffs, sacrificing how well one criterion is met to improve others. To help sort through how to use them, Table 14.2 sorts the criteria by three characteristics:

When trying to decide whether your model is defensible, think ahead to how you will describe it in a paper. If you find yourself tied in knots trying to justify why you chose the model you did, you should see if you can find a model which is either simpler, more general, or more physical.

Table 14.2 Characteristics of Criteria for Judging Fits

Criterion	Comparison Only?	Critical?	Threshold?
Statistical quality	Yes	No	Yes
Closeness of fit	No	Sort of	Yes
Precision	No	Yes	No
Size of data ranges	No	No	No
Agreement outside fitted range	No	No	No
Stability	No	Yes	Yes
Physical results	No	Yes	Yes
Defensible model	No	Yes	No

- *Comparison only?* Statistical measures should be used only for comparing fits.
- *Critical?* The criteria listed as critical must be met; otherwise the fit should be rejected, regardless of how well other criteria are met.
- *Threshold?* Criteria with a threshold reach a point of being 'good enough.' Criteria without a threshold are ones we always wish were better, no matter how well they are met.

When choosing between fits, it is helpful to first reject those that fail to meet critical criteria; for example, reject a fit if the results are nonphysical or the fit is wildly unstable.

Next, if threshold criteria reach the 'good enough' level for all fits being compared, then that criterion is no longer useful. For example, if one fit has an R factor of 0.016, and another has an R factor of 0.011, then R factor doesn't weigh in favor of either fit, since they're both good enough (i.e., under 0.02). But if one fit has an R factor of 0.016 and another has an R factor of 0.035, then the R factor should be considered along with other criteria.

That will still leave you with several criteria to trade off. How you choose to do so depends on your own priorities for the scientific study you are undertaking; we'll look at how some experts have done it in Chapter 19.

WHAT I'VE LEARNED IN CHAPTER 14, BY SIMPLICIO

There are eight criteria for evaluating how good a fit is: χ_ν^2 (used for comparison between two fits on the same data), R factor, precision of parameters of interest, size of the data ranges, agreement above the fitting ranges, stability, whether the results are physically possible, and the defensibility of the model.

- The criteria must be balanced against each other. Some of the criteria, such as the requirement that results be physically possible, can disqualify a fit regardless of what the others say.
- Errors can be divided into *random* and *systematic*. Random errors can be determined experimentally, but it is difficult to know the extent of systematic errors.

■ We should, therefore, make a preliminary assumption that errors are purely random. Once we have settled on our final fit, we then switch to assuming that the fit is good to estimate the uncertainties in fitted parameters.

■ The Hamilton test should be used to determine if one model can be rejected in favor of another.

References

Bevington, P. R., and D. K. Robinson. 2003. *Data Reduction and Error Analysis for the Physical Sciences*, 3rd ed. Boston: McGraw-Hill.

Chantler, C. T., B. A. Bunker, H. Abe, M. Kimura, M. Newville, and E. Welter. 2018. A call for a round robin study of XAFS stability and platform dependence at synchrotron beamlines on well defined samples. *J. Synch. Rad.* 25:935–943. DOI:10.1107/S1600577518003752.

Downward, L., C. H. Booth, W. W. Lukens, and F. Bridges. 2006. A variation of the F-test for determining statistical relevance of particular parameters in EXAFS fits. *X-Ray Absorption Fine Structure—XAFS13: 13th International Conference. AIP Conf. Proc.* 882:129–131. DOI:10.1063/1.2644450.

Kelly, S. D., S. R. Bare, N. Greenlay, G. Azevedo, M. Balasubramanian, D. Barton, S. Chattopadhyay, S. Fakra, B. Johannessen, and M. Newville. 2009. Comparison of EXAFS foil spectra from around the world. *J. Phys. Conf. Ser.* 190:012032. DOI:10.1088/1742-6596/190/1/012032.

IXS Standards and Criteria Committee. 1998. A survey of error analysis procedures used by existing XAFS software packages. http://ixs.iit.edu/survey/errors/index.html.

IXS Standards and Criteria Committee. 2000. Error reporting recommendations: A report of the standards and criteria committee. http://ixs.iit.edu/subcommittee_reports/sc/err-rep.pdf.

Nyquist, H. 1928. Certain topics in telegraph transmission theory. *Trans. Am. Inst. Electric. Eng.* 47:617–644. DOI:10.1109/T-AIEE.1928.5055024.

Stern, E. A. 1993. Number of relevant independent points in x-ray-absorption fine-structure spectra. *Phys. Rev. B Condens. Matter.* 48:9825–9827. DOI:10.1103/PhysRevB.48.9825.

Chapter 15

The Process of Fitting

The Main Event

The main event? But this is such a short chapter!	
Yes, friend Simplicio, but that is only because we've already laid so much ground work.	
A short chapter deserves a short introduction. Enough of the chitchat: let's get started!	

15.1 Identify Your Questions

Before beginning to fit, you should decide on the primary scientific questions you are trying to answer (see Section 5.1 of Chapter 5). It is often sensible to identify secondary questions as well: information that would be nice to know, but which is not necessary for the analysis to be considered a success.

We will next present two examples, which we will follow for the rest of this chapter.

15.1.1 Example: Which Ligand?

Consider an unknown iron compound in a soil sample, for which we have collected EXAFS data at the iron K edge. It is known that there is a fairly high sulfur content in the soil, and there is plenty of oxygen as well. As a researcher, we are interested in discovering whether the iron is bonded primarily to sulfur, to oxygen, or to a mix of

Even when fitting a known substance to practice how to fit, one should imagine a scientific question one might be trying to answer. For example, 'does this material show local distortions that are not reflected in the structure derived from x-ray diffraction?'

DOI: 10.1201/9780429329555-19

the two; that is our primary question. (A mix would not necessarily mean that an individual iron atom had ligands of both types—it might mean that some iron was present in the form of a sulfide and some in the form of an oxide.)

Secondary questions might include a more specific identification of the iron phases that are present (e.g., which of the many possible iron oxides are present?), and an estimate of the fraction of each iron mineral present, if it does turn out there is a mixture.

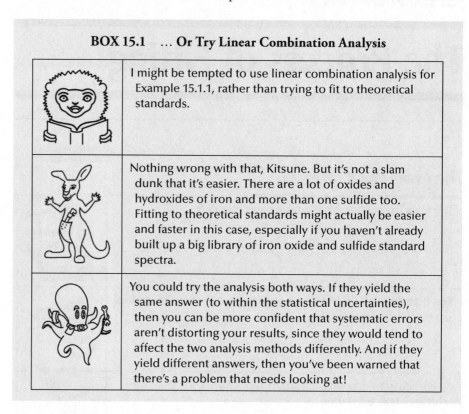

BOX 15.1   … Or Try Linear Combination Analysis

	I might be tempted to use linear combination analysis for Example 15.1.1, rather than trying to fit to theoretical standards.
	Nothing wrong with that, Kitsune. But it's not a slam dunk that it's easier. There are a lot of oxides and hydroxides of iron and more than one sulfide too. Fitting to theoretical standards might actually be easier and faster in this case, especially if you haven't already built up a big library of iron oxide and sulfide standard spectra.
	You could try the analysis both ways. If they yield the same answer (to within the statistical uncertainties), then you can be more confident that systematic errors aren't distorting your results, since they would tend to affect the two analysis methods differently. And if they yield different answers, then you've been warned that there's a problem that needs looking at!

15.1.2 Example: Where's the Dopant?

Consider a thin zinc oxide film doped with 1% manganese. Data were collected on both the zinc and the manganese K edges. It is not known whether the manganese substitutes for the zinc in the lattice, is situated interstitially, or forms distinct manganese oxide clusters. Deciding between those possibilities is the primary question. Important secondary questions depend on the answer to the primary question. If the manganese substitutes directly for zinc, do manganese atoms tend to cluster near each other in the lattice? If they are interstitial, how is charge balance maintained (zinc vacancies, interstitial oxygens, etc.)? If distinct manganese oxide clusters are formed, what is the typical size of those clusters?

15.2 Prepare Your Data

15.2.1 Transform to $\chi(k)$

This was covered in Chapter 7. It is not important at this stage to make sure you have the 'perfect' normalization and background subtraction, as long as it's reasonable. At the end of the fitting process, you can check the effect of your data reduction choices.

15.2.2 Choose k *Weighting*

The choice of k-weight is one of those things that shouldn't end up making a significant difference to your final fit. In addition to choosing a single k-weight, some fitting software provides you the option of choosing multiple k-weights. If, for instance, you choose k-weights 1, 2, and 3, the software will perform three separate Fourier transforms (one for each k-weight) on the data and on the theoretical standard. Each $\chi(R)$ resulting from the data is then treated as if it were a separate data set, but with one set of free parameters being simultaneously applied to all three sets.

BOX 15.2   What k-Weight Do You Use?

	I usually pick a k-weight that makes $\chi(k)$ look like it has a roughly constant amplitude.
	I prefer a less subjective technique, my dear Dysnomia. If I think my nearest-neighbor scatterers are lighter than krypton, I begin with a k-weight of 3. If between krypton and barium, I choose a k-weight of 2. And if heavier than barium, a k-weight of 1 is my initial preference. This tends to accomplish your goal of 'roughly constant amplitude' in a systematic way.
	I begin with k-weight 2, my friends. That is a simple compromise.
	Beginning with all three k-weights simultaneously might reduce correlations. Why not start by using all of the information?
	But Kitsune! Fitting all three k-weights simultaneously takes more computation! If your fitting model becomes complicated, that could end up slowing you down! It's also a little harder to explain the statistics behind a multiple k-weight fit to someone unfamiliar with it, although that is becoming less of an issue as the technique becomes more common.
	The truth is that you are all just making choices you find convenient. The systems you are studying don't have a 'correct' k-weight hidden inside them, and as long as you transform your data and your theoretical standard in the same way, a good fit shouldn't depend on the k-weight chosen.

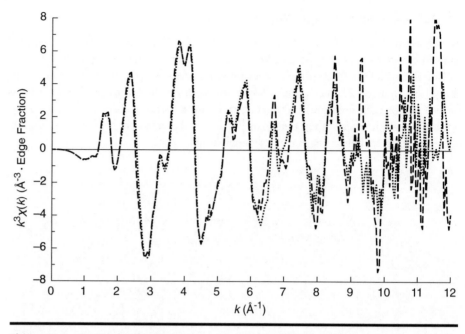

Figure 15.1   Ca *K*-edge $k^3\chi(k)$ for two scans of a modified calcium carbonate sample.

15.2.3 Choose k Range

Choose an initial range of $\chi(k)$ data to fit. You usually want to start out conservative and try expanding the range later in the fitting process once you have a reasonable model in place.

The high end of the $\chi(k)$ range is generally the easier one to choose because the problem at the high end is the signal-to-noise ratio. Most of us can tell when the signal begins to get lost in the noise and can choose the high end of the range accordingly.

For example, Figure 15.1 shows two scans of the calcium edge of a sample of modified calcium carbonate.

The two scans stop following each other after about 9 Å⁻¹, so if these were the only two scans 8 or 9 Å⁻¹ would be a good value for the high end of the range to be used for fitting. Averaging more than two scans might allow that value to be pushed a bit higher.

The lower end of the $\chi(k)$ range is a more difficult question. One reason is that the low-*k* part of the data tends to be more sensitive to the choice of background. Figure 15.2, for example, shows $\chi(k)$ data from the selenium *K* edge of a copper indium selenide solar cell plotted using two different 'acceptable' backgrounds (i.e., these are the *same* data with two different backgrounds subtracted).

Notice that the different background subtractions primarily affect the data below 6 Å⁻¹ and have a more dramatic effect below 3 Å⁻¹. If you weren't sure which background was the appropriate one, 6 Å⁻¹ would be a conservative choice for the lower end of the data range to be fit, and below 3 Å⁻¹ would certainly be unwise.

In addition, Section 13.1.5 of Chapter 13 taught us that the low-*k* part of the data is more sensitive to questions surrounding E_o. If your sample contains absorbing atoms with different oxidation states, the difference in E_o between them has a more significant effect at low *k*. If you think it is appropriate to use different values of E_o for different scattering shells from the same absorber, perhaps to correct for errors in the theoretical standards (Section 12.2 of Chapter 12), then that will have the greatest effect at low *k* as well.

Saying that two scans 'follow' each other doesn't mean they have to lie smack on top of each other. You can see differences between the scans starting at around 5 Å⁻¹, but those kinds of differences are why we average multiple scans in the first place. By the time you get above 9 Å⁻¹, though, you've often got one scan going up when the other is going down, and the structure is pretty much lost in the noise.

Be sure you understand the difference between what we're showing in Figures 15.1 and 15.2! Figure 15.1 shows two scans that have been processed in the same way; the differences are due to noise. Figure 15.2 shows data after it has been averaged, but with two different choices of background subtraction. The differences seen in Figure 15.2 are not due to noise; they instead indicate the systematic error that may be present due to choices made during background subtraction.

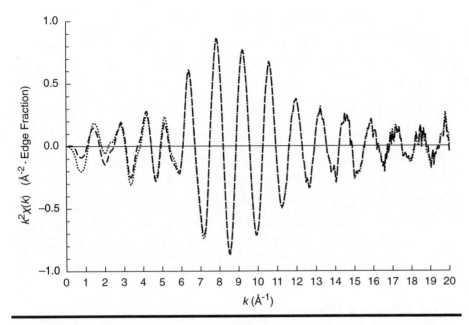

Figure 15.2 Se *K*-edge $k^2\chi(k)$ for a CuInSe₂ solar cell, computed using two different background subtractions.

Similarly, broadening due to limited energy resolution, either because of intrinsic effects such as the core-hole lifetime or instrumental effects such as those caused by finite slit height (Section 8.5.2 of Chapter 8), will have the greatest effect at low *k*.

In Section 13.2.5 of Chapter 13, we saw that the mean free path also affects the data more strongly at low *k*. The simplifying assumption that S_o^2 is independent of *k* is least accurate at low *k* as well (Section 13.1.4 of Chapter 13).

Finally, multiple-scattering paths are particularly important at low *k* (Section 12.3.2 of Chapter 12).

The choice of the minimum value of *k* for which you want to fit your data is thus as much a choice about how much complexity you want to include in your model as a question of data characteristics. If you are prepared to wrestle with triangle paths, multiple E_o parameters, discriminating between choices for the background, accounting for energy resolution, and correcting for errors in your theoretical standards, then you could choose to begin fitting at 2 Å⁻¹ or even lower. If you want to minimize the effect of all those complications, then a choice like 6 Å⁻¹ is more appropriate. Values in between, of course, represent a compromise.

BOX 15.3   The Transition from EXAFS to XANES

Once again, I will remind you that there is no physical distinction between XANES and EXAFS. Over the years, scientists have discovered that there are many approximations that work pretty well in the range of 6–20 Å⁻¹ or so: expanding in terms of scattering paths, using a single energy origin for all the photoelectrons, employing a mean free path formalism, finding a 'smooth background' through the data, and so on. Those approximations are never perfect, but they become less and less accurate as we extend toward lower and lower *k*.

All those low-*k* approximations make me nervous. So I usually start fitting at 5 Å⁻¹ and then check later to see if I can push that lower without messing up my fit.

That is probably because you are an industrial chemist, Dysnomia, so you usually have pretty good data to work with. My data are often very noisy because some of my samples are present in such low concentrations. If I did not start my fitting range until 5 Å⁻¹, then sometimes I would have almost nothing to fit at all! So I have to start my fitting lower than that, knowing that it may make my background subtraction and modeling a bit more complicated.

15.2.4 Choose k *Window*

As discussed in Section 7.5.5 of Chapter 7, this decision should be considered largely arbitrary. Some researchers prefer to always start with the same window function, while others make decisions based on data quality.

15.3 Plan Your Strategy

In Section 12.4 of Chapter 12, we discussed two broad strategies for fitting *bottom-up* and *top-down*.

15.3.1 Example: Which Ligand?

What a nice example, friends, of how starting simple can lead to success!

(A continuation of the example introduced in Section 15.1.1.)

In this case, a bottom-up strategy seems warranted. A reasonable plan would be to first try to fit the near neighbors with oxygens, choosing coordination number, bond length, MSRD, and E_o as free parameters. To constrain S_o^2, we could invoke chemical transferability (Section 13.1.4.7 of Chapter 13) and use a value found from a related standard. After the fit using near-neighbor oxygens was performed, we would then try fitting with sulfur near neighbors instead. Using the criteria from Chapter 14, it is possible that only one of the two would yield a reasonable fit.

If so, then we would be well on the way to answering the primary question for this experiment, although mixtures would still need to be investigated. If things went well, eventually we could move on to the secondary questions.

15.3.2 Example: Where's the Dopant?

(A continuation of the example introduced in Section 15.1.2.)

Unlike the example in the previous section, a top-down approach is probably best here. After all, we might spend quite a bit of effort in the bottom-up approach confirming that the nearest-neighbor environment for the zinc atoms is something like 5 ± 2 oxygens, which is not particularly helpful for answering our primary question.

But a little research shows that there are a lot of possibilities. 'Zinc oxide' could adopt the *wurtzite* structure, the *zincblende* structure or even occasionally the *rocksalt* form. 'Manganese oxide' offers an even longer list of possible structures. And that's before we even consider clustering and nanoscale phases. A top-down approach would suggest we choose one and try to see if it comes close. If the first one we choose fails, we then try another, and so on. But which one do we start with?

BOX 15.4 Start Under the Light

This may be a good time to remind yourself that EXAFS is not the only arrow in your quiver. X-ray diffraction, for instance, while it probably won't tell you where the manganese is located, may be able to identify which zinc oxide structure has been formed. And examination of the manganese XANES using the methods of Chapter 9 is likely to narrow down the possibilities for manganese oxide species. Other tools, such as x-ray photoemission spectroscopy, could also be very helpful. And while theory and the existing literature on the system aren't always right, they can at least identify particularly likely structures.

Kitsune has helped you narrow the list down, but you'll still have choices. After all, if you're down to just one possibility, then why do we need EXAFS? (Well, it never hurts to have confirmation, but still.) Which structure should you try first? I say try the easy ones first. Interstitials, for example, are hard because it takes some thought to figure out where there's space for an atom and even then it's likely to distort the structure around it more than a substitution would. And some structures, like Mn_3O_4, have more than one crystallographic site for the manganese atom and that's a pain in the tail. Clustering and nanoscale phases sound annoying too, so I wouldn't start with those unless other evidence already pointed toward them.

I once heard a joke about someone at night looking for their car keys under a street light because that's where the light was. Isn't that what you're doing, Dysnomia? Looking where it's light because it's easier?

You missed the point of that joke, Simplicio. The original joke is funny because the guy knows he dropped his keys somewhere else. If he's not sure where he dropped them, it makes perfect sense to start by looking in places where they'd be easy to find. After all, if he finds them, he's done, and he never has to look in the hard places, like under a car.

Be careful, Dysnomia! How do you know the choices you're not checking don't also fit well?

You're right, Mr. Handy. I do have to know that the choices I'm not investigating thoroughly would produce spectra different enough from the one that works that they couldn't also work. But I can find that out either by just trying a quick-and-dirty fit on each of the candidates or because I have a good enough theoretical sense of how spectra look. For example, if an oxide fits well according to the criteria from Chapter 14, then I don't need to bother to try a metallic form because the nearest-neighbor distance would be so different.

Since zinc oxide is the predominant phase, the first part of the task would be to determine what its structure was in this sample. It turns out that the EXAFS of the wurtzite and zincblende structures look quite similar, as both feature a zinc atom in an oxygen tetrahedron which connects to 12 identical tetrahedra through the vertices.

But wurtzite is hexagonal while zincblende is cubic, so it's a cinch to tell them apart by x-ray diffraction.

Once the correct zinc oxide phase was identified, we could follow Dysnomia's advice and try the easiest choice for the manganese edge first, which probably means substituting a single manganese absorbing atom for a zinc atom in the zinc oxide structure we had established.

Initial fits could use very few parameters. As a first simple model, we could approximate the effect of the manganese atom as changing the size of the local tetrahedron without disturbing atoms farther out in the lattice much. That would mean fitting S_o^2, E_o, a half-path length for the nearest-neighbor oxygens, an MSRD for the nearest-neighbor oxygens, and an MSRD for more distant paths. The key to this approach is to fit out well beyond the nearest-neighbor peak in the Fourier transform from the start, without giving much freedom to the structure to move around. If the structure is wrong because, for instance, the manganese is interstitial, the peaks beyond the nearest-neighbor oxygen scatterers will not even be close. If, on the other hand, the initial simple model is on the right track, there may be some mismatches in both amplitude and phase, but the major peaks of the fit will qualitatively follow the data.

If the structure we first try appears to be completely wrong, the plan is to move on to another structure. If it's on the right track, we can then gradually make the model more realistic, perhaps introducing features from our secondary questions, such as clustering of manganese atoms on nearby sites.

Hopefully, this plan will eventually yield an extremely convincing fit according to the criteria of Chapter 14. At that point, you can be fairly confident that you have the right structure. But, as Dysnomia mentioned in Box 15.4, it might still be helpful to make simple models for some of the other possible structures that you haven't tried yet as that will strengthen your argument as to which structure is the correct one.

The strength of this plan is the complexity of the theoretical standard—there are loads of paths and tons of wiggles in the Fourier transform—matched with the simplicity of a model with just a few free parameters. If a fit like that works, you can feel pretty good that you've nailed down the right basic structure.

15.4 Fit!

15.4.1 *Choice of* R *Range*

The low end of the R range is usually chosen based on the background subtraction. While different software packages use different schemes for background subtraction, there is often a value of R below which the data are compromised by the background subtraction algorithm; that value, or a few tenths of an angstrom above it, is generally a good lower end for our fitting range.

While we refer to Fourier transforms throughout this chapter, all advice given also applies for wavelet transforms, with the additional requirement of choosing ηs (Section 7.6.3 of Chapter 7).

If it is not clear from your software documentation how to establish that value, take the Fourier transform after performing different, but reasonable, background subtractions and choose the point where the background stops having a strong effect on the Fourier transform. For example, look back at Figure 7.16 in Chapter 7. In that figure, the two reasonably appropriate background subtractions yield Fourier transforms that don't differ much above around 1.1 Å. In that case, therefore, 1.1 Å would be a reasonable choice for the low end of the fitting range in R-space.

Choosing the high end of the fitting range is a more involved question, which we will address in the context of the next two examples.

15.4.1.1 Example: Which ligand?

In Chapter 13, we used the scattering off of the nearest-neighbor sulfurs in pyrite as the example path for our graphs. Since pyrite is one of the candidates for the iron species present in the soil sample of the example from Section 15.1.1, let's continue that example by examining, in Figure 15.3, the magnitude of the Fourier transforms of the first four paths in pyrite.

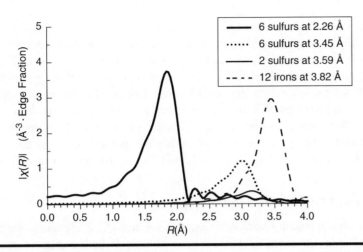

Figure 15.3 · Magnitude of the Fourier transform of the first four single-scattering paths in the structure of pyrite, FeS$_2$.

Suppose we were adopting a bottom-up strategy and we wanted to try to fit just the first sulfur path. An examination of the nominal half path lengths of the first four paths of pyrite, shown in the legend of Figure 15.3, might make you think that $R = 3.0$ Å would be a reasonable top end of our fitting range as it is well above the $D = 2.26$ Å of the first path and well below the $D = 3.45$ Å of the next one.

But inspection of the graph shows this would not be a good choice, as below $R = 3.0$ Å in the Fourier transform there is still a large contribution from the second-nearest neighbors and even a substantial contribution from the more distant iron scatterers ($D = 3.82$ Å).

This reminds us of two facts we should consider when choosing the R range over which to fit:

1. As discussed in Box 1.4, the Fourier transforms of paths peak at a value of R well below their half path length D.
2. The Fourier transforms of paths are not sharp spikes—they have substantial width. This is in part because of disorder but is also because of the effect of taking a Fourier transform on a finite range of data.

It's not surprising, therefore, that a path with half path length $D = 3.45$ Å peaks around $R = 3.0$ Å and has substantial amplitude well below that.

If we knew we were dealing with pyrite and only wanted to fit the first path, Figure 15.3 shows us that we would do well by choosing the top end of our fitting range to be around $R = 2.1$ Å.

Of course, in this example, we wouldn't know whether the material was pyrite when we began fitting. But if the nearest-neighbor atoms were all sulfurs, then the next-nearest neighbors couldn't be a whole lot closer than the $D = 3.45$ Å found in pyrite—there just isn't room. That means that, while the Fourier transform for a different iron sulfide mineral—pyrrhotite (Fe$_{1-x}$S), for example—might not bear much resemblance to Figure 15.3 in the region around $R = 3$ Å, it is still likely that the first path could be reasonably well isolated by choosing a fitting region that stopped at $R = 2.1$ Å.

But what if we did begin to suspect we were dealing with a pyrite-type structure and wanted to extend the fit to include the sulfurs at $D = 3.45$ and 3.59 Å but were not yet ready to try to include the iron scattering at $D = 3.82$ Å? That would be a bit more challenging, as Figure 15.3 shows that the sulfur paths we want to include overlap significantly with the iron path that we do not. The best we could do might be to choose the top end of the fitting range around $R = 2.9$ Å, thus including a significant portion of the sulfur paths but avoiding most of the iron.

In agreement with the graphs in Chapter 13, Figure 15.3 shows an FEFF calculation for the iron K edge of pyrite, using half path lengths and degeneracies drawn from the crystallographic structure and using the following parameters for all paths: $\sigma^2 = 0.003$ Å^2, $S_0^2 = 0.90$, $E_0 = 7112$ eV, $C_3 = C_4 = 0$. The Fourier transforms are taken from 3 to 15 Å^{-1} using a k-weight of 2 and Hanning windows with sills of 0.5 Å^{-1} on each side.

But, 2.1 Å would put the top edge of our fitting range *below* the D for the first path! That doesn't seem right!

Get over it, Simplicio! The Fourier transform is not a radial distribution function. The sooner you come to grips with that, the better.

It's also worth noting that the sulfur path at $D = 3.45$ Å and the path at $D = 3.59$ Å could initially be lumped together into a single sulfur-scattering path since they overlap substantially and feature the same scattering species. In fact, if the material is structurally *similar* to pyrite but is modified in some way, it's not unlikely that the details of how many sulfurs are at what distances could be quite different than in the pure pyrite that we are using for a model. Whether or not particular data are fit better by a split shell with two sulfur distances, or by a single shell with one, is the kind of thing that can be addressed as the fitting process progresses. (This is a great chance to try the Hamilton test, described in Sections 10.4.5 and 14.2.2 of Chapters 10 and 14, respectively!)

15.4.1.2 Example: Where's the dopant?

For a top-down approach like that we began in Section 15.1.2, the choice for the high end of the Fourier transform is somewhat more arbitrary. If, for instance, we are attempting to model the manganese edge by using a manganese atom substituted into a zinc site in a wurtzite structure, we actually know more about what the structure should look like far away from the absorber than near it. After all, the manganese substitution will introduce distortions near the site, but those distortions may be smaller farther away.

Nevertheless, there are limits to how high in the Fourier transform we can fit. For one thing, both the $1/D^2$ and the $e^{-2D/\lambda}$ in the EXAFS equation will result in a decrease in the signal-to-noise ratio as R increases.

Besides the loss of signal at high R, high-R data are more sensitive to errors in theory and the fitting model. For example, suppose that at some value of k within the data range the mean free path should be 9 Å, but the model is using a value of 10 Å. For a path at $D = 2.0$ Å, the error in the amplitude, as given by the $e^{-2D/\lambda}$ factor, would be about 5%. But for a path at $D = 6.0$ Å, the error in amplitude would be 14%. High-D paths also oscillate much more rapidly in k-space, causing them to be more sensitive to a proper choice of energy origin. Thus, the correct treatment of E_o and energy broadening are more important for high-D paths. (As another example of a high-R effect, Kitsune will show us how an oversimplified fitting model can affect high-R data more than low-R data in Box 15.5.)

Geometrical considerations also dictate that multiple-scattering paths become more numerous relative to single-scattering paths as D increases. For many materials, it can be difficult to know what constraint scheme is most appropriate for multiple-scattering paths (see Section 17.5 of Chapter 17), and as their relative importance increases, the question becomes more pressing. (Mandelbrant will show a graphical example of this in Box 15.5.)

Finally, extending fits to higher R (and thus including paths at higher D) is computationally taxing. With today's computers, this doesn't matter much for a single fit. But it could be important for automated fitting of large amounts of data, such as can sometimes arise in connection with quick-XAFS experiments.

BOX 15.5   How High?

 The phrase 'somewhat more arbitrary' at the start of Section 15.4.1.2 and the discussions of theory don't actually tell me how to choose the top end of the range. How do I decide how far up to fit in R when I'm using a top-down strategy?

At some point as you push R higher and higher, the fit usually starts to get messed up. This is one place where reduced χ^2 (Section 14.1.3 of Chapter 14) is really useful: just keep moving the top of your fitting range up until the reduced χ^2 starts to go up.

But Dysnomia, why does it get worse like that? I've never fully understood that. From the point of view of the EXAFS equation, both $1/D^2$ and $e^{-2D/\lambda}$ should damp out the signal relative to the noise, yes. And for simple cases like copper metal that's more or less what happens. But for the kinds of projects you work on, Dysnomia, such as the doped zinc oxide film we're considering, I've noticed that the fit often degrades fairly rapidly around 5 Å. I don't like not understanding why that happens. Do you have any ideas?

Perhaps I can help. Copper metal adopts a cubic space group. But the wurtzite structure (Figure 12.4) is hexagonal, and crystallography has shown the c/a ratio is about 2% smaller than the 'ideal' ratio. In terms of local structure, that means the zinc atoms sit inside oxygen tetrahedra that are a tiny bit 'squashed,' so that one oxygen is a different distance from the zinc atom than the other three. That's a subtle effect for nearest neighbors, and the scattering can probably be handled by a single zinc–oxygen path with a fitted MSRD and bond length. The first shell of zinc–zinc scattering is also mildly split, with the in-plane distance being a few hundredths of an angstrom different from the out-of-plane distance. Once again, this scattering could likely be modeled by a single path if desired. But as we move to paths at higher and higher D, the difference of about 2% between the vertical and the horizontal axes becomes a greater and greater splitting when measured in angstroms (2% of 6 Å is three times as much as 2% of 2 Å, for instance). At the very least, this will show up as an increase of MSRD for high-D paths, but it may even cause a splitting that was not resolved at low R to begin to show up at high R. Figure 15.4 compares calculated spectra for the crystallographic and ideal (undistorted tetrahedra) cases. It's easy to see that at low R the shapes of the two curves are qualitatively similar, while at high R they increasingly diverge.

Well put, friend Kitsune. But that is not the only problem we must face. Multiple-scattering paths have a greater and greater influence on the spectrum as we move to higher R. As we shall see in Section 17.5 of Chapter 17, we are often in doubt as to how to constrain multiple-scattering paths. At high R, those paths that fill us with uncertainty begin to dominate and our fits degrade. We can see this in Figure 15.5, which shows the same 'crystallographic' calculation friend Kitsune used, along with one in which the multiple-scattering paths are constrained to have double the MSRD as the single-scattering paths.

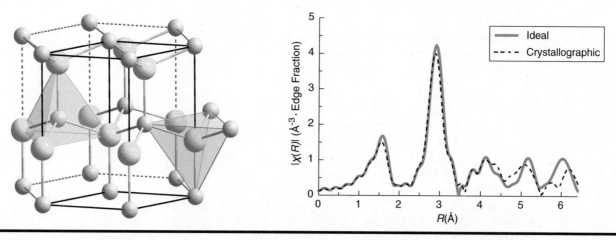

Figure 15.4 (left) Wurtzite structure. Larger spheres represent oxygen atoms; smaller spheres zinc atoms. (right) Magnitude of the Fourier transform of the zinc *K* edge of wurtzite, calculated for both the ideal and crystallographic *c/a* ratios. Details of structural parameters (e.g., the value for the MSRDs) and transform are the same as those used in Chapter 13 and Figure 15.3, except that $E_o = 9659$ eV.

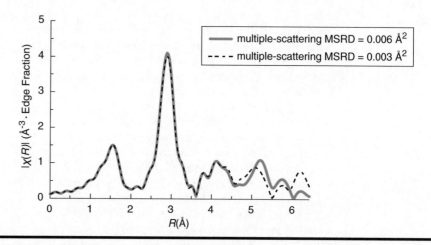

Figure 15.5 Dashed line is the 'crystallographic' graph from Figure 15.4. Solid line is the same calculation, except that the MSRD for multiple-scattering paths is constrained to be twice that of single-scattering paths.

Does it matter what causes the fits to mess up when we go higher in *R*? It could be signal-to-noise ratio, or the geometry of the model, or multiple-scattering paths, or something else.

Regardless of the cause, the, χ^2_ν of the fits will tell us when we are going past what our data and model can handle.

15.4.2 Art of Fitting

As Field Marshall Helmuth Carl Bernard Graf von Moltke wrote, 'No plan of operations extends with any certainty beyond the first contact with the main hostile force' (Hughes and Bell 1995). What is true of war is also true of fitting: the results of the first fit you attempt might send you on a trajectory that you would not have expected at the outset. We cannot, therefore, tell you 'how to fit' in the same way that we could tell you how to synthesize a particular chemical or bake a particular kind of cake. The best we can do, for now, is to ask our panel for advice (Box 15.6) and then, in Chapter 19 of this book, show you some examples of what experts have done.

BOX 15.6   Fitting Advice

	My take is: *be fearless.* Nobody's hovering over your shoulder when you're doing fits; you don't have to have a reason for the stuff you try. Follow hunches, try wacky things, explore the fitting space. Once you start to understand how to fit a set of data, there will be plenty of time to confirm your conclusions, stress test your fit, and construct a convincing argument.
	One should *keep proper notes.* Much as it pains me to admit, Dysnomia and the Field Marshall are correct: much of fitting is improvisation. Because of that, it is very important to take notes as one fits, both on the attempts one is making and on what results arise.
	Continuously evaluate your results and your constraints, using the methods of Sections 14.7 and 14.8 of Chapter 14. You can waste a lot of time by constructing a fit where the fitted model matches the data beautifully—and makes no physical sense.
	I suggest you *do more than just fit.* As you begin to learn more about your system through EXAFS analysis, you may think of aspects of the system that you or your colleagues could investigate in other ways. Perhaps, in a soil sample, the fits suggest the presence of a compound you did not expect. Before continuing too far, consult with a soil scientist who knows something about your sample to see what they think. Or maybe a fit raises questions about morphology that a look at transmission electron microscopic images would address.
	One way or another, friends, we should *start simple,* either by beginning with very few paths or by starting with very few free parameters. A fit with too many free parameters early in the process is only likely to cause confusion. We can always add complexity as we go.
	I've got one! *Don't be afraid to make mistakes.* I bet I can learn as much about my system (and how to get better at fitting) from things that don't work as things that do.

15.4.3 Perfecting Your Fit

Often, you will find that early in the fitting process the fit seems very fragile—perhaps it only works for a certain *k*-weight or when a certain path is excluded. But as the process continues and you get a feel for the data and the model, it will seem to become more robust. By the time that you're satisfied that you have a good fit, you are likely to have several good fits, in the sense that while details of the fitting model or ranges vary, the results are consistent. You will naturally pick the 'best' from this set, using the criteria of Chapter 14. But it's also usual at this stage to systematically explore the fitting space a little bit around the fit you have chosen. Here are some things you could try:

- Move the lower end of the k range up or down, perhaps by 0.5 Å^{-1} at a time
- Move the upper end of the k range up or down
- Move the lower end of the R range up a bit (perhaps 0.2 Å at a time). You can't move it down, though, if you've already set it at the minimum determined by the background subtraction procedure.
- Move the upper end of the R range up or down a bit
- Try a different (but reasonable) background subtraction.
- Try relaxing a constraint that you're not sure about. For example, if you've forced all MSRDs beyond the nearest neighbor to adopt the same value, try splitting off the next-nearest neighbors.
- Try adding a constraint. For example, if two related MSRDs adopt values that are consistent to within their uncertainties, try constraining them to each other.
- Try adding multiple-scattering paths that you'd previously considered negligible or paths that have most of their amplitude above the range over which you are fitting.

While you don't have to do all those things, each one that you do try provides you more fits to choose from, and more of an opportunity to use the approaches of Chapter 14 to pick the one you like best.

15.4.4 Stressing Your Fit

As the Nobel Prize-winning physicist Richard Feynman said, 'The first principle is that you must not fool yourself—and you are the easiest person to fool' (Feynman 1997). So once you have a fit that you really like, the next step is to try to prove it wrong.

The process for this is very similar to the process we just described in Section 15.4.3 except that now instead of looking for the best fit, you're looking for other perfectly reasonable fits that yield different results: in other words, you're checking the 'stability' criteria from Section 14.6 of Chapter 14. Picture yourself as a rival researcher trying to knock down your argument and try to do things to the fit that give a different result that is also plausible.

In the end, you will often find that you have a set of conclusions that are quite robust. In the example, we introduced in Section 15.1.2, for example, you might come to a firm conclusion that the manganese is substituting for the zinc in a wurtzite-type structure. You might also have conclusions that are fairly robust but leave some room for doubt—perhaps you have indirect evidence for clustering of the manganese atoms but can't quite rule out some other kind of distortion. And you may have some questions you end up unable to answer at all because you couldn't get an unambiguous result when you stressed your fits.

But that mix works just fine for a paper, a presentation, or a dissertation, as we'll see in Chapter 19. Making clear to your readers which of your conclusions are firm and which have some wiggle room increases their confidence in your work overall.

WHAT I'VE LEARNED IN CHAPTER 15, BY SIMPLICIO

■ Before beginning to fit, we should identify the primary and secondary questions we're trying to answer.

■ We should use signal-to-noise ratio to choose the high end of our data range in k-space.

■ We should choose the low end of our data range in k-space, in part, based on where $\chi(k)$ becomes independent of reasonable background choices. We should also consider the trade-off between incorporating additional data and having to deal with theoretical and analytical approximations that become less valid at low k.

■ We should choose the low end of our fitting range in the Fourier transform based on the background subtraction we used. The high end should be based on excluding substantial contributions from paths we are *not* including in our model, rather than on trying to include the entire contribution from paths we *are* including.

■ When we have found a fit we're satisfied with, we should stress it in a variety of ways to understand its limitations.

■ We should be fearless, keep proper notes, continuously evaluate our results, do more than just fit, start simple, and not be afraid to make mistakes.

References

Feynman, R. P. 1997. Cargo cult science. In *Surely You're Joking, Mr. Feynman*, ed. R. P. Feynman and R. Leighton, and A. R. Hibbs, 338–346. New York: W. W. Norton.

Hughes, D. J. and H. Bell. 1995. *Moltke on the Art of War: Selected Writings*. New York: Presidio Press.

Chapter 16

Starting Structures

Wrong Is Better Than Nothing

Theoretical standards are, of course, different for different edges! Make sure, when you are using software to calculate a theoretical standard, that you specify the correct edge!

This one I understand! I start wrong all the time, and yet that's the way I find out what's right!

In Chapter 12, we discussed how theoretical standards are calculated. To calculate a theoretical standard, however, we must have a starting structure. This chapter is about choosing that starting structure.

16.1 Crystal Structures

Frequently, we begin with a known crystal structure. As a simple example, we might be dealing with a modified version of a bulk crystal—perhaps a thin film or a nanoparticulate powder. Beginning with the structure of the bulk crystal and using a top-down strategy (Section 12.4.2 of Chapter 12) is a good way to go.

But crystal structures can also be good starting points for amorphous materials. Section 16.1.1 will explain why.

16.1.1 Cluster Size and EXAFS

XAFS can also be dependent on the interaction between the polarization of the incident x-ray and the orientation of the material being measured (Ankudinov and Rehr 1997; Bunker 2010). For powders, liquids, gases, and the like, polarization effects will average out because those samples exhibit all orientations. But for single crystals (if not being spun), layered solids, and other 'textured' materials, one should consider polarization when computing the theoretical standard. This is particularly crucial for edges such as L_2 and L_3 that promote electrons from orbitals that are not spherically symmetric.

Consider a set of theoretical standards based on the iron(II) oxide structure. Iron(II) oxide adopts the rock salt structure and thus consists of alternating iron and oxygen atoms on a cubic lattice, as shown in Figure 16.1.

Suppose we measure the iron K-edge EXAFS and want to investigate the nearest-neighbor oxygen scattering. To be accurate, the theoretical standard needs to be able to construct a potential (most likely a muffin tin, as in Section 12.2.1 of Chapter 12). Depending on the algorithm used, the calculation may incorporate aspects such as charge transfer between atoms, overlap of the atomic potentials, and values for the interstitial potential. If we only feed the calculation the absorbing iron atom and the six oxygen atoms immediately surrounding it, those aspects are likely to be approximated poorly; that is, we would be calculating a theoretical standard for an FeO_6 molecule, which is significantly different from an FeO crystal.

DOI: 10.1201/9780429329555-20

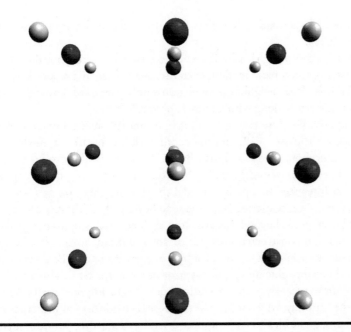

Figure 16.1 **Iron(II) oxide structure. Iron atoms are shown darker than oxygen atoms.**

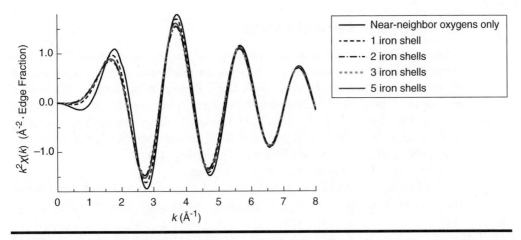

Figure 16.2 **(See color insert.) Theoretical standards for the nearest-neighbor scattering path of iron(II) oxide using clusters of different sizes. The calculation, for example, for 'two iron shells' also includes all oxygen shells closer to the absorber. Calculations were made using FEFF6L with default settings.**

How many more shells would we need to include in the calculation to get the first shell nearly right? To see, examine Figure 16.2.

As expected, including only the near-neighbor oxygen atoms in the calculation yields a noticeably different result from using a large cluster, especially at low k. But it is intriguing that even when the second shell of iron scatterers (and all atoms interior to them) is included, the amplitude is still off by nearly 5% for the peak just below 4 Å^{-1}. When we consider what a 5% amplitude error means to, for example, the determination of coordination numbers we can see that this is a significant source of systematic error.

Convergence has essentially been achieved, however, by the time the third shell of iron atoms is added. For the peak just below 4 Å^{-1}, the amplitude of the 'three iron

Please notice, friends, that Figure 16.2 shows the theoretical standard for the nearest-neighbor scattering path only. This is not an issue that we can solve by Fourier filtering since we're already looking at only one path!

shells' theoretical standard only differs from that of the 'five iron shells' theoretical standard by 0.3%.

It is also worth noting that by 7.5 Å⁻¹, the systematic error introduced by using a smaller cluster is much more modest: even the 'one iron shell' amplitude is only off by about 0.5%. (The 'near-neighbor oxygen atoms only' standard, however, still exhibits a 5% amplitude error, along with a modest phase shift.)

Now suppose we were trying to model an amorphous iron oxide and we think the iron valence is close to +2, and the coordination is mostly octahedral. Since the material is amorphous, the radial distribution function is likely to be quite different from that of crystalline iron(II) oxide. We would not use a top-down approach—we are really just interested in a path to model the near-neighbor oxygen scattering, and perhaps the iron–iron scattering from the shell beyond that. The exact locations of the oxygen and iron atoms beyond that first iron shell are irrelevant to our analysis.

And yet, if we constructed a cluster of an iron atom surrounded by six oxygen atoms, or even an iron atom surrounded by 6 oxygen atoms and 12 iron atoms beyond that, we could be introducing significant error into the nearest-neighbor path. Instead, we would want to create a cluster extending out several additional shells. As long as we had to do that anyway, why not just use the iron(II) oxide crystalline structure? That's why it's often a good idea to use a similar crystal as a starting point for analyzing an amorphous material, even if we plan to use a bottom-up approach (Section 12.4.1 of Chapter 12).

As we said in Section 15.2.3 of Chapter 15, you avoid a lot of complications by not trying to fit too low in *k* range. Sometimes your data aren't good enough to start higher, but if you've got data out to 12 or 15 Å⁻¹ or something good like that, do yourself a favor: don't try to start at 2 Å⁻¹!

16.1.2 Cluster Size and XANES

When learning XAFS analysis, it is easy to get the impression that XANES depends primarily on the absorbing atom and its nearest neighbors; after all, those are the characteristics most often identified by XANES fingerprinting (Sections 9.2 and 9.3 of Chapter 9). But Figure 16.2 suggests that cluster size has a greater impact at energies closer to the edge, and Section 12.3.2 of Chapter 12 tells us that multiple scattering also plays a greater role. In addition, we will not have the benefit of Fourier filtering; any XANES analysis necessarily involves all scattering directly.

Figure 16.3 demonstrates this slow convergence.

Unlike for the EXAFS shown in Figure 16.2, even the inclusion of atoms out to the third shell of iron atoms is not enough to ensure good convergence, while the

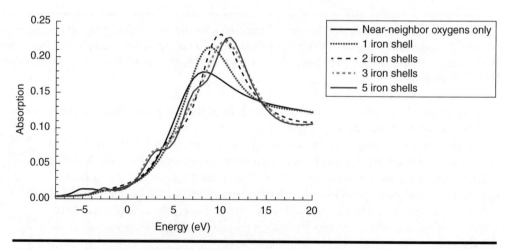

Figure 16.3   Theoretical standards for the x-ray absorption near-edge structure of iron(II) oxide using clusters of different sizes. The calculation, for example, for 'two iron shells' also includes all oxygen shells closer to the absorber. Calculations were made with FDMNES using the finite difference method and including quadrupole contributions.

inclusion of only the nearest-neighbor oxygen atoms and the first shell of iron scatterers is a rough enough approximation that it looks like a different species from the 'five iron shells' model. Considering that the latter model could take an hour or more to compute on a laptop, it is clearly computationally intensive to produce quantitatively accurate XANES standards.

16.1.3 Sources for Crystal Structures

While it's possible to find crystal structures by searching the primary literature, it is often efficient to first search databases such as the Crystallography Open Database (Gražulis et al. 2012), which includes many published structures for small- to medium-sized unit cells. Structures for macromolecules can be found in the Protein Data Bank (Berman et al. 2002).

16.2 Calculated Structures

Many techniques now exist for optimizing the structures of small molecules, atomic clusters, crystals, and solutions. It is, therefore, usually possible to take a rough guess as to the arrangement of atoms in a material and compute a set of stable atomic positions, perhaps even including MSRDs (Poiarkova 1999) and cumulants. Such techniques are still specialized to particular classes of materials and often require specialized knowledge to use reliably. We therefore will not detail them in this book, except to mention that they constitute a valid method of generating starting structures, and to include a few examples in our case studies (Chapter 19).

Realize that the process described in Section 16.2 is generally a two-step process: first you use one piece of software to generate the coordinates for a reasonable starting structure and then another (such as FEFF) to compute a theoretical standard that corresponds to that structure.

BOX 16.1   A Little More

	Wait! We're just going to leave it there? No references? No examples? No details?
	We've got to leave it there, Simplicio. Calculating structures for a particular material is about as discipline-specific as it gets. It is really a question for textbooks on the individual systems being studied, not for a book on XAFS.
	We can at least provide a few examples from the literature, even if the list is in no way comprehensive: if you're interested, you can check out Choi et al. (2003), Dimakis and Bunker (2001), Haumann et al. (2006), and Molenbroek et al. (2001).

16.3 Mixtures

What if your sample comprises a mixture of materials contributing to the measured edge, such as was the case in the examples from Chapters 10 and 11? Then, just as in those chapters, we can treat the spectrum as a linear combination of standards, only this time, they will be theoretical standards.

The EXAFS equation could then be written as

$$\chi(k) = \sum_j \left(X_j S_o^2 \sum_i N_{ji,pure} \frac{f_{ji}(k)}{kD_{ji}^2} e^{-\frac{2D_{ji}}{\lambda(k)}} e^{-2k^2\sigma_{ji}^2} \sin\left(2kD_{ji} + \delta_{ji}(k)\right) \right) \quad (16.1)$$

where the outer summation is over the different constituents j of the mixture, each with fraction X_j.

In principle, a mixture can now be fit in the same manner as a pure-phase sample. There is more than one theoretical standard, each weighted by X_j. The X_js can be fitted or constrained to values determined in other ways. At least one constraint can certainly be applied:

Remember that there are lots of ways to measure percent composition that are not the same as the XAFS fractions! Look back to Box 10.1 of Chapter 10 if you need a refresher!

$$\sum_j X_j = 1 \quad (16.2)$$

While Equation 16.1 is probably the most straightforward way of writing the EXAFS equation for mixtures, it does require introducing new parameters in the form of the X_js. That's not friendly to software that expects you to be expressing your models in terms of the standard, single-phase EXAFS equation.

One way to modify Equation 16.1 to express it in terms of the standard parameters alone is to define

$$S_{oj}^2 = X_j S_o^2 \quad (16.3)$$

and thus write

$$\chi(k) = \sum_j \left(S_{oj}^2 \sum_i N_{ji,pure} \frac{f_{ji}(k)}{kD_{ji}^2} e^{-\frac{2D_{ji}}{\lambda(k)}} e^{-2k^2\sigma_{ji}^2} \sin\left(2kD_{ji} + \delta_{ji}(k)\right) \right) \quad (16.4)$$

Alternatively, the degeneracy N_{ji} could be considered to be the average degeneracy of that path for the atoms of the element, leading to

$$\chi(k) = S_o^2 \sum_j \sum_i N_{ji} \frac{f_{ji}(k)}{kD_{ji}^2} e^{-\frac{2D_{ji}}{\lambda(k)}} e^{-2k^2\sigma_{ij}^2} \sin\left(2kD_{ji} + \delta_{ji}(k)\right) \quad (16.5)$$

Equation 16.5 is elegant, in that N_{ji} has a natural extension of the usual meaning of N_i and that the sum over paths is now clearly generalized by being a sum over the paths of each theoretical standard. With this definition,

$$N_{ji} = X_j N_{ji,pure} \quad (16.6)$$

Whichever method is used in the computation, S_o^2 and the X_js are the quantities that should actually be reported to avoid confusion between quantities such as S_{oj}^2 and S_o^2, or between N_{ji} and $N_{ji,pure}$.

BOX 16.2   An Example

Perhaps an example will be useful, friends. Suppose we have a material where 24% of the cadmium is present as monteponite and 76% as greenockite, as in Section 10.3 of Chapter 10. Formally, we would wish to take each path in the theoretical standard for monteponite and multiply it by 0.24 and add the result to each path in the theoretical standard for greenockite multiplied by 0.76. This is what Equation 16.1 tells us to do and is the same principal as is used in linear combination analysis.

But a lot of XAFS software won't let us do that. Instead, we've got to work with the regular path parameters. So here's what Equations 16.3 and 16.4 tell us we can do. Suppose, the S_o^2 for cadmium in the measurement is 0.87. (Remember, S_o^2 shouldn't be different for the same edge of different constituents.) We could use the paths for monteponite, but make the S_o^2 for monteponite $0.24 \times 0.87 = 0.21$, and use the paths for greenockite but make its $S_o^2 = 0.76 \times 0.87 = 0.66$. To a fitting program, it looks like one big theoretical standard with a lot of paths, some of which have different values for S_o^2. It doesn't make any physical sense, but it works.

Hrumph. I would prefer something with a more physical basis. And that can be done using Equations 16.5 and 16.6. Monteponite takes on the rock salt structure (Section 16.1.1), so an absorbing cadmium atom in pure monteponite has six nearest-neighbor oxygen atoms. Since 24% of the cadmium atoms are in that environment, the average degeneracy for that path is $0.24 \times 6 = 1.44$. The next-nearest-neighbor path in monteponite is made up of 12 cadmium atoms, so in our mixture that path would have an average degeneracy of $0.24 \times 12 = 2.88$. Greenockite, on the other hand, takes on the wurtzite structure (Section 15.4.1 of Chapter 15). In pure greenockite, there would be four nearest-neighbor sulfur atoms. In our mixture, the average degeneracy of that path would be $0.76 \times 4 = 3.04$. All paths in both theoretical standards, including multiple-scattering paths, can be handled analogously and doing it this way gives the parameters a physical meaning.

In either method, the fraction of each constituent in the mixture could be a free parameter of the fit.

Whichever method one uses, however, the results should be reported in the same way. Dysnomia would report S_o^2 as 0.87, not 0.21 or 0.66, and Carvaka would specify the coordination number of the near-neighbor oxygen atoms in monteponite as 6, not 1.44.

16.4 Inequivalent Absorbing Sites

Some chemical compounds or crystal structures have the same element present in different environments within the same structure. For example, in Section 9.2 of Chapter 9, we discussed the spinel structure, in which cations can occupy tetrahedral or octahedral sites. Consider magnetite, a spinel with chemical formula Fe_3O_4. In this material, one-third of iron atoms are in tetrahedrally coordinated sites, while two-thirds are in octahedrally coordinated sites.

For EXAFS, this can be thought of as if it were a mixture of two constituents and the methods of Section 16.3 applied.

16.5 Histogram Methods

Up until now, we have accounted for thermal and static disorders by relying on the MSRD in the EXAFS equation, possibly with the modifications introduced by higher-order cumulants (Section 13.2 of Chapter 13). In some systems, however, the cumulant expansion does not provide a good description of the actual distribution of distances (Ravel et al. 2006). As an example, nanoparticles may include atoms in the cores, atoms on faces, atoms on edges, and atoms at vertices, resulting in a multimodal distribution of near-neighbor bond lengths (Price et al. 2012).

For these cases, one approach is to divide nearest-neighbor bond lengths into ranges (e.g., 1.90–1.91 Å, 1.91–1.92 Å, 1.92–1.93 Å …) and calculate a theoretical standard for a configuration corresponding to each. This is sometimes known as a *histogram approach* (Price et al. 2012). This, however, introduces at least one new parameter for each bin, specifying how much weight is to be given to that bin (analogous to the X_js of Section 16.3). While this approach may also remove the need for a few parameters, such as bond length and cumulants, there will nevertheless generally be a dramatic increase in the number of parameters needed to describe the system.

This in turn means that the parameters must be highly constrained. This can be done by creating a theoretical simulation of the distribution of bond lengths, perhaps by application of density functional theory (Ravel et al. 1999), molecular dynamics (Price et al. 2012), or Monte Carlo techniques. The distribution can then be used to dictate the weightings.

BOX 16.3   What's the Point?

	This approach is new to me, and I'm a little confused as to the point. If I am using a theoretical method to calculate the distribution of bond lengths, why I am using EXAFS at all? What information am I extracting from the measurement itself?
	Fitted parameters can still be introduced. For example, an average bond length could be fit by introducing a parameter that changes all the bond lengths in the histogram by the same fraction. Or an additional MSRD adjustment can be applied, allowing for more or less disorder than in the theoretical calculation. Or the results of the calculations can be used to fit the distribution as a superposition of a small number of Gaussian peaks, each of which could then be treated as an ordinary path with the usual adjustable parameters.

The conventional EXAFS equation makes the assumption that the distribution of lengths for a given path can be modeled well by a Gaussian. The histogram method can be thought of as simply allowing one to substitute a more appropriate distribution function.

 And we should not forget a simple application, my friends. Even without free parameters, we can use the EXAFS to confirm or refute the theoretical computation. This in itself provides us valuable information about our system.

16.6 Multiple-Edge Fits

Often, it is possible to measure more than one edge of a sample (e.g., the nickel *K* edge and the iron *K* edge). Both edges can then be fit simultaneously.

In Sections 16.3 through 16.5, different environments around absorbing atoms of a particular element were contributing to the same set of data. For multiple-edge fits, in contrast, the environment around each kind of element is contributing to different sets of data.

What, then, makes this a 'simultaneous' fit? It must be that constraints are used to couple the two edges to each other. Consider, for instance, an alloy of copper and zinc. The theoretical standard for the copper *K* edge will include a path for scattering by nearby zinc atoms, while the theoretical standard for the zinc *K* edge will include a path for scattering by nearby copper atoms. It is geometrically necessary that those paths be the same length, have the same MSRD, and exhibit the same cumulants.

This is not quite as straightforward as it may sound, however. We are asking a fitting routine to minimize the sum of the χ^2 values for each edge *m*:

$$\chi^2_{total} = \sum_m \chi^2_m \qquad (16.7)$$

It is *not* true, though, that the degeneracies have to be the same!

But as we learned in Section 14.1.3 of Chapter 14, we don't generally know how to compute χ^2! This is a significant problem, as it determines the relative weighting of the edges in the fit. The problem is particularly acute if different *k*-weightings or different *k* ranges are applied to the data sets associated with the different edges.

Despite the frequent appearance of multiple-edge EXAFS fits in the literature, this issue has not, to our knowledge, been adequately addressed. Fortunately, there is a solution which is relatively easy to implement.

Recall from Section 14.3.1 of Chapter 14 that when we are done fitting, we generally assume we have a good fit and use the preliminary value of χ^2_ν to estimate ε. If we follow the procedures from Chapter 14, then for each edge

$$\varepsilon_m = \varepsilon_{random,m} \sqrt{\chi^2_{\nu,m}} \qquad (16.8)$$

While fitting two (or more) edges simultaneously adds substantially to the number of independent points by the Nyquist criterion, the information is likely to fall substantially short of ideal packing (Box 14.1 of Chapter 14).

where we use one of the usual ways to estimate $\varepsilon_{random,m}$ (see Section 7.2.2 of Chapter 7) for each edge. We then rerun the fit using our new estimates of ε_m for each edge. If necessary, we can repeat the process iteratively until the estimates for ε_m stabilize.

The right reason to use multiple-edge fitting is to take advantage of constraints between the edges. If, instead, you're tempted to fit more than one edge simultaneously so that you can 'borrow' independent points from one edge to help fit the other, you're abusing the fact that your data aren't ideally packed!

16.7 Site Occupancy

We are often faced with materials that adopt a crystal lattice but with *substitutional disorder*. For example, an iron–nickel alloy might adopt the nickel crystal structure (i.e., face-centered cubic) but with roughly 20% of the sites occupied by iron atoms. If the iron atoms don't form a regular superstructure but are scattered about more or less at random, we would say that the material adopts the nickel structure with site occupancy of 80% nickel and 20% iron. The lattice constant of the alloy would differ from pure nickel, and there might be short-range distortions such as a modest difference between the nearest-neighbor distance of a nickel–nickel pair and a nickel–iron pair, but the face-centered cubic nickel structure would provide an excellent starting structure for calculating a theoretical standard.

Substitutional disorder can also involve vacancies, that is, site occupancies that total to less than 100%.

16.7.1 Vacancies

Vacancies are the easiest to model, so we'll address them first. They can be modeled as a reduction in path degeneracy. For example, holmium arsenide adopts the rock salt structure we discussed in Section 16.1 but can include arsenic vacancies (Taylor et al. 1974). If the vacancy fraction were f, then the degeneracy of single-scattering paths involving arsenic scatterers would be scaled by $(1 - f)$. If the vacancies were randomly distributed, then multiple-scattering paths would be scaled by $(1 - f)^n$, where n is the number of arsenic scattering sites involved in the path.

Notice that it does not matter what the *absorbing* atom is. A single-scattering path at the arsenic edge with a holmium scatterer would *not* be reduced due to arsenic vacancies as data reduction has already normalized the data to the size of the arsenic signal.

If the absorbing atom is found in more than one site or constituent, as in Sections 16.3 and 16.4, then vacancies can change the fractions X_j. But Equation 16.2 still holds, friends—every absorbing atom must be somewhere!

16.7.2 Treating as a Mixture

When a site may be occupied by more than one element, one strategy is to compute a theoretical standard for each possible occupant. For example, let us return to the example from the beginning of this section of an alloy, that is, 20% iron and 80% nickel. When analyzing the nickel edge, one standard could be computed from a face-centered cubic structure with every site occupied by nickel, while another could be computed from a face-centered cubic structure with every scattering site occupied by iron. (The absorbing site should *always* be occupied by the atom corresponding to the edge being analyzed.) The problem can then be treated in the manner of Sections 16.3 and 16.4.

Personally, I prefer the method of Section 16.7.3 to that of Section 16.7.2. The structures where all of the scattering sites are occupied by one element may not correspond to physically realizable compounds, and I thus question how accurate a theoretical standard will be. In addition, multiple scattering will not be properly reflected in this method.

16.7.3 Creating a Mixed Model

An alternative to the method in Section 16.7.2 is to use a model structure that already contains atoms of both types. For example, consider attempting to model the nickel edge of the 20% iron, 80% nickel alloy we have been discussing. In the face-centered cubic structure, the nickel absorber has 12 nearest neighbors. A model could be built with three of those near neighbors occupied by iron atoms and the remaining nine by nickel atoms. Since that makes 25% of the neighboring sites iron, rather than 20%, the degeneracy of each of the iron-direct scatterers would need to be multiplied by $f/25 = 20/25 = 0.80$, while the degeneracy of each of the nickel-direct scatterers would need to be multiplied by $(100 - f)/75 = 80/75 = 1.07$. More distant scatterers can be treated analogously. For example, a triangle path involving scattering off of a nickel atom followed by an iron atom would have its degeneracy multiplied by $\left(\dfrac{f}{25}\right)\dfrac{(100-f)}{75}$, while one involving scattering off of two different nickel atoms would have its degeneracy multiplied by $\left[\dfrac{(100-f)}{75}\right]^2$.

16.7.4 Creating Multiple Mixed Models

The most accurate approach is to create theoretical standards based on multiple mixed models, with the substituted atoms placed in various possible configurations. To take advantage of this level of detail, it is probably best to use theoretical tools to predict the distribution of substitutional disorder, much as theory is used to improve the modeling of absorber–scatterer distances in the histogram method (Section 16.5).

Yeah, I might be a little nervous to use it with an alloy of say copper and mercury because they have very different atomic weights and so the scattering is very different. But iron and nickel? The scattering differences are subtle anyway. The method of Section 16.7.2, while an approximation, would probably be good enough there, and it's a lot less work than figuring out all the proper ratios required by the method in Section 16.7.3.

WHAT I'VE LEARNED IN CHAPTER 16, BY SIMPLICIO

- Crystal structures are good starting points, even if we're not working with crystals.
- We should use a cluster size that goes quite a bit further than the furthest direct scattering path we're planning to analyze.
- Mixtures and compounds with multiple inequivalent absorbing sites can be handled in a way analogous to linear combination analysis.
- If the distribution of absorber–scatterer distances associated with a given path is not fit well by a Gaussian, then we can consider using histogram methods.
- The weighting for multiple-edge fitting should be determined by an iterative process of improved estimates of the measurement uncertainty ε for each edge.
- Systems with fractional site occupancy can be modeled by creating a separate theoretical standard for each kind of element in the site (easiest but least accurate), creating a theoretical standard with some sites occupied by each kind of element and then scaling the degeneracies appropriately, or creating multiple theoretical standards that include different distributions of elements among the sites.

References

Ankudinov, A. L. and J. J. Rehr. 1997. Relativistic calculations of spin-dependent x-ray absorption spectra. *Phys. Rev. B.* 56:R1712–R1715. DOI:0.1103/PhysRevB.56.R1712.

Berman, H. M., T. Battistuz, T. N. Bhat, et al. 2002. The protein data bank. *Acta Crystallogr. D.* 58:899–907. DOI:10.1107/S0907444902003451.

Bunker, G. 2010. *Introduction to XAFS.* New York: Cambridge University Press.

Choi, S. H., B. R. Wood, J. A. Ryder, and A. T. Bell. 2003. X-ray absorption fine structure characterization of the local structure of Fe in Fe-ZSM-5. *J. Phys. Chem. B.* 107:11843–11851. DOI:10.1021/jp030141y.

Dimakis, N. and G. Bunker. 2001. Chemical transferability of single- and multiple-scattering EXAFS Debye-Waller factors. *J. Synchrotron Radiat.* 8:297–299. DOI:10.1107/S0909049500019269.

Gražulis, S., A. Daškevič, A. Merkys, D. Chateigner, L. Lutterotti, M. Quirós, N. R. Serebryanaya, P. Moeck, R. T. Downs, and A. Le Bail. 2012. Crystallography Open Database (COD): an open-access collection of crystal structures and platform for worldwide collaboration. *Nucleic Acids Res.* 40:D420–D427. DOI:10.1093/nar/gkr900.

Haumann, M., M. Barra, P. Loja, S. Löscher, R. Krivanek, A. Grundmeier, L.-E. Andreasson, and H. Dau. 2006. Bromide does not bind to the Mn_4Ca complex in its S_1 state in Cl(–)— Depleted and Br(–)—Reconstituted oxygen-evolving photosystem II: Evidence from x-ray absorption spectroscopy at the Br K-edge. *Biochemistry* 45:13101–13107. DOI:10.1021/bi061308r.

Molenbroek, A. M., J. K. Nørskov, and B. S. Clausen. 2001. Structure and reactivity of Ni-Au nanoparticle catalysts. *J. Phys. Chem. B.* 105:5450–5458. DOI:10.1021/jp0043975.

Poiarkova, A. V. 1999. *X-Ray Absorption Fine Structure Debye-Waller Factors*. Seattle, WA: University of Washington.

Price, S. W. T., N. Zonias, C.-K. Skylaris, T. I. Hyde, B. Ravel, and A. E. Russell. 2012. Fitting EXAFS data using molecular dynamics outputs and a histogram approach. *Phys. Rev. B.* 85:075439(14). DOI:10.1103/PhysRevB.85.075439.

Ravel, B., E. Cockayne, M. Newville, and K. M. Rabe. 1999. Combined EXAFS and first-principles theory study of $Pb_{1-x}Ge_xTe$. *Phys. Rev. B.* 60:14632–14642. DOI:10.1103/PhysRevB.60.14632.

Ravel, B., Y.-I. Kim, P. M. Woodward, and C. M. Fang. 2006. Role of local disorder in the dielectric response of $BaTaO_2N$. *Phys. Rev. B.* 73:184121(7). DOI:10.1103/PhysRevB.73.184121.

Taylor, J. B., L. D. Calvert, J. G. Despault, E. J. Gabe, and J. J. Murray. 1974. The rare-earth arsenides: Non-stoichiometry in the rocksalt phases. *J. Less Common Metals* 37:217–232. DOI:10.1016/0022-5088(74)90038-1.

Chapter 17

Constraints

The Multi-Tool of Modeling

Why do we need constraints? Shouldn't we just let the fit sort everything out?	
There are usually not enough independent points in the data (Section 14.1.1 in Chapter 14) to let all one's parameters be free, my dear Simplicio.	
And constrained fits are usually more stable.	
There may be physical reasons that a constraint must be true. For example, the distance from atom A to B must be the same as the distance from atom B to A. Not including constraints that enforce those kinds of truths withholds knowledge from the model.	
You may get information about your system from other sources, such as other experimental probes. This should be fed into the model as well.	

DOI: 10.1201/9780429329555-21

Constraints can help you troubleshoot! If your fit is insisting that an MSRD should be negative, you can try forcing it to adopt some reasonable value and see what it does to the rest of the fit. Then, when you identify the cause of the problem, you might be able to fix it and remove the constraint!

Constraints help us to start with a simple model and then add complexity: in bottom-up fits by adding them (Section 12.4.1 in Chapter 12), and in top-down fits by taking them away (Section 12.4.2 in Chapter 12).

Degeneracy, S_o^2, and E_o do not have to be the same for A to B as B to A, though!

Be careful about symmetry arguments! If you only know about symmetry from long-range methods such as XRD, and it is not confirmed by short-range methods such as Mössbauer, infrared spectroscopy, or XANES, then it is possible that the symmetry is only present on average but is broken locally.

Sometimes, though, it is helpful to try removing a rigorous constraint as a check on fit stability (see Section 14.6 in Chapter 14). If, for example, a multiedge fit does not report that the distance from A to B is the same as from B to A (within the specified uncertainties, of course), it is a warning flag that there may be other problems with the fit.

17.1 Rigorous Constraints

Some constraints express information about the modeled system that is rigorously true or nearly so. Constraints of this type should always be included, if present. A few examples:

- The distance from A to B is the same as from B to A. This can be used, for example, to constrain the path from A to B back to A (measured at the A edge) to be the same length as the path from B to A back to B (measured at the B edge). This also means that the MSRDs and higher cumulants of those two paths must be the same.
- Geometrical constraints can be based on the symmetry of the material. If, for example, all iron sites in a crystal are known to be equivalent, and the length of iron–iron path B is twice that of iron–iron path A in the model structure, then if path A is lengthened by an amount x, path B must be lengthened by an amount $2x$.
- The fraction of the atoms of an element that are present in each constituent or site must add up to 1.
- Certain multiple-scattering constraints are also rigorously true; we will cover those in Section 17.5.

17.2 Constraints Based on a Priori Knowledge

You may have knowledge about your system before you begin your EXAFS analysis. One source for this kind of information is other characterization techniques. Two examples are as follows:

1. The stoichiometric ratio of the elements in the sample is often known. If so, this constrains the ratio of elements present in various sites or constituents (see Section 19.4.6 in Chapter 19 for a worked example of this kind of constraint).
2. Diffraction may give lattice parameters, or the average position of atoms. Be careful, however; while this constrains *average* positions, there may be noncrystallographic local distortions to which EXAFS is sensitive but not x-ray diffraction (XRD) (see Section 19.2 in Chapter 19 for a case study examining this kind of distortion).

Another possibility is that you have a strong theoretical belief in characteristics of your material. Box 13.2 in Chapter 13 provides an example in which ligands are expected to be sulfur or nitrate, and thus, the degeneracies of the paths modeling the nitrate ligands can be constrained.

Finally, there are constraints based on other kinds of XAFS analysis, particularly XANES. As an example, suppose you wish to learn about the chromium species present in a soil sample. Because chromium (VI) exhibits a strong, distinctive preedge feature (this can be seen in Figure 10.4 of Chapter 10), it is fairly straightforward to estimate the ratio of chromium (VI) to chromium (III) in a sample by using semi-quantitative fingerprinting methods (Szulczewski et al. 1997) or by linear combination analysis using empirical standards (Kappen et al. 2008). That ratio could then be used to constrain the relative contribution of chromium (VI) and chromium (III) theoretical standards in a model (see Section 16.3 in Chapter 16).

Often, your a priori knowledge may be somewhat uncertain. In such cases, it might be better to use a restraint (see Section 17.6.1) than a constraint.

17.3 Constraints for Simplification

At the start of this chapter, our panel provided us several different reasons why constraints are useful. Those reasons can broadly be divided into two categories: constraints add information to a fit, and they simplify a model.

Sections 17.1 and 17.2 provided examples of constraints that add information. We almost always want to use constraints of this type when possible.

Constraints that are intended to simplify a model, on the other hand, have pros and cons. On the positive side, they increase the degrees of freedom of a fit, increase a fit's stability, make statistical tests more likely to show significance, reduce the error bars on fitted parameters, and can sometimes make a model more defensible. On the other side, they can introduce bias into a fit, reduce our ability to detect aspects of our material's structure, increase a fit's R factor, and can sometimes make a model less defensible. Choosing how much to use constraints of this kind is thus intimately connected to deciding how to weight the criteria for a good fit we discussed in Chapter 14.

17.3.1 Constraints Based on Grouping

Usually, simplifying constraints involve grouping parameters together. For example, consider a measurement on the iron K edge of a mineral incorporating iron, oxygen, and sulfur. A theoretical model might include many single-scattering paths, including both sulfur and oxygen near neighbors; sulfur, oxygen, and iron next-nearest neighbors; and more distant paths. There are several ways these paths could be grouped, as suggested by the Venn diagram in Figure 17.1.

There are thus multiple ways to constrain the MSRDs of the paths in this theoretical standard:

- Force the MSRD for all paths to be the same, giving only one fitted parameter
- Allow one fitted MSRD for near neighbors, and another for all other paths
- Allow one fitted MSRD for scatterers with an oxidation state of −2 (i.e., oxygen and sulfur), and one for iron scatterers
- Allow one fitted MSRD for each scattering element (making three in all)
- Allow one fitted MSRD for near neighbors, one for next-nearest neighbors, and one for more distant paths
- Allow one fitted MSRD for near-neighbor oxygens, one for near-neighbor sulfurs, and one for all other paths
- Allow one fitted MSRD for near-neighbor oxygens, one for near-neighbor sulfurs, one for more distant oxygens, one for more distant sulfurs, and one for irons

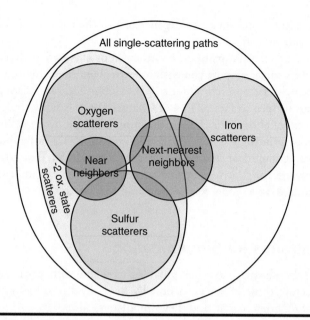

Figure 17.1 Venn diagram for paths in a model of a compound containing iron, oxygen, and sulfur, measured at the iron edge.

■ Allow one fitted MSRD for near-neighbor oxygens, one for near-neighbor sulfurs, one for more distant scatterers with an oxidation state of –2, and one for irons

■ Allow one fitted MSRD for near-neighbor oxygens, one for near-neighbor sulfurs, one for next-nearest-neighbor irons, one for more distant irons, and one for all scatterers with an oxidation state of –2 that are not near neighbors

■ Allow one fitted MSRD for near-neighbor oxygens, one for near-neighbor sulfurs, one for next-nearest-neighbor oxygens, one for next-nearest-neighbor sulfurs, one for next-nearest-neighbor irons, one for more distant oxygens, one for more distant sulfurs, and one for more distant irons

We have not exhausted all of the reasonable possibilities for ways to group the MSRDs, but you get the idea! To decide which to use, you need to evaluate fits using the criteria from Chapter 14. All of the listed possibilities are defensible, so other criteria will have to come into play.

One trade-off to consider is the number of parameters to allocate to a category that can be grouped in various ways. In the above example, it is likely that one MSRD for all single-scattering paths will not work very well; most structures require more flexibility than that. On the other hand, the last option in the list, which allocates eight free parameters to MSRDs, probably has more detail than the fit can support. Successful constraint schemes are likely to lie somewhere in between.

As a second example of constraints that simplify by grouping, consider a mixture of iron metal, an amorphous iron oxide, and an iron carbide. How should the E_os be constrained in the theoretical standards? Some possibilities are as follows.

■ Force E_o for all paths to be the same, giving only one fitted parameter
■ Allow one fitted E_o for each constituent
■ Allow one fitted E_o for the metal and the carbide, since they are both zerovalent, and one for the oxide
■ Estimate the difference between the E_o for the oxide and the E_o for the zerovalent forms, using either fitting to standards or theoretical computations, and use

the estimate to constrain the difference in E_o between constituents. This requires only one free parameter, but does account for the effect of oxidation.

- Allow one E_o parameter for the oxygen scatterers in the oxide, one for the iron scatterers in the oxide, one for the carbon scatterers in the carbide, one for the iron scatterers in the carbide, and one for the metal

Again, this list is not comprehensive, but does give a sense of some of the trade-offs and options.

17.3.2 Constraints Based on Estimates

Another type of simplifying constraint is that based on an estimate. For example, S_o^2 might be constrained to some reasonable value such as 0.9; bond lengths might be fixed at the values provided by XRD or 'typical' values for the bond in question; or vacancies and surface effects might be neglected (thus fixing oxidation numbers to those of a bulk crystal).

When using constraints of this type, it is crucial to modify your reported uncertainties accordingly. For example, suppose you set S_o^2 to 0.9 and then fit the coordination number of the nearest neighbors in a single-shell fit and your fitting software, using an algorithm based on the method described in Section 14.3 of Chapter 14, reports a coordination number of 5.7 ± 0.6. The use of an estimate like 0.9 for S_o^2 might be reasonably assumed to have an uncertainty of 15% (see Section 13.1.4.4 of Chapter 13). Since coordination number is entirely coordinated with S_o^2 for a single-shell fit, the 15% uncertainty applies to the coordination number as well. 15% of the best-fit value of 5.7 is 0.9. This uncertainty of ± 0.9 is independent of the ± 0.6 that came from the fitting algorithm, so they should be added in quadrature, giving an overall uncertainty of $\sqrt{0.9^2 + 0.6^2} = 1.1$. Thus, the coordination number found by the fit should be reported as 5.7 ± 1.1.

Of course, the calculation in the previous paragraph depended on the two parameters being completely coordinated, while in most cases (e.g., S_o^2 and σ^2), the correlation is only partial. In those cases, the uncertainty due to the simplifying constraint can be ascertained by forcing the constrained parameter to the two ends of its range and observing the effect on the best-fit values of other parameters. For example, if constraining S_o^2 to 0.75 gives a best-fit value of 0.012 ± 0.001 Å^2 for σ^2 and constraining S_o^2 to 0.95 gives 0.016 ± 0.002 Å^2, then reporting σ^2 as 0.014 ± 0.003 Å^2 is reasonable.

17.3.3 Constraints Based on Standards

Another simplifying strategy is to fit a related standard, and use the results to constrain values for the sample. In fitting the zinc K edge of an aluminum-doped zinc oxide thin film, for instance, it is reasonable to measure the zinc K edge of a pure zinc oxide thin film of known structure (i.e., an empirical standard), fit the standard with S_o^2 as a free parameter, and then constrain the S_o^2 for the doped sample to have the same value.

While this strategy is very commonly used for S_o^2, it can also be used for other parameters. For example, if your sample is unlikely to be in a different oxidation state from that of your standard, it might make sense to constrain E_o in this way.

Sometimes, the constraints may involve the relationship between parameters rather than individual values. For example, suppose a sample consists of rutile titanium oxide doped with terbium, and is measured at the titanium K edge. The pure rutile structure is shown in Figure 17.2.

Fits are most sensitive to small difference in E_o at low k. Sometimes, it's worth it to start your fit a little higher in k if it means you can get by with fewer free parameters.

Constraining a correlated parameter to an estimated fixed value and then not including the uncertainty in that estimate in the error budget of correlated parameters is one of the most common errors in the literature! If you are reviewing a paper that does this, make sure to point it out and ask for the problem to be fixed before publication.

You may be wondering how we got an uncertainty of 0.003 Å^2 in the last example shown in Section 17.3.2. The answer? We eyeballed it. It doesn't really matter what fancy statistical technique you use; if the software is reporting uncertainties of a thousandth of a square angstrom or two, and changing the correlated parameter causes the fit to change by up to two thousandths of a square angstrom from the reported value, then it's pretty clear that the overall uncertainty is about 0.003 Å^2. It's not necessary to get too picky with these uncertainty estimates, as long as you're not lowballing them by leaving off altogether the uncertainty due to correlation with an estimated parameter.

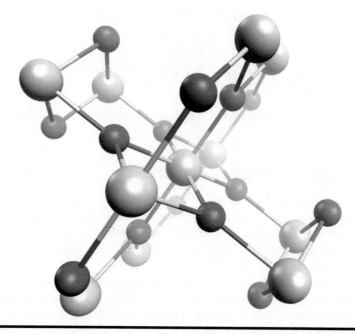

Figure 17.2   The rutile structure. Titanium atoms are shown larger and lighter; oxygen atoms are smaller and darker.

Don't forget to account for the uncertainty in the fitted value taken from the standard when reporting the uncertainties in parameters fitted for the sample! This is done in essentially the same way as we discussed in Section 17.3.2.

In this section, we're discussing using standards to constrain fits to samples. But one thing you should almost never do is use previous fits to a sample to constrain values of parameters for subsequent fits to that sample! For example, it is not acceptable to fit the first shell, and then constrain the parameters for the first shell to their best-fit values before fitting the second. Doing so builds in systematic biases and wreaks havoc with statistics, including uncertainties!

Well, I can think of a couple of exceptions, Mr. Handy. For example, if I'm not sure whether the coordination number of my nearest neighbors is four or six, and I convince myself from a single-shell fit that it's six, I might then constrain it to that value, because that's more a choice between structures than something like a bond length or an E_0 would be. But your basic point is good.

For the sake of discussion, let's label the single-scattering paths in this way:

■ Call scattering from the six near-neighbor oxygens Path O1
■ The next-nearest atoms to the absorber are the two titanium atoms along the *c*-axis (the *c*-axis runs roughly perpendicular to the page in Figure 17.2). Call the single-scattering path to those atoms T1
■ Call the path to the eight remaining titanium scattering atoms in Figure 17.2 T2
■ Call the path to the eight remaining oxygen scattering atoms in Figure 17.2 O2

Note that the length of paths T2 and O2 are very similar; they will show substantial overlap in the Fourier transform. Path T2 will probably show the greater amplitude because titanium has an atomic number that is significantly higher than oxygen.

Let's suppose that a process like that described in Section 17.3.1 has resulted in three MSRDs being fit for the pure rutile standard: one for O1, one that is shared by T1 and T2, and one for O2.

This might work fine for the pure rutile. But compared to the standard, the doped sample might exhibit more disorder, resulting in higher MSRDs. This might make the contribution from O2, while still enough to affect the spectrum, very difficult to stably fit, particularly because of the overlap with T2. In addition, the presence of an atom of terbium somewhere in the structure must also be considered, adding more free parameters to the fit.

Under those circumstances, it's highly desirable to find a way to constrain the MSRD for O2 in the doped sample. But, just forcing it to adopt the value from the standard won't work, because the sample is much more disordered than the standard.

One solution is to assume the disorder affects the MSRD for O1 and O2 to the same extent, so that

$$\frac{\sigma^2_{O2,doped}}{\sigma^2_{O2,pure}} = \frac{\sigma^2_{O1,doped}}{\sigma^2_{O1,pure}} \qquad (17.1)$$

Rearranging gives

$$\sigma^2_{O2,doped} = \frac{\sigma^2_{O2,pure}}{\sigma^2_{O1,pure}} \sigma^2_{O1,doped} \qquad (17.2)$$

We can thus use the results from a fit on the pure sample to express the MSRD for O2 in the doped sample as a multiple of O1, reducing the number of free parameters by one.

BOX 17.1   Standards in Multiple Data-Set Fits

	The method discussed in Section 17.3.3 sets up a distinction between standard and sample, which is, in some ways, artificial. Why should the standard get to decide what the value of S_o^2 or E_o or any other parameter is on behalf of the sample?
	Because we know more about the standard, Carvaka. And sometimes the signal-to-noise is better for a standard, too.
	We know more about the standard, but we don't know everything about it—if we did, we wouldn't need to fit it at all. Instead of fitting the standard and using the results to constrain the sample fit, we could fit them both simultaneously, as a multiple data-set fit. The parameters that we would take from the standard and apply to the sample if we were following the method discussed in Section 17.3.3, we can instead force to be the same, but refine simultaneously. If the sample and standard have different signal-to-noise ratios, we can use them to weight the data in the fits, so that the standard fit gets weighted more heavily. If we do that, we're only giving the standard the extra weight it merits, and not more.
	The method proposed by Carvaka has an additional advantage, in that it obviates the need for taking into account the uncertainty in the parameters borrowed from the standard when reporting the uncertainty in parameters fitted for the sample.
	Oh, right! That error propagation stuff is a pain. I'm totally sold, guys!

17.4 Some Special Cases

There are a couple of special cases that come up frequently enough that we should discuss them here.

17.4.1 Lattice Scaling

If a crystal has a unit cell that is cubic, rhombohedral, or hexagonal with an ideal *c/a* ratio (1.633), then the lattice size is determined by only one parameter, *a*. It can, therefore, be reasonable to parameterize all path lengths as being modified by the same percentage compared to the theoretical standard. Substituting a dopant or changing temperature or pressure, for example, might expand or contract the lattice by, say, 5%, but that change to the lattice scale should make *every* path longer or shorter by the same percentage.

That's not to say that atoms can't change relative positions within the unit cell under those circumstances, necessitating different free parameters for the half path length of different paths. But a uniform lattice scaling can serve as a useful first approximation for changes relative to the theoretical standard when pursuing a top-down strategy (see Section 12.4.2 in Chapter 12).

A uniform lattice scaling can also work fairly well, particularly for relatively short paths, when the unit cell is only slightly distorted from cubic (e.g., there is a modest tetragonal distortion) or ideal hexagonal (e.g., *c/a* is slightly smaller or larger than 1.633). It is unlikely to work, however, when the starting structure for the theoretical standard is a molecule rather than a crystal; such systems often contort under stress rather than expand or contract uniformly.

17.4.2 Correlated Debye Model

Several attempts have been made to theoretically predict MSRDs for systems fulfilling certain special requirements (Dalba and Fornasini 1997). In most cases, this has been done primarily to test the accuracy of either EXAFS theoretical standards or the theory used to make the prediction, rather than as a practical method for generating constraints for MSRDs for use in fitting unknown samples. The correlated Debye model (Beni and Platzman 1976), however, has been shown to have utility for an important class of materials, namely, cubic monatomic metals where the atoms reside on a Bravais lattice such as face- or body-centered cubic metals (Dalba and Fornasini 1997). According to that model,

$$
\sigma^2 = \frac{3\hbar^4}{2k_b^3 \mu \Theta^3} \int_0^{\frac{k_b\Theta}{\hbar}} \coth\left(\frac{\hbar\omega}{2k_bT}\right)
$$

$$
\left[1 - \frac{\sin\left(D\sqrt[3]{\frac{6\pi^2}{V}}\frac{\hbar\omega}{k_b\Theta}\right)}{\left(D\sqrt[3]{\frac{6\pi^2}{V}}\frac{\hbar\omega}{k_b\Theta}\right)} \right] \omega d\omega \tag{17.3}
$$

where V is volume per atom, k_b is Boltzmann's constant, Θ is the Debye temperature of the material, μ is the projected reduced mass (Poiarkova 1999) of the atoms in the path, ω is a variable of integration, and D is the half path length.

While the Debye model only applies quantitatively to a relatively small fraction of the materials studied by XAFS, it is interesting to examine its dependence on several parameters of interest.

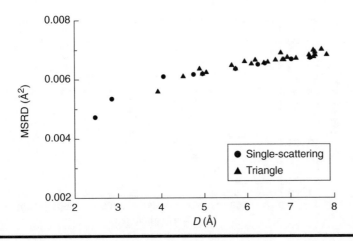

Figure 17.3   MSRD as a function of the half path length for several paths in iron metal, as predicted by the correlated Debye model. Calculations were performed by IFEFFIT 1.2.12, using a measurement temperature of 300 K and a Debye temperature of 470 K.

Let us begin by examining the dependence on the half path length D. Figure 14.3 shows the MSRD predicted by the Debye model as a function of D for several paths in metallic iron.

While the details of the plot are particular to the correlated Debye model, some of the qualitative features apply to a wide variety of systems:

- The nearest-neighbor scattering path has an MSRD that is significantly smaller than paths further out. This is because bonded atoms tend to move in a correlated manner, making the bond length less variable than would be expected if the atoms moved independently.
- Most of the distance dependence is found in the first few shells. Although a gradual increase with D can be seen over the entire range in Figure 17.3, the change from 4 to 8 Å is small compared to typical uncertainties in fitted MSRDs.
- Triangular multiple-scattering paths exhibit similar MSRDs as single-scattering paths with the same half path length.

Next, let's consider the temperature dependence, shown in Figure 17.4.

The temperature dependence is roughly parabolic well below the Debye temperature, and linear above it. This functional dependence on temperature is observed in a broad range of systems, even when the Debye model itself does not apply (Dalba and Fornasini 1997).

If data are collected on a sample as a function of temperature, therefore, it may be worth seeing if they can be fit with a quadratic function at low temperatures, a linear function at high temperatures, or both. The spectra for multiple temperatures can be fitted simultaneously, and the simple functional dependence, if present, can thus be used to reduce the number of free parameters in the fit.

In addition, observing the temperature dependence of the MSRD can give an estimate of how much would remain at 0 K. While this residual MSRD is partly due to zero-point thermal motion (that is, the origin of the zero-point value shown in Figure 17.4), a comparison to well-ordered standards can yield an estimate of how much of the MSRD is due to static disorder (Section 13.1.1.3 in Chapter 13).

You may be wondering why anyone would bother using EXAFS to analyze a plain-vanilla metal such as copper or iron. After all, we already know their structure! While that's true, they play a role in more interesting systems, often as constituents in a mixture. Examples may be found among nanostructured materials, cathodes, and biogeochemical systems.

Spectra of the same edge of the same sample taken at different temperatures are certainly not independent, and the number of independent points should not be taken to be that given by the Nyquist criterion (Section 14.1.1 in Chapter 14) multiplied by the number of spectra! A good rule of thumb is to make sure that each spectrum has enough independent points by the Nyquist criterion to fit all the parameters that directly impact the fit to that spectrum.

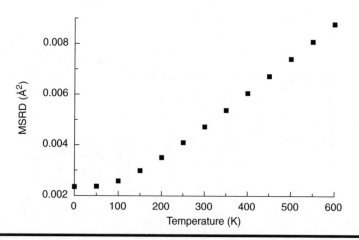

Figure 17.4 MSRD as a function of temperature for the nearest-neighbor path in iron metal, as predicted by the correlated Debye model. Calculations were performed by IFEFFIT 1.2.12, using a Debye temperature of 470 K.

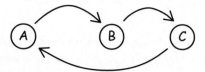

Figure 17.5 A focused path.

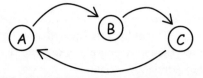

Figure 17.6 A partially focused path.

17.5 Multiple-Scattering Paths

17.5.1 Focused Paths

In Box 12.3 in Chapter 12, we introduced the concept of a *focused* path. For convenience, we reproduce Figure 12.4 in Chapter 12 here as Figure 17.5.

While atom B affects the amplitude and phase shift of the path starting at A, it does not affect its half path length, MSRD, or higher cumulants, which are the same as for the single-scattering path from A to C back to A. This provides a constraint on those parameters; a focused path of this type does not add any free parameters to a fit.

What if the path starts at B instead? The path is still geometrically identical to the path from A to C back to A. If A or C is an absorbing atom in our model, either because one of them is the same element as B or because we are performing a multiple-edge fit for which A or C is an absorbing atom, then we can still constrain the geometry for this path rigorously. If not, we have to handle it more like a triangle path (see Section 17.5.4).

While Figure 17.5 shows a perfectly linear path, we can still consider a path to be focused if it is almost linear. For example, Figure 17.6 shows a partially focused path. The geometry is very nearly the same as the path shown in Figure 17.5 and to a good approximation can be again treated as the same as the single-scattering path A to C to A.

BOX 17.2 How Small Is Small?

How good is the approximation? How far off can B get from the midline before we shouldn't use it?

It is based on the small-angle approximation, dear Kitsune. The triangle ABC has two small angles, one with its vertex at A and one at C. To the extent that those angles are small, to first order the distance from A to C is the same as the sum of the distances from A to B and B to C. An estimate of the accuracy of this approximation can be determined by seeing how much cos θ varies from 1 for the small angles. For example, if the angle with the vertex at C is 10°, cos 10° = 0.985, suggesting the approximation is good to something like 1% or 2%. Even at 30°, the approximation is good to about 10%–15%.

And that's the error in the difference from the starting structure, not the error in the half path length! For example, the half path length for the multiple-scattering path in the starting structure might be 2.45 Å. That number is exact for the starting structure, and accounts for the geometry. If fitting finds that the single-scattering path from A to C to A should be made 0.04 Å longer, then assuming that the multiple-scattering path should change length by the same amount represents an error of only around 15% of 0.04 Å, or less than 0.01 Å. That's probably OK for most fits.

Thanks—that tells me the accuracy of constraining half path lengths. What about MSRDs?

Let us find out, friend Kitsune. To estimate this, let us start with a set of simple assumptions. We'll treat the three atoms as if they move in an uncorrelated, isotropic, harmonic manner, each with the sample amplitude. While this is certainly not the case, it will help us understand how sensitive the MSRD is to the geometry. In the interest of simplicity, let us also place B equidistant from A and C. We can then place A at the origin, C at the coordinates $(D,0)$, and B at the location $(D/2,y)$. The half path length D_{MS} is then

$$D_{MS} = \frac{1}{2}\left(D + \sqrt{D^2 + 4y^2}\right) \tag{17.4}$$

As a check, at 30° Equation 14.4 gives us a half path length for the focused path 8% longer than that for the direct path. (That's a bit closer than Robert's estimate above. The case where the focusing atom is equidistant from the others is relatively favorable for keeping the deviations from the single-scattering case small.) Differentiating with respect to D gives

$$\frac{\partial D_{MS}}{\partial D} = \frac{1}{2}\left(1 + \frac{D}{\sqrt{D^2 + 4y^2}}\right) \tag{17.5}$$

and with respect to y gives

$$\frac{\partial D_{MS}}{\partial y} = \frac{1}{2}\left(\frac{4y}{\sqrt{D^2 + 4y^2}}\right) \tag{17.6}$$

Because we have assumed the atomic motions are uncorrelated, isotropic, and the same for all atoms,

$$\sigma_{MS}^2 = \left[\left(\frac{\partial D_{MS}}{\partial D}\right)^2 + \left(\frac{\partial D_{MS}}{\partial y}\right)^2\right]\sigma^2 \tag{17.7}$$

where σ^2 is the MSRD for the single-scattering path from A to C back to A. Since we are treating the focused case, $y \ll D$. After a bit of mathematics we get, to lowest order in y/D,

$$\sigma_{MS}^2 = \left[1 + 4\left(\frac{y}{D}\right)^2 \right] \sigma^2 \qquad (17.8)$$

This provides us the estimate we sought. If the angle were 30°, then $y/D = 1/2 \tan 30° = 0.29$, and Equation 14.8 tells us that the MSRD is about a third larger than the corresponding single-scattering path. At 10°, in contrast, the difference is only about 3%.

In reality, the MSRD for the focused path is likely to be even closer to that of the single-scattering path than is shown in your simple model, Mandelbrant. You've treated the motions as uncorrelated, but when atom B moves down, shortening the path, atoms A and B are likely to move apart, lengthening it. When atom B moves up, the reverse will occur. So, the addition of atom B to the path causes smaller variations in path length than your uncorrelated model assumes.

The bottom line, then, is that for almost all purposes we're OK constraining a path where the small angles of the triangle are 10° to have the same changes to path length and MSRD as the corresponding single-scattering path. If the small angles get up to 30°, though, the approximation (particularly for the MSRD) is getting a bit iffy. Fortunately, the amplitude of paths with angles that big also tends to be smaller!

17.5.2 Double Paths

Go high enough in D, and you're likely to encounter a path that looks like Figure 17.7.

This strange-looking path involves the photoelectron traveling from the absorber A to a scatterer B, then back to A, back to B again, and finally back to A once again. There's no widespread terminology for this kind of path, so we'll call it a *double* path.

Unlike focused paths, double paths are not particularly high amplitude, but the double path off of the nearest neighbor can sometimes have an effect on your fit. Fortunately, double paths can be rigorously constrained: the half path length is twice that of the corresponding single-scattering path, and the MSRD is four times as great. Even the third cumulant can be rigorously constrained for this kind of path; it is eight times larger than that for the corresponding single-scattering path.

Don't confuse a *double path* with *double scattering*, which is a reasonable description of a triangle path (see Section 17.5.4)!

17.5.3 Conjoined Paths

Figure 17.8 shows another multiple-scattering path, which in this text we will call *conjoined*.

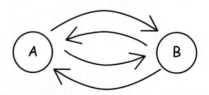

Figure 17.7　A double path.

This is, essentially, two single-scattering paths in sequence: first A to B and back to A, and then A to C and then back to A. (Of course, the reverse order works equally well!)

The half path length for this path can be rigorously constrained; it is the sum of the half path lengths of the two related single-scattering paths.

The MSRD is more difficult, however. If the single-scattering paths were uncorrelated, then the MSRD for the conjoined path would be the sum of the MSRDs for the two single-scattering paths. But the paths are not uncorrelated; aside from the possible effects that the motion of B and C might have on each other, A is certainly common to both! If A moves a little to the left in Figure 17.8, for instance, the distance from A to B gets larger, whereas that from A to C gets a bit smaller. On the other hand, if Figure 17.8 had been drawn so that triangle CAB were acute rather than obtuse, then when the distance from A to B got larger the distance from A to C would also tend to get larger.

Box 17.2 anticipated this issue: Mandelbrant provided a derivation that accounted for the atom in common, but did not account for the fact that the motions were likely to be correlated, and Carvaka then provided an argument as to the qualitative effect of the correlation. Short of simulating the atomic motion of the entire structure (difficult, since the structure is at least partially unknown if we're investigating it!), a rigorous estimate of the MSRD for this kind of path is difficult. Fortunately, the amplitude of paths of this kind is generally small.

17.5.4 Triangles, Quadrilaterals, and Other Minor Multiple-Scattering Paths

This brings us to *triangle* paths, such as the one shown in Figure 17.9.

If your model allows the average positions of the atoms to be calculated (e.g., it is parameterized in terms of crystallographic parameters, or in terms of bond lengths and angles), then the half path length can be rigorously constrained. If not (most commonly because the model is parameterized in terms of distances from absorber to scatterers only, without specifying bond angles or positions), a reasonable estimate can probably still be made.

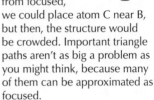

The paths shown in Figures 17.8 and 17.9 aren't too far off from being focused. To make it far from focused, we could place atom C near B, but then, the structure would be crowded. Important triangle paths aren't as big a problem as you might think, because many of them can be approximated as focused.

There is still a problem, Dysnomia! What if the path starts at A, and neither B nor C are atoms for which we've measured the edges? Even if the atoms were collinear, we couldn't constrain them in terms of an existing single-scattering path, because the path from B to C back to B wouldn't be part of our theoretical standards! While your idea that some of these paths can be thought of as focused is helpful, sometimes we need other solutions.

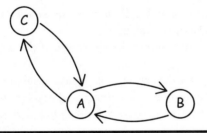

Figure 17.8   A conjoined path.

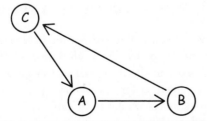

Figure 17.9   A triangle path.

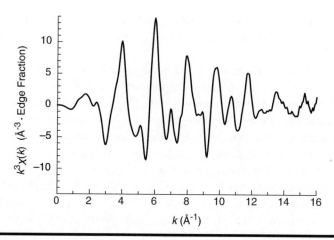

Figure 17.10 k^3-weighted $\chi(k)$ for nickel K edge of nickel (II) fluoride.

But as with conjoined paths, it is not immediately clear how to constrain MSRDs for triangle paths (or related paths such as quadrilaterals).

The correlated Debye model gives us a hint as to how to proceed with this kind of path, as well as other low-amplitude multiple-scattering paths. In Figure 17.3, the triangle paths were assigned MSRDs similar to the single-scattering paths of the same length. Does that work in other kinds of materials?

To find out, we will now provide an example using nickel (II) fluoride. The k^3-weighted $\chi(k)$ data are shown in Figure 17.10.

Nickel (II) fluoride adopts the rutile structure (Section 17.3.3), with nickel in place of titanium and fluorine in place of oxygen. This structure does not feature a lot of short collinear paths, but does provide several possibilities for triangular paths. The highly polar bonds between nickel and fluorine make it quite different from the monatomic metals to which the correlated Debye model applies, and should thus provide a good test of the effect of small-amplitude multiple-scattering paths in a material where the Debye model does not hold.

For our model, we will use the following free parameters:

■ S_o^2 and E_o
■ The lattice parameters a and c. These two parameters determine the position of all nickel atoms relative to each other.
■ A parameter u that determines the placement of the fluorine atoms within the lattice.
■ Three MSRDs: one for the paths to the six near-neighbor fluorine atoms, one for the paths to the 10 nearest nickel atoms, and one for all other single-scattering paths. This is the same parameterization that was used as an example in Section 17.3.3.

Fits are performed on the data from 5 to 14 Å⁻¹, fitting from 1.4 to 5.0 Å in the Fourier transform. This gives 20 independent points by the Nyquist criterion, which is plenty to fit the 8 parameters above.

As a baseline fit, we will include all important single-scattering paths through $D = 7.0$ Å but no multiple-scattering paths. The result is shown in Figure 17.11.

The fit appears fairly close to the data, and in fact gives an R factor of 0.0112, well within our informal guidelines for satisfactory agreement (see Section 14.2.1 in Chapter 14). The fitted parameters are shown in Table 17.1.

All parameters have adopted reasonable values.

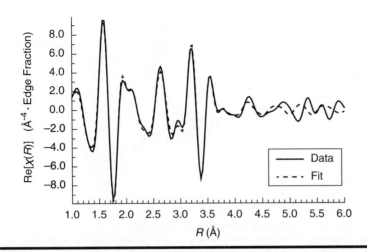

Figure 17.11 Real part of the Fourier transform for the nickel (II) fluoride data and fit using no multiple-scattering paths.

Table 17.1 Results of NiF$_2$ Fit Using No Multiple-Scattering Paths

Parameter	Best Fit[a]	Crystallographic Values[a,b]
S_0^2	0.88(6)	—
E_0	8346.3(9) eV	—
a	4.68(1) Å	4.65(0) Å
c	3.09(1) Å	3.08(0) Å
u	0.309(3)	0.301(1)
Near-F MSRD	0.005(1) Å^2	—
Near-Ni MSRD	0.008(1) Å^2	—
Other MSRD	0.020(4) Å^2	—

[a]Uncertainties in the last digit are given in parentheses.
[b]Crystallographic values are from Baur and Khan (1971).

In fact, this fit generally satisfied the criteria provided in Chapter 14, with one notable exception: the fit does not agree with the data above the fitted range (Section 14.5 in Chapter 14). In fact, examination of Figure 17.11 shows that the misfit begins well below 5.0 Å, which is the top of the fitted range.

Nevertheless, this is probably an acceptable fit for most purposes. Would the inclusion of multiple-scattering paths make it better?

To answer that, we added one multiple-scattering path at a time, starting with the highest amplitude path (as determined by curved wave amplitudes by FEFF6L; see Zabinsky et al. 1995) and working our way down. The MSRD for focused paths starting at a nickel atom was constrained to that of the related single-scattering path and double paths were handled correctly, but all 'difficult' paths (triangle, quadrilateral, focused stating on a fluorine atom) were given MSRDs equal to each other. Each fit was run in two ways: in one set, the MSRDs of these multiple-scattering paths were constrained to be equal to the 'Other MSRD' already defined in the fitting model; in the second, they were assigned a new free parameter.

The resulting R factors are shown in Figure 17.12.

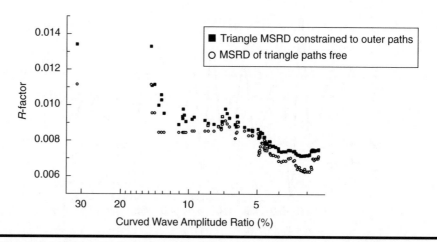

Figure 17.12 *R* factors of fits to nickel (II) fluoride data as multiple-scattering paths are added. The *x*-axis is the amplitude of the last multiple-scattering path added as compared to the amplitude of the nearest-neighbor path, as computed by FEFF6L.

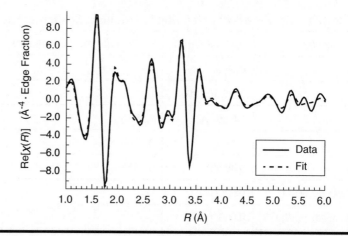

Figure 17.13 Real part of the Fourier transform for the nickel (II) fluoride data and fit using 102 multiple-scattering paths but no additional free parameters.

While the fit was already satisfactory without multiple-scattering paths, adding them improved the closeness of the fit, even without increasing the number of free parameters. To get a sense of the level of significance, let's use the Hamilton test to compare the fit with no multiple-scattering paths to the one with the maximum number shown in Figure 17.12 (102, as it turns out). For the case where the MSRD of triangle paths was constrained to be the same as the MSRD for outer paths we haven't added or changed any of the existing free parameters, and so we'll use the minimum, 1, for the Hamilton test. The *R* factor of the fit with all 102 multiple-scattering paths but no new free parameters is 0.00752, as compared to 0.0112 for the fit with no multiple-scattering paths, giving an r of $0.00752/0.0112 = 0.67$; $a = (20-8)/2 = 6$, and we've agreed to make $b = 1/2$. This gives $I_{0.67}(6,0.5) = 0.03$. So we would have accepted the fit as *significantly* better, even if we had to use a free parameter to achieve it!

This can also be seen by observing the Fourier transform of the fit with all of the multiple-scattering paths (Figure 17.13), but with the MSRD for triangle paths constrained to be the same as for outer paths.

The mismatch at the top of the fitting range seen in Figure 17.11 is largely gone. There has been a qualitative improvement to the fit, not just a quantitative one.

The *R* factors are a bit better yet if we do use a free parameter for the triangle MSRDs, though. Is the improvement due to that free parameter significant?

This time, we can use the Hamilton test without hesitation, as we're adding a parameter. We'll compare the fit with all 102 multiple-scattering paths but no additional free parameter (that's the one shown in Figure 17.13) to the equivalent fit with the new MSRD parameter. The latter had an R factor of 0.00713, giving an r of $0.00713/0.00752 = 0.95$ and $I_{0.95}(6,0.5) = 0.44$—that is, the improvement in closeness of fit seen by allocating a free parameter to the multiple-scattering paths is no better than what would be expected to happen by chance. We, therefore, do *not* need to allocate a free parameter to the multiple-scattering paths in this case.

This combination of results, which is quite common in EXAFS analyses, is worth reiterating. **Including multiple-scattering paths with a simple constraint scheme that does not add any free parameters provided a significant improvement over leaving the paths off altogether, while a more flexible constraint scheme that allocates multiple-scattering a free parameter did not provide any additional statistical improvement.**

Figure 17.12 holds some additional lessons as well. It is not the case that adding each multiple-scattering path provided an improvement over the fit before. While the general trend is toward improvement, adding an individual multiple-scattering path often increased the R factor. This does not mean that the multiple-scattering path in question is 'bad'! Since $\chi(k)$ can take on positive or negative values, it often happens that multiple-scattering paths partially cancel each other. Thus, adding one path may make the fit worse, even though adding a group that includes that path makes the fit better. No individual multiple-scattering path should be evaluated as improving or degrading the fit; it is only in combination that the improvement can be seen. In the case of our nickel (II) fluoride fits, the multiple-scattering path with the highest amplitude makes the fit quite a bit worse when added on its own—the R factor jumps from 0.0112 to 0.0134. But with a few more paths the combination improves the fit, and the improvement continues, in fits and starts, nearly through the end of the 102 paths we attempted.

When we fit EXAFS, therefore, we should not concern ourselves too much with the effect of individual multiple-scattering paths. A good procedure is to add paths in groups, perhaps by choosing all that are above a certain amplitude.

We'll close this section by repeating Table 17.1 but with the results corresponding to the fit shown in Figure 17.13 added to form Table 17.2.

For seven of eight parameters, the uncertainty is smaller in the fit with the multiple-scattering paths than the one without. The additional paths help to reduce

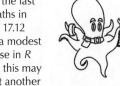

While the last few paths in Figure 17.12 show a modest increase in R factor, this may be just another blip upward like the ones seen near 13%, 10%, 7%, and 6%. If we continued to add paths with smaller and smaller amplitudes, the R factor might continue to drop on average. On the other hand, at some point we will reach the level where errors in the model, systematic error, and noise limit the ability to obtain additional improvement from adding paths!

Table 17.2   Results of NiF₂ Fits with and without Multiple-Scattering Paths

Parameter	No Multiple Scattering[a]	All Multiple Scattering[a]	Crystallographic Values[a,b]
S_0^2	0.88(6)	0.91(5)	—
E_0	8346.3(9) eV	8346.4(7) eV	—
a	4.68(1) Å	4.68(1) Å	4.65(0) Å
c	3.09(1) Å	3.09(1) Å	3.08(0) Å
u	0.309(3)	0.306(4)	0.301(1)
Near-F MSRD	0.005(1) Å²	0.006(0) Å²	—
Near-Ni MSRD	0.008(1) Å²	0.008(0) Å²	—
Other MSRD	0.020(4) Å²	0.017(3) Å²	—

[a]Uncertainties in the last digit are given in parentheses.
[b]Crystallographic values are from Baur and Khan (1971).

correlations that are otherwise present between parameters. And while none of the parameters shifted by more than the original uncertainties, several shifted by amounts nearly that large.

BOX 17.3 Multiple-Scattering Paths Are Your Friends

	Multiple-scattering paths still kind of scare me.
	C'mon, Simplicio, they're awesome! You can add them to a fit and, if the basic model is right, the fit should get closer and the uncertainties smaller, all without using up any free parameters!
	They also make your model better reflect what is happening physically in the material.
	I use them as additional evidence that a model is correct. If one's model is poor, then there is no reason that adding groups of multiple-scattering paths should result in substantial improvement in the closeness of the fit. That improvement could be added to the list from Chapter 14 as a ninth criterion.
	Well, even so, maybe I'll just leave the triangle paths off. I'm afraid of constraining them wrong.
	'Leaving them off' is a constraint, and a very poor one—you're constraining their amplitude to zero, which you know isn't right! A great thing about small-amplitude multiple-scattering paths is that they're relatively insensitive to reasonable choices of constraints.
	Of course, the more physically accurate your constraint scheme for multiple-scattering paths is, the better it should fit. We might be able to do even better with the nickel (II) fluoride fit if we used a more realistic model to constrain the MSRDs of multiple-scattering paths. For example, we could construct a model that accounted for the multiple-scattering paths that incorporate legs from near neighbors, since the MSRD for the near neighbors is different than that for the outer paths.

Sure, Carvaka, but there's a point of diminishing returns. I don't want to go to all the trouble of sorting out the geometry of 102 multiple-scattering paths, just for a little additional improvement!

Perhaps you can write some code to do what you're suggesting, Carvaka? Then we'd only have to put the effort in once, and we could automatically apply it to future systems of study.

This is an interesting discussion, my friends, but it is time to move on. We have learned enough to incorporate multiple-scattering paths into our fits and have the tools to go further if we wish.

17.6 Alternatives for Incorporating A Priori Knowledge

In Sections 17.2 and 17.3, we discussed incorporating a priori knowledge, estimates, or information gained from fitting a standard into a model by using constraints. Often, however, priori knowledge is uncertain, and likewise we might expect the values of parameters for a sample to be similar to, but not identical to, those for a corresponding standard. Several methods have been developed to address the issue of partial knowledge.

17.6.1 Restraints

One method for incorporating uncertain a priori information is to penalize the fit to the extent that the fitted value of the parameter deviates from the a priori value, just as the fit is penalized to the extent that the spectra for data and fit don't match. If that is done, the condition is called a *restraint* (Ellis and Freeman 1995).

For example, suppose we believe that the nearest-neighbor distance in a material is 2.02 ± 0.03 Å. This means that we would be a bit surprised to find that the bond length is actually 1.98 or 2.06 Å, and quite surprised if it were 1.95 or 2.09 Å.

In accord with the suggestion from the IXS that we compute uncertainties in fitted parameters by specifying how much they would need to change to increase χ^2 by 1 (see Section 14.3.1 in Chapter 11), we apply the same principle here, forcing χ^2 to increase by 1 if the restrained parameter differs from the a priori value by an amount equal to the uncertainty in that value.

Since we don't know the measurement uncertainty until we achieve a good fit, however, we don't know how to weight the restraint initially. We, therefore, recommend the following procedure for weighting restraints:

1. Set the weighting by trial and error for preliminary fits, so that the restrained parameter seems to generally stay within the desired range while still being able to vary within it.
2. Once a satisfactory fit (as judged by the criteria discussed in Chapter 14) is achieved, observe the value of χ_ν^2 reported by the fitting software. Weight the restraint so that changing the parameter by its uncertainty penalizes the fit by

Restrained parameters reduce degrees of freedom, just as free parameters do! Restraints are not a way of getting around statistical restrictions.

For a discussion of using restraints to incorporate the concept of bond valence sums, see Newville (2005).

The weighting we suggest in step 2 is equivalent to dividing the uncertainty in the a priori assumption by $\sqrt{\chi_v^2}$; depending on your software, this may be the way the weighting is adjusted.

an amount equal to χ_v^2; this corresponds to changing the χ^2 of the rescaled fit by 1, as desired.

3. Of course, reweighting the constraint may cause the reported value of χ_v^2 to change. Repeating step 2 in an iterative process should allow the fit to converge within a few iterations. If it does not, it is an indication that the restraint may not be compatible with the fit and should be reexamined.

17.6.2 Bayes–Turchin Analysis

In the conventional approach to curve fitting to a theoretical standard, it is up to the judgment of the person doing the fitting to choose what parameters to constrain and which to vary freely. One contributing factor in that decision is the importance of a parameter to the fit; if its value has little effect on the fitted spectrum, perhaps because the path it is associated with has low amplitude, then it is not worth assigning a free parameter to. The process of choosing parameters and constraints is generally iterative, with the scientist deciding on constraint schemes based in part on the behavior of previous fits.

The subjective nature of this process is unsettling, and requires considerable experience to do well. While it has been applied successfully for decades, it still seems desirable to find a way to make the progress more rigorous.

In *Bayes–Turchin analysis* (Krappe and Rossner 2009), in contrast, the scientist's best guess and estimated uncertainty is provided a priori for all parameters. In this way, Bayes–Turchin analysis is similar to a fit in which all parameters are restrained, as opposed to free or constrained.

If that were all there was to the approach, it would fail: for multishell fits, there are usually not enough independent points in the data to support varying that many parameters. But Bayes–Turchin analysis mathematically selects the parameters that most strongly affect the fit, in effect automating the process of choosing free and constrained parameters traditionally undertaken by the scientist doing the analysis. The merits of this technique are summarized effectively by Krappe and Rossner (2009):

'The Bayes–Turchin approach makes it possible to start the data analysis with a very large model-parameter space. The method itself, rather than a mere guess, yields the subspace where the data determine the outcome of the fit. This allows the simultaneous analysis of atomic-like background and structure parameters.'

The method also has a notable disadvantage, however. While a scientist can choose constraint schemes that make physical sense, the Bayes–Turchin method is driven only by the ability to reproduce the fit.

Consider, for example, the Bayes–Turchin fit to copper metal provided by Krappe and Rossner (2000). This fit uses 158 potential free parameters, considerably larger than the 55 independent points available according to the Nyquist criterion. As a test of the procedure, they assigned a priori values for the MSRDs according to the correlated Debye model at 291 K but measured the data at 10 K.

Depending on the criterion chosen for significance, the Bayes–Turchin method identified no more than 40 of the 158 potential free parameters as having a significant impact on the fit. So far, this is similar to what a scientist might do using the traditional method. But the Bayes–Turchin method assigned values more than two standard deviations different from the a priori value for the first, second, third, fifth, and sixth single-scattering paths, and not for the fourth or seventh. The MSRD also differed by more than two standard deviations for eight of the multiple-scattering paths, distributed throughout the path list.

BOX 17.4 Opinions on Bayes–Turchin

	I can't say I care for this method much. If a person proposed a fit where the first, second, third, fifth, and sixth direct-scattering paths of copper were allowed to have independent values for the MSRD, but the fourth and seventh did not, I would say that they have an unphysical model and suggest they try again.
	But it is more objective than the traditional method, dear Carvaka. The way a person chooses constraints is haphazard, and leaves more room for biasing the results. I like this method.
	There's no reason we can't do both. The Bayes–Turchin method, for example, could be applied with the Debye temperature of copper and the temperature of the measurement as potential free parameters, rather than the MSRD for each path. That would prevent the kind of unphysical result Carvaka is concerned about, while still increasing the objectivity of the procedure.

WHAT I'VE LEARNED IN CHAPTER 17, BY SIMPLICIO

- *Constraints* can be used to add information to a fit, or to simplify a model.
- Quantitative constrains and *restraints* can be based on a priori experimental or theoretical knowledge about the sample, or the results of a fit to a standard. Quantitative constraints should not, however, be based on the results of preliminary fits to the same standard.
- Constraints that group parameters together are usually selected by trial and error, using the criteria discussed in Chapter 14 to evaluate which schemes work best. An alternative to this trial and error is the *Bayes–Turchin approach*.
- Most focused and double multiple-scattering paths can be constrained rigorously.
- Multiple-scattering paths that cannot be constrained rigorously, such as triangle paths, should usually be included in the fitting model. The exact constraint scheme is not crucial, and usually does not need to involve new free parameters.
- Restraints can be used to incorporate uncertain *a priori* knowledge to a fit.

References

Baur, W. H. and A. A. Khan. 1971. Rutile-type compounds. IV. SiO_2, GeO_2 and a comparison with other rutile-type structures. *Acta. Crystallogr. B.* 27:2133–2139. DOI:10.1107/S0567740871005466.

Beni, G. and P. M. Platzman. 1976. Temperature and polarization dependence of extended x-ray absorption fine-structure spectra. *Phys. Rev. B.* 14:1514–1518. DOI:10.1103/PhysRevB.14.1514.

Dalba, G. and P. Fornasini. 1997. EXAFS Debye-Waller factor and thermal vibrations of crystals. *J. Synchrotron Radiat.* 4:243–245. DOI:10.1107/S0909049597006900.

Ellis, P. J. and H. C. Freeman. 1995. XFIT—An interactive EXAFS analysis program. *J. Synchrotron Radiat.* 2:190–195. DOI:10.1107/S0909049595006789.

Kappen, P., E. Welter, P. H. Beck, J. M. McNamara, K. A. Moroney, G. M. Roe, A. Read, and P. J. Pigram. 2008. Time-resolved XANES speciation studies of chromium on soils during simulated contamination. *Talanta.* 75:1284–1292. DOI:10.1016/j.talanta.2008.01.041.

Krappe, H. J. and H. H. Rossner. 2000. Error analysis of XAFS measurements. *Phys. Rev. B.* 61:6596–6610. DOI:10.1103/PhysRevB.61.6596.

Krappe, H. J. and H. H. Rossner. 2009. The Bayes-Turchin approach to the analysis of extended x-ray absorption fine-structure data. *Physica Scripta* 79:048302(7). DOI:10.1088/0031-8949/79/04/048302.

Newville, M. 2005. Using bond valence sums as restraints in XAFS analysis. *Physica Scripta T.* 115:159–161. DOI:10.1238/Physica.Topical.115a00159.

Poiarkova, A. V. 1999. *X-ray Absorption Fine Structure Debye-Waller Factors.* Seattle: University of Washington.

Szulczewski, M. D., P. A. Helmke, and W. F. Bleam. 1997. Comparison of XANES analyses and extractions to determine chromium speciation in contaminated soils. *Environ. Sci. Technol.* 31:2954–2959. DOI:10.1021/es9701772.

Zabinsky, S. I., J. J. Rehr, A. Ankudinov, R. C. Albers, and M. J. Eller. 1995. Multiple-scattering calculations of x-ray-absorption spectra. *Phys. Rev. B.* 52:2995–3009. DOI:10.1103/PhysRevB.52.2995.

XAFS IN THE LITERATURE

Chapter 18

Communicating XAFS

Mind Your χ's and Error Bars

Why do we need a chapter on how to write about and present XAFS research? We've just had a whole book explaining everything!	
And how many pages did it take us?	
Oh. I guess I can't spend hundreds of pages talking about my latest result, huh?	
Only in a dissertation, dear Simplicio. And yet one must *always* provide information that is concise, clear, and accurate. Thus, this chapter.	

18.1 Know Your Audience

As with any writing or presentation, it's important to tailor your paper or talk to the audience. $\chi(k)$ graphs, for instance, would be meaningless to a group unfamiliar with EXAFS, but their absence would be a serious flaw to those who are.

Of course, in many situations the audience is mixed, comprising people with a variety of knowledge levels. With presentations, that is not such a big deal—specialists

DOI: 10.1201/9780429329555-23

One example where it might not be practical to graph *all* the data is for quick-XAFS experiments, since you might have hundreds or thousands of scans. In a case like that, graph a representative subset of the data.

For presentations, you can use the 'hidden slide trick' as if it were supporting data. For an audience with mixed levels of knowledge, for example, you can put the $\chi(k)$ data on a slide after your conclusions and only flip to it if someone in the audience asks a question that needs it.

can ask their questions afterward. But with papers it often used to be an awkward problem, particularly in formats where there were length limits.

In recent decades, however, the increased use of *supporting information* by journals has helped address the problem. This allows information of interest to specialists to be available online without cluttering up the main article.

18.2 Experimental Details

The following should be included in a paper or its supporting information:

- The beamline on which the measurements were made
- The edges that were measured
- The method for harmonic rejection (mirror, percentage of detuning)
- Special environmental conditions, such as temperature (if not room temperature)
- Measurement mode: transmission, fluorescence, electron yield, quick-XAFS, microprobe, and so on. For fluorescence, the type of detector should also be identified.
- The identity and mode of measurement of the energy reference: measured simultaneously with the sample, between each measurement, only at the start of the experiment, and so on
- Some information on sample form. At a minimum, indicate whether it is powder on tape, a pellet, a thin section, and so on. Additional information, such as a summary of the method for assuring small particle size and the number of absorption lengths above the measured edge can also be helpful.
- It is helpful to know if scans were averaged to produce the data

While that seems like a long list of items, it can generally be accomplished in a paragraph.

Unless they are unusual, scan parameters are often not reported; the signal-to-noise ratio visible in $\chi(k)$ speaks for itself.

18.3 Data

18.3.1 What to Include

If a paper includes EXAFS, it is necessary to provide $\chi(k)$ graphs (with your chosen k-weighting) for all the data analyzed, if practical. In short papers for audiences less familiar with EXAFS, some or all of these can be presented in the supporting information.

For XANES analyses, normalized XANES graphs are needed. Always include at least a few representative spectra in the main article; the rest can go in supporting information.

Aside from those requirements, you can use some discretion as to what to present graphically.

For an audience relatively unfamiliar with EXAFS, for instance, the magnitude of the Fourier transform can be useful, as long as you make it clear that while it's correlated to a radial distribution function, it's not the same thing. For a more savvy group, the real part of the Fourier transform can be more informative.

Supporting information also allows data to be uploaded as text files. This can be a great boon to those interested in working further with your results. We heartily endorse this practice!

18.3.2 Labeling Graphs

This is a more contentious subject than you might suspect.

First off, there are the graphs we have called *unnormalized absorption* and labeled $M(E)$. Unnormalized data appear in publications relatively rarely, but when they do there is the question of how to label the y-axis. Options such as $\mu(E)$, $x\mu(E)$, and $\ln(I_o/I_t)$ have been used, but none are unambiguously accurate. If you use $M(E)$, on the other hand, no one will know what you're referring to unless they've read this book or unless you take the time to explain it in the text. While there is no ideal solution to this quandary currently, it is at least a question worth putting some thought into before creating your graph.

Next is the question of the y-axis units. $M(E)$ is dimensionless. Normalized data are not just dimensionless; they've been scaled so that information from $M(E)$ has been lost—we no longer know the edge jump. Some have chosen to label the units on the y-axis of a normalized absorption plot as 'arbitrary units' or 'a.u.' In this text, we label the unit as *edge fraction*, which is somewhat more specific.

Since the convention of normalizing so that the edge jump is one is firmly established, numerical values and tick marks should be provided on the y-axis of normalized data. In this way, readers or viewers can see your choice of normalization for themselves.

It is acceptable, and often advisable, to plot 'stacked' data; that is, to shift spectra up or down to provide separation between them, as we've done in Figure 11.5 of Chapter 11, so long as the shift is noted in the caption.

k-weighted $\chi(k)$ creates another conundrum. Unweighted $\chi(k)$ is like normalized data: it has been scaled and should thus have numerical values on the y-axis but no units (or arbitrary units). But k has units—probably Å^{-1}, nm^{-1}, or pm^{-1}. So what happens when we multiple arbitrary units by Å^{-1}? Many would say that the result is Å^{-1}. But if that were the case, how come the numbers come out differently if we change the normalization? While the convention of normalizing to 1 is very strong, in theory we could have normalized the edge jump to, say, 100. That would increase the numerical values of $k\chi(k)$ by a factor of 100 as well, and feels as if it were a change of units, from 'edge fraction' to 'edge percentage.'

In this book, we solve that issue by carrying the 'edge fraction' unit through, labeling the units of the y-axis of $k^3\chi(k)$ graphs, for instance, as 'Å^{-3}•edge fraction.'

The units of Fourier transforms are related to the units of the k-weighted $\chi(k)$, but with one more factor of k since the Fourier transform integral involves multiplying the k-weighted $\chi(k)$ by dk.

The units of wavelet transforms depend on the normalization used. For the normalization used in Section 7.6 of Chapter 7, the units are related to the units of the k-weighted $\chi(k)$ but with one more factor of $k^{1/2}$.

One more issue with units: angstroms are not part of the SI system. Some journals, and some referees, prefer the use of nanometers or picometers.

When discussing wavelet transforms, it is not necessary to provide units for η and s, and many authors choose not to do so. Although η should have units of length and s units of inverse length, only their dimensionless product ηs affects the shape of the wavelet transform.

18.3.3 Estimate of Noise

Using one of the methods from Section 7.2.2 of Chapter 7, provide an estimate of $\varepsilon_{\text{random}}$, the measurement uncertainty due to noise.

18.4 Data Reduction

If the procedures followed are not too far out of the ordinary, only a few sentences need to be allocated to data reduction. Citing the software used for normalization and background subtraction, for instance, will provide most of the information as to how they were done.

Most people provide the k range as part of the text, which is fine. But it's very convenient for readers if you show the range directly on the $\chi(k)$ graphs. The same comment can be made about specifying the fitting range in the Fourier transform.

k-weighting and the range of data in k-space used for Fourier transforms should be specified. If wavelet transforms are being used, a description of the parent function (e.g. 'Morlet wavelet') and the value of its parameters (in the case of a Morlet wavelet, ηs) should also be provided.

18.5 Models and Standards

18.5.1 Curve Fitting to Theoretical Standards

There are no hard-and-fast rules as to how much detail should be provided on models used for fitting, as it will be strongly influenced by the nature of the article or presentation and the nature of the model.

At a minimum, however, the following should be provided (based in part on the suggestions of IXS Standards and Criteria Committee, 2000):

- The software package used for analysis (with citation, if available)
- The software used to compute theoretical standards (with citation)
- How S_o^2 was determined or constrained (this is a specific suggestion of IXS Standards and Criteria Committee, 2000)
- Whether the fit was performed in k-space, the Fourier transform, a wavelet transform, or the back-transform
- The fitting range
- An estimate of the number of independent points and how it was arrived at (e.g., by the Nyquist criterion)
- The number of free parameters (if the fit is in some way iterative, *all* free parameters must be reported, not just those in the final iteration). The degree of detail concerning the free parameters depends on the nature of the article or presentation.
- The number of degrees of freedom of the fit (computed as independent points − free parameters)

The IXS recommendations from 2000 include some requirements that we didn't list, like providing the function that is being minimized by the fitting routine, because those requirements are usually taken care of by citing the software package that performs the fit.

18.5.2 Linear Combination Analysis and Principal Component Analysis

We suggest that when linear combination is employed in a paper, the following be reported:

- The list of standards successfully used in fits or as target transforms. A list of standards or targets considered but not used in final fits is also helpful but not required.
- The fitting space on which the analysis was performed: normalized XANES, derivative of normalized XANES, $k^2\chi(k)$, and so on
- The range over which the analysis was performed
- For linear combination analysis, whether energy shifts were allowed

Don't forget to say something about how the standards were prepared and measured in your experimental section, especially if they were measured in a different mode from the samples!

18.6 Results

As with the model, the detail provided in the results will depend on the nature of the paper or the presentation.

18.6.1 Graphs of Fits

It is typical to show graphs of a few representative fits along with the corresponding data, whether modeling, linear combination, or target transform is being used.

18.6.2 Closeness of Fit

An *R* factor should be reported for all EXAFS fits to theoretical standards. Since there are several different definitions in use (Section 10.4.3 of Chapter 10), don't forget to indicate which you are using!

The International XAFS Society (IXS Standards and Criteria Committee, 2000) recommends that χ^2 be reported as well, but we prefer an alternative: reporting an estimate of the measurement uncertainty ε_{sys} due to systematic error. Using the logic from Sections 10.4.6 and 14.1 of Chapter 10 and 14, respectively,

$$\varepsilon_{sys} = \varepsilon_{random} \left(\sqrt{\chi_\nu^2} - 1 \right) \tag{18.1}$$

In a paper, make sure to indicate that this was the method used to compute ε_{sys}.

Linear combination analyses feature the combination of being easy to interpret, difficult to characterize statistically (the measurement uncertainty problem again), and dominated in terms of closeness of fit by the edge. Accordingly, the issue of reporting quantitative measures of closeness of fit for linear combination analyses has garnered less attention than for EXAFS modeling. We recommend reporting an *R* factor for any fits that are not graphed. (Fits that are graphed provide the same information visually, so the *R* factor is optional.) Alternatively, if there are multiple spectra for which linear combination fits were performed and all were acceptable, then the one with the highest *R* factor can be graphed and the rest reported as having smaller residuals.

Reporting the measurement uncertainty due to systematic error provides the same information as the IXS suggestion of reporting χ^2, but with less potential for confusing those not familiar with the field.

Principal component analysis should report how much variance is accounted for by the included components. While we recommend reporting this as a fraction of the variance after the first component, it is more common to include the first component in the total. Make sure you're clear as to which you are reporting!

SPOIL plays a role for target transforms analogous to *R* factor for fits to theoretical standards. With that in mind, we recommend providing SPOIL values for all target transforms, including rejected ones. In addition, it is very helpful to show graphically the poorest fit for a target transform that was accepted, as that will allow readers to see where, if anywhere, the components failed to reproduce the target.

18.6.3 Uncertainties

It is vitally important to report uncertainties with all fitted parameters.

For curve fitting to theoretical standards, the method given in Section 14.3 of Chapter 14 works well as it accounts for both random and systematic errors. Be sure to provide an indication of how the uncertainties were computed—a citation to software is usually sufficient. Do *not* just use an arbitrary rule of thumb, such as 'EXAFS can determine coordination numbers to within such-and-such percent.' What EXAFS can do depends intimately on the particular parameter being fit for the particular data using a particular model.

For linear combination analysis, the uncertainty is often dominated by systematic error, particularly associated with normalization. The uncertainties reported by software do not account for systematic error and are thus underestimated.

That statement is so important, it should be repeated: it is vitally important to report uncertainties with all the fitted parameters.

One good method for estimating uncertainties in linear combination analysis, analogous to the established method used for curve fitting, is given in Section 10.4.6 of Chapter 10. Alternatively, a reasonably good estimate can be made by estimating

For example, suppose you think the error in your linear combination analysis is dominated by the normalizations of your spectra, which you estimate are uncertain by 10%. If linear combination analysis indicates that a sample is 20% constituent A, 50% constituent B, and 30% constituent C, it's reasonable to report the percentages as 20 ± 3%, 50 ± 6%, and 30 ± 4%. I make the uncertainties a little *more* than 10% of the values because normalization isn't the only source of error. The bottom line is that you don't have to be too picky about how you calculate the uncertainties as long as you're conservative—just make the uncertainty high enough that you believe it covers all the sources of error!

systematic uncertainties, such as that due to the edge jump during normalization of the sample and standards, and incorporating the estimate into the reported uncertainties.

18.7 Conclusions

Be honest with the strength of your conclusions. If two models can be distinguished by using the Hamilton test on fits, you can with confidence select one over the other. If not, you may still be able to make a conclusive argument by bringing together multiple lines of argument; for example, XANES fingerprinting and EXAFS modeling, or EXAFS modeling and nuclear magnetic resonance.

If results are only suggestive and not conclusive, that should be made clear.

Much of the time, the same paper or presentation will include both kinds of conclusions.

WHAT I'VE LEARNED IN CHAPTER 18, BY SIMPLICIO

- For EXAFS analyses, we should always include $\chi(k)$ graphs somewhere—either in the main text or in supplementary information.
- We should provide an estimate of the measurement uncertainty due to noise. When modeling EXAFS, we should also provide an estimate of the measurement uncertainty due to systematic error.
- We should report R factors for both modeling and linear combination fits.
- All fitted parameters should be reported with uncertainties.

Reference

IXS Standards and Criteria Committee. July 26, 2000. Error reporting recommendations: A report of the standards and criteria committee. Accessed January 12, 2013. http://ixs.iit .edu/subcommittee_reports/sc/err-rep.pdf.

Chapter 19

Case Studies

The Real Deal

At last! I get to see how XAFS is actually done! But have we covered enough? Will these papers be too advanced?	
Nope. In fact, no paper really uses every single trick and technique we've talked about. The cool thing about looking at case studies is that you get a feel for what's *really* important in a good study.	
And how to cope when things go wrong!	
Yes, dear colleagues, but one can also observe the commonalities; those represent the best practices that all good papers follow.	

19.1 Introduction to the Case Studies

In this chapter, we will examine a number of actual studies from the literature. These studies come from multiple disciplines and use the full range of techniques we have learned about in this book.

DOI: 10.1201/9780429329555-24

367

In order to get the most out of this chapter, it's best if you peruse a copy of the article being discussed, although you'll still be able to follow the gist even if you do not. Many of these articles are freely available on the Web, while others are in commonly held journals.

In each case, we discuss what can actually be gleaned from the papers, sometimes using a little detective work to deduce quantities that aren't explicitly stated in the article. We offer a few critiques as well—no paper is perfect, if for no other reason than that the field has advanced since the paper was written. But the eight papers chosen here stand as exemplars in their fields, demonstrating the effective use of XAFS to solve scientific problems.

When discussing these case studies, we sometimes simplify the scientific questions for the sake of brevity. You can refer to the full articles for a more careful treatment.

19.2 Lead Titanate, a Ferroelectric

19.2.1 The Paper

This paper (Sicron et al. 1994) was published in 1994. This is the oldest paper we will use as a case study, and it's fair to say that the field has advanced considerably since it was written. But it still stands as an excellent example of the use of EXAFS to solve a problem in materials science and of the differences between local and average structure.

19.2.2 The Scientific Question

Lead titanate is a ferroelectric material up to 763 K. At the time Sicron et al. wrote this paper, the phase change at 763 K was widely thought to be *displacive*, meaning that the distance between atoms changes sharply across the phase transition. Figure 19.1 shows a two-dimensional example of a displacive transition.

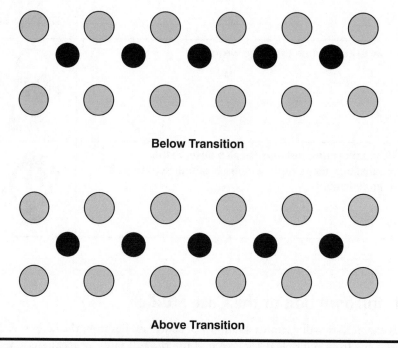

Below Transition

Above Transition

Figure 19.1 Below the transition, the black dots are closer to two nearby gray dots and further from two others. Above the transition, the black dots are equidistant from their four nearest gray neighbors. This is an example of a displacive transition.

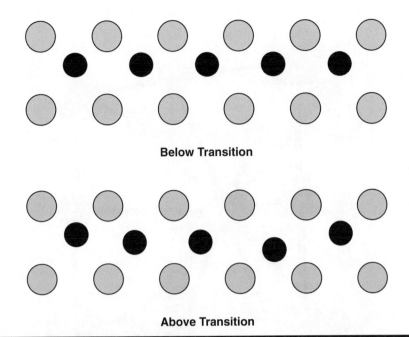

Below Transition

Above Transition

Figure 19.2 Below the transition, the black dots are always close to two nearby gray dots and further from two others and are always oriented the same way. Above the transition, the black dots are still closer to two nearby gray dots and further from two others, but the orientation is random. This is an example of an order–disorder transition.

Some other ferroelectric materials, however, showed evidence of undergoing *order–disorder* transitions. In an order–disorder transition, the local structure does not change abruptly across the phase transition, but the longer range structure becomes less ordered. Figure 19.2 shows a two-dimensional example of each kind of transition.

Sicron et al. chose to investigate the structure of lead titanate as a function of temperature through the transition, to see whether it was purely displacive or whether it also exhibited aspects of an order–disorder transition.

19.2.3 Why XAFS?

EXAFS, as a probe of local order, is well suited to the changes accompanying a displacive transition. But 'local' goes well beyond the nearest neighbor, and EXAFS may also be able to detect features of an order–disorder transition. The same, however, can be said of tools such as x-ray diffraction (XRD), which could in principle reveal either kind of change (the difference between the bottom structures in Figures 19.1 and 19.2 would show up as a change in the Debye–Waller factor).

But what if the truth is messier than one or the other? If the transition has aspects that are displacive and aspects that are order–disorder, then the picture might be more difficult to sort out. Interpreting either EXAFS or XRD would be fraught with pitfalls. The answer is to make an interpretation informed by both kinds of probes, along with other experimental and computational techniques.

These authors used FEFFIT (a predecessor to IFEFFIT) to analyze their data.

Both lead L_3 (13,035 eV) and titanium K (4966 eV) edges are readily available to many beamlines and are well separated.

19.2.4 The Structure

Lead titanate adopts the *perovskite* structure. In the ideal form of this structure (Figure 19.3), each lead atom occupies the center of a cube with titanium atoms at the

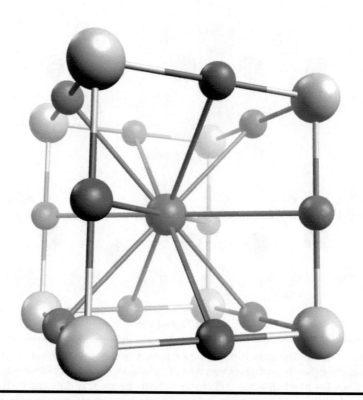

Figure 19.3   Ideal perovskite structure. For lead titanate, the atom shown in the center is lead, those on the corners are titanium, and those on the edges are oxygen.

corners and oxygen atoms at the center of each edge. This can also be thought of as each titanium atom occupying the center of a cube with lead atoms at the corners and oxygen atoms at the center of each face. Thus, the lead atoms have 12 near-neighbor oxygens and the titanium atoms have 6. The lattice constant for the unit cell is roughly 4 Å, meaning that the lead–lead and titanium–titanium distances should be within the range of EXAFS, at least at low temperatures.

The ideal perovskite structure commonly undergoes several kinds of distortion. On the basis of diffraction and previous studies of perovskites, Sicron et al. anticipated the following possibilities:

It turned out that the titanium data were too noisy to analyze the titanium–titanium scattering. But in principle, at the start of the experiment, it was a possibility!

■ A tetragonal distortion, in which the unit cell is stretched along one axis. At 150 K, diffraction revealed that the long (*c*) axis was 4.17 Å, whereas the short (*a* = *b*) axis was 3.90 Å. (Below 183 K, there is a slight orthorhombic distortion as well, so that *a* ≠ *b*. The orthorhombic distortion was considered negligible by the authors.)

■ A displacement of the lead atoms along the *c* axis. This breaks the 12 near-neighbor oxygens into the following three groups: the four on the edges of the face the lead atom is displaced toward, the four on the edges of the opposite face, and the remaining four in the middle. This also breaks the eight nearby titanium atoms into the following two groups relative to the lead: the four on the corners of the face the lead atom is displaced toward and the four on the corners of the opposite face.

■ A displacement of the titanium atoms along the *c* axis. This breaks the six near-neighbor oxygens into three groups as follows: the one the titanium atom is displaced toward, the one it is displaced away from, and the four in the middle. The eight nearby lead atoms split into two groups.

19.2.5 *Experimental Considerations*

The measurement of the lead L_3 edge was fairly straightforward. The sample was finely ground, mixed with graphite, and pressed into a pellet. The dilution fraction and thickness of the pellet were chosen to make an edge jump of around one.

The titanium measurement, however, was more challenging. Sicron et al. point out that the absorption length of lead titanate above the titanium edge is only 2 μm. It would be difficult to grind the powder down to be small compared to that length, so they decided to measure in fluorescence.

BOX 19.1   Sample Thickness

We can use the information the authors provide in their paper to estimate the total absorption of their sample. We begin by computing the absorption length of lead titanate just above the lead L_3 edge. At 13.1 keV, the mass absorption coefficient of $PbTiO_3$ is about 120 cm²/g. Using 7.5 g/cm³ as the density of lead titanate (this density is readily available on the Web), we arrive at an absorption length of 11 μm, which matches the number provided by the authors. So far, so good, my friends!

Just below the lead L_3 edge, at 13.0 keV, we get about 55 cm²/g. Thus, the edge jump is just a little over half of the total absorption, and an edge jump of one should correspond to a total absorption of more than two.

But we have not yet accounted for dilution. The paper does not specify the thickness of the pellet used for transmission, but it does say the pellet 'could be easily handled.' Let us suppose, then, that it were 1 mm thick. At 13 keV, the absorption length of graphite is about 4 mm, so the graphite would add about another 0.2 absorption lengths to their sample.

Taken together, we can estimate that the total absorption of their sample above the lead L_3 edge was most likely between 2 and 2.5.

That's a little higher than the ideal we recommended in Section 6.3.4 of Chapter 6, but well within the range of the recommendation of experts given in Table 6.3.

Yes, and the authors of the paper take pains to note that they detuned 75% to be especially vigilant about suppressing harmonics. Note the sign of skilled experimenters: a slightly thick sample is paired with aggressive detuning!

However, fluorescence raises its own problems. We can easily calculate, using the methods of Section 6.3.1 of Chapter 6, that the mass absorption coefficient of the sample just above the titanium edge would be 610 cm²/g, of which 110 cm²/g, or almost 20%, is attributable to the titanium. That is enough to cause considerable self-absorption. Nevertheless, the authors chose to take their self-absorption lumps and measured a 1 mm thick undiluted pellet in fluorescence.

BOX 19.2   What Else Could They Have Done?

	Why didn't they just dilute the fluorescence sample?
	It wouldn't do any good; they switched to fluorescence because they couldn't grind the particles to well below 2 μm, as they would have needed to in order to get undistorted transmission data! If there were particles several absorption lengths thick embedded in a graphite matrix, they still would have seen almost as much self-absorption (because the local concentration was just as high) but would have degraded signal to noise (because the average concentration was lower)!
	Besides, Simplicio, their main interest was in displacements, which are phase variables, not amplitude ones. Phase variables aren't affected much by self-absorption. If they wanted to fit changing coordination numbers or perform a XANES linear combination fit, then their choice wouldn't be so good, but as it is, it's not a big deal.
	Alternatively, they could have made the choice to measure in transmission and taken the distortion that comes with uneven thickness. But that would have been somewhat more inconvenient in terms of sample preparation; a millimeter of graphite is more than four absorption lengths at energies near the titanium edge, so they couldn't have used their pellet method. They could have used powder on tape, although even tape absorbs noticeably at those energies. They would also have had to choose between making the edge jump fairly small or the sample really thick, because although the titanium absorbs enough to cause self-absorption, it's still a small fraction of the absorption compared to the lead. This is a 'tweener' sample—it's somewhat inconvenient to measure, no matter which strategy you use.
	They could have tried sedimentation to get the fine particle size they needed.
	Yeah, they could have, but they didn't really need to. They were transparent about how they did the measurement, described and attributed the distortions they observed, and got the answers to the scientific questions they were looking for. That's a win!

One more experimental aspect is worth discussing. Because they were using a Si(111) crystal in the monochromator, there was no second harmonic. The third harmonic above the titanium edge (5.0 keV), however, would be around 16 keV. Because the L_1 edge of lead is at 15.9 keV and the L_2 edge at 15.2 keV, any intensity in the third harmonic would at least exacerbate the fluorescent background and might additionally contaminate the titanium scans with lead EXAFS. Mitigating these issues was another reason that the authors detuned I_o by 75%.

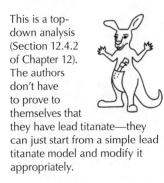

Presumably, based on the year the paper was written, Sicron et al. were using a Lytle-type detector to measure fluorescence. With an energy-discriminating detector, the contamination from the lead L edges would have been less of a concern.

19.2.6 The Model

The authors collected spectra on both the lead and the titanium edges at a variety of temperatures above and below the known tetragonal to cubic phase transition. Although a modern analysis might include a simultaneous refinement of all spectra, these features were not widely available in EXAFS modeling software in the early 1990s, in part because such fits would strain the capabilities of typical computers of that period. Therefore, each edge and each temperature was refined separately. Although this prevented the implementation of some of the efficiencies described in Chapter 17, it yielded redundant information on the scientific questions they were trying to address, increasing their confidence in their results.

In this case study, we will focus on their analysis of the lead L_3 edge.

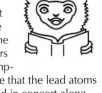

This is a top-down analysis (Section 12.4.2 of Chapter 12). The authors don't have to prove to themselves that they have lead titanate—they can just start from a simple lead titanate model and modify it appropriately.

19.2.6.1 Paths

The model comprised 11 single-scattering paths as follows:

- Three paths for the three groups of nearby oxygen scatterers (four atoms each)
- Two paths for the titanium scatterers (four atoms each)
- Two paths for the lead scatterers (one with two atoms and one with four)
- Four paths for more distant oxygens: four oxygen atoms in the $+c$ direction from the $+c$ titanium atoms, eight oxygen atoms in neighboring unit cells parallel to the $+c$ face, eight oxygen atoms in neighboring unit cells parallel to the $-c$ face, and four oxygen atoms in the $-c$ direction from the $-c$ titanium atoms. The furthest of these oxygens was located at a distance of $D = 5.38$ Å in the model.

The only one of these path sets that is not clear from the geometry is the lead. It appears that the assumption was made that the lead atoms were displaced in concert along the c axis, with adjacent stacks being displaced in the opposite sense. No discussion is provided in the paper for this choice.

In addition, the paper indicates they included 34 double-scattering and 14 triple-scattering paths.

Above room temperature, the signal degraded (as expected, due to increased MSRDs), so they only fit the near-neighbor oxygen and titanium scatterers. To make sure this didn't introduce an artifact into their analysis, they also tried the shorter range fit on the low-temperature spectra and confirmed that it yielded consistent values for the fitted parameters.

19.2.6.2 Free parameters

Free parameters included the following:

- S_o^2 and E_o.
- The displacement of the lead atoms along the c axis and the displacement of the titanium atoms along the c axis.
- MSRDs. One for each of the following direct scattering groups: the four nearest oxygen atoms, the eight other near oxygen atoms, the four nearest titanium atoms, the other four near titanium atoms, the lead atoms, and the more distant oxygen atoms.
- A third cumulant for the four nearest oxygen atoms.

But the conclusions they end up drawing in the paper don't depend on this choice, as they end up being mostly based on the behavior of the lead–oxygen, lead–titanium, and titanium–oxygen scattering.

It's clear from what they wrote that the authors tried other free parameters in preliminary fits, probably including lattice parameters for the unit cell and a third cumulant for the titanium atoms.

19.2.6.3 Constraints

Perhaps the most interesting constraints are geometric. Rather than using absorber–scatterer distances as free parameters, Sicron et al. used displacement of individual atoms relative to their lattice sites. From there, half path lengths can be readily calculated.

For example, consider the path from the lead absorber to the nearest titanium scatterers. If we follow the nomenclature from the paper and call the displacements of the lead and titanium atoms from their ideal positions d_{Pb} and d_{Ti}, then the distance from a lead atom to the nearest titanium atoms is given by

$$D_{Pb-Ti_1} = \sqrt{(0.5a)^2 + (0.5a)^2 + (0.5c - d_{Pb} + d_{Ti})^2} \tag{19.1}$$

Other half path lengths, including those for multiple-scattering paths, could be expressed in similar ways. This form of geometric constraint dramatically reduces the number of free parameters needed for the fit.

Other constraints included the following:

In calling this the distance to the nearest titanium atoms, we have assumed that $d_{Pb} > d_{Ti}$ and that the two displacements are in the same direction. If $d_{Pb} < d_{Ti}$, then Equation 19.1 represents the distance to the further titanium atoms. Displacement of the two atoms in opposite directions could be represented by a negative value of d_{Ti} in Equation 19.1.

- For the lead L_3 edge, a and c were constrained to have the values they adopt, according to XRD, at 150 K. The authors justify this by referring to the results from their titanium EXAFS analysis, which shows that these parameters are nearly constant across the temperature range measured.
- Above room temperature, the MSRD for the two groups of near oxygen atoms that are not the nearest (i.e., paths 2 and 3 when the single-scattering paths are sorted from short to long) was constrained to be 28% larger than for the nearest neighbors. This ratio was taken from the results found at lower temperatures. This constraint was imposed because, as the temperature was raised, the contribution from these oxygen paths decreased.
- The MSRDs for multiple-scattering paths were constrained in terms of the MSRDs for direct scattering paths. Although the exact method of constraint is not provided in the paper, Section 17.5.4 of Chapter 17 tells us that the precise method of constraint is not likely to be crucial.
- S_o^2 and E_o were constrained to be the same for all lead L_3 edge fits, with values chosen 'so as to optimize the fits at all temperatures.' In a modern study, this probably would have been done by refining spectra from multiple temperatures simultaneously. Although the process used by the authors was doubtless somewhat more laborious than that, the effect is similar.

19.2.6.4 Degrees of freedom

Because MSRDs increase with temperature, the range of usable $\chi(k)$ tends to shrink with temperature. In addition, the shells further out in the Fourier transform were not fit at higher temperatures. For these reasons, the number of independent points according to the Nyquist criterion (Section 14.1.1 of Chapter 14) ranged from 24 at their lowest temperature measurement to only 10 at their highest temperature. Their constraint scheme reflected this loss of information, allowing for nine free parameters at the lowest temperature and only six at the highest temperatures. (In addition, temperature-independent values of S_o^2 and E_o were extracted from the entire data set.)

The results for the lattice parameters from the titanium edge appear at first glance to conflict with the diffraction results. The titanium edge results show a and c differ from each other by about 0.2 Å even well above the point where diffraction shows that they are equal. The authors speculate that this means that, above the phase transition, 'the crystal consists of correlated nanoregions of distorted cells with different distortion orientations.' This would yield the cubic structure on average, as indicated by diffraction, and the tetragonal structure locally, as indicated by EXAFS.

19.2.7 Drawing Conclusions

EXAFS is best at making choices between alternatives. There are enough arbitrary choices made during data reduction and modeling that it is difficult to know how much to conclude from a single fit.

In this paper, Sicron et al. presented not only their primary fits but also some important alternatives. Because the diffraction data indicated that the average structure above the transition was cubic, they tried to fit the lead L_3 data with a model that was locally cubic (i.e., one with a structure like Figure 19.3). This model had four free parameters as follows: E_o, an MSRD for the 12 near oxygen atoms, an MSRD for the 8 near titanium atoms, and a 'phase shift correction' for the titanium scatterers (the S_o^2 was presumably extracted from the lower temperature data, which the diffraction and EXAFS agreed showed tetragonal distortion). This was actually generous to the cubic fit; there is no a priori reason E_o should differ from the lower temperatures, and the 'phase shift correction' for titanium, presumably taken to address some possible error in the theoretical standards, is the kind of fudge factor that would attract scrutiny if used in their preferred fit.

Even with the two questionable parameters allowed to vary, the least-squares residual was five times that for the primary fit.

Let's apply the Hamilton test to these results. Going from the cubic to the tetragonal high-temperature fit adds four free parameters: two displacements, a third cumulant for the nearest oxygens, and an additional MSRD for the titanium atoms. Thus, $b = 4/2 = 2$. The number of degrees of freedom of the tetragonal fit is roughly $10 - 6 = 4$ (S_o^2 and E_o are almost entirely determined by the spectra from lower temperatures), so $a = 4/2 = 2$. r, according to the paper, is $1/5 = 0.2$. Computing, $I_{0.20}(2,2) = 0.10$, which is not quite enough to say the tetragonal fit is better at the 95% confidence level.

That's not the only evidence they had, however. They also had the fits to the titanium edge. Although we have not given the details of those models here (consult the paper if you're interested!), the high-temperature tetragonal fit used three free parameters and seven independent points, giving four degrees of freedom. The cubic fit, utilizing one fewer free parameters, yielded a least-squares residual that was twice as high. Thus, for a Hamilton test on the titanium data, $b = 1/2 = 0.5$, $a = 4/2 = 2$, and $r = 1/2 = 0.5$. $I_{0.50}(2,0.5) = 0.12$. Once again, on its own, this result is not conclusive.

The results were, however, determined essentially independently. If for the lead L_3 edge the tetragonal model has a 10% probability of being as much better as the cubic model as was seen by chance, and for the titanium K edge the figure is 12%, the probability of them both being that much better by chance is approximately $0.12 \times 0.10 = 1\%$. The cubic fit can, therefore, be rejected in favor of the tetragonal fit.

And with that, the core of the scientific question is answered. If the local structure is still tetragonal above the transition while the average structure becomes cubic, then there must be some order–disorder aspect to the phase change.

If the two edges had been refined simultaneously, then the conclusion would presumably have been the same. The higher degrees of freedom of the combined data set would have likely yielded a result that passed the Hamilton test.

19.2.8 Presentation

The paper includes figures giving the k-weighted $\chi(k)$ for all data. Fits for both edges and several representative temperatures are presented with the real part, imaginary part, and magnitude of the Fourier transforms. There is also a plot of the magnitudes of the Fourier transforms of each of the path groups from the fitted model at the lowest temperature. Fitted parameters are presented, with error bars, as a function of temperature.

BOX 19.3 What We Learned and Liked

	Let us follow Simplicio's lead, my friends, and each say something we learned from this paper, or perhaps simply something we liked. We should like that they provided enough information so that we could reconstruct what they did, even when they didn't explicitly spell it all out—notice how we deduced both the sample thickness and the geometric constraints!
	I like that the free parameters are based on what is happening physically in the material. The two displacement parameters are very easy to interpret and well worth the trouble of having to work out the geometry for the half path lengths.
	I am impressed by how they responded to adversity! The titanium measurement was clearly challenging, and their data on that edge ended up being, by their own admission, a bit noisy. And yet it was a key and convincing part of their overall story!
	I learned from how they used previous diffraction results in combination with EXAFS to reach their conclusions.
	I admire their attention to detail, as in this excerpt: 'Care was taken at every stage of the experiment to avoid systematic distortions to the data. The sample was pure and homogeneous. A homogeneous spot in the beam was chosen and the width of the slits defining the beam was optimized for resolution. Gases were properly chosen for linear response in both detection chambers. The sample was repositioned in the beam frequently during the course of the experiment. This frequent realignment compensated for any changing conditions in the experiment or in the beam.'
	I learned from the paper and from our discussion that there's more than one way to do things right. They could have refined the data simultaneously (at least if the paper had been written today), but they did it separately, and that worked. They could have tried to measure and analyze the difficult titanium edge several different ways, but what they did worked. They were always clear about what they did, and always thoughtful about what it meant, resulting in a compelling paper.
	I love the unapologetic focus on what works! Can't measure one way? Try another! Your low-temperature data are less noisy than the high-temperature ones? Use a more constrained model at high temperature! Titanium edge shows the lattice parameters are only very weakly dependent on temperature? Use that to help constrain the lead edge! Great stuff!

19.3 An Iron–Molybdenum Cofactor Precursor

19.3.1 The Paper

This paper was published by Corbett et al. (2006). We'll use it as an example of applying XAFS analysis to a biological question, and as a clever way of addressing a difficult problem.

19.3.2 The Scientific Question

Bacteria reduce nitrogen with the help of an enzyme called iron–molybdenum cofactor or FeMoco for short. The structure of this enzyme is known (Einsle et al. 2002), but the mechanism of its synthesis within the bacteria is not.

One step in this process transfers a precursor compound from a protein called NifB to a protein complex called NifEN. Although the precursor was known to contain iron and sulfur, the structure was not known. The goal of Corbett et al., then, was to determine, or at least narrow down, the structure of the precursor.

19.3.3 Why XAFS?

The precursor has not been isolated, but is only found bound to one of the proteins. Determining its structure by diffraction would therefore require crystallizing an entire protein complex and then solving its structure, a very challenging task. Because the precursor is built around iron atoms, XAFS can provide structural information localized to the precursor.

19.3.4 A Challenge and a Solution

These authors used XFIT to reduce their data and EXAFSPAK for analysis.

One of the biggest challenges Corbett et al. faced was that NifEN contains 'permanent' iron atoms even before the precursor binds to it. XAFS measured at the iron K edge, then, would be a linear combination of the signals from the precursor iron and the permanent iron. Finding the structure of the precursor is challenging enough; having to find the structure of other regions of the protein at the same time would be formidable.

So Corbett et al. devised a different plan. They had a sample of the NifEN protein without the precursor bound (referred to as $\Delta nifB$ NifEN) and another where it was bound. They knew, from previous work, that $\Delta nifB$ NifEN contains 8.5 ± 1.0 atoms of iron per molecule, whereas NifEN (with the precursor) contains 16.1 ± 2.4 atoms of iron per molecule (Hu et al. 2005). This means that roughly half of the $\chi(k)$ at the iron K edge comes from the precursor.

Because proteins are big molecules, the local structure around the permanent iron atoms is probably unchanged by the addition of the precursor. Corbett et al., therefore, applied the concepts of linear combination analysis in reverse: knowing how much of each constituent they had, the spectrum of the 'mixture' (protein + precursor), and the spectrum of one of the constituents ($\Delta nifB$ NifEN), they could subtract the known fraction of $\Delta nifB$ NifEN from the spectrum of the precursor + protein, rescale the residual by the fraction of precursor, and thus arrive at the $\chi(k)$ for just the precursor.

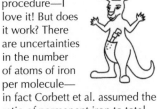

This is a daring procedure—I love it! But does it work? There are uncertainties in the number of atoms of iron per molecule—in fact Corbett et al. assumed the ratio of permanent iron to total iron was 1:2, which is not quite the best fit values of 8.5:16.1. Throw in the usual uncertainties associated with normalization, and there's the potential for the derived spectrum to be a bit off. To address this concern, Corbett et al. first tested the method on a similar system where all the structures were known. Showing that the method works on a similar system is a crucial part of their paper.

19.3.5 Possible Structures

The structure of the iron cluster in FeMoco is shown in Figure 19.4.

At the time Corbett et al. undertook their study, it was known that the precursor contributed iron and sulfur atoms to the final structure shown in Figure 19.4, but it wasn't known whether it also incorporated a molybdenum atom, and the arrangement of the iron and sulfur atoms was not known. Were the iron atoms tetrahedrally coordinated to sulfur, as is often the case? Or was the unusual geometry of the final product, in which several of the iron atoms feature a coordination to sulfur that is a trigonal pyramid so flattened as to be nearly trigonal planar, also present in the precursor?

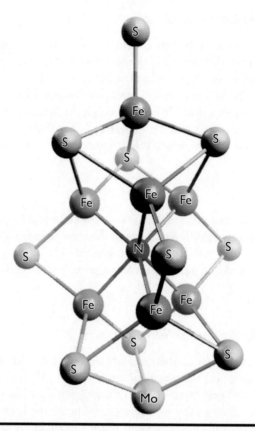

Figure 19.4   Structure of the iron cluster in FeMoco (additional organic structures bonded to the molybdenum atom and the top sulfur are not shown).

19.3.6 *Experimental Considerations*

The concentration of iron in the protein as prepared is given in the paper as 5 mM for NifEN and 3 mM for Δ*nifB* NifEN. Using the atomic mass of iron, we can calculate that 5 mM corresponds to 0.28 mg/mL. Iron has a mass absorption coefficient of roughly 200 cm²/g above its *K* edge, while carbon (a proxy for the other elements in the protein and accompanying materials) is roughly 6 cm²/g. Assuming that the rest of the components have a density of about 1,000 mg/mL, we can compute the fraction of absorption due to iron as on the order of (200 × 0.28)/ (200 × 0.28 + 6 × 999.7) = 0.9%. Thus, self-absorption is not a significant concern.

Proteins are fragile samples, vulnerable to beam damage and other sources of degradation. To help maintain the samples, they were added to a solution comprising a Tris•HCl buffer, imidazole, dithionite, and glycerol. This solution was flash frozen and measured in fluorescence at 10 K. The low temperature served the following two purposes: it minimized the risk of beam damage and allowed data to be collected to high *k* by minimizing the thermal contribution to MSRDs.

With the additional dilution due to the protective solution, the concentration of iron was fairly low, and so a 30-element germanium detector was used for the fluorescence measurement. The biggest source of fluorescent background in this case was likely scatter, because the majority of the atoms in the sample were of low atomic weight. Corbett et al., therefore, used a manganese filter and Soller slits (Section 8.7.1 of Chapter 8).

The signal-to-noise ratio in their resulting data was impressive, allowing them to use the range of $\chi(k)$ from 2 to 16 Å⁻¹ for analysis.

19.3.7 Fingerprinting

Corbett et al. used fingerprinting on the preedge features to provide a preliminary understanding of the structure of the precursor. They used pseudo-Voigt functions (Section 9.3.2 of Chapter 9) to model the features, allowing the energies, widths, and amplitudes to vary. The rise of the edge itself, which acted as a background for the preedge features, was also modeled by a pseudo-Voigt function.

As it turned out, the differences between the precursor and other tested protein-bound iron–sulfur clusters were clear; although almost all other clusters showed a preedge peak or peaks centered around 7112.2 eV, the precursor had an extra preedge peak centered at 7113.6 eV. The only other substance tested that showed this extra peak was FeMoco itself.

Because the other tested materials featured iron tetrahedrally coordinated to sulfur, the second peak suggests that the precursor, like the final FeMoco product, had a substantial contribution from the flattened trigonal pyramidal geometry.

19.3.8 The Models for EXAFS

Given the evidence from fingerprinting, Corbett et al. decided to try fitting the EXAFS data using models based on the structure of FeMoco itself.

One candidate would be the structure shown in Figure 19.4, but with different organic structures attached to the ends of the cluster. If that were the case, the iron *K*-edge EXAFS of the precursor should look very similar to FeMoco, but that was not the case—for one thing, the ratios of the amplitudes of the peaks in the Fourier transform were significantly different.

Another possibility would be that the molybdenum atom was substituted by an iron atom (the '8Fe' model) or were simply missing/replaced by organic structures (the '7Fe' model). Finally, both the molybdenum and the iron atom furthest from it could be missing (we'll call that the '6Fe' model, in agreement with Corbett et al.'s naming scheme).

At this point, it is helpful to examine the FeMoco structure more closely, to examine characteristics of the radial distribution function around iron in preparation for choosing paths.

Each iron atom has three or four sulfur near neighbors. Most of them also have a nitrogen neighbor, but Corbett et al. decided to disregard the nitrogen on the basis that the scattering would be weak compared to the sulfur and iron.

Beyond that, the alignment of the two central triangles of iron atoms creates an unusual distribution of iron–iron distances. There are 'short' iron–iron distances along the sides of the triangles, between an atom in one triangle and the corresponding atom in the other triangle, and between the apical iron atoms, if any, and the iron atoms in the triangle near it. There are also 'long' iron–iron distances, from one atom of a triangle to one of the noncorresponding atoms in the other triangle. In addition, the FeMoco structure is not entirely symmetric, with the result that the short iron–iron distances range from 2.58 to 2.67 Å, whereas the long distances range from 3.69 to 3.72 Å.

In the 6Fe model, each iron atom has only three neighbors at the short iron–iron distance, whereas in the 7Fe model or FeMoco itself that rises to four neighbors for the three atoms in the triangle nearest the apical iron atom. The authors essentially use a fingerprinting technique to reject the 6Fe model, saying 'the intensity of the short-range iron–iron scattering precludes a six-iron cluster' (Corbett et al. 2006).

That leaves the 7Fe and 8Fe models to try. They are very similar, so we'll detail them together.

One could look askance at such a statement. Were the authors basing that statement on the amplitudes of peaks in the magnitude of the Fourier transform? That can be dangerous, considering the effects of multiple scattering, truncation, and interference between paths.

Sure, Robert, I'd 'look askance' if they didn't back it up. But they *do* back it up—they go on to model the 7Fe and 8Fe structures, and as we'll see, the 8Fe structure ends up fitting better. Because the three models form a nice sequence structurally, it's probably OK for them to dismiss the 6Fe model.

19.3.8.1 Paths

The model included single-scattering paths for the near-neighbor sulfurs and the short and long irons. The short and long iron atoms were allowed two paths each, to reflect the kind of modest spread of distances seen in the FeMoco crystal structure. There were thus a total of five single-scattering paths. An examination of the structures suggests that the contributions from multiple-scattering paths with half path lengths shorter than 4 Å is modest; there are no focused paths, and with the exception of those involving the apical sulfur, there are no short triangle paths. Corbett et al. therefore chose to neglect multiple scattering in their fits.

19.3.8.2 Free parameters

Free parameters comprised the following:

- E_o
- Half path length and MSRD for each path (10 parameters total)
- The fraction of the short iron–iron scattering assigned to the shorter of the short paths
- The fraction of the long iron–iron scattering assigned to the shorter of the long paths

This gives a total of 13 free parameters.

19.3.8.3 Constraints

The most interesting constraint is on the coordination numbers. For each model, the coordination number was fixed at the average specified by the model structure.

For example, consider the coordination of the short iron–iron path in the Fe7 model. The apical iron has three iron neighbors in that range; the iron atoms in the triangle below it each have four iron neighbors in that range; and the iron atoms in the lower triangle again each have three iron neighbors in that range. The average coordination for the short iron path in the Fe7 is therefore $(1 \cdot 3 + 3 \cdot 4 + 3 \cdot 3)/7 = 3.4$. In contrast, the Fe8 model would have $(2 \cdot 3 + 6 \cdot 4)/8 = 3.8$.

As discussed under 'paths,' however, the short iron path was split into two, and the fraction of the scattering assigned to each was a free parameter. If we call that parameter x, then the degeneracy for the shorter of the short iron paths under the Fe7 model is $3.4x$ and the degeneracy for the longer of the short iron paths under that model is $3.4(1-x)$.

The other constraint of note is S_o^2, which was constrained to be 1.0.

BOX 19.4   The S_o^2 Constraint

1.0 is a little high for S_o^2—wouldn't it have been better to constrain it to something like 0.9?

Better yet, the authors could have extracted S_o^2 from their test fit to FeMoco—the coordination numbers for that structure are known.

If they were fitting coordination numbers, that would mean the coordination numbers they found would come out too low by the same ratio as the overestimate in S_o^2. But within each model, they weren't fitting coordination numbers, so the effect would more likely show up as fitted values for MSRDs that were a little too large. On the other hand, the difference between the 7Fe and the 8Fe models does depend on coordination numbers! So their fits will have a modest bias toward the less-coordinated model; for example, the 7Fe.

 As we'll see, the systematic error introduced by this choice actually works opposite the direction of their final conclusion. So they made a conservative choice. When in doubt, make the conservative choice; it makes your conclusion stronger in the end!

19.3.8.4 Degrees of freedom

In this paper, Corbett et al. chose to fit $\chi(k)$ directly, rather than fitting the Fourier transform. This is a reasonable strategy for this compound—the Fourier transform shows very little structure beyond the chosen paths, in agreement with the structures under consideration. There is thus no need to use Fourier filtering to reduce the amount of modeling that has to be done.

Of course, when fitting in $\chi(k)$, there is no explicit R range to substitute into the Nyquist criterion (Section 14.1.1 of Chapter 14). But it's reasonably easy to get an estimate. Inspection of Figure 2 in Corbett et al. (2006) suggests the meaningful data are concentrated between 1.5 and 4 Å; this is also consistent with the interatomic distances in their single-scattering paths. Using the k range of 2–16 Å^{-1} over which they fit, this gives a Nyquist estimate of 22 independent points. With 13 free parameters, they have a comfortable 9 degrees of freedom.

19.3.9 Drawing Conclusions

Corbett et al. use the alternate definition of the R factor from Equation 10.4 in Chapter 10, which they call F. This is just the square root of our usual definition of R. To avoid confusion, we'll square their values of F and report them as R for the remainder of this discussion.

Their fit to the 7Fe and 8Fe models yields R values of 0.049 and 0.044, respectively. Those values, according to our informal guidelines in Table 14.1, signal that the 'model has some details wrong or data are low quality. Nevertheless, consistent with a broadly correct model.' That is to be expected. This is a very challenging problem, and systematic error will accumulate despite the best efforts of the authors. The reverse linear combinational analysis, the constraint of S_o^2, the grouping of single-scattering paths into a few bins, and the neglect of multiple-scattering paths and scattering beyond the first few shells will all take their toll. But they allowed the authors to address a problem that had long resisted solution.

Let's apply the Hamilton test, and see if the 8Fe fit is *significantly* better than the 7Fe one.

It's not entirely clear how to define b in this case, because of the way paths are averaged. So let's start by assigning b the minimum value we can, 0.5, and see if the Hamilton test shows a significant difference in that case. a is half the number of degrees of freedom, or around 5. And r is $0.044/0.049 = 0.90$. $I_{0.90}(5, 0.5) = 0.32$, so the 8Fe is not statistically better at the 95% confidence level.

Although they do not explicitly perform the Hamilton test, the authors make the point clearly:

 Notice that if we had used the statistic the authors called F, we would have had to square it to apply the Hamilton test!

'Although the 7Fe model cannot be excluded on this basis, the 8Fe model is considered to be the more likely structure because it is a better match both to the EXAFS fit results and previous biochemical studies.'

The previous studies provide additional evidence that, along with the EXAFS results, gives the edge to the 8Fe model.

In the supporting information, the authors also report results of several fits using several other models for the paths beyond the short iron, such as using another shell of sulfur atoms rather than the long iron paths. Each of these fits had a larger *R* factor than the 8Fe model, although again the differences did not reach the level of statistical significance. Nevertheless, this exploration of the fitting space is helpful because it shows that the authors considered multiple possibilities, and the 8Fe model still appears preferable.

What then, can be concluded?

Remember—the choice of S_o^2 as 1.0 probably biased the results slightly toward the 7Fe model. The fact that the 8Fe model still fits better is a point in its favor.

- It is very likely that the iron–sulfur cluster in the precursor is structurally similar to the cluster in FeMoco, in particular favoring flattened trigonal pyramidal coordination over solely tetrahedral.
- It is quite unlikely that the precursor includes molybdenum.
- It is not settled whether the precursor comprises seven iron atoms or eight, although the evidence leans more toward eight.

That set of conclusions represents a considerable advance from the state of knowledge before Corbett et al. conducted their study.

19.3.10 Presentation

One interesting wrinkle in the way the data were presented is that E_o was given relative to 7130 eV, an energy somewhat higher than the white line. This causes the fitted E_o shifts to be tabulated as values around −10 eV—usually a red flag! But in this case, because the reference energy was chosen so high, they are reasonable.

The paper includes figures giving k^3-weighted $\chi(k)$ for all data. Normalized XANES is included in the supporting information, with the detail of the preedge included in the published article. Also in the published article, second derivative plots of the preedge features are shown to emphasize the multiple peaks. Key fits are presented both in k^3-weighted $\chi(k)$ and the magnitude of the Fourier transform.

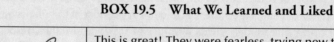

BOX 19.5   What We Learned and Liked

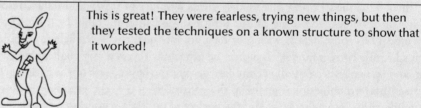	This is great! They were fearless, trying new things, but then they tested the techniques on a known structure to show that it worked!
	The authors were very careful to provide thorough information, so we could understand precisely what they had done. This is particularly important when developing new techniques.
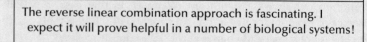	The reverse linear combination approach is fascinating. I expect it will prove helpful in a number of biological systems!

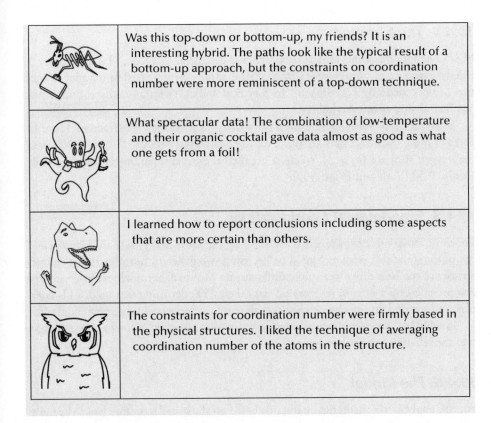

	Was this top-down or bottom-up, my friends? It is an interesting hybrid. The paths look like the typical result of a bottom-up approach, but the constraints on coordination number were more reminiscent of a top-down technique.
	What spectacular data! The combination of low-temperature and their organic cocktail gave data almost as good as what one gets from a foil!
	I learned how to report conclusions including some aspects that are more certain than others.
	The constraints for coordination number were firmly based in the physical structures. I liked the technique of averaging coordination number of the atoms in the structure.

19.4 Manganese Zinc Ferrite, an Example of Fitting Site Occupancy

19.4.1 The Paper

This paper was written by Calvin et al. (2002). It is an example of determining site occupancy in spinels, a topic we already touched on in Section 9.2 of Chapter 9.

19.4.2 The Scientific Question

Mixed metal ferrites such as manganese zinc ferrite (MZFO) have important magnetic and electrical properties, which depend crucially on how the cations are distributed between tetrahedral and octahedral sites. Nanoparticulate ferrites were thought to have different characteristics than their bulk counterparts, some of which were thought to be associated with different cation distribution.

For this study, MZFO nanoparticles were prepared using two different protocols and compared to three differently processed bulk MZFO samples. The goal was to compare the site occupancies of these samples.

19.4.3 Why XAFS?

Because the atomic numbers of manganese and iron differ by only one, they appear nearly indistinguishable to XRD. In addition, the diffraction peaks for the nanoparticulate samples are broadened. The combination means that XRD cannot tell us the site occupancy of these samples.

Mössbauer spectroscopy could provide the distribution of the iron cations, but manganese does not have a Mössbauer-active isotope.

XAFS, on the other hand, is element specific. By collecting and simultaneously refining data on the manganese, iron, and zinc *K* edges, the site occupancy of each can be determined.

As one might suspect, Calvin who wrote this paper is the same person who wrote this book. We have tried to use a variety of examples, from a variety of fields, over the course of this text. So you will forgive our author, I trust, if he indulges in an example drawn from one of his own papers.

Sure, why not? But if there's something I don't like in this paper, I'm still gonna call him on it!

These authors used FEFFIT (a predecessor to IFEFFIT) to analyze their data.

19.4.4 *The Structure*

MZFO adopts the spinel structure, which was discussed in Section 9.2 of Chapter 9. Briefly, cations can occupy sites that are tetrahedrally coordinated to oxygen, or sites that are octahedrally coordinated. In a typical spinel, twice as many cations sit in octahedral sites as tetrahedral ones.

More precisely, most spinels adopt the space group Fd$\bar{3}$m . The fractional coordinates of the tetrahedral sites are (0, 0, 0), the octahedral sites are (5/8, 5/8, 5/8), and the oxygen sites are (*u*, *u*, *u*), where *u* is called the oxygen parameter and is usually around 0.38 (Smit and Wijn 1959).

Fd$\bar{3}$m is a space group that can be described by more than one setting. Thus, the oxygen parameter is sometimes defined so that it is 0.125 less than in our definition.

19.4.5 *Experimental Considerations*

The most remarkable experimental aspect of this study is that two of the samples and one of the standards were measured twice, using two different beamlines. The characteristics of the beamlines were quite different; for instance, harmonic rejection on one was provided by a mirror, whereas on the other, 25% detuning was used. This comparison provided an interesting check on the measurement and subsequent analysis.

In addition, for each protocol, two samples were prepared, allowing investigation into the reliability of the synthesis.

The paper indicates that the samples were from one to four absorption lengths thick at all three edges. That's a bit thick, and, what's more, the detuning was only 25%. Fortunately, the comparison to a beamline with a harmonic rejection mirror helps us to see if harmonics are causing a problem.

19.4.6 *The Model*

For this analysis, six theoretical standards based on the spinel structure had to be built: a tetrahedral and an octahedral absorber for each of the three cation edges. All edges were weighted equally in the simultaneous refinement; that is, the measurement error ε was assumed to be the same for the purposes of finding a fit.

The site occupancy of the scattering atoms was addressed by using the mixed model technique from Section 16.7.3 of Chapter 16.

BOX 19.6   Was a Mixed Model Necessary?

Did Calvin et al. really need to create a mixed model of scatterers? After all, manganese, iron, and zinc all have atomic numbers within five of each other.

It is sometimes said that EXAFS cannot distinguish between elements with atomic numbers within five of each other. But that statement needs to be understood in the proper context. It is true that it is difficult to distinguish between elements with atomic numbers that close to each other in a fit in which bond length and coordination number are free to vary. But that is in part because the bond length parameter can compensate for small differences in the phase shift due to the scattering element, and the coordination number parameter can compensate for small differences in the scattering amplitude. Thus, it might be hard to discriminate between a manganese and a zinc scatterer in such a fit. But using the wrong scatterer introduces systematic errors into both bond length and coordination number, even if the atomic number of that scatterer is within five of that of the actual element. Thus, whenever possible, one should use the actual scattering element in theoretical standards.

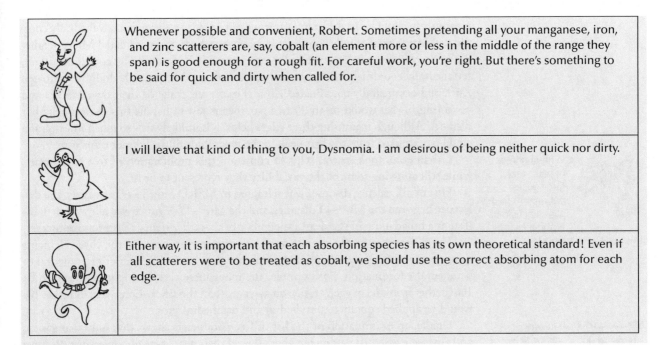

Whenever possible and convenient, Robert. Sometimes pretending all your manganese, iron, and zinc scatterers are, say, cobalt (an element more or less in the middle of the range they span) is good enough for a rough fit. For careful work, you're right. But there's something to be said for quick and dirty when called for.

I will leave that kind of thing to you, Dysnomia. I am desirous of being neither quick nor dirty.

Either way, it is important that each absorbing species has its own theoretical standard! Even if all scatterers were to be treated as cobalt, we should use the correct absorbing atom for each edge.

19.4.6.1 Paths

The number of paths was not specified in the paper, but encompassed 'multiple shells of cations' as well as multiple-scattering paths. The fitting range extended from 1.0 to 5.5 Å in R.

On the basis of the spinel structure and the fitting range, the number of single-scattering paths would have been on the order of 10 for each theoretical standard if a mixed model were not being used. The mixed model increases this to perhaps 20 per theoretical standard. (The cation scatterers would have an individual path for each cation, but the oxygen paths would need no modification.) That suggests a total of roughly a hundred single-scattering paths. Add in multiple-scattering paths and the total number of paths was likely in the hundreds.

With so many paths, it's understandable that Calvin et al. did not provide an enumeration. But in retrospect, it might have been helpful had they provided the number of single- and multiple-scattering paths in each theoretical standard, so that readers could know what level of detail was present in the models.

19.4.6.2 Free parameters

The free parameters used were as follows:

- S_o^2 and E_o for each edge (six parameters)
- A lattice parameter
- The oxygen parameter
- The fraction by which the distance to the nearest oxygen differs from the average for the manganese cations, and a similar parameter for the zinc cations (two parameters).
- MSRDs: nearest-neighbor manganese absorber to oxygen; nearest-neighbor iron absorber to oxygen; nearest-neighbor zinc absorber to oxygen; and all more distant paths (four parameters)
- Tetrahedral site occupancy for zinc and manganese ions (two parameters)

This yields a total of 16 parameters.

19.4.6.3 Constraints

The list of free parameters implicitly includes rather dramatic constraints. For example, both manganese and iron cations in spinels exhibit *valence disorder*; that is, they are likely to exist as a mixture of +2 and +3 ions in a spinel (zinc should be only +2). This means there could be 10 different nearest-neighbor oxygen distances: tetrahedrally coordinated Mn^{2+}, octahedrally coordinated Mn^{2+}, tetrahedrally coordinated Mn^{3+}, octahedrally coordinated Mn^3, tetrahedrally coordinated Fe^{2+}, octahedrally coordinated Fe^{2+}, tetrahedrally coordinated Fe^{3+}, octahedrally coordinated Fe^{3+}, tetrahedrally coordinated Zn^{2+}, and octahedrally coordinated Zn^{2+}. If each were assigned their own MSRD and bond length, that would mean 20 free parameters just to handle this nearest-neighbor distance! Although measuring three edges helps, it is unlikely that so much information could be extracted from the first peak in the magnitude of the Fourier transform.

Calvin et al. took several steps to constrain this proliferation of free parameters, while still allowing some of the variability they represent to be fit.

First of all, valence disorder was relegated to MSRD only. In other words, the difference between the Mn^{2+}–O distance and the Mn^{3+}–O distance was allowed to manifest as a broadening of the nearest-neighbor peak, rather than as distinct distances.

The differences in the nearest-neighbor oxygen distances for the three elements were treated as if it were a fractional difference that was independent of tetrahedral or octahedral coordination. For example, the manganese cations might be found to be 1% further from the oxygen atoms, on average, than the zinc cations were, but this 1% would be applied equally to tetrahedral and octahedral sites.

Finally, an examination of the list of free parameters shows that only manganese and zinc were assigned parameters that allowed their nearest-neighbor oxygen distance to vary from the average. Why not iron as well?

The answer is that, despite jokes to the contrary, not everything can be above average. If we know how much the zinc and manganese deviate from the average, and if we know the relative amount of zinc, manganese, and iron, then we can calculate the deviation of the iron from the average.

As it turns out, Calvin et al. did know the relative amount of the three cations; there are a number of methods for finding the stoichiometry of a compound (Calvin et al. used inductively coupled plasma atomic emission spectroscopy). Thus, they reduced the number of free parameters related just to the nearest-neighbor oxygen distances to five: the fractional changes in half path length for manganese and zinc and one MSRD for each element. (Most of the other free parameters affect the nearest-neighbor scattering as well, but also affect more distant paths.) Data from three edges are likely to support fitting five parameters related just to the nearest-neighbor peak.

Another important constraint relates to the site occupancy. Although the fraction of manganese ions and zinc ions present in tetrahedral sites are both free parameters, the fraction of iron atoms is then determined by the stoichiometry of the compound and the sites that are left available.

An example of such calculation is worth doing in detail. Suppose the fraction of manganese, zinc, and iron cations in tetrahedral sites is T_M, T_Z, and T_F, respectively. Similarly, we can designate the fractions in octahedral sites as O_M, O_Z, and O_F. Finally, let's designate the mole fraction of each element in the sample as X_M, X_Z, and X_F. We know that each cation must be in one site or the other, thus,

$$T_M + O_M = 1; \quad T_Z + O_Z = 1; \quad T_F + O_F = 1 \tag{19.2}$$

We also know, because we have a spinel, that the total occupancy in the octahedral sites must be twice that of the tetrahedral sites (neglecting vacancies!):

$$X_M O_M + X_Z O_Z + X_F O_F = 2(X_M T_M + X_Z T_Z + X_F T_F) \tag{19.3}$$

Substituting Equation 19.2 in Equation 19.3 and solving for TF yields:

$$T_F = \frac{X_M + X_Z + X_F - 3(X_M T_M + X_Z T_Z)}{3 X_F} \tag{19.4}$$

With hundreds of paths and three edges, but only 16 free parameters, this is about as top-down as it gets, friends!

If this paper was written now, some sort of histogram method (Section 16.5 of Chapter 16) might have been applied to better model the distribution of bond distances.

Finally, multiple-scattering paths were constrained using a heuristic method that added no new free parameters (see Section 17.5 of Chapter 17).

19.4.6.4 Degrees of freedom

Not only were three edges used, but somewhat different ranges of k-space were used on different samples, consistent with the trade-offs discussed in Chapter 14. All edges were fit from 1.0 to 5.5 Å in the Fourier transform.

For the manganese edge, the k range was typically around 3–8 Å^{-1}, yielding about 14 points by the Nyquist criterion. For the iron and zinc edges, the k ranges were generally somewhat larger.

Because separate edges represent independent data sets for the purpose of computing degrees of freedom, there were typically 50 or more independent points per analysis, considerably more than the 16 free parameters.

19.4.7 Drawing Conclusions

The purpose of this paper was somewhat different than the two examined in the previous sections of this chapter. Rather than attempting to choose between models, this study sought quantitative analysis. Thus, the primary answers are provided by the values of the fitted parameters, rather than the statistical quality of the fits. The repetition of the analysis on two beamlines helped confirm the reliability of the determinations, and their agreement with magnetic measurements spoke to their validity.

In order to reach conclusions about the values of fitted parameters, it is important to examine the uncertainties in those parameters. One of the samples, for example, yielded 42 ± 11% of the manganese in tetrahedral sites when measured on one beamline and 50 ± 11% on the other. Those values are consistent, as their error bars overlap. The corresponding values for the standard measured on both beamlines were also consistent, 15 ± 5% on one and 18 ± 4% on the other. But the values for the sample and the standard lie far outside of each others' uncertainties, indicating that the cation distributions are significantly different.

Another interesting result was the MSRDs of the nearest-neighbor oxygens. In every case (all standards, all samples, and all beamlines), the MSRD for the manganese–oxygen near-neighbor path was larger than zinc–oxygen, usually significantly so. And in nearly every case, the iron–oxygen MSRD fell in between. The paper speculates that this may be a marker for valence disorder, as zinc ions should always be of a single valence, whereas manganese ions would be expected to be present as a mixture of +2 and +3.

19.4.8 Presentation

The paper includes graphs of k-weighted $\chi(k)$ for all data and the real part of the Fourier transform for all data and all fits. An example of unnormalized absorption is presented. To facilitate comparison with fingerprinting studies in the literature, the magnitudes of the Fourier transforms of the manganese edge of two of the samples are also shown.

A detailed discussion of sources of systematic error within the model is provided.

BOX 19.7   What We Learned and Liked

I like seeing some of these older papers. They're still convincing, but we can think of ways to do the same study even better now!

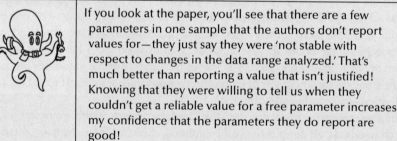

	The list of simplifying assumptions that went into their model was much appreciated. It was thorough and direct.
	The site occupancy constraints were very elegant and helped couple together the data from the three edges.
	Me? I liked the near-neighbor parameters and constraints. The authors found a way to reflect some of the distortions caused by the different cations without using so many free parameters that the fit became unstable.
	I like that the MSRDs were given a physical interpretation.
	If you look at the paper, you'll see that there are a few parameters in one sample that the authors don't report values for—they just say they were 'not stable with respect to changes in the data range analyzed.' That's much better than reporting a value that isn't justified! Knowing that they were willing to tell us when they couldn't get a reliable value for a free parameter increases my confidence that the parameters they do report are good!

19.5 Sulfur XANES from the Wreck of the Mary Rose

19.5.1 The Paper

On July 19, 1545, the English warship *Mary Rose* sank during a battle with the French navy. In 1982, the wreck of the *Mary Rose* was raised and the difficult work of preserving the remains began (Figure 19.5). One of the greatest risks was posed by sulfur compounds in the wood as they could potentially convert to sulfuric acid now that the ship had been freed from the seafloor.

This paper by Sandström et al. (2005) used XANES, along with a battery of other techniques, to determine the form in which sulfur was present in the wood from the *Mary Rose*.

19.5.2 The Scientific Question

Although several questions were addressed by the paper, the main focus was on the identity and amounts of sulfur compounds as a function of depth in wood cores taken from the *Mary Rose*.

Figure 19.5 The wreck of the *Mary Rose*. (Photo courtesy of the Mary Rose Trust. This photo is licensed under the Creative Commons Attribution-Share Alike 3.0 Unported license.)

19.5.3 Why XAFS?

This is a problem especially well suited for XAFS. Slices from the heterogenous wood cores could be measured *in situ* through microprobe, and because XAFS is element-specific, the sulfur-containing species present could be identified even though they are a small percentage of the total sample.

19.5.4 Experimental Considerations

The sulfur concentration was around 1% by mass in most of the measured wood, except for 3.5% in a sample taken from a wooden gunshield. Iron was also found, although the amount showed considerable variation, with typical values around 0.2% by mass (2.8% in the gunshield). Approximating the rest of the material as carbon, and using a mass absorption coefficient above the sulfur K edge of 2000 cm²/g for sulfur, 900 cm²/g for iron, and 160 cm²/g for carbon, we can calculate that the fraction of absorption due to sulfur from the gunshield is roughly (2000 × 0.035)/(2000 × 0.035 + 900 × 0.028 + 160 × 0.937) = 29%! Even the more typical samples would have about 10% of the absorption due to sulfur.

One option, then, would be powdering the samples and making transmission measurements; one absorption length of the wood corresponds to roughly 50 μm above the sulfur K edge.

Another option would be to measure fluorescence in the thin limit (see Section 6.4 of Chapter 6); this is the choice Sandström et al. made. They therefore ground

One advantage of measuring in the thin limit of fluorescence is that homogeneity is not much of an issue. Unlike in transmission, the sample can be filled with pinholes, as long as it's thin.

segments from the wood cores into very fine powder and brushed them on sulfur-free tape. At the sulfur K edge, the absorption length of air is only about 3 cm, so the measurement region was filled with helium gas, which has an absorption length an order of magnitude larger.

A sodium thiosulfate energy reference was measured before and after each sample, assuring the careful alignment necessary for principal component and linear combination analyses.

Sandström et al. also measured sections of the wood using a microprobe technique, to see where in the wood's cellular structure the sulfur compounds were concentrated. For this purpose, they cut slices a few microns thick from the wood (once again, ensuring the thin limit) and measured in fluorescence using an energy-discriminating detector.

19.5.5 Principal Component Analysis

The paper mentions a preliminary principal component analysis (PCA) of the data from the wood cores. (As described in Section 11.10 of Chapter 11, PCA of this kind sometimes doesn't even make it into the final paper!) The PCA did not include the gunshield, which had a significantly different elemental composition than the wood cores. In this case, Sandström et al. find 'at least' six significant components.

19.5.6 Linear Combination Analysis

Rather than using solid compounds as standards, Sandström et al. preferred to use dilute solutions, finding that they serve as a better analogue to the materials as found in wood. The standards they used were meant to be representative of various classes of sulfur compounds, rather than exact matches, and are shown in Table 19.1.

Not counting the two forms of cystine separately (two forms were included partially as a test of which was more appropriate), there were seven standards. Linear combination analysis found each in at least one core sample, consistent with the large number of constituents suggested by PCA.

For the gunshield, melanterite ($FeSO_4 \cdot 7H_2O$) was used instead of sodium sulfate as a sulfate standard, because it was presumably closer to the form present in the iron-rich gunshield.

Table 19.1   Sulfur Standards Used for Linear Combination Analysis of Wood Cores in Sandström et al. (2005)

Standard	Form	Class
Cystine	Solid	Disulfide
Cystine	Aqueous	Disulfide
Cysteine	Aqueous	Thiol
Elemental sulfur	In xylene	Elemental sulfur
Methionine sulfoxide	Solid	Sulfoxide
Sodium methylsulfonate	Aqueous	Sulfonate
Sodium sulfate	Aqueous	Sulfate
Pyrite	Solid	Pyrite

19.5.7 *Drawing Conclusions*

Like the previous case study, this is not a study attempting to choose between two different possibilities. Preservation of cultural artifacts depends on the details, and this study provides them in the form of the distribution of sulfur species as a function of depth of core and choice of wood. Information on the distribution of the sulfur compounds on the microscale was provided by microprobe XANES, and the results of the linear combination analysis were corroborated by x-ray photoelectron spectroscopy.

19.5.8 *Presentation*

Normalized XANES spectra for all data are provided, some with the main paper and some in the supporting information. A few representative linear combination fits are shown, with the rest being tabulated. Microprobe maps are shown at different energies to reveal distribution of sulfur-bearing species; once again, some maps are included in the main paper and some in the supporting information.

BOX 19.8   What We Learned and Liked

	So many kinds of data! I could read this paper over and over and over again, and keep learning more about the system!
	I appreciate how careful they were to maintain energy calibration (crucial for linear combination analysis!) and to explain how they did it.
	I appreciate that they showed a picture of the gunshield, friends. That sounds like a small thing, but it helps put an unfamiliar artifact in context and also gives a visual impression of its condition.
	The test of solid versus aqueous cystine as a standard for linear combination analysis was one of many interesting features!
	I liked seeing how PCA can be used as a preliminary probe.

I am not a biologist, nor do I analyze archaeological artifacts. But Sandström et al. made the chemistry of the different portions of the wood very clear to me.

Sometimes I feel that XAFS papers can lose sight of the system being studied and get bogged down in mechanics. In this case, though, the authors always kept their eye on the ball: the role of sulfur in the wood recovered from the *Mary Rose*.

19.6 Identification of Manganese-Based Particulates in Auto Mobile Exhaust

19.6.1 The Paper

Ressler, Wong, Roos, and Smith (2000) published a paper identifying the species of manganese-based particulates emitted by automobiles that were using fuel with the additive (methylcyclopentadienyl)manganese tricarbonyl (abbreviated as MMT).

This paper is an important part of the literature not just because of the conclusions it drew about automobile exhaust, but because it was a clear and early exemplar of the use of XANES PCA to attack a difficult speciation question.

19.6.2 The Scientific Question

In order to understand the health impacts of manganese particulates in automobile exhaust, it is crucial to know the speciation.

The primary question addressed by this study is this speciation. Secondary questions were the effect on this speciation of different simulated driving patterns.

19.6.3 Why XAFS?

Ressler et al. address this question directly in the introduction to their paper. They describe an attempt to get XRD data, and wryly state the results: 'Studies of four relatively heavily loaded samples (292–4244 μg) showed only background from the filter substrate material.' X-ray photoelectron spectroscopy (XPS) did suggest that oxygen, phosphorus, and sulfur were associated with the manganese, but did not provide more definite speciation.

The low concentration, small sample size, and possible poor crystallinity led to the use of XAFS.

19.6.4 Experimental Considerations

These authors used WinXAS to analyze their data.

The samples were, in essence, soot collected from automobile exhausts during testing. We can use the mass given for the 'heavily loaded' diffraction samples to estimate the number of absorption lengths in the soot layer.

Let's start with the upper end of the range at 4,000 μg. Presuming the sample to be the diameter of an exhaust pipe on the 1997 Ford Taurus sedans used in the study, the area would be about 20 cm². Soot from auto mobile exhaust is mostly carbon—the manganese species, while the focus of the study, were likely to be present at low concentration. Above the manganese *K* edge, carbon has a mass absorption coefficient of around 8 cm²/g. The number of absorption lengths is thus approximately $(8 \text{ cm}^2/\text{g})(4{,}000 \text{ µg})/(20 \text{ cm}^2) \approx 0.002$. That's a thin sample! Even if the sample were pure manganese, a similar computation only arrives at around 0.1 absorption lengths. Fluorescence can be used without fear of self-absorption.

The samples were measured using a Lytle detector. An energy-discriminating detector was not called for in this case; because the sample was thin, the amount of scatter was probably small. And because the other elements in the sample were dominated by those with low atomic numbers, air would act to filter out much of the background.

Because the samples were thin, it was straightforward to measure a reference foil simultaneously in transmission.

Samples were obtained from two different cars (but identical models), at two different mileages, using four different simulated driving patterns. Twelve of the sixteen possible combinations of these variables were used.

Twelve measurements were made in part not only to gain information about the dependence on the variables that had been changed but also to provide enough related spectra to perform robust PCA.

Ressler et al. made their measurements using quick-XAFS, even though the experiment wasn't time resolved. It says good things about the reliability of quick-XAFS that it can be a good choice even when it's not needed!

19.6.5 Principal Component Analysis and Target Transforms

PCA on XANES of the twelve samples revealed that three components were sufficient to recreate the spectra to within 1%.

Guided by the indications of phosphorous, sulfur, and oxygen from the XPS measurements, target transforms were attempted on 10 different standards: MnO, Mn_3O_4, Mn_2O_3, β-MnO_2, α-MnO(OH), MnS, $MnSO_4 \cdot H_2O$, $MnPO_4$, MnP_2O_7, and $Mn_5(PO_3OH)_2(PO_4)_2 \cdot 4H_2O$ (hureaulite). Only Mn_3O_4, $MnSO_4 \cdot H_2O$, and hureaulite provided reasonably good matches to the target transform.

Because three constituents were expected from PCA, the identification of three by target transform indicates that the complete set has been found.

A cluster plot of the second component versus the first revealed a natural division of the samples into two groups. Intriguingly, this division into groups does not correspond with any of the experimental parameters, except that the less-populated group included spectra from only one of the two cars. This suggests there is a 'hidden variable' controlling the mix of constituents in the exhaust.

Although Ressler et al. measured their samples in fluorescence, they measured their standards in transmission (with 50% detuning), allowing them to achieve good signal to noise without self-absorption. It is not necessary to measure standards and samples in the same mode to perform linear combination analysis!

19.6.6 Linear Combination Analysis

With the constituents identified, linear combination analysis using the XANES became straightforward. This analysis revealed that all three constituents were present in all of the samples, but to different extents. The spectra in the smaller of the two groups identified on the cluster plot were distinguished by having more than a quarter of their spectral weight provided by Mn_3O_4.

Ressler et al. use Equation 10.5 for *R*, rather than the version we've been using throughout this text. Thus, when they set a threshold of 1%, it corresponds to something like 0.01% of the variance (including the first component).

19.6.7 Fingerprinting

Ressler et al. also used fingerprinting to estimate the average manganese valence of the samples. They chose to use an arctangent function to fit the edge and found that the standards established a linear relationship between edge energy and valence. By comparing with the edge energy of the standards, they were able to conclude that all of the standards exhibited a manganese valence of roughly 2.1 to 2.2. Because manganese in Mn_3O_4, $MnSO_4 \cdot H_2O$, and hureaulite have a valence of +3, +2, and +2, respectively, this can be compared to the fractions found from linear combination analysis. For example, according to linear combination analysis, 43% of the manganese in sample

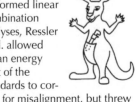

When they performed linear combination analyses, Ressler et al. allowed for an energy shift of the standards to correct for misalignment, but threw out any standards that needed to be shifted by large amounts as unphysical.

In this paper, the conversion between the percentages found directly by linear combination analysis and percent by weight would seem a bit convoluted to a synthetic chemist because the discussion of the conversion is mixed together with an empirical demonstration of the accuracy of the linear combination approach for a sequence of known mixtures. But different audiences respond to different ways of presenting information, and this paper is written for environmental scientists, not synthetic chemists!

12 was present as Mn_3O_4, 24% as $MnSO_4 \bullet H_2O$, and 33% as hureaulite. This implies an average manganese valence of $0.43 \bullet 3 + 0.24 \bullet 2 + 0.33 \bullet 2 = 2.4$. Although slightly higher than the value extracted from fingerprinting the position of the edge, this is reasonably good agreement for a comparison of this type.

In addition to XANES, qualitative fingerprinting was also applied to the magnitude of the k^3-weighted Fourier transform of the EXAFS. The Fourier transform of the Mn_3O_4 standard shows a prominent peak at about 3.0 Å, which is absent in the other two linear combination standards. This peak is also seen in the group of samples with high Mn_3O_4 concentrations, but not in those with lower concentrations.

19.6.8 Drawing Conclusions

The target transforms leave little doubt that Mn_3O_4, $MnSO_4 \bullet H_2O$, and hureaulite, or very similar compounds, account for nearly all of the manganese species produced in the auto mobile exhaust studied.

The finding of two different groups of samples, one higher in Mn_3O_4 concentration than the other, is an interesting result. Perhaps surprisingly, which group a sample falls into did not depend in any obvious way on parameters such as driving pattern or mileage.

Thus, the primary scientific question—the identities of the manganese species admitted—has been answered. The secondary question as to what controls this speciation led to a mystery that would require further studies to solve.

19.6.9 Presentation

Normalized XANES spectra of all samples are provided. A few examples of the magnitudes of the Fourier transforms are provided for use in fingerprinting. Examples of linear combination fits and target transforms are provided. Importantly, two examples of rejected target transforms are shown, similar to Figure 11.21 of Chapter 11. A discussion of error analysis is provided to quantify the uncertainty in the linear combination analyses.

BOX 19.9   What We Learned and Liked

	The use of quick-XAFS in a non-time-resolved study is fun! Hey, if it works, it works!
	The authors took the time to lay out the method of PCA. Their care and clarity has helped make this an important paper in the XAFS literature.
	The technique of allowing constituent spectra in linear combination to vary in energy calibration, but throwing them out if they showed unphysical shifts, shows a good use of a 'reality check.'

	Verifying their linear combination technique on a series of mixtures of known proportions lent credence to their results.
	Those are interesting samples. I'd never thought of collecting and measuring a sample like that!
	I like surprises! And the results showing two groups of samples were a surprise!
	The way they used XPS to help choose target transforms for PCA, and then used those results to perform linear combination analyses, is an excellent example of how techniques can be chained together.

19.7 *In Situ* Investigation of Cobalt Molybdenum Sulfide Catalysts

19.7.1 The Paper

Plais et al. (2021) published this paper in part to demonstrate the application of modern XAFS analysis techniques to the use of cobalt molybdenum sulfide catalysts. While they chose relatively straightforward model systems for their investigation, that doesn't mean everything about them were already known, and their experiment yielded some novel and important results.

19.7.2 The Scientific Question

Their paper investigates a number of related systems and processes, but for this case study we'll focus on Plais et al.'s measurement of the sulfidation of a catalyst produced by calcination of an alumina-supported CoNiMo precursor. They wished to discover the cobalt sulfide species that evolve when the catalyst is heated under a mixture of hydrogen sulfide and hydrogen gas. Understanding the species formed during this process has implications to the stability of cobalt molybdenum sulfide catalysts.

19.7.3 Why XAFS?

XAFS lends itself well to time-resolved *in situ* catalysis experiments, particularly since catalytic materials often have nanoscale order. While other techniques such as Raman spectroscopy can also help characterize the process, XAFS is able to analyze many species that are difficult to identify using other techniques.

It is important to note that the precursor material is *not* a pure compound! Based on linear combination fitting, the authors found that the cobalt was primarily present as a mixture of $CoAl_2O_4$, $Co(H_2O)_6^{2+}$, and CoAl-based layered double hydroxide. Raman spectroscopy suggested the molybdenum was primarily present as MoO_3 or related species.

These authors used Demeter and IFEFFIT to reduce and analyze their data, along with a MATLAB toolbox for MCR-ALS developed by Jaumot et al. (2005).

The beamline capabilities gave them lots of flexibility. If they wanted better signal to noise, they could have merged more scans at once, at the cost of some time resolution. Or they could have chosen to measure only one edge per heating cycle, but then they wouldn't be sure their measurements of the cobalt and molybdenum edges were on a consistent sample. As an industrial chemist, I study a lot of systems that evolve over the time scale of hours. While I can measure those kinds of processes on standard stepping-mode beamlines, it's really nice when I get beam time on a quick-XAFS line!

19.7.4 *Experimental Considerations*

The authors used the ROCK quick-XAFS beamline at the SOLEIL synchrotron (La Fontaine et al. 2020). Using a flow cell to facilitate the *in situ* measurement (La Fontaine et al. 2013), they measured powder in transmission, which allowed the use of a reference foil.

For the *in situ* measurements, they ramped the temperature from room temperature to 400°C at a rate of 3°C per minute, leading to a total heating cycle time of more than two hours. In quick-XAFS mode, they could potentially collect one EXAFS spectra every 250 ms, but the time scale of the process they were studying did not require such fine resolution. Accordingly, they chose to only use scans taken in one direction on the monochromator, avoiding artifacts that could be introduced by including scans taken in the reverse direction. In addition, they merged scans in sets of 20 to improve signal to noise. The result allowed them to measure a merged spectrum in 10 seconds, which was still more than adequate for their experiment. Finally, roughly every 30 seconds they would switch edges from cobalt to molybdenum or vice versa, a process which itself took 30 seconds. In each 2-minute cycle they would therefore collect three merged spectra at the cobalt edge and three merged spectra at the molybdenum edge. The effective time resolution for this experiment was therefore two minutes, but in that time, they were collecting EXAFS data for two different edges. During each 2-minute cycle, the temperature would increase by 6°C.

For their EXAFS analyses, they used a k range from 2.8 to 10.6 Å^{-1}, with the top of that range dictated by noise in the spectra.

19.7.5 *Principal Component Analysis and Blind Source Separation*

For the remainder of this case study, we'll focus on the cobalt edge data. The authors also analyzed data from the molybdenum edge, and collected, but did not thoroughly analyze, data from the nickel edge. In this system, each element is present as a number of different species, many of which include only one type of metal atom. Therefore, the analyses of different edges might reveal different numbers of species.

At the cobalt edge, the scree plot clearly suggested that there were only three significant components.

The authors then attempted blind source separation. They employed evolving factor analysis to provide the initial guesses, followed by MCR-ALS (see Section 11.8.3 of Chapter 11).

For initial analysis of the results, they used fingerprinting, comparing the spectra achieved from blind source separation with the spectra of likely constituents. The results for the cobalt edge were initially difficult to interpret. One constituent appeared to be similar to the precursor mixture, and one appeared to be a sulfide. But the third did not appear similar to any known spectrum of a pure material or to a mixture of materials thought to coevolve. Instead, it had some features of an oxide spectrum and some of a sulfide. While it was possible that a sulfide and oxide species appeared simultaneously at some point during calcination and then had their concentrations rise and fall together throughout the rest of the temperature ramp, that did not seem likely to the authors. Using their knowledge of the system, they strongly believed there would be two separate species contributing to this 'constituent,' and that the two species should be evolving to some extent independently. This would mean that their choice of three significant spectra from the scree plot was wrong, and that they should have picked four.

19.7.6 Column-Wise Augmentation

One way of trying to tease out additional constituents from PCA data is to change the data set in some way. For example, in section 11.6 of Chapter 11 we discussed how if the data set contains too many very similar spectra, that PCA might begin to consider one or two that contain a different constituent to be just due to noise. PCA is not 'wrong' in that respect: if one were to increase by a factor of ten the number of spectra in the set used as data for a PCA analysis, for example, it becomes much more likely that noise causes one of the spectra to show a hint of a constituent that is not present.

One possibility would be to cut down on the number of very similar spectra used for PCA. But another method is to add some spectra to the set which are thought to include only constituents present in the experimental spectra, but perhaps with different ratios. The additional spectra will help 'teach' the PCA, and then the MCR-ALS, what the spectra of some of the actual individual constituents look like, allowing it to avoid spurious solutions. This process is called column-wise augmentation (CWA), described in Ruckebusch et al. (2006). In this case, the authors added spectra from one of the other experiments they had done during the run: a dried cobalt on alumina sample that they had determined to be a mixture of CoS_2 and Co_9S_8, which were expected to be present in the calcined CoNiMo on alumina sample as well. They included a roughly equal number of spectra from the calcination of the CoNiMo sample and from the calcination of the dried cobalt on aluminum sample. In addition, they imposed a constraint on the MCR-ALS that the spectra from the CoS_2 and Co_9S_8 sample could be built out of at most two of the constituents.

The scree plot on the new data set unambiguously showed at least four significant spectra, with a hint there might even be a fifth. MCR-ALS using four significant spectra yielded one that was similar to the precursor, one similar to CoS_2, one similar to Co_9S_8, and one which could be some sort of CoMoS phase.

This is why I do not like relying only on the scree plot to determine the number of significant components. I would prefer to supplement it with one or more of the numerical criteria described in Section 11.3.5 of Chapter 11.

That's a good idea, Robert, but as described in Section 11.4 of Chapter 11, it's not foolproof either.

Please note that the authors did not simply add a spectrum of a standard collected at another beamline in a different sample holder. Such a choice would have introduced additional variables that would have made their results difficult to interpret. Instead, they added data from the same experimental visit, using the same beamline, with the same sample holder, and the same scan parameters.

BOX 19.10   What Happened to the Oxide?

	I thought the three-component MCR-ALS showed that the XANES of one of the constituents was a mixture of oxide and sulfide. But then the four-component MCR-ALS gave us another sulfide. What happened to the oxide?
	It was never there. The three-component fit couldn't handle four components, of course. It might have 'borrowed' a little of the precursor oxide constituent to fit some other feature in some of the sulfide spectra, creating the illusion of a mixture distinct from the precursor.
	This illustrates the risk of trusting MCR-ALS in the absence of other evidence.
	Fortunately, in this case, there was other evidence for the four-constituent solution, both theoretical and experimental, some of which we'll see in the remainder of this case study.

The initial model Plais et al. used is very simple. If there had been much cobalt–cobalt scattering in this constituent, then the initial model would not have fit well, and the authors would have tried modifying it until they found a model that made sense and fit well. Since their model without cobalt–cobalt scattering yielded a good fit, however there was no need to try more complicated models.

I don't really know this particular family of materials, so I don't know if they're likely to be leaving out any focused paths. If the main goal of their study was to find the values of variables linked to EXAFS amplitude, I might be nervous about that possibility. But since they're mainly trying to determine if the unknown constituent might be a CoMoS compound, just the two direct scattering paths should be good enough.

For the suspected CoMoS phase, Plais et al. next turned to curve fitting to a theoretical standard.

19.7.7 EXAFS Modeling

For the purposes of this study, Plais et al. just wanted to verify that the unknown constituent could plausibly be a CoMoS phase. They therefore proceeded with a simple, flexible model, a bit like the first step in a bottom-up fitting approach (see Section 12.4.1 of Chapter 12). They couldn't start with *just* the sulfur neighbors, though, since the strong peak from the nearest shell of molybdenum would have significant overlap with the contribution from the nearest-neighbor sulfur atoms. For this model, they assumed there were no cobalt–cobalt interactions within the R range they chose to fit, a choice presumably driven by their knowledge of related systems.

19.7.7.1 Paths

The model included only a single-scattering Co-S path and a single-scattering Co-Mo path.

19.7.7.2 Free parameters

Free parameters comprised the following:

- E_o
- Half path length, coordination number, and MSRD for each path (six parameters total)

This gives a total of seven free parameters.

19.7.7.3 Constraints

Based on analysis of CoO and CoS_2 standards measured under similar conditions to the samples, S_o^2 was chosen to be 0.71.

19.7.7.4 Degrees of freedom

Plais et al. fit in back-transform space, using a k range of 2.8 to 10.6 Å^{-1} and an R range of 1 to 3 Å. Under the Nyquist criterion, this yields 10 independent points. With 7 free parameters, this yields 3 degrees of freedom, which is acceptable.

19.7.7.5 Results

The fit yielded an R factor of 0.003 (reported as 0.3% in the paper). Absorber–scatterer distances, MSRDs, and E_o adopted physically reasonable values. The coordination number for the nearest-neighbor sulfurs was 4.7 ± 0.6, and for the cobalt–molybdenum scattering it was 0.9 ± 0.4, a reasonable value for the nanoscale compounds often found in studies of catalysts. Using the criteria from Chapter 14, the fit is a good one, confirming the presence of a nanoscale CoMoS compound during the calcination.

19.7.8 Drawing Conclusions

While the study in the paper by Plais et al. is wide-ranging and draws multiple conclusions, the piece we've focused on in this case study primarily demonstrates that a CoMoS compound is formed during calcination of the alumina-supported CoNiMo precursor. In addition, their analysis allows them to evaluate the time-resolved percentages of precursor, CoS_2, Co_9S_8, and CoMoS compound during calcination.

19.7.9 Presentation

This paper studies multiple systems using a variety of techniques—too much to include all of the required detail in 13 pages! Therefore, it relies heavily on including information in the supporting information, which runs an additional 26 pages.

Continuing our focus on the cobalt edge of the CoNiMo calcination data, the paper itself provides all normalized XANES spectra for the series, plotted on a single graph using spectral colors to indicate the time evolution, with 'deep blue' at the start of the series and 'deep red' at the end.

The supporting info also provides a representative plot of k^3-weighted $\chi(k)$ data.

For the main paper, they chose to highlight their fitting process using the following figures and tables:

■ A comparison of the k^3-weighted $\chi(k)$ of one of the components identified by MCR-ALS after column-wise augmentation with two spectra obtained during other parts of the experiment that were thought to be CoS_2, showing that the component is likely to represent CoS_2.

■ A comparison of the k^3-weighted $\chi(k)$ of one of the components identified by MCR-ALS after column-wise augmentation with two spectra obtained during other parts of the experiment that were thought to be Co_9S_8, showing that it is plausible that the component is Co_9S_8. The experimental data is quite noisy at high k.

■ A graph of the k^3-weighted $\chi(k)$ of one of the components identified by MCR-ALS after column-wise augmentation. This component was the one thought to be a CoMoS compound, which was not closely similar to any other measurement in their experiment. Therefore, no comparison to experimental spectra was provided.

■ A graph of the time evolution of the four identified constituents (the precursor, CoS_2, Co_9S_8, and the CoMoS compound). The x-axis is labelled as temperature, rather than time.

■ A table of results for the EXAFS fit to a theoretical standard of the suspected CoMoS spectrum, providing coordination number, absorber–scatterer distance, and MSRD for both the cobalt–sulfur scattering and cobalt–molybdenum scattering, along with the fitted shift in E_o, the EXAFS R factor (expressed as a percentage), and the reduced χ^2. Uncertainties are provided for all of the fitted parameters except the E_o shift.

The supporting information provided additional information:

■ A graph illustrating the linear combination fit used to determine the cobalt species in the CoNiMo precursor. The experimental spectrum and the linear combination fit are shown as normalized XANES, while each constituent contributing to the fit is shown scaled to the amount of its contribution. The residual (effectively, the 'misfit') is also shown.

■ The scree plot for the cobalt edge data *without* column-wise augmentation

■ The normalized XANES spectra and concentration profiles for the constituents found by MCR-ALS from the data *without* column-wise augmentation.

■ The scree plot for the cobalt edge data *with* column-wise augmentation

■ The normalized XANES spectra and concentration profiles for the constituents found by MCR-ALS from the data *with* column-wise augmentation

■ The normalized XANES spectra and concentration profiles using the constituent spectra found by MCR-ALS with column-wise augmentation, but applied only to the unaugmented data. This was done so that they could provide an apples-to-apples comparison of statistics for the four-constituent fits relative to the three-constituent fits.

The printed version of the plot of the normalized XANES spectra is in grey-scale, but refers readers to the online, color version, which also includes a high-resolution version for download. In addition to being information-packed, it is quite pretty. While conveying information is the primary purpose for graphs in a scientific paper, pleasing aesthetic qualities are a nice bonus!

It was crucial for the authors to provide the details of their temperature ramp rate in the text of the article. While it is understandable that they would choose to label the x-axis of their time evolution graph with temperature, the system would also have shown evolution if left at the same elevated temperature for an extended time. The graph, therefore, does not depict what would happen if the system were raised to a certain temperature and held there, but rather what constituents were present as the system passed through a given temperature while being ramped at 3°C per minute.

- The magnitude of the Fourier transform of the fit and constituent for the modelling described in Section 19.7.8
- The k^3-weighted $\chi(k)$ of the fit and constituent for the modelling described in Section 19.7.8

BOX 19.11   WHAT WE LEARNED AND LIKED

	While not entirely reflected in our abbreviated case study here, in the full paper the authors work their way from one system to the next, using the results of simpler systems to interpret more complicated ones.
	I like that they were willing to share their 'mistakes'! They thoroughly discussed the three-component fit, rather than just jumping straight to the four-component, column-wise augmented one. It makes their process easier to follow, and more convincing.
	I am glad that they were careful to follow good reporting practices, such as sharing statistical information and uncertainties.
	I like that they made ample use of findings that had been previously established about these systems to guide their investigations and analyses.
	I agree, Kitsune. They did not allow themselves to get distracted by their analytical techniques. It would have been easy to trust spurious MCR-ALS results, for example, but they always made sure they were backed up by theory, previous results, and other experimental data.
	This was an excellent use of a quick-XAFS line! They juggled the trade-offs between time resolution, signal to noise, and monitoring multiple edges well.
	I love how they jumped from technique to technique, always picking out what worked best for their next question. It's rare to find a single paper that includes fingerprinting, linear combination analysis, principal component analysis, and curve fitting to a theoretical standard—but this one does!

19.8 Speciation of Gold in Hydrothermal Fluids

19.8.1 The Paper

This paper was published by Trigub et al. in 2017. It features an array of advanced techniques, both experimentally and analytically.

19.8.2 The Scientific Question

The transport of gold in hydrothermal fluids is influenced by factors such as temperature, pH, redox potential, and chloride concentration. This study focused on determining the gold-bearing species present under four sets of conditions:

- A reduced environment with sulfur in the form of hydrogen sulfide (with a smaller amount of sodium hydroxide to adjust the pH to 8.2) at 450°C
- A partially oxidized environment with sulfur in the form of sodium thiosulfate at 350°C
- A partially oxidized environment with sulfur in the form of sodium thiosulfate at 450°C
- A partially oxidized environment with sulfur in the form of sodium thiosulfate, enriched with sodium chloride, at 400°C

19.8.3 Why XAFS?

Because the species present could easily change with conditions, it was important to perform the characterization *in situ*. In this case, that did not mean taking samples from the field to the synchrotron, but rather recreating the conditions desired for study at the synchrotron. Concentrations, temperatures, etc. were therefore dictated more by geochemical considerations than experimental ease.

XAFS is particularly well suited to measuring species in solution. Because it is element-specific, the environment and state of gold ions can be measured without interference from other sulfur and chloride species.

19.8.4 Experimental Considerations

The gold L_3 edge is at 11.9 keV. Since the gold L_2 edge is not until 13.7 keV, that makes the L_3 edge well suited to EXAFS, suffering no interference from lower-energy edges while allowing for good estimation of the postedge background.

This study, however, relies heavily on XANES, with its sensitivity to oxidation state and symmetry. The core-hole lifetime broadening of the gold L_3 edge is more than 5 eV (Krause and Oliver 1979), significantly impacting XANES spectra. Data was therefore collected in a HERFD mode, allowing for much better energy resolution.

The measured solubility of gold under similar conditions was at most about 0.1 moles per kg of water, depending on the other species present. In all cases, the solutions also included concentrated sodium and sulfur, and in one case chlorine, generally in concentrations of several moles per kg of water. The ratio of sodium, sulfur, and/or chlorine ions to gold was therefore on the order of 50 to 1. The mass absorption coefficient of gold just above the L_3 edge is on the order of five times larger than the other elements. On its face, this could suggest some self-absorption effects could be present, but that's only at the upper limit of the solubility of gold (which is strongly dependent on conditions), and even then the effects would be modest.

While one might expect self-absorption not to have a substantial effect on the spectra collected in this experiment, I would have preferred if the authors had included a direct measure of the amount of gold present in the solutions they measured, such as the relative strength of fluorescent lines of different elements. This would have allowed readers to verify whether self-absorption effects are a concern in this experiment.

19.8.5 HERFD-XANES Results

One of the most important results of the experiment was particularly simple: the XANES of all four experimental solutions was very similar, but quite different from the XANES of gold(I) thiolate and gold(I) sulfide. This suggests that the local environment of the gold species formed was largely independent of the redox potential under the conditions studied.

It was also clear that the species formed was neither of the two reference species. The next question, therefore, would be to identify the species.

19.8.6 Theoretical XANES Standards

In order to model different possible structures of the local environment around the gold ions in solution, ab initio molecular dynamics (AIMD) simulations were used.

For the thiosulfate solution at 350°C, for instance, the authors ran AIMD simulations for $Au(HS)_2^-$, $Au(HS)S_3^-$, and $Au(S_3)_2^-$. They then used the finite difference method (see Section 12.5.1 of Chapter 12) to compute theoretical XANES standards and compared them to the experimental spectrum. The three calculations differed significantly, particularly in the relative amplitudes of the white line and the first peak above it. Importantly, the authors did not simply assert that the theoretical standard that matched the experimental spectrum most closely must be the species that was actually present. The experimental and theoretical uncertainties were too large for that, and the match between experiment and standard was only semi-quantitative in all cases. Instead, they used the substantial differences between the theoretical standards, and the large degree of similarity between the measured spectra, to assert that the experimental spectra could not consist of *more than one* of the materials from the theoretical standards. If, for example, one of the experimental solutions was dominated by $Au(HS)_2^-$, then it must be the case that all of them are.

19.8.7 EXAFS Data

The authors chose to weight their $\chi(k)$ data by k^2. With that weighting, noise was clearly visible starting at about 6 Å^{-1}, with signal to noise gradually degrading as k increased. To determine the k range to use for the Fourier transform, they plotted transforms using ranges of 3–10 Å^{-1}, 3–11 Å^{-1}, and 3–12 Å^{-1}. For the thiosulfate solution at 350°C, the latter two ranges showed similar structure up through 5.0 Å^{-1}, but when the data was cut off at 10 Å^{-1}, there were unresolved peaks in the Fourier transform. They therefore chose to use a range from 3–11 Å^{-1}, reasoning that the 3–11 Å^{-1} transform was adding structural information relative to the 3–10 Å^{-1} one, but that extending the $\chi(k)$ range further was likely adding noise without providing more structural insight. They used similar approaches for the other spectra, resulting in cutoffs of either 10 or 11 Å^{-1} in each case.

19.8.8 EXAFS Analysis

Since the system was expected to be highly disordered, the authors employed three different methods for analyzing the EXAFS.

In one method, they used traditional fitting to theoretical standards using IFEFFIT.

For another, they used ab initio molecular dynamics (AIMD) to estimate the range of atomic positions for various starting structures. They then computed theoretical standards based on these calculated positions, comparing the experimental spectra to those standards (a form of fingerprinting, since no fitting was involved).

Finally, they employed a reverse Monte Carlo (RMC) method, in which the positions of the atoms in a starting cluster are shifted directly. While considerably more computationally intensive than the traditional method, it has several significant advantages:

- Even when the atomic positions vary substantially from the initial guess, the quality of the theoretical calculation does not degrade.
- RMC is inherently similar to a histogram approach, allowing non-Gaussian distributions of absorber–scatterer distances to be accounted for without the need for a cumulant expansion.
- The thorny question of how to parameterize multiple-scattering paths in terms of single-scattering ones (see Section 17.5 of Chapter 17) is avoided.

19.8.8.1 AIMD

Perhaps unsurprisingly, they found their AIMD method to fit the experimental spectra least well, since the method involves no fitting. Nevertheless, the AIMD method was a useful supplement to their other analysis because it helped clarify the role different scattering atoms were likely to play in the structures. Since the atoms in the AIMD structures necessarily occupied physically reasonable positions, there was no risk of finding false minima (see Section 14.6 of Chapter 14).

19.8.8.2 Models

For traditional fitting of the thiosulfate solution at 350°C, they tried two different models. The first, meant to accommodate $Au(HS)_2^-$, included only nearest-neighbor sulfur atoms. S_o^2 was fixed at 0.9. A single triangle path was used for multiple scattering, with a degeneracy of two (clockwise and counterclockwise).
The following parameters were allowed to vary:

- E_o
- Half path length, coordination number, and MSRD for the nearest-neighbor sulfur (three parameters total)
- Half path length and MSRD for the multiple-scattering path (two parameters total)

This gives a total of 6 free parameters.

While the authors chose to fit directly in k-space, the number of independent points can still be estimated using the Nyquist criterion by observing that they attribute features from about 1.5 to 4.5 Å to correspond to structural features. Using these values yields 15 independent points, so they clearly had enough information to fit 6 parameters.

They also tried a second model, meant to represent $Au(HS)S_3^-$. This model was similar to the first, except that it added a second shell of sulfur atoms at a greater distance, with a fixed coordination number of 1. It thus had eight free parameters, still well within the number of independent points provided by the data.

19.8.8.3 Results for traditional fits

The two fits produced the same R factor, 0.05, indicative of a model which might be broadly correct, but with some significant errors in details (see Section 14.2.1 of Chapter 14). The second shell of sulfur atoms in the second model had a large and uncertain MSRD (0.025 ± 0.03 Å^2), suggesting that the model was 'zeroing out' the additional path, thus making it essentially the same as the first model, and explaining why the addition of more free parameters did not improve the *R* factor of the fit.

19.8.8.4 RMC

For the RMC method, the authors chose three starting materials: Au(HS)$_2^-$, Au(HS) S$_3^-$, and Au(S$_3$)$_2^-$. All three models fit reasonably well, although the Au(S$_3$)$_2^-$ was a bit of a closer fit at high *R*.

19.8.9 Drawing Conclusions

In this case study, we've focused primarily on just one of the samples, the thiosulfate solution at 350°C. The entire paper, of course, also includes analyses of the other samples described in Section 19.8.2.

For the thiosulfate solution at 350°C, neither XANES-HERFD nor any of the EXAFS analysis approaches they tried yielded an unambiguous structure for the gold species present. That does not mean they did not reach important conclusions, however! The XANES-HERFD showed that all of the samples they studied had very similar, or perhaps identical, structures, suggesting that within the range of conditions studied the gold speciation did not depend on pH, redox potential, or chloride concentration. The EXAFS analyses were productive as well, ruling out certain kinds of structures, such as ones with highly-ordered shells beyond the nearest-neighbor sulfurs.

19.8.10 Presentation

As with the paper highlighted in Section 19.7, the authors relegated much of the detailed information to the supporting information.

In the main paper they provided:

- HERFD-XANES spectra of the samples and empirical standards, as well as several theoretical standards
- The magnitude of the Fourier transform for the same two samples, showing the effects of different choices of *k* range
- A table of results for traditional fits of the EXAFS data, including the *R* factor and uncertainties in fitted parameters
- k^2-weighted $\chi(k)$ spectra for two of the samples, along with the theoretical standards generated from AIMD and the best fits generated from RMC. (Analogous graphs for the traditional fits were included in the supporting information.)
- The magnitude of the Fourier transform for the same two samples, along with the theoretical standards generated from AIMD and the best fits generated from RMC

BOX 19.12   What We Learned and Liked

I like that they were clear about what their data could and couldn't show. They didn't try to overinterpret their XANES and EXAFS data, and yet reached important conclusions about their systems.

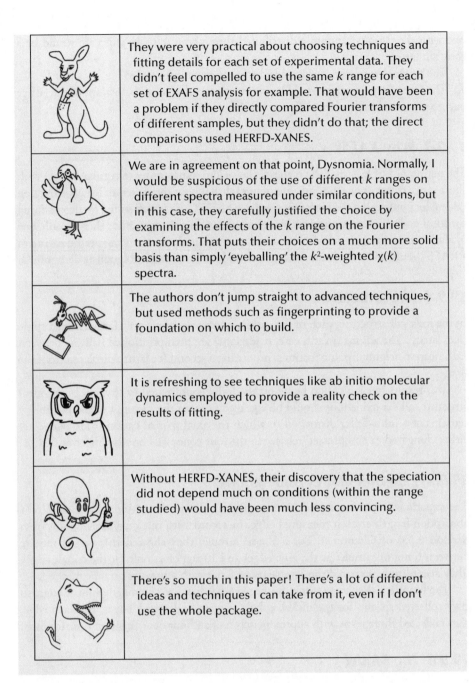

They were very practical about choosing techniques and fitting details for each set of experimental data. They didn't feel compelled to use the same *k* range for each set of EXAFS analysis for example. That would have been a problem if they directly compared Fourier transforms of different samples, but they didn't do that; the direct comparisons used HERFD-XANES.

We are in agreement on that point, Dysnomia. Normally, I would be suspicious of the use of different *k* ranges on different spectra measured under similar conditions, but in this case, they carefully justified the choice by examining the effects of the *k* range on the Fourier transforms. That puts their choices on a much more solid basis than simply 'eyeballing' the k^2-weighted $\chi(k)$ spectra.

The authors don't jump straight to advanced techniques, but used methods such as fingerprinting to provide a foundation on which to build.

It is refreshing to see techniques like ab initio molecular dynamics employed to provide a reality check on the results of fitting.

Without HERFD-XANES, their discovery that the speciation did not depend much on conditions (within the range studied) would have been much less convincing.

There's so much in this paper! There's a lot of different ideas and techniques I can take from it, even if I don't use the whole package.

19.9 Local Structure of an Entropy-Stabilized Oxide

19.9.1 The Paper

We chose this paper, published by Rost et al. in 2017, in part because they cited the first edition of *XAFS for Everyone*, and clearly followed its precepts in the conduct of their experiment and the presentation of the information in their paper.

19.9.2 The Scientific Question

Entropy-stabilized oxides (ESOs) are materials in which the different species of cations are distributed randomly throughout the lattice. The authors synthesized a $Mg_{0.2}Ni_{0.2}Co_{0.2}Cu_{0.2}Zn_{0.2}O$ powder, with the intention of producing an ESO; that is,

Of course, there are lots of good sources of best practices for EXAFS, many of which Rost et al. cite in their paper! Just because there's a practice we endorse and they follow, doesn't mean they got it from our first edition.

one with the cations randomly distributed throughout what in this case would be a rock salt lattice.

The primary question, then, was whether they had succeeded.

The secondary question would be to investigate what local distortions in the lattice occurred around each of the different species of cations.

19.9.3 Why XAFS?

The nature of ESOs is that the material shows long-range order (a regular lattice) with short-range distortions randomly distributed throughout the material. X-ray diffraction is ideal for confirming the long-range order, while EXAFS is well suited for determining the local environment around each cation species. In this case, since there is only one oxidation state for each metal present in the compound, the element-specific nature of EXAFS should allow the local structure around each to be unambiguously determined.

19.9.4 The Structure

In the rock salt structure, each of the cations adopts positions on a face-centered cubic (fcc) lattice. The anions (in this case, oxide ions) are then positioned halfway between each nearest-neighbor pair of cations, producing a second fcc lattice displaced from the first.

Because each element absorbs so strongly just above its K edge, calculations show that the edge jump would be in the range of 0.1 to 0.7, depending on the edge. With modern beamlines and no need for time-resolved data, this is consistent with excellent signal to noise in transmission mode.

This results in each cation being octahedrally coordinated to oxygen. In the ideal structure, all six oxide ions should be equidistant from the cation, but the possibility remains of a Jahn-Teller distortion in which the axial pair of cation-oxide bonds are either elongated or compressed relative to the four equatorial bonds (Halcrow 2013).

19.9.5 Experimental Considerations

The experiment was conducted in transmission. The authors aimed for a total of 1 absorption length at each measured edge, in accord with our recommendation from Section 6.3.4 of Chapter 6. For a 1 cm² sample, they showed this would require between 6 mg of sample (at the zinc edge) and 10 mg of sample (at the cobalt edge). They accordingly prepared the sample using 8 mg with a powder on tape method.

They only detuned by 20%. If they had aimed for greater edge jumps it could have introduced distortion due to harmonics.

The beamline they were using did not extend down to the magnesium K edge, so they collected spectra for the nickel, cobalt, copper, and zinc K edges. For each edge, they collected three scans with approximately half an hour of acquisition time for each.

19.9.6 The Model

For their initial analysis, the authors used a model of an ESO with the rock salt structure and all oxygens equidistant (no Jahn-Teller distortion). For cations for which Jahn-Teller distortions were suspected, either due to a poor fit to the undistorted model or for theoretical reasons, a model with a Jahn-Teller distortion was also tested. Finally, models with varying degrees of clustering of cations were also tested.

19.9.6.1 Paths

For the fits without Jahn-Teller distortion, a single nearest-neighbor oxygen path was used, with degeneracy set to 6. In the rock salt structure, the next-nearest neighbors would be 12 equidistant cations on the fcc lattice. To simulate an ESO, the authors chose a model with one cation of the same species as the absorbing cation, and two or three of each of the other cations, so that the total came to 12. For instance, the nickel cation was given one nickel neighbor, two copper neighbors, and three each of zinc, cobalt, and magnesium. To simulate clustering, the number of neighbors of the same species as the absorbing cation was increased to 2, 3, 4, and so on, up to a maximum of 8, with the number of cations of other species being reduced to keep the total at 12. Structures with anti-clustering, so that there were no neighbors of the same species as the absorbing cation, were also tested.

For fits with Jahn-Teller distortion, two different nearest-neighbor oxygen paths were used, one with degeneracy 2 and the other with degeneracy 4.

Multiple-scattering paths were incorporated—the paths that use an oxide ion to focus scattering off of a cation are particularly important!

19.9.6.2 Free parameters

The free parameters used for each edge were as follows:

- E_o
- The distance to the nearest-neighbor oxygens (1 parameter for the standard fits; 2 for those allowing for Jahn-Teller distortion)
- The distance to the next-nearest neighbor shell of cations (1 parameter)
- The MSRD for the near-neighbor oxygens (1 parameter, even if Jahn-Teller distortion was allowed)
- The MSRD for the next-nearest neighbor cations (1 parameter)

This yields a total of 5 or 6 parameters, depending on whether a Jahn-Teller distortion was allowed.

19.9.6.3 Constraints

The following constraints were applied:

- S_o^2 was constrained to 1.0
- In the models allowing the Jahn-Teller effect, the same MSRD was used for both axial and equatorial oxygens
- All cation–cation scattering was constrained to be at the same distance and to have the same MSRD

19.9.6.4 Degrees of freedom

A k range of 3.25 to 11.9 Å^{-1} was used for each edge, with a fitting range from 1.0 to 3.4 Å, yielding 13 independent points according to the Nyquist criterion.

19.9.7 Drawing Conclusions

Only the copper edge spectrum showed an improvement in fit when allowing for the Jahn-Teller distortion, but in that case, it was necessary for producing a good fit.

For all cations, high degrees of clustering was soundly rejected by the fits, with R factors increasing from less than 0.02 to above 0.8 (!) when seven or more neighbors of the same species as the cation were included. The zinc and cobalt edges favored anti-clustering, with only the fit with no same-species neighbors producing R factors below 0.02. For nickel and copper, a random distribution was favored, with one or two same-species neighbors.

All edges yielded a cation–cation distance of approximately 2.99 ± 0.01 Å, with copper at the low end of that range. This is within 0.2% of the value implied by x-ray diffraction measurements of the sample.

In the paper, the authors say the Nyquist criterion yields 11 independent points, rather than 13. It is possible that the discrepancy is related to how the sills of the Hanning window are accounted for. Regardless, it is clear that they had enough independent points to support the number of free parameters in their fit.

The combination of the consistency of the cation–cation distances found by fits at different edges, the consistency with the distance implied by x-ray diffraction, and the close fits produced with relatively simple models provides very strong evidence that the cations, sit, on average, at the lattice points of the rock salt structure. In combination with the failure of models with high degrees of clustering to produce good fits, it seems very likely that the cations are distributed through the lattice randomly, as expected for an ESO, or perhaps even with a modest degree of anti-clustering.

This does not, however, preclude some amount of local distortion, randomly distributed so as not to affect the x-ray diffraction measurement. In addition to the slightly lower cation–cation distance found for the copper edge, the MSRDs for the cation–cation scattering ranged from 0.008 to 0.011 Å^2, relatively high values for a crystalline substance measured at room temperature. In addition, the Jahn-Teller distortion of the oxygen atoms around copper yielded an equatorial copper–oxygen distance of 1.99 ± 0.01 Å and an axial copper–oxygen distance of 2.22 ± 0.02 Å. Finally, the cation–oxygen distances for the other cation absorbers are slightly lower than the 2.12 Å implied by the x-ray diffraction measurement. Taken together, this suggests local distortions of the oxide lattice.

19.9.8 Presentation

The paper includes graphs of k^2-weighted $\chi(k)$ for all four measured edges, and both the imaginary part and the magnitude of the Fourier transform for all data and the best fit to each edge. The supplementary information includes the normalized XANES for each edge, and the magnitude of the Fourier transform for the fit to copper using the model which did not allow for Jahn-Teller distortion.

BOX 19.13   What We Learned and Liked

	I like that they showed the copper-edge fit using the model without Jahn-Teller distortion, so that we could see that it didn't work.
	Each graph was appropriately and clearly labelled— something that is not always true in XAFS literature! They also showed individual points in the Fourier transforms of the data and fits as well as the usual smoothed curve, in order to emphasize the limited spatial resolution.
	In addition to the EXFAS analysis, the authors used density functional theory to simulate what might be happening physically in their structure. The results were broadly consistent with their EXAFS data.
	I appreciate that they described clearly how they prepared their samples, including details such as the thickness of Kapton tape they used.

	The authors also showed us how they calculated the amount of sample to use.
	The comparison to x-ray diffraction data really strengthened their argument.
	I have to say, they had beautiful data, with their *k* ranges limited not by signal-to-noise, but by the presence of four relatively closely spaced edges.

19.10 The Next Case Study: Yours

We have now come to the end of the book, and I will once again, for the first time since the introduction, speak to you directly, without my cast of characters as intermediaries. If you have read through to this point, and thought carefully about what you have read, you now have the tools to use XAFS effectively in your own scientific projects and to interpret and critique its use by others.

Of course, you won't be alone. There is a community of others using XAFS for their work, and like any community, we can all help each other. The website xafs.org is an excellent starting point for learning about the resources that are available.

There is also another sense in which I hope you won't be alone. In the preface, I introduced the characters who have accompanied us through this text and explained how they are a composite of the important characteristics of every successful scientist who understands XAFS. That's you too, now.

So the next time you try to use XAFS to analyze a system that is of interest to you, ask yourself, when you are stuck and feeling overwhelmed, 'what would Dysnomia do to shake things up?' When you are pleased that the wiggly lines on your graphs are matching up, imagine Carvaka asking, 'yes, but what does it mean?' and Kitsune asking 'yes, but why do we care?' Mandelbrant can help you take an imposing challenge one step at a time, and Mr. Handy can keep you from walking off a cliff. Once you have the perfect solution, let Robert remind you that it needs to be understandable to the rest of the community. And, of course, Simplicio is around to tell you that you shouldn't be afraid of being wrong.

Goodbye, then, from me and from them. May your data be good and your analyses sound!

References

Calvin, S., E. E. Carpenter, B. Ravel, V. G. Harris, and S. A. Morrison. 2002. Multiedge refinement of extended x-ray absorption fine structure of manganese zinc ferrite nanoparticles. *Phys. Rev. B.* 66:224405(13). DOI:10.1103/PhysRevB.66.224405.

Corbett, M. C., Y. Hu, A. W. Fay, M. W. Ribbe, B. Hedman, and K. O. Hodgson. 2006. Structural insights into a protein-bound iron-molybdenum cofactor precursor. *Proc. Natl. Acad. Sci. USA.* 103:1238–1243. DOI:10.1073/pnas.0507853103.

Einsle, O., F. A. Tezcan, S. L. Andrade, B. Schmid, M. Yoshida, J. B. Howard, and D. C. Rees. 2002. Nitrogenase MoFe-protein at 1.16 Å resolution: A central ligand in the FeMo-cofactor. *Science.* 297:1696–1700. DOI:10.1126/science.1073877.

Halcrow, M.A. 2013. Jahn-Teller distortions in transition metal compounds, and their importance in functional molecular and inorganic materials. *Chem. Soc. Rev.* 42:1784–1795. DOI:10.1039/c2cs35253b.

Hu, Y., A. W. Fay, and M. W. Ribbe. 2005. Identification of a nitrogenase FeMo cofactor precursor on NifEN complex. *Proc. Natl. Acad. Sci. USA.* 102:3236–3241. DOI:10.1073/pnas.0409201102.

Jaumot, J., R. Gargallo, A. de Juan, and R. Tauler. 2005. A graphical user-friendly interface for MCR-ALS: A new tool for multivariate curve resolution in MATLAB. *Chemometr. Intell. Lab.* 76:101–110. DOI:10.1016/j.chemolab.2004.12.007.

Krause, M. O. and J. H. Oliver. 1979. Natural widths of atomic K and L levels, $K\alpha$ x-ray lines and several KLL Auger lines. *J. Phys. Chem. Reference Data.* 8:329–338. DOI:10.1063/1.555595.

La Fontaine C., S. Belin, L. Barthe, O. Roudenko, and V. Briois. 2020. ROCK: A beamline tailored for catalysis and energy-related materials from ms time resolution to μm spatial resolution. *Synch. Rad. News* 33:20–25. DOI:10.1080/08940886.2020.1701372.

La Fontaine C., L. Barthe, A. Rochet, and V. Brios. 2013. X-ray absorption spectroscopy and heterogeneous catalysis: Performances at the SOLEIL's SAMBA beamline. *Catal. Today* 205:148–158. DOI:10.1016/j.cattod.2012.09.032.

Plais, L., C. Lancelot, C. Lamonier, E. Payen, and V. Briois. 2021. First *in situ* temperature quantification of CoMoS species upon gas sulfidation enabled by new insight on cobalt sulfide formation. *Catal. Today* 377:114–126. DOI:10.1016/j.cattod.2020.06.065.

Ressler, T., J. Wong, J. Roos, and I. L. Smith. 2000. Quantitative speciation of Mn-bearing particulated emitted from autos burning (methylcyclopentadienyl)manganese tricarbonyl-added gasolines using XANES spectroscopy. *Environ. Sci. Technol.* 34:950–958. DOI:10.1021/es990787x.

Rost, C. M., Z. Rak, D. W. Brenner, and J.-P. Maria. 2017. Local structure of the $Mg_xNi_xCo_xCu_xZn_xO(x=0.2)$ entropy-stabilized oxide: An EXAFS study. *J. Am. Ceram. Soc.* 100:2732–2738. DOI:10.1111/jace.14756.

Ruckebusch, C., A. De Juan, L. Duponchel, and J. P. Huvenne. 2006. Matrix augmentation for breaking rank-deficiency: A case study. *Chemometr. Intell. Lab.* 80:209–214. DOI:10.1016/j.chemolab.2005.06.009.

Sandström, M., F. Jalilehvand, E. Damian, Y. Fors, U. Gelius, M. Jones, and M. Salomé. 2005. Sulfur accumulation in the timbers of King Henry VIII's warship *Mary Rose*: A pathway in the sulfur cycle of conservation concern. *Proc. Natl. Acad. Sci. USA.* 102:14165–14170. DOI:10.1073/pnas.0504490102.

Sicron, N., B. Ravel, Y. Yacoby, E. A. Stern, F. Dogan, and J. J. Rehr. 1994. Nature of the ferroelectric phase transition in $PbTiO_3$. *Phys. Rev. B.* 50:13168–13180. 10.1103/PhysRevB.50.13168.

Smit, J. and H. P. J. Wijn. 1959. *Ferrites.* New York: Wiley.

Trigub, A. L., B. R. Tagirov, K. O. Kvashnina, S. Lafuerza, O. N. Filimonova, and M. S. Nickolsky. 2017. Experimental determination of gold speciation in sulfide-rich hydrothermal fluids under a wide range of redox conditions. *Chem. Geol.* 471:52–64. DOI:10.1016/j.chemgeo.2017.09.010.

Index

A

Absorber–scatterer distance, 262
Absorbing atom, 9–16, 266, 285, 334
Absorption coefficient, 63–65
Absorption edge, *see* Edge
Absorption length, 64–66, 76–77, 81–82, 138–139
Agreement outside the fitted range, 304–306
Amplitude reduction factor (S_o^2), 14–15, 268, 271–274, 338, 380, 403
ATHENA, 101, 102
Attenuation, 208–209
Auger-Meitner electrons, 8, 36, 50, 138, 155
Automation, 56–57

B

Back-transforms, 121–124, 216
Bayes–Turchin Analysis, 356
Beamline, 30, 141
Beamline equipment, 37
Beamline for Materials Measurement (6-BM), 56
Beam optimization
    aligning, 142–143
    pre-I_o vertical slit width, 143–145
    reducing harmonics, 145–149
Beamtime applications, 38–39
Bending magnets, 30, 32
Bottom-up strategy, 252–254, 316
Bouguer's Law, 64–65, 195
Bragg diffraction, 146, 171–172, 222
Bragg's Law, 142–143, 145
Brilliance, 32

C

Calculated structures, 329
Central atom, 285
Characterization questions, 56
Child function, 125–126, 128
Closeness of fit, 299–302, 309, 365
Cluster analysis, 231–233
Column-wise augmentation (CWA), 397–398
Combinatoric fitting, 206–207
Compton scattering, 79, 154
Conjoined paths, 348–349
Constraints, 253

alternatives for incorporating a priori knowledge, 355
    Bayes–Turchin Analysis, 356
    restraints, 355–356
based on a priori knowledge, 338–339
multiple-scattering paths
    conjoined paths, 348–349
    double paths, 348
    focused paths, 346–348
    triangles, quadrilaterals and minor, 349–355
rigorous constraints, 338
for simplification, 339–343
some special cases, 344
    correlated Debye Model, 344–346
    lattice scaling, 344
Constraints for simplification, 339
    estimates, 341
    grouping, 339–341
    solar slits (*see* Soller slits)
    standards, 341–343
Convergence of the path expansion, 248–250
Coordination charge, 188–190
Core-hole lifetime broadening, 15, 37, 50–52, 143–144, 162, 247, 315, 401
Correlated Debye Model, 344–346
Counting statistics, 67–68, 95–96, 137–138, 155, 162, 164, 209, 223
Crysotats, 87
Crystal/diffractive analyzers, 37, 140
Crystal structures, 326
    cluster size and EXAFS, 326–328
    cluster size and XANES, 328–329
    sources for, 329
Cumulant expansion, 16
Cumulants, 16–17, 277–278
Current amplifier, 150
Current-mode semiconductor detectors, 36
Curve fitting to theoretical standards
    EXAFS fitting strategy, 252
        bottom-up strategy, 252–253
        top-down strategy, 253–254
    fitting, 244–245
    overview, 244
    the path expansion, 248
        convergence, 248–250
        full multiple scattering, 250–252
    theoretical EXAFS standards, 245–246

final state rule, 247
  losses, 247–248
  muffin-tin potentials, 246–247
theoretical XANES standards, 254
  finite difference method (FDM), 254
  real-space multiple-scattering
    (RSMS), 254
Cyclic voltammetry, 38, 58, 164

D

Dead time, 152, 208–209
Defensible model, 307–308
Degeneracy (fitting parameter, *N*), 12, 253,
    266–269, 274, 380, 403, 406–407
  in mixtures, 330–331
  when substitutional disorder is present,
    334–335
  when vacancies are present, 334
Degrees of freedom, 204
Detail wavelet transforms, 130
Detector choice, 134–135
  energy-discriminating fluorescence detectors,
    139–140
  signal-to-noise ratio, fluorescence, 138–139
  signal-to-noise ratio, transmission, 135–138
  wavelength-dispersive fluorescence
    detectors, 140
Detectors
  current-mode semiconductor detectors, 36
  electron yield, 37
  energy-discriminating fluorescence detectors,
    36, 74, 97, 135, 139, 152–153, 155–158
  ionization (ion) chambers, 36, 135, 149–153
  wavelength-dispersive fluorescence detectors,
    37, 53
Detuning, 145–147
Distribution tails, 63
Double paths, 348

E

Edge, 1, 9, 17
Edge fraction, 363
Edge jump, 19, 69–70, 99–103
Elastic scattering, 153
Electronics out of range, 174–175
Electron wave, 9
Electron yield, 36–37, 88
Electropolishing, 88
Empirical standards, 25, 183
Energy alignment, 207–208
Energy-discriminating fluorescence detectors, 36,
    74, 97, 135, 139, 152–153, 155–158
Energy-dispersive XAFS, 43–44
Energy misalignment, 228–229
Energy-recovery linacs, 30
Energy resolution, 50–52, 209
E$_o$, 11, 17, 19, 97–98, 212, 275–278
Étendue, conservation of, 32
Evolving Factor Analysis (EFA), 237–238
Ex situ experiments, 59
Extrinsic background, 98

F

Fast Fourier transforms, 115
Fermi's Golden Rule, 9, 15
Ferrites, 185
Fifth and higher cumulants, 283
Final state rule, 247
Fingerprinting, 25
  matching empirical standards, 183–185
  semiquantitative fingerprinting, 186–187
    fitting features, 189–191
    vanadium XANES, 187–189
  spectral features, 185–186
  theoretical XANES standards, 192–193
Finite difference method (FDM), 254
First-generation SLS, 30
Fluorescence measurement, 36
Fluorescent background, 152–153
  assumptions, 158
  low-energy peaks, 156
  quantitative analysis, 157
  scatter peaks, 153–156
Focused paths, 250, 346–348
Fourier Transform, 22, 108–109
  back-transforms, 121–124
  choice of windows, 116
  complex function, 117–120
  of components, 224
  corrected, 120
  finite data ranges, 109–114
  windows, 114–115
  zero padding, 115–116
Fourth cumulant (C_4, $\sigma^{(4)}$), 280–282
Fourth-generation SLS, 30
Free electron lasers (FEL), 33–34, 45
Full multiple scattering, 250–252
Full width at half maximum (FWHM), 152

G

Gaussian factor, 125
Gaussian function, 190
Glitches (monochromator), 97, 147, 162,
    166–167, 209
Graphs of fits, 365
Grazing entry, 80
Grazing exit, 80

H

Half path length (D), 169, 262–265, 269, 271,
    277, 288–289, 319, 344–345, 374; *see
    also* Symbol *D*
  for multiple-scattering paths, 347–349
Hamilton test, 203–208, 211, 213–214, 300–
    302, 352–353, 366, 375, 381
Hanning window, 116
Harmonics, 63, 208–209
  rejection mirror, 145
  suppression
    detuning, 145–147
    rejection mirror, 145
    testing, 147–149

Heisenberg uncertainty principle, 50
High-energy-resolution fluorescence-detected
    x-ray absorption spectroscopy
    (HERFD-XAS), 52–53, 401–402
Histogram methods, 332–333

I

Inelastic scattering, 79
Inequivalent absorbing sites, 332
Inhomogeneous transmission samples, 208–209
Insertion devices
    bending magnets, 30, 32
    undulators, 33, 34, 135
    wigglers 32–33
In situ experiments, 59
Integration time, 135
International Union of Crystallography (IUCr)
    definitions, 14, 17–18
Intrinsic background, 98
Isosbestic points, 218–219
Iterative Target Transform Factor Analysis
    (ITTFA), 236–237
IXS Standards and Criteria Committee
    recommendations, 96, 203, 224, 293,
    296–299, 302–303, 355, 364

K

Kaiser–Bessel window, 116
k weighting, 20, 71, 313

L

Lattice scaling, 344
Less common fitting parameters, 277
    cumulants, 277–278
    fifth and higher cumulants, 282
    fourth cumulant, 280–282
    mean free path, 283–285
    third cumulant, 278–280
Linear combination analysis (LCA),
    25–26, 312
    choosing data range and space, 210
        back-transform of EXAFS, 216
        EXAFS in energy space, 211
        EXAFS in $\chi(k)$, 211–215
        XANES in derivative space in energy
            space, 210–211
        XANES in energy space, 210
    combinatoric fitting, 206–207
    example of, 197–200
    sources of systematic error, 207
        attenuation, 208–209
        background, 208
        energy alignment, 207–208
        energy resolution, 209
        glitches, 209
        noise, 209–210
    statistics of fitting, 200
        degrees of freedom, 204
        degrees of freedom and statistically
            distinguishable fits, 202–203

        the Hamilton Test, 204–205
        normalization: a source of systematic error,
            200–202
        quantifying fit mismatch, 203–204
        uncertainties, 205–206
    when doesn't work, 197
        nonuniform samples in transmission, 197
        surface gradients in thick fluorescence
            samples, 197
    when works
        intimate mixtures in fluorescence,
            196–197
        intimate mixtures in transmission, 196
        simple example, 194–195
Local field effect, 247
Lorentzian function, 190
Losses, 247–248
Low-density polyethylene (LDPE), 87
Lytle detector, 94

M

Malinowski's *factor indicator function*, 227
McMaster correction, 108, 269
Mean free path ($\lambda(k)$), 15, 248–249, 270, 272,
    283–285, 315, 320
Mean square radial displacement, *see* Mean
    square relative displacement (σ^2, C_2)
Mean square relative displacement (σ^2, C_2), 16,
    169, 171, 265, 268–271, 278, 307,
    321–322, 329, 332, 387–388, 403, 408
    constraints on, 306, 308, 318, 321, 324,
        333, 338–340, 342–354, 356–357,
        373–375, 385–386, 403, 407
    correlations with 265, 268, 271, 274, 282
Microprobe beamlines, 38, 45–46
Mixtures, 329–331
Monochromators, 35, 63, 141
Morlet wavelet, 125
Moving average, 26
Muffin-tin potentials, 246–247
Multibend achromat (MBA) lattice, 30
Multidimensional experiment, 60
Multielectron excitations, 170–171
Multielement semiconductor detector, 36
Multimodal experiment, 60
Multiple-edge fits, 333
Multiple scattering, 13
    paths
        conjoined paths, 348–349
        double paths, 348
        focused paths, 346–348
        triangles, quadrilaterals and minor paths,
            349–355
Multivariate curve resolution—alternating least
    squares (MCR-ALS), 237–238

N

Near-edge x-ray absorption fine structure
    (NEXAFS), 18
Noise, 67–68, 134, 209–210
Normalization source of systematic error, 200–202

Normalized absorption
 choosing E*o*, 97–98
 deadtime correction, 97
 deglitching, 97
 normalization, 98–103
 self-absorption correction, 103–104

O

Offset, 152
Operando catalysis, 58, 164
Operando experiments, 59–60, 93
Overabsorption, 75
Oversampling, 162
Overview wavelet transforms, 130

P

Parent function, 125
Passivated implanted planar silicon (PIPS), 134
Path expansion, 248
 convergence, 248–250
 full multiple scattering, 250–252
Periodic extension, 110
Photoelectron, 9–10
Pinholes, 66
Planning beamtime, 140–141
Plates, 259–262
Poisson statistics, *see* Counting statistics
Polychromator, 43–44
Polyethylene oxide (PEO), 3
Precision, 302
 calculations of uncertainties by analysis
   software, 302–303
 correlations, 303
 finding uncertainties in fitted parameters, 302
 how precise?, 303
Preedge, 18–19
Principal component analysis (PCA), 26–27, 223
 blind source separation, 235–236
   Evolving Factor Analysis (EFA), 237–238
   Iterative Target Transform Factor Analysis
     (ITTFA), 236–237
   SIMPLe-to-Use Interactive Self-Modeling
     Mixture Analysis (SIMPLISMA),
     238–239
   Transformation Matrix (TM), 236
 cluster analysis, 231–233
 components, 222–223
   appearance of, 223
   compare to measurement error, 224–225
   Fourier Transform, 224
   objective criteria, 226–227
   scree, 225–226
 constituents
   coupled, 230
   energy misalignment, 228–229
   relationship to number of
     components, 228
   structural free parameters, 229–230
 of EXAFS, 239
 formalism, 230–231

 idea of, 219–222
 isosbestic points, 218–219
 is used, 239–241
 literature example, 218
 overview, 218
 target transforms, 233–235
Pseudo-Voigt function, 190
Pump-probe technique, 44–45
Pyrite (FeS$_2$), 1, 2

Q

Quantifying fit mismatch, 203–204
Quenching, 42
Quick XAFS (QXAFS), 43, 140

R

Radial distribution function, 23
Real-space multiple-scattering (RSMS), 254
Resolving power, 50
Resonant inelastic x-ray scattering (RIXS), 53–54
Resonant Raman spectroscopy, 53
Resonant x-ray emission spectroscopy (RXES), 53
Resonant x-ray fluorescence spectroscopy
     (RXFS), 53
Restraints, 355–356

S

Sample
 environmental conditions, 62, 89
 gases, 89
 powder, 82–83
   air-sensitive, 86–87
   fillers, 83
   pressing into pellet, 86
   sedimentation, 83
   spreading on tape, 84–86
 in situ and *operando*, 89–90
 solid metals, 88
 solutions and liquids, 88–89
 thin films, 88
 transmission
   absorption lengths, 64–65
   edge jump, 69–70
   mass, 64–65
   noise, 67–68
   optimum thickness, 68–69
   thickness, 64–65
   uniformity, 66–67
 type, 62
Sample wheels, 56–57
Scan parameters
 EXAFS region, 162
 number of scans, 163–164
 postedge region, 162
 preedge region, 161
 time-resolved studies, 164–165
 XANES region, 161–162
Scattered photons, 53, 62
Scattering parameters, 285–290

Scree plot, 226
Second-generation SLS, 30
Self-absorption, 75, 135, 208–209
Self-absorption correction, 103–104
Self-amplified spontaneous emission (SASE), 33
Semiquantitative fingerprinting, 186–187
  fitting features, 189–191
  vanadium XANES, 187–189
Shot noise, *see* Counting statistics
Signal-to-noise ratio, 135
  in fluorescence, 138–139
  in transmission, 135–138
SIMPLe-to-Use Interactive Self-Modeling
    Mixture Analysis (SIMPLISMA),
    238–239
Simultaneous probes, 38
Single scattering, 13, 248–250, 346
Single-shell fit, 252–253
Singular value decomposition (SVD), 230
Site occupancy, 334
  creating a mixed model, 334–335
  creating multiple mixed models, 335
  treating as a mixture, 334
  vacancies, 334
Size of data ranges, 303–304
Slicing, 45
Soller slits, 155
Sources of systematic error for linear combination
    analysis, 207
  attenuation, 208–209
  background, 208
  energy alignment, 207–208
  energy resolution, 209
  glitches, 209
  noise, 209–210
Speciation questions, 56
Spherical waves, 13–14
SPOIL, 233–234
Spot size, 38
Standard, 25
Starting structures
  calculated structures, 329
  crystal structures, 326
    cluster size and EXAFS, 326–328
    cluster size and XANES, 328–329
    sources for crystal structures, 329
  histogram methods, 332–333
  inequivalent absorbing sites, 332
  mixtures, 329–331
  multiple-edge fits, 333
  site occupancy, 334
    creating a mixed model, 334–335
    creating multiple mixed models, 335
    treating as a mixture, 334
    vacancies, 334
Static disorder, 16
Statistics of linear combination fitting, 200
  degrees of freedom, 204
    and statistically distinguishable fits,
      202–203
    the Hamilton Test, 204–205
    normalization: a source of systematic error,
      200–202

quantifying fit mismatch, 203–204
  uncertainties, 205–206
Storage-ring light sources (SLS), 30
Stroboscopic technique, 44
Substitutional disorder, 334
Symbol D
  common constraints, 264–265
  correlations, 265
  effect on Fourier transform, 264
  effect on $\chi(k)$, 264
  nomenclature, 262
  physical interpretation, 262–264
  typical values, 264
Symbol E_o
  common constraints, 276
  correlations, 276
  effect on Fourier transform, 276
  effect on $\chi(k)$, 276
  nomenclature, 275
  physical interpretation, 275–276
  typical values, 276
Symbol N
  common constraints, 266–267
  correlations, 268
  effect on Fourier transform, 266
  effect on $\chi(k)$, 266
  nomenclature, 266
  physical interpretation, 266
  typical values, 266
Symbol S_o^2
  common constraints, 273–274
  correlations, 274
  effect on Fourier transform, 273
  effect on $\chi(k)$, 273
  nomenclature, 271
  physical interpretation, 271–273
  typical values, 273
Symbol σ^2, C_2
  common constraints, 270–271
  correlations, 271
  effect on Fourier transform, 270
  effect on $\chi(k)$, 269
  nomenclature, 269
  physical interpretation, 269
  typical values, 269
Synchrotron light sources, 30

T

Target transforms, 233–235
Theoretical EXAFS standards, 245–246
  final state rule, 247
  losses, 247–248
  muffin-tin potentials, 246–247
Theoretical standards, 25
Theoretical XANES standards, 254
  finite difference method (FDM), 254
  real-space multiple-scattering (RSMS), 254
Thermal disorder, 16, 263
Third cumulant (C_3, $\sigma^{(3)}$), 270, 278–280, 308,
    348, 373–375
Third-generation SLS, 30
Time-resolved experiments

energy-dispersive XAFS, 43–44
   freezing time, 42
   pump-probe, 44–45
   QXAFS, 43
   scale of 15 minutes, 42
   timescale of months/years, 42
Top-down strategy, 253–254
Total-yield detectors, 156, 158
Transformation Matrix (TM), 236
Transition from EXAFS to XANES, 315–316
Transmission geometry, 35, 195
Triangle and quadrilateral multiple-scattering
   paths, 349–355

U

Uncertainties, 205–206, 365–366
Uncertainty principle, 127–128
Undesirable photons, 62–63
Undulators, 33–34, 135
Unnormalized absorption, 93–94, 98, 363

V

Voigt function, 190

W

Wavelength-dispersive fluorescence detectors, 37
Wavelet, 125

Wavelet transforms
   vs. Fourier transform, 125–127
   parent function, 128–130
   uncertainty principle, 127–128
Wavenumber, 11
White line, 18, 50, 219–220
Wigglers, 32–33

X

X-ray absorption edge, *see* Edge
X-ray absorption near-edge structure (XANES),
   9, 17–18, 42
   derivative space in energy space, 210–211
   energy space, 210
   modeling software, 255
X-ray fluorescence (XRF), 46, 88
X-ray microprobes, 46
X-ray nanoprobes, 46
X-ray sources, 29
   energy-recovery linacs, 30
   synchrotron light sources, 30
   tabletop sources, 30, 32

Z

Zero padding, 115–116

For Product Safety Concerns and Information please contact our
EU representative GPSR@taylorandfrancis.com Taylor & Francis
Verlag GmbH, Kaufingerstraße 24, 80331 München, Germany